Radios

of the Baby Boom Era

1946 to 1960

VOLUME 1

Admiral to Clearsonic

Imprint of Howard W. Sams & Company
Indianapolis, Indiana 46214-2012

For radio enthusiasts everywhere,
keeping the memories alive

Special thanks to Doug Heimstead, whose knowledge of and enthusiasm for old radios was instrumental in creating this series. Thanks also to George Fathauer, Jr., of Antique Electronic Supply, and John Terrey of Antique Radio Classified, for their helpful suggestions.

PHOTOFACT® is a registered trademark of Howard W. Sams & Company

Cover Illustration: Ned Shaw Studios
Cover Production: Woodfield Hi-Tech Prepress
Database Design: Kevin Etter, Bill Skinner, Bill Smith WB9TUI, Barry Buchanan
Database Input: Kelly Jenkins, Deloris Bibb
Text: Brian McCaffrey, Wendy Ford
Technical Advisers: Bill Fink, Leon Lewis, Stan Scott, Dave Stitt, George Weliver
Editing: Dave Crouch, Ken Cobb
PHOTOFACT Research: Demetrice Bruno, Angelina Cooper, Lisa Gorman, Crystal Johnson, Rose Smith
Production Assistance: Jim Beever, Lynne Clark, Alejandro Heyartz, Michele Lambert, Jonathan Mangin

FIRST EDITION

PROMPT Publications
An Imprint of Howard W. Sams & Company
2647 Waterfront Parkway East Drive, Suite 300
Indianapolis, IN 46214-2012

ISBN 0-7906-1002-7

Library of Congress Catalog Number: 91-61691

Printed in the United States of America

CONTENTS

SAMS PHOTOFACT® PICTURE GUIDES

Radios of the Baby Boom Era is only one of a whole series of picture references to your favorite home entertainment devices of the years 1946 to 1960.

We have assembled thousands of photographs of old and collectible *televisions, car radios,* and *hi-fi components* from the PHOTOFACT files and bound them in handy book form.

Each book presents technical data, helpful indexes, and hints for restoring these devices to working order.

Look for Sams PHOTOFACT Picture Guides at your local Sams authorized distributor or your nearest bookstore.

INTRODUCTION

The baby boom years, 1946 to 1960...a time of bebop, rhythm-and-blues, and rock-and-roll, a time that forever transformed the views and values of America's youth...and radio played a central role. Radio makers produced thousands of different brands and models to satisfy eager consumers, from consoles to table models to portables.

Whether you're looking for one particular radio that sparks fond memories or are building a wide-ranging collection, this book can help you identify your radio and obtain the information needed to bring it back to operating condition.

The period covered by the *Radios of the Baby Boom Era* series produced profound changes in American culture, and radio was instrumental in bringing them about. From Frank Sinatra and the bobby soxers to Elvis and beyond, radio is intertwined with people's memories of the best years of the 1940s and 1950s. Radio is a key element of the nostalgia for this period, and the great variety of radios produced during those years are now becoming a nationwide focus for collectors.

Radio in the Baby Boom Years

Radio itself underwent a metamorphosis during the years after World War II, not because of dramatic changes in technology, but because of how people came to use it in their daily lives.

From its beginning, radio had wrought significant changes in our culture. Through the 1920s and 1930s it was a daily indulgence of millions, bringing sports, drama, comedy, news, and music to virtually every corner of the nation. Network broadcasting, with its standardized live programming, created a common thread which helped unify the nation.

When war came, involving the United States in a massive military effort that sealed its place as the world's leading economic and military power, radio played a key role in creating awareness of the situation. America learned of Japan's attack on Pearl Harbor through President Roosevelt's stirring "Day of Infamy" speech. Throughout World War II, radio was a vital communications and entertainment medium...a constant companion for civilians and soldiers alike, bringing news and entertainment which rallied the country to the cause and saw it through to victory.

The end of the war brought a new world order and an audience with different wants, needs, and values. This was a major crisis for radio, a crisis deepened by the emergence of television and its quick rise into the information and entertainment medium of choice. Radio was about to go through a profound change. Some even said that radio would die.

But the predictions of radio's death were greatly exaggerated. While TV became the focal point of entertainment for the family, radio became the companion to the individual. It provided a portability which TV couldn't match, with music and news as the predominant programming. Radio experienced a metamorphosis that reflected the new America...mobile, affluent, and commercialized.

Radio Manufacturers

Consumers were eager to buy radios, and manufacturers obliged by producing a wide assortment of models, some simple, some elegant, some whimsical, some disguised as other objects. Radio manufacturers came and went, others made fortunes. Some of the leading names were:

- *Admiral* Started during the depression on a $3,400 investment, it rose to be one of the largest corporations in America because it dared to be innovative.
- *Arvin* Which began by making automobile heaters and became one of a very few manufacturers to produce a metal cabinet radio.
- *Fada* Highly prized by collectors because of their distinctive cabinet designs by Catalin Corporation.

- *Philco* Rose from near-bankruptcy as a producer of batteries to become one of the top three names in radio.
- *RCA* Another of the "Big Three" in radio whose productivity in developing and manufacturing vacuum tubes was a major factor in developing the entire radio-electronics industry.
- *Zenith* The third of the "Big Three" whose involvement in radio spans the history of the industry from 1918 to the present.

Howard W. Sams and Radio

While manufacturers turned out thousands of radios, Howard W. Sams built a company to supply the people who repaired them--service technicians. In those days the average tube radio needed servicing six times a year, and technicians used Sams service data to make the job easier. That service data was called Sams PHOTOFACT, because it included a cabinet photo along with basic facts to help the repairer: a schematic, a list of vacuum tubes and other replaceable parts, power supply information, tuning ranges, adjustment and alignment guidelines, and troubleshooting techniques. Soon home handymen and hobbyists discovered PHOTOFACT and used them to repair their own radios at home.

The pictures and index data in this book are taken from those early PHOTOFACT sets, which are available to you today to help restore your old radio. The details of how to order a specific PHOTOFACT are found with the repair information at the back of this book.

Collecting Classic Radios

Collecting has been a distinctly American passion. It's a passion partly to possess things which have significant meaning to us, partly for financial gain, and partly for the thrill of the hunt. The amount of money invested or, indeed, the overall worth of the collection are strictly side issues. The real thrill of collecting is to possess a piece of the past. The real value of a collection is assessed in the satisfaction gained through the collecting process.

For years, collectors have focused on pre-World War II radios, some of which brought great emotional and financial rewards to those who pursued them. Unlike the pre-World War II sets, most radios of the baby boom era are still in relatively plentiful supply, creating the opportunity to get in on the ground floor and build an extremely diverse collection which will bring years of pleasure.

To get started in collecting, you can join a national club such as the Antique Wireless Association or the Antique Radio Club of America, or find a local antique radio club and start haunting their swap meets.

Subscribe to publications such as *Antique Radio Classified* and *Radio Age,* and get on the mailing lists for Antique Electronic Supply and for Vintage TV & Radio.

How to Use This Book

Radio Photos The old radios themselves are the "stars" of this series, and you'll learn a lot just by browsing through the hundreds of pictures. If you know what a radio looks like but not the brand name or model number, perhaps you'll recognize it among these photos. With each picture, we furnish a description and basic technical information that helps in classifying the radio.

Indexes The indexes provide a wealth of useful information cross-referenced to the pictured radios. If you know the model number of a radio, you can look it up to find out on which page it or a similar model is pictured, which tubes it uses, and which tubes would be suitable replacements. If you know the manufacturer of a radio but not the brand name, there's an index that summarizes all the manufacturers and brand names of the period.

Fix-Up Information Because the purpose of this series is not only to list radios, but also to help in restoring them to operating condition, we provide a basic troubleshooting guide for working on tube radios. We also help you find sources for parts, collector's information, and radio restoration aids. There's a schematic of a typical 5-tube radio to help you understand the circuitry.

We hope you enjoy reading and using this book as much as we did bringing it to you. For those of us who appreciate what radio has accomplished over the years, it was a labor of love. Happy dialing!

THE RADIOS

From Admiral to Clearsonic, these pictures from the Sams PHOTOFACT library represent the radios that were America's constant companions during the baby boom years. From the living room to the kitchen and the bedroom to the beach, they provided continuous entertainment whether we were at work, at rest, or at play.

About the Brands in This Volume

Volume 1 shows 56 brand names, with pictures of 581 models and more than 1100 other models listed in the index. Some of the more noteworthy brands and models include:

- *Air King* Model A-600 (p.55)   A highly collectible radio because its body is made by Catalin.
- *Airline* Model O5GCB-1541A (p.134)   The "Lone Ranger" radio.
- *Andrea* (pp.168-171)   By Frank Angelo d'Andrea, who also made the famous Fada line.
- *Automatic* Tom Thumb Camera/Radio (p.227)   A unique combination.
- *Capehart* (pp.263-273)   Whose reputation for quality led to the company slogan, "The World's Most Luxurious Musical Instrument."

Names to look for in Volumes 2 through 6 include Crosley, General Electric, Philco, Remler, and Zenith.

The radio pictures are grouped together by brand name. Within brand names, they are organized roughly by historical sequence as determined by when their specific PHOTOFACT set was published.

TOP TEN NETWORK RADIO PROGRAMS OF 1946	TOP TEN NETWORK RADIO PROGRAMS OF 1957
1. Jack Benny	1. Our Miss Brooks
2. Fibber McGee and Molly	2. Edgar Bergen
3. Bob Hope	3. Two for the Money
4. The Charlie McCarthy Show	4. Dragnet
5. Fred Allen	5. News from NBC
6. Radio Theatre	6. Gene Autry
7. Amos 'n' Andy	7. The Great Gildersleeve
8. Walter Winchell	8. You Bet Your Life
9. Red Skelton	9. Gunsmoke
10. The Screen Guild Players	10. People Are Funny

TOP TEN DAYTIME RADIO PROGRAMS OF 1946	TOP TEN DAYTIME RADIO PROGRAMS OF 1957
1. When a Girl Marries	1. Wendy Warren
2. Young Widder Brown	2. Arthur Godfrey
3. Our Gal, Sunday	3. Helen Trent
4. Portia Faces Life	4. Guiding Light
5. Kate Smith Speaks	5. Young Dr. Malone
6. Ma Perkins	6. Our Gal Sunday
7. Breakfast in Hollywood	7. The Second Mrs. Burton
8. Aunt Jenny	8. The Road of Life
9. Right to Happiness	9. Aunt Jenny
10. The Romance of Helen Trent	10. My True Story

ADMIRAL

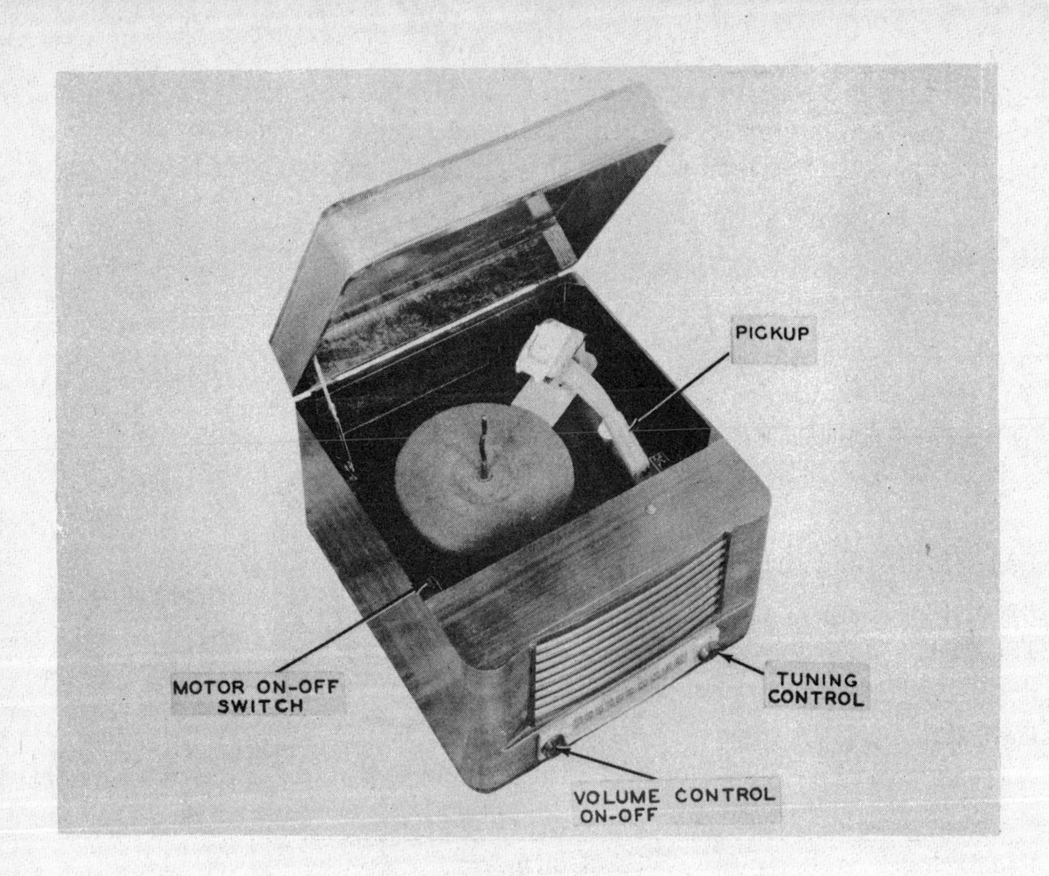

MODEL PICTURED
6RT43

AC superheterodyne radio & auto phono, self-contained loop antenna

TUBES
5

POWER SUPPLY
117 volts AC

TUNING RANGE
540-1630KC

MFR/SUPPLIER
Admiral Corp.

PHOTOFACT SET
4-24

PUBLISHED
1946

ADMIRAL

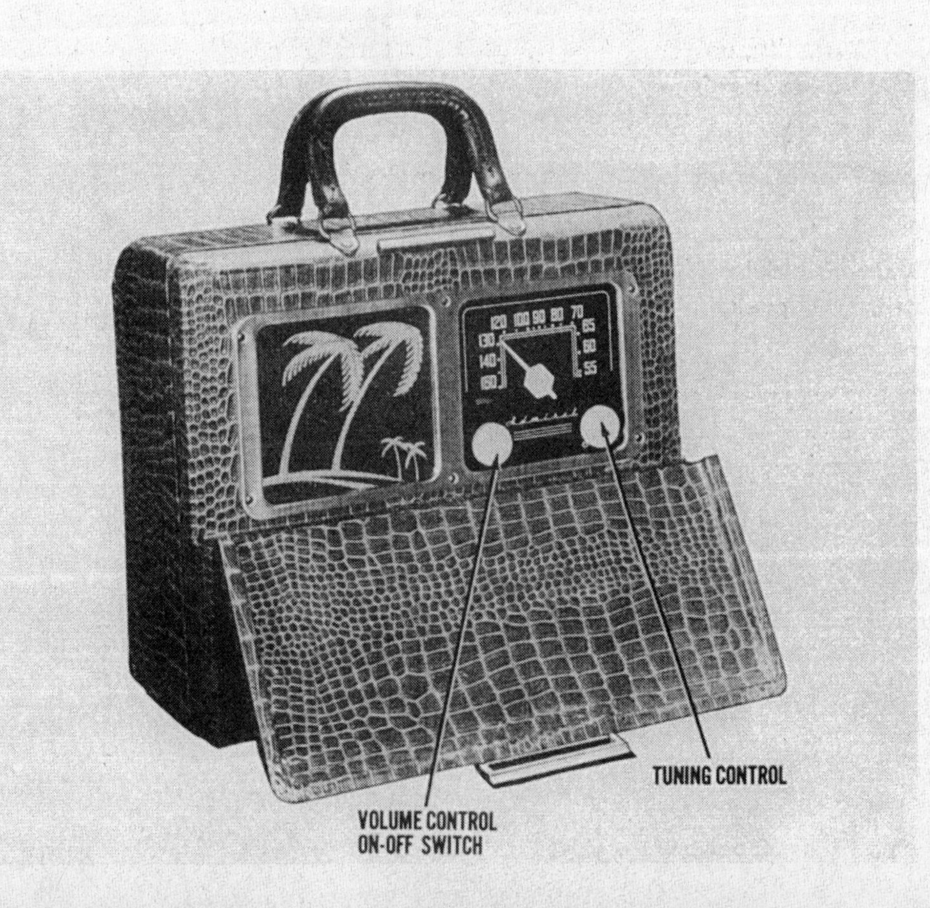

MODEL PICTURED
6P32

Three-power portable AC/DC battery superheterodyne, self-contained loop antenna

TUBES
6

POWER SUPPLY
105-125 volts AC/DC or 90 volts B supply, 9 volts A supply in pack form

TUNING RANGE
535-1620KC

MFR/SUPPLIER
Admiral Corp.

PHOTOFACT SET
6-1

PUBLISHED
1946

ADMIRAL

MODEL PICTURED
6RT42A

AC operated phono-radio combo superhet with loop antenna

TUBES
5

POWER SUPPLY
110-120 volts AC

TUNING RANGE
540-1630KC

MFR/SUPPLIER
Admiral Corp.

PHOTOFACT SET
18-1

PUBLISHED
1947

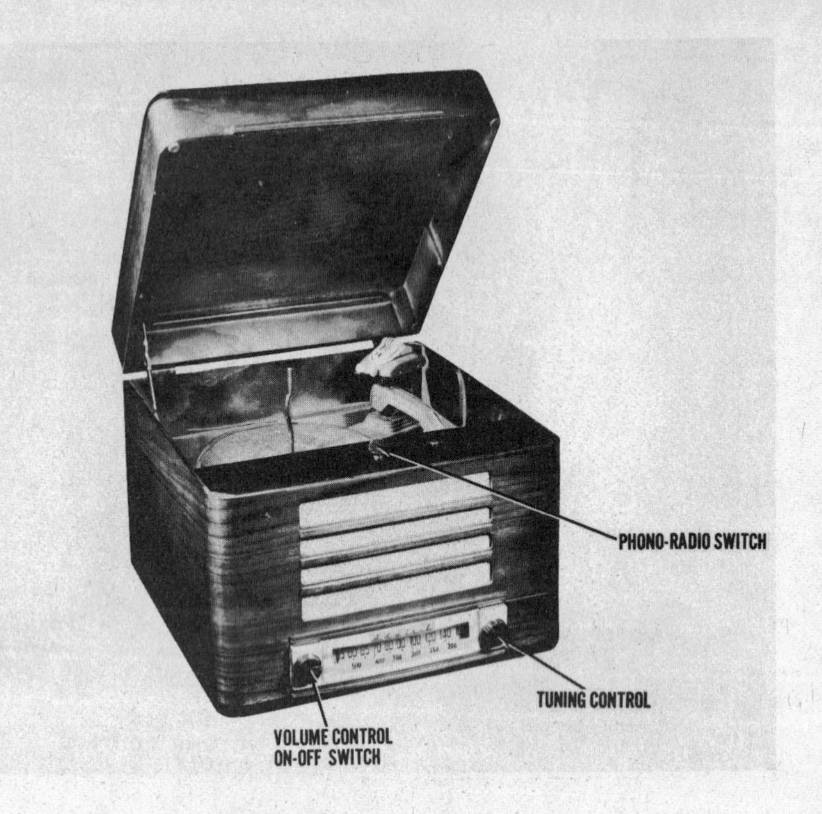

ADMIRAL

MODEL PICTURED
6RT44

AC operated phono-radio combo superheterodyne with loop antenna

TUBES
7

POWER SUPPLY
110-120 volts AC

TUNING RANGE
531-1650KC, 8.9-12.5MC

MFR/SUPPLIER
Admiral Corp.

PHOTOFACT SET
18-2

PUBLISHED
1947

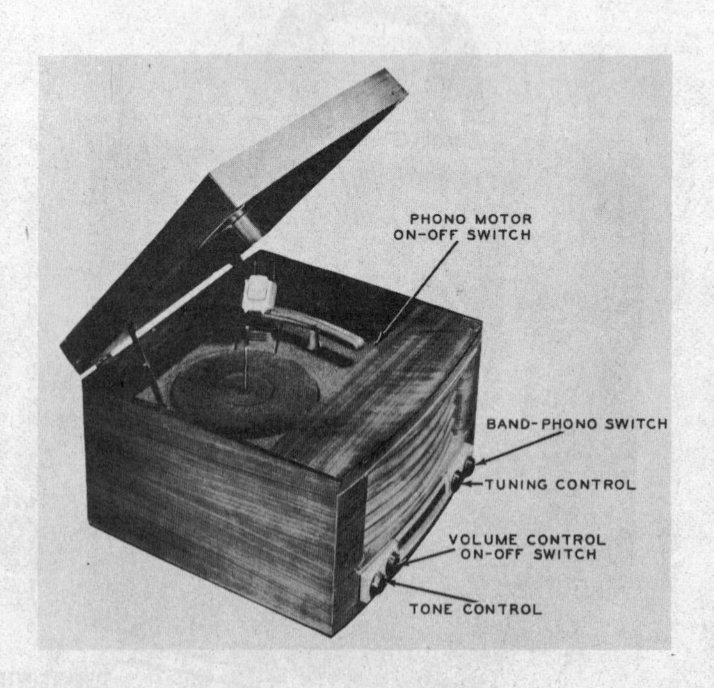

ADMIRAL

MODEL PICTURED
7T12

Battery operated superheterodyne receiver

TUBES
4

POWER SUPPLY
1.5 volts A & 90 volts B supply in pack form

TUNING RANGE
535-1630KC

MFR/SUPPLIER
Admiral Corp.

PHOTOFACT SET
24-1

PUBLISHED
1947

ADMIRAL

MODEL PICTURED
7P33

Three-power portable superheterodyne receiver with loop antenna

TUBES
5

POWER SUPPLY
105-125 volts AC/DC, or 9 volts DC A & 90 volts DC B in pack form

TUNING RANGE
540-1620KC

MFR/SUPPLIER
Admiral Corp.

PHOTOFACT SET
26-1

PUBLISHED
1947

ADMIRAL

MODEL PICTURED
7RT42

AC operated combo
radio-phono
superheterodyne receiver

TUBES
6

POWER SUPPLY
110-120 volts AC

TUNING RANGE
540-1630KC

MFR/SUPPLIER
Admiral Corp.

PHOTOFACT SET
26-2

PUBLISHED
1947

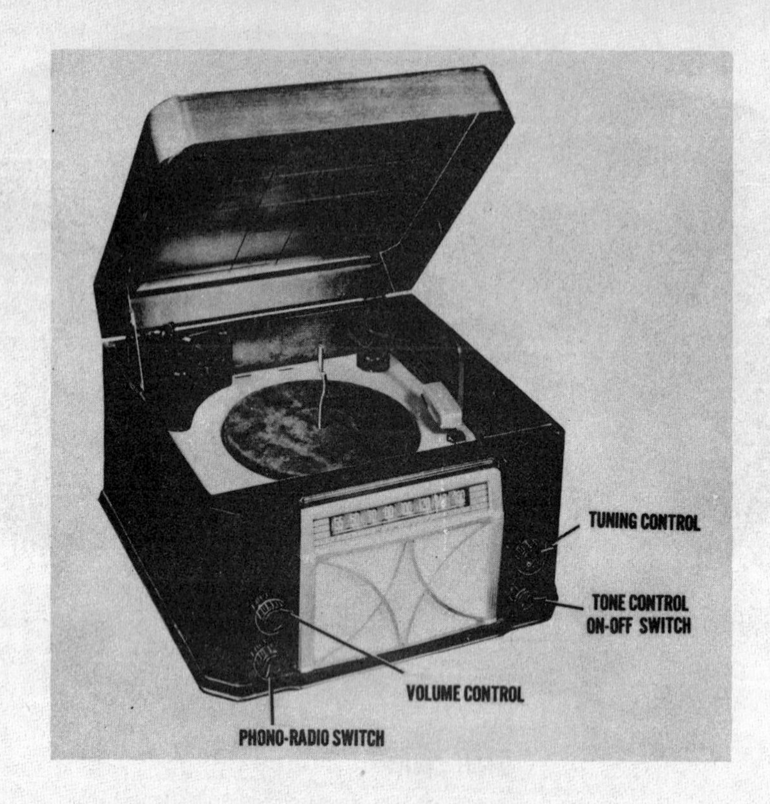

ADMIRAL

MODEL PICTURED
7T10

AC/DC operated
superheterodyne receiver
with loop antenna

TUBES
5

POWER SUPPLY
110-120 volts AC

TUNING RANGE
540-1630KC

MFR/SUPPLIER
Admiral Corp.

PHOTOFACT SET
30-1

PUBLISHED
1947

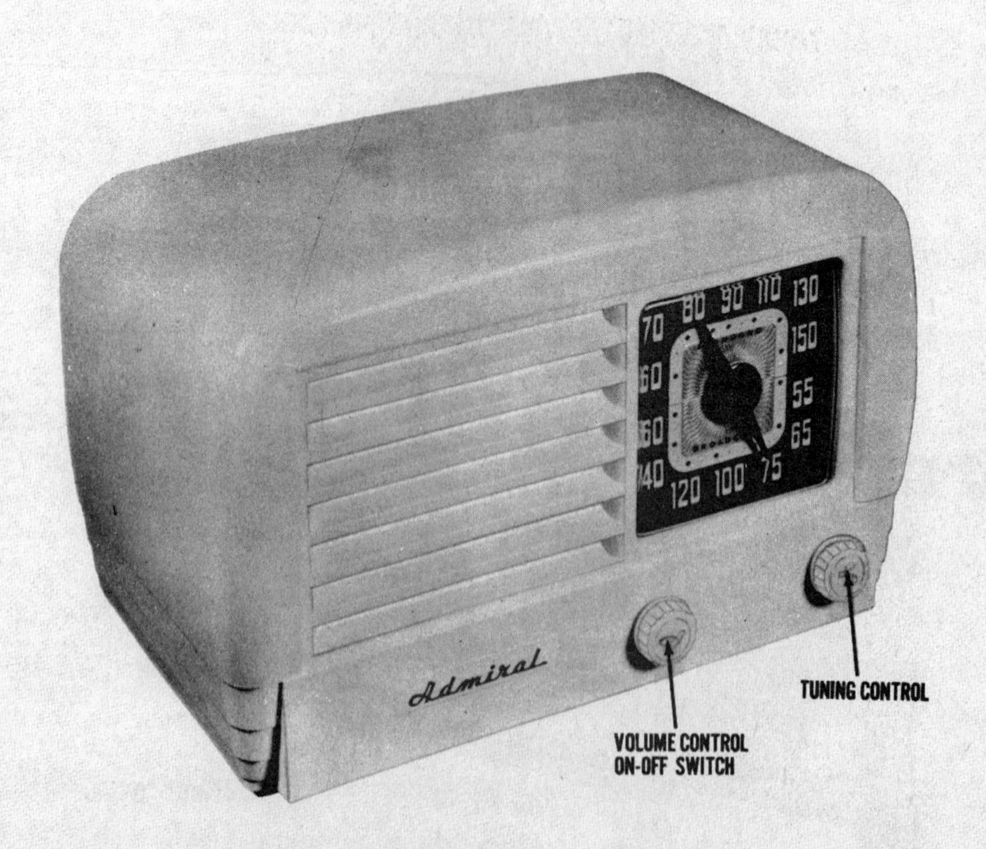

7T01M-UL

7T04-UL

ADMIRAL

MODEL PICTURED
7T01 and 7T04

AC/DC operated superheterodyne receiver with loop antenna

TUBES
5

POWER SUPPLY
110-120 volts AC/DC, .25 amp @117 volts AC

TUNING RANGE
540-1630KC

MFR/SUPPLIER
Admiral Corp.

PHOTOFACT SET
31-1

PUBLISHED
1948

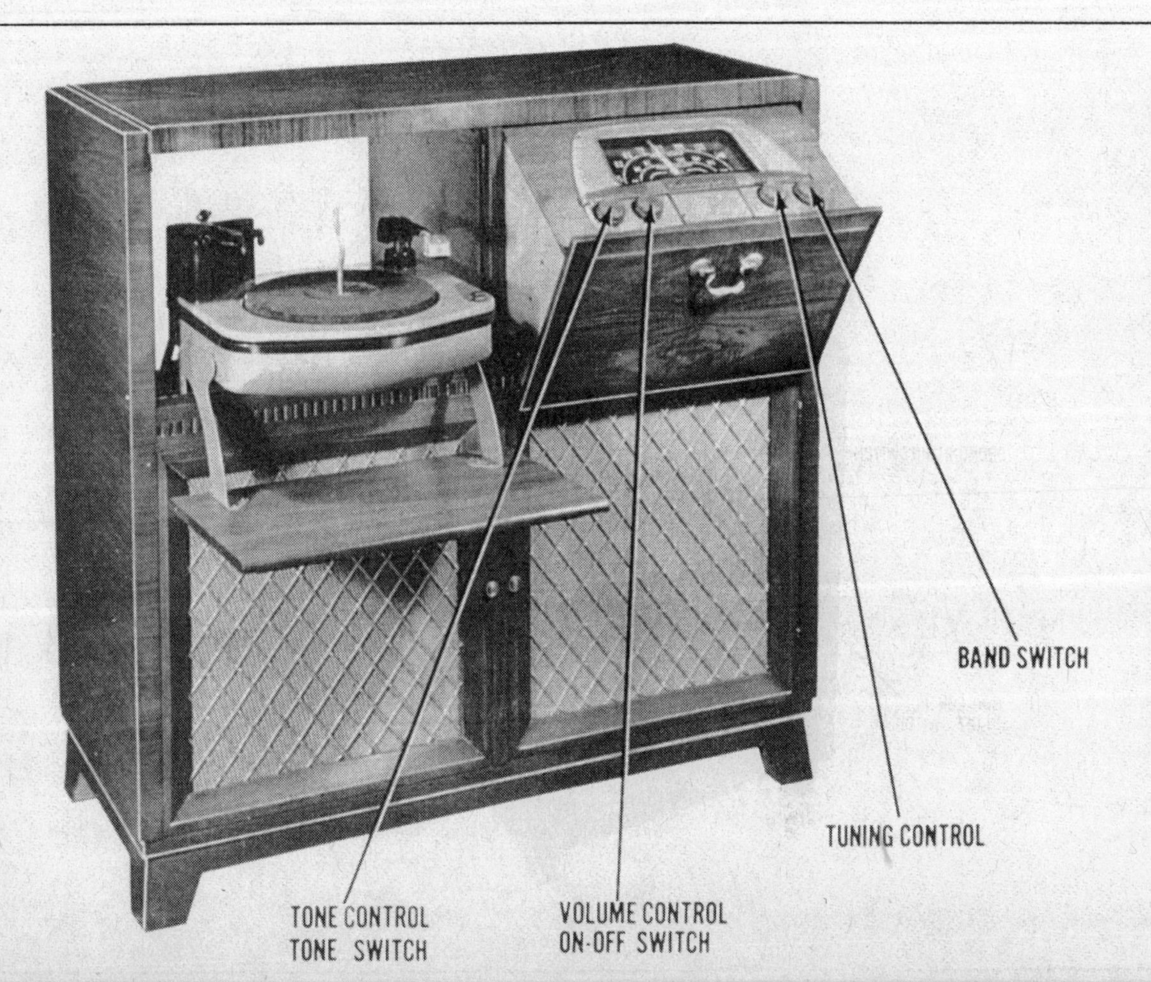

ADMIRAL

MODEL PICTURED
7C73

AC operated combo phono-radio AM/FM superheterodyne with loop antenna

TUBES
9

POWER SUPPLY
110-120 volts AC

TUNING RANGE
540-1605, 87.5-108.5MC

MFR/SUPPLIER
Admiral Corp.

PHOTOFACT SET
32-1

PUBLISHED
1948

ADMIRAL

MODEL PICTURED
7C65W

AC operated combo phono-radio superheterodyne receiver with loop antenna

TUBES
7

POWER SUPPLY
110-120 volts AC

TUNING RANGE
540-1630KC

MFR/SUPPLIER
Admiral Corp.

PHOTOFACT SET
36-1

PUBLISHED
1948

ADMIRAL

MODEL PICTURED
7C60M

AC operated phono-radio superhet receiver with loop antenna

TUBES
6

POWER SUPPLY
110-120 volts AC

TUNING RANGE
540-1630KC

MFR/SUPPLIER
Admiral Corp.

PHOTOFACT SET
48-2

PUBLISHED
1948

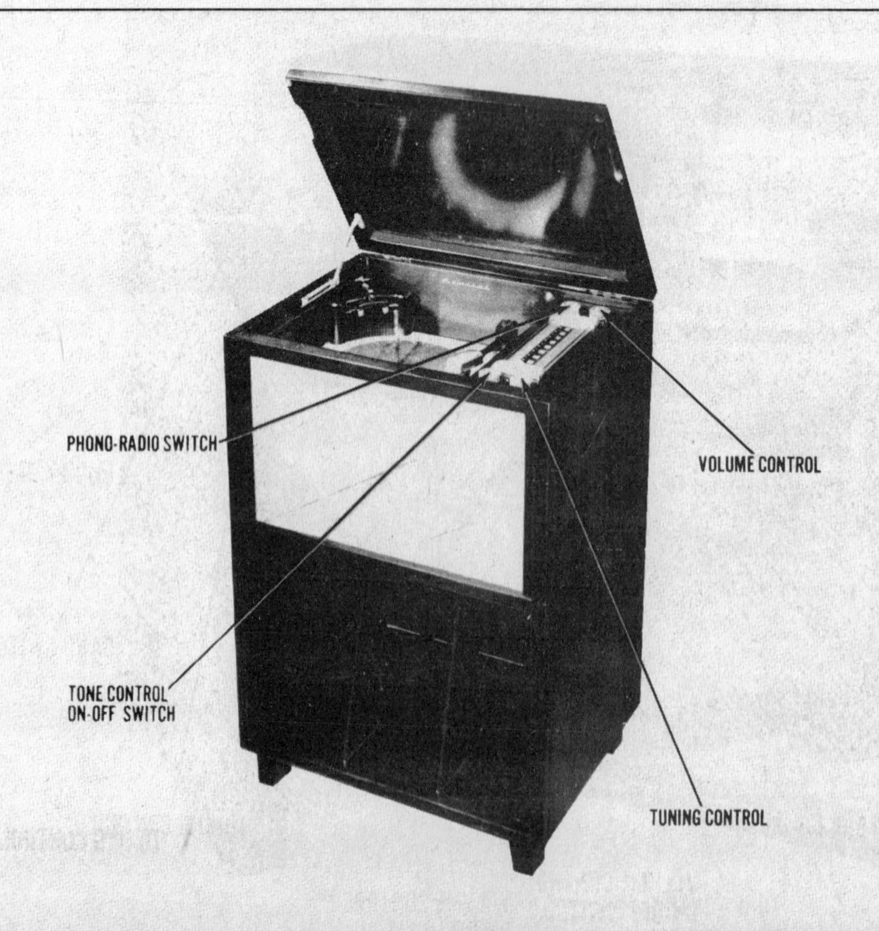

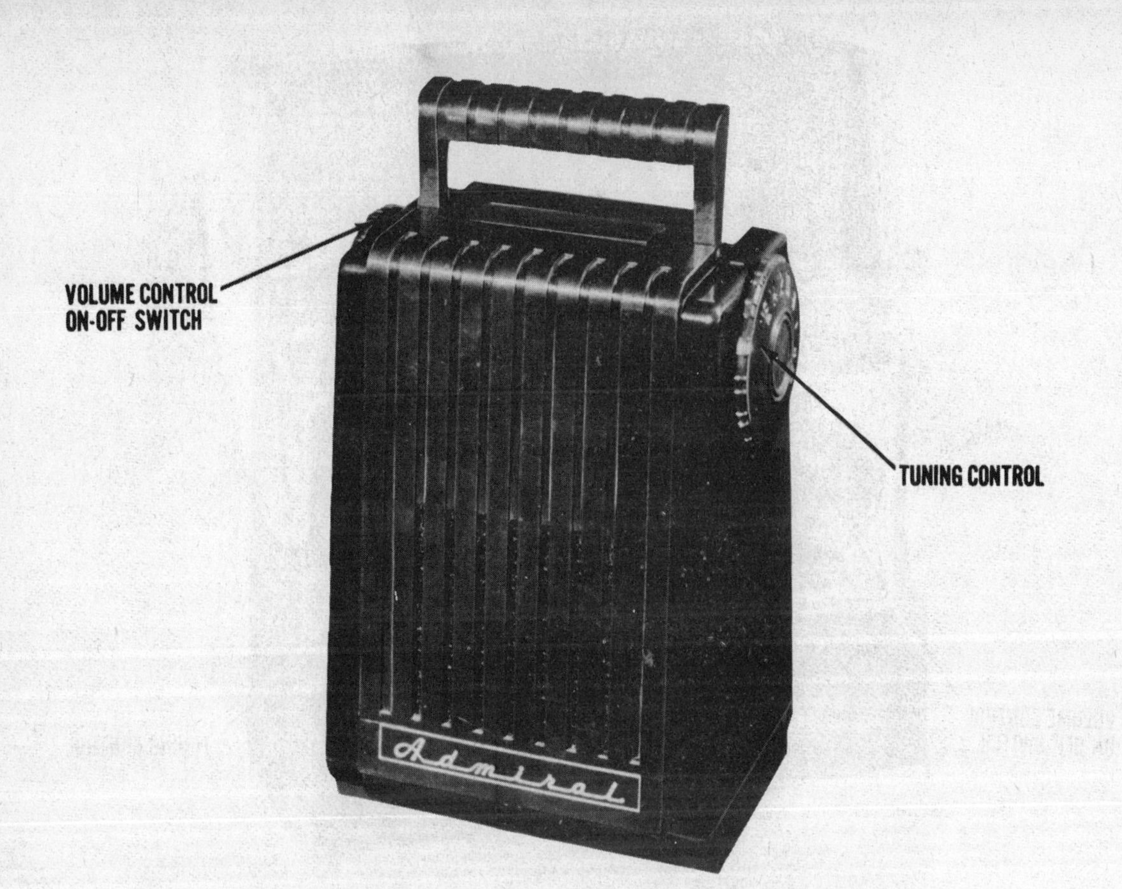

ADMIRAL

MODEL PICTURED
4D11

Battery operated portable superheterodyne receiver with loop antenna

TUBES
4

POWER SUPPLY
1.5 volts A & 67.5 volts B supply

TUNING RANGE
540-1630KC

MFR/SUPPLIER
Admiral Corp.

PHOTOFACT SET
49-1

PUBLISHED
1948

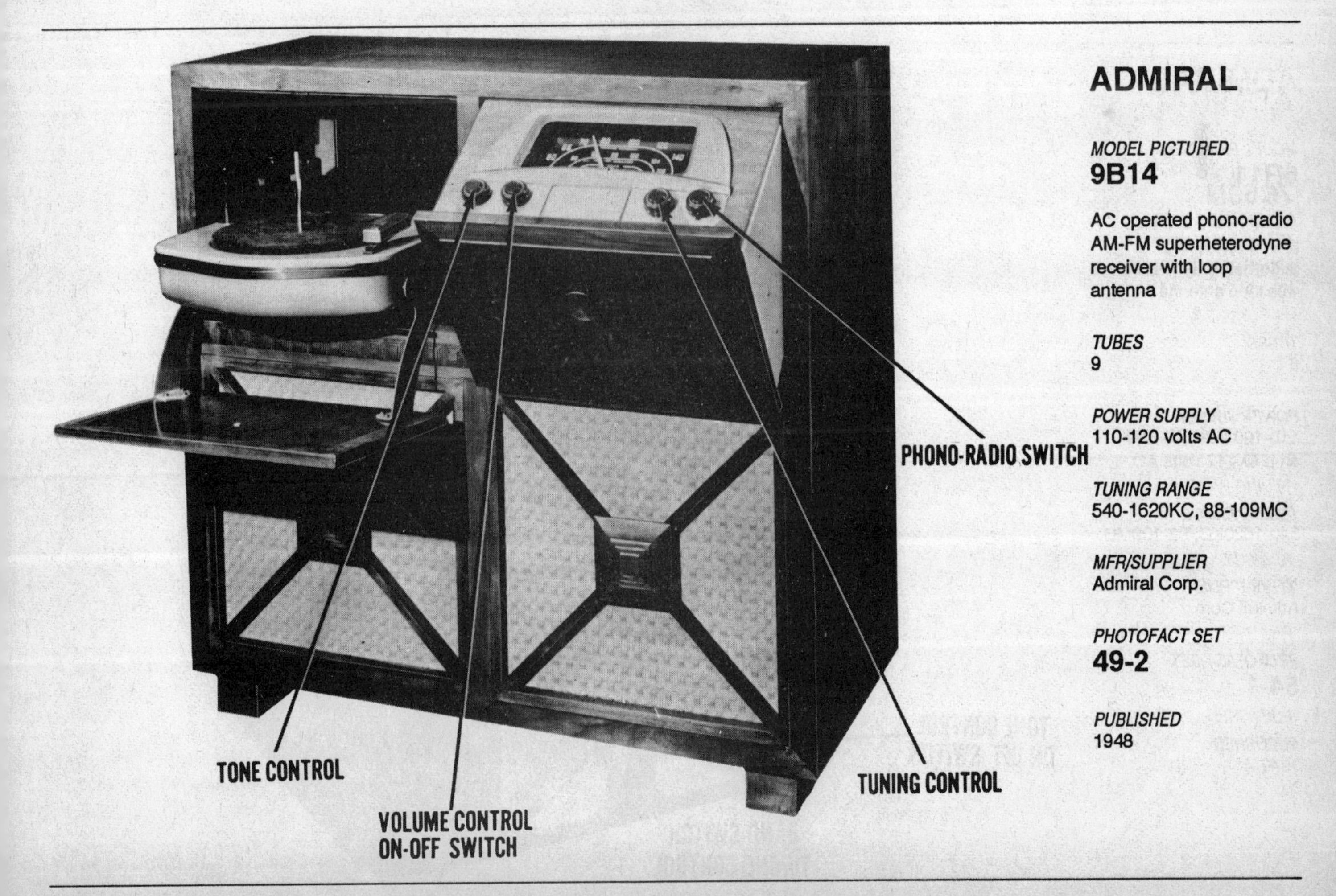

ADMIRAL

MODEL PICTURED
9B14

AC operated phono-radio AM-FM superheterodyne receiver with loop antenna

TUBES
9

POWER SUPPLY
110-120 volts AC

TUNING RANGE
540-1620KC, 88-109MC

MFR/SUPPLIER
Admiral Corp.

PHOTOFACT SET
49-2

PUBLISHED
1948

ADMIRAL

MODEL PICTURED
6C11

Three-power operated portable superheterodyne receiver with loop antenna

TUBES
5

POWER SUPPLY
110-120 volts AC/DC or 9 volts A supply & 90 volts B supply in pack form

TUNING RANGE
540-1620KC

MFR/SUPPLIER
Admiral Corp.

PHOTOFACT SET
53-1

PUBLISHED
1949

ADMIRAL

MODEL PICTURED
6R11

AC operated combo phono-radio, AM/FM superheterodyne receiver with loop antenna

TUBES
6

POWER SUPPLY
110-120 volts AC, .39 amp @ 117 volts AC

TUNING RANGE
540-1620KC, 88-109MC

MFR/SUPPLIER
Admiral Corp.

PHOTOFACT SET
54-1

PUBLISHED
1949

ADMIRAL

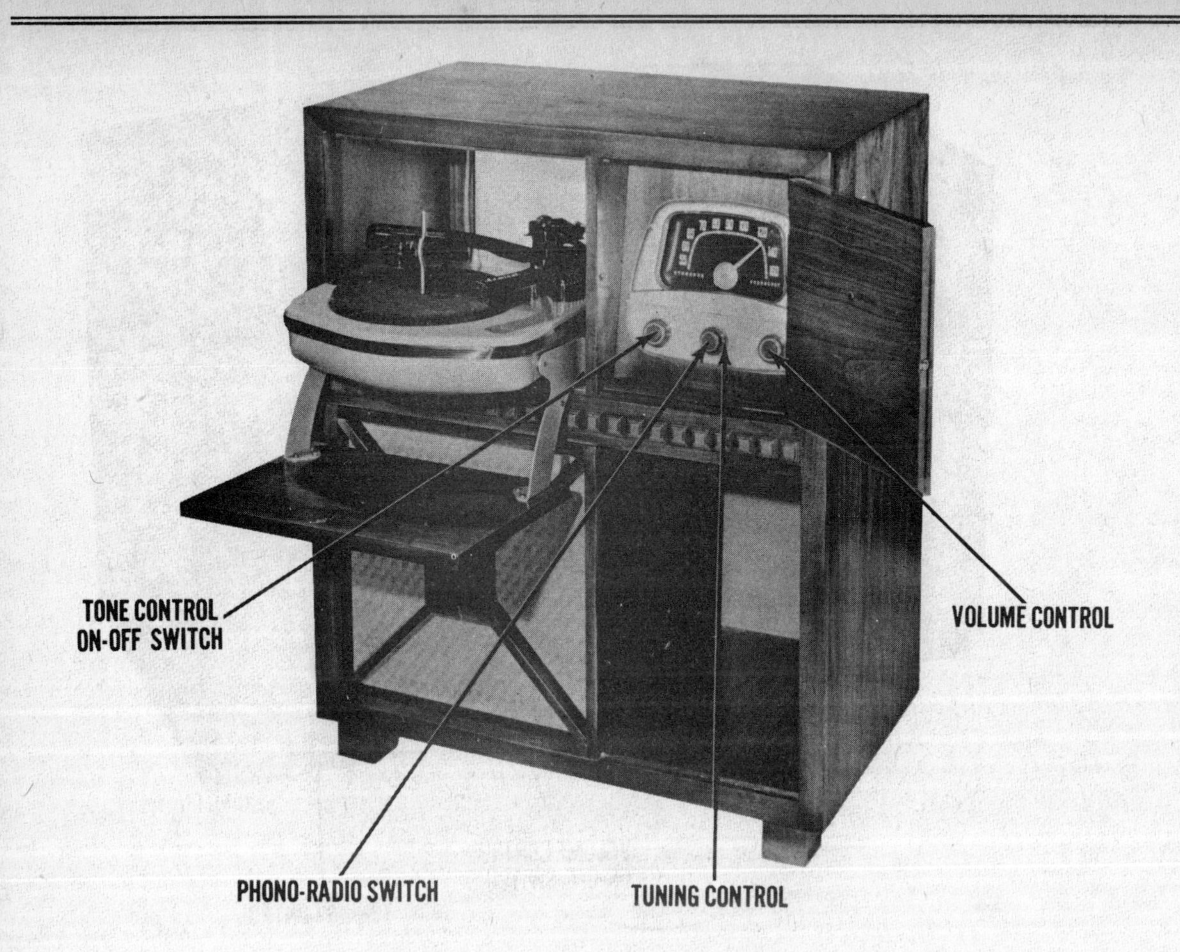

MODEL PICTURED
7G14

AC operated combo phono-radio superheterodyne receiver with loop antenna

TUBES
7

POWER SUPPLY
110-120 volts AC

TUNING RANGE
540-1620KC

MFR/SUPPLIER
Admiral Corp.

PHOTOFACT SET
54-2

PUBLISHED
1949

ADMIRAL

MODEL PICTURED
5F11

Three-power operated portable superheterodyne receiver with loop antenna

TUBES
4

POWER SUPPLY
110-120 volts AC/DC or 7.5 volts A & 67.5 volts B battery

TUNING RANGE
535-1620 KC

MFR/SUPPLIER
Admiral Corp.

PHOTOFACT SET
57-1

PUBLISHED
1949

ADMIRAL

MODEL PICTURED
5R11

AC/DC operated superheterodyne receiver with loop antenna

TUBES
5

POWER SUPPLY
110-120 volts AC/DC, .27 amp @ 117 volts AC

TUNING RANGE
530-1620KC

MFR/SUPPLIER
Admiral Corp.

PHOTOFACT SET
59-1

PUBLISHED
1949

ADMIRAL

MODEL PICTURED
6V12

AC operated phono-radio superheterodyne receiver with loop antenna

TUBES
6

POWER SUPPLY
110-120 volts AC

TUNING RANGE
540-1620KC

MFR/SUPPLIER
Admiral Corp.

PHOTOFACT SET
62-1

PUBLISHED
1949

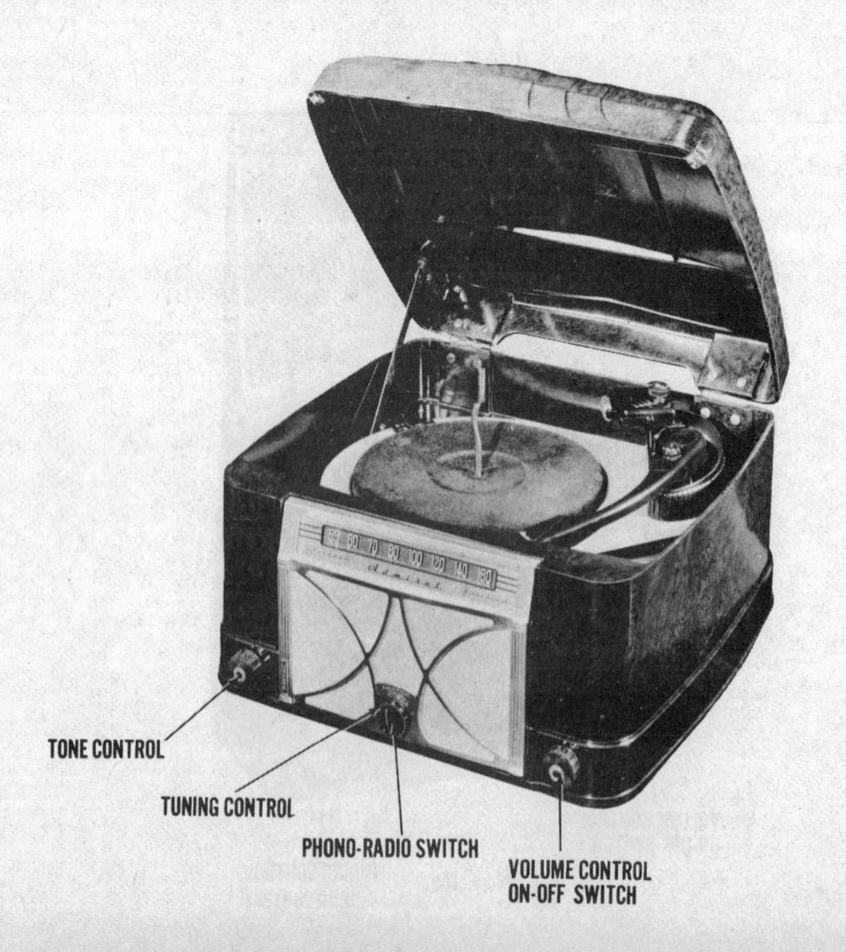

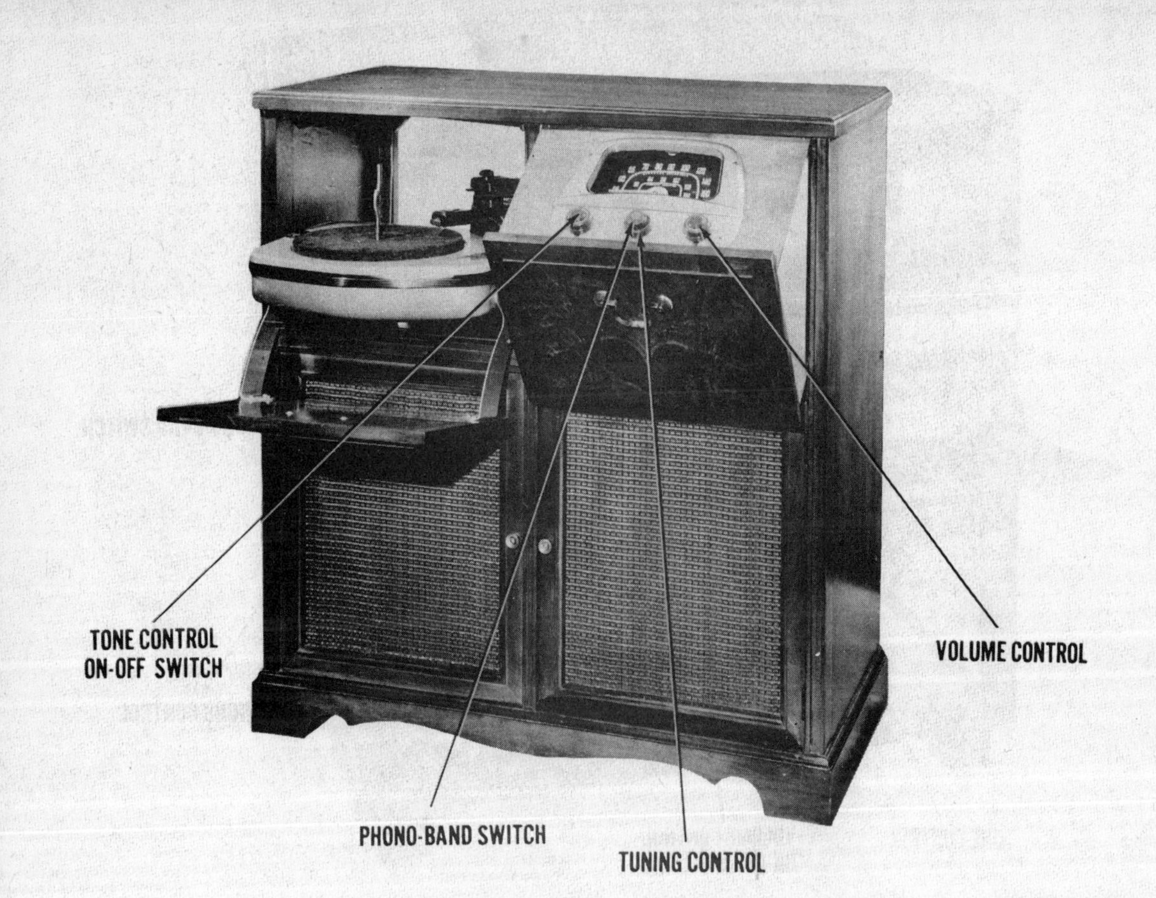

ADMIRAL

MODEL PICTURED
8D15

AC operated phono-radio, AM-FM superheterodyne receiver with loop antenna

TUBES
8

POWER SUPPLY
110-120 volts AC

TUNING RANGE
530-1620KC, 88-108MC

MFR/SUPPLIER
Admiral Corp.

PHOTOFACT SET
67-1

PUBLISHED
1949

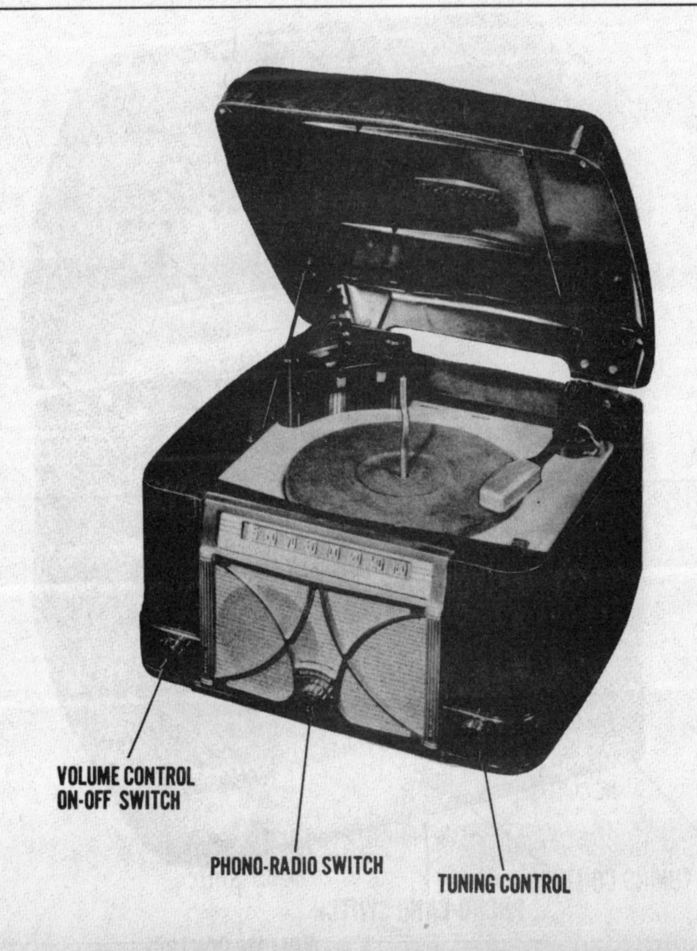

ADMIRAL

MODEL PICTURED
5T12

AC operated combo phono-radio, superheterodyne receiver with loop antenna

TUBES
5

POWER SUPPLY
110-120 volts AC, .25 amp @ 117 volts AC

TUNING RANGE
540-1620KC

MFR/SUPPLIER
Admiral Corp.

PHOTOFACT SET
68-1

PUBLISHED
1949

ADMIRAL

MODEL PICTURED
9E15

AC operated combo phono-radio, AM/FM superheterodyne receiver with loop antenna

TUBES
9

POWER SUPPLY
110-120 volts AC, .75 amp @ 117 volts AC

TUNING RANGE
540-1600KC, 88-108MC

MFR/SUPPLIER
Admiral Corp.

PHOTOFACT SET
68-2

PUBLISHED
1949

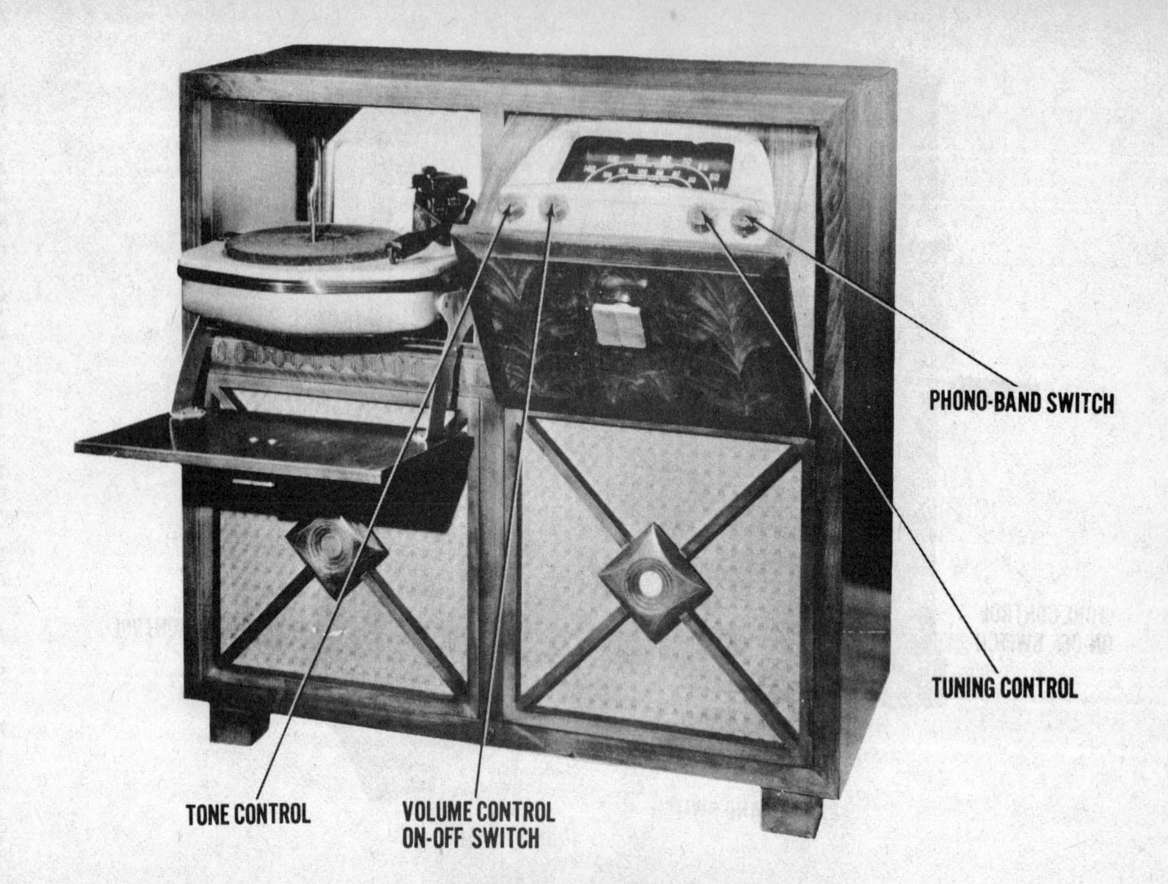

ADMIRAL

MODEL PICTURED
6W12

AC operated combo phono-radio, AM/FM superheterodyne receiver with loop antenna

TUBES
6

POWER SUPPLY
110-120 volts AC

TUNING RANGE
540-1620KC, 88-108MC

MFR/SUPPLIER
Admiral Corp.

PHOTOFACT SET
71-1

PUBLISHED
1949

ADMIRAL

MODEL PICTURED
6Y18

Three-power operated portable superheterodyne receiver with loop antenna

TUBES
5

POWER SUPPLY
110-120 volts AC/DC or 9 volts A & 90 volts B in pack form

TUNING RANGE
540-1620KC

MFR/SUPPLIER
Admiral Corp.

PHOTOFACT SET
75-1

PUBLISHED
1949

ADMIRAL

MODEL 5X13

MODEL 5X12

MODEL PICTURED
5X12

AC/DC operated superheterodyne receiver with loop antenna

TUBES
5

POWER SUPPLY
110-120 volts AC/DC

TUNING RANGE
550-1600KC

MFR/SUPPLIER
Admiral Corp.

PHOTOFACT SET
76-3

PUBLISHED
1949

ADMIRAL

MODEL PICTURED
6Q12

AC/DC operated AM/FM superheterodyne receiver with loop antenna

TUBES
6

POWER SUPPLY
110-120 volts AC, .26 amp @ 117 volts AC

TUNING RANGE
540-1620KC, 88-108MC

MFR/SUPPLIER
Admiral Corp.

PHOTOFACT SET
78-1

PUBLISHED
1949

ADMIRAL

MODEL PICTURED
5W12

AC operated combo phono-radio superheterodyne receiver with loop antenna

TUBES
5

POWER SUPPLY
110-120 volts AC

TUNING RANGE
550-1620KC

MFR/SUPPLIER
Admiral Corp.

PHOTOFACT SET
79-2

PUBLISHED
1949

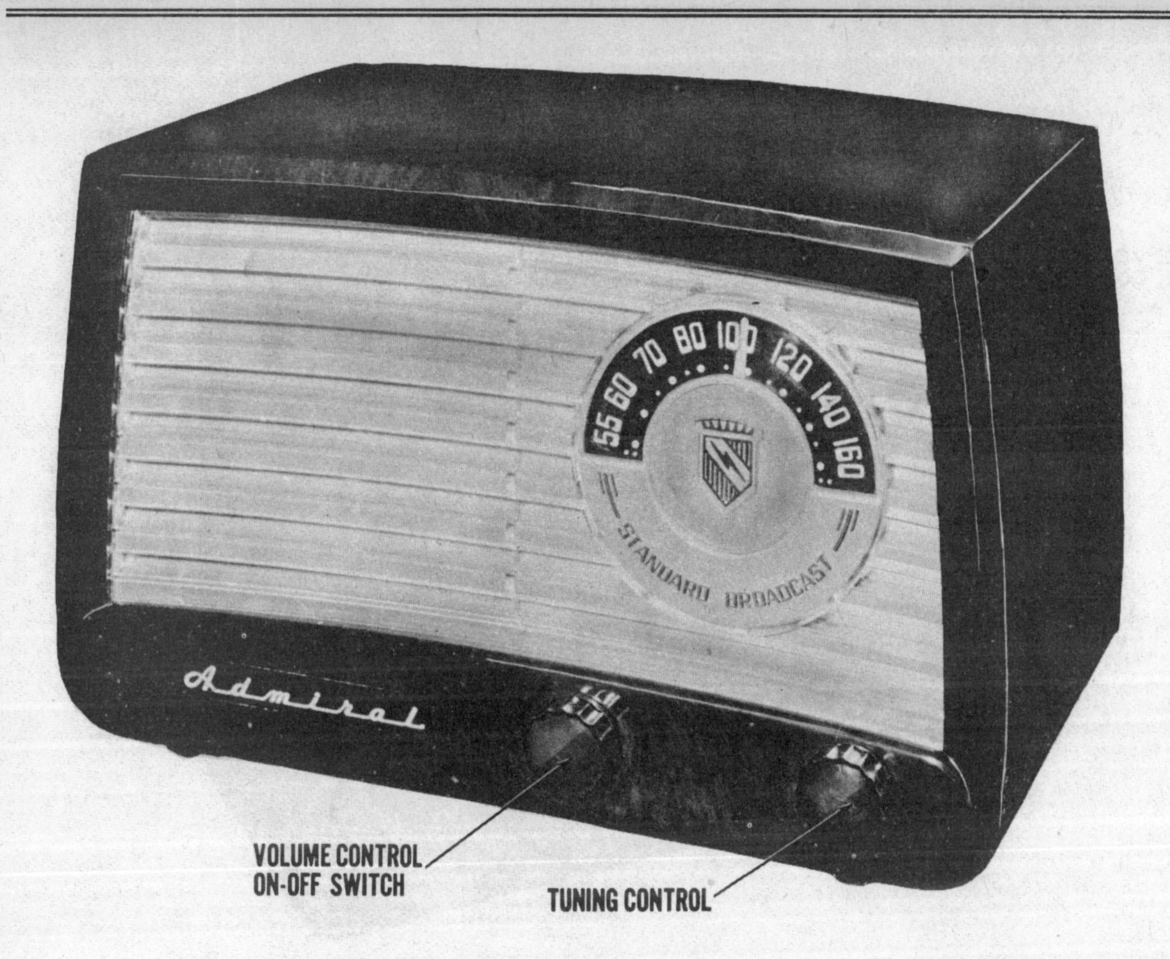

ADMIRAL

MODEL PICTURED
6A22

AC/DC operated superheterodyne receiver with loop antenna

TUBES
6

POWER SUPPLY
110-120 volts AC/DC

TUNING RANGE
540-1600KC

MFR/SUPPLIER
Admiral Corp.

PHOTOFACT SET
103-1

PUBLISHED
1950

ADMIRAL

MODEL PICTURED
6S12

AC operated phono-radio superheterodyne receiver with loop antenna

TUBES
6

POWER SUPPLY
110-120 volts AC

TUNING RANGE
540-1620KC

MFR/SUPPLIER
Admiral Corp.

PHOTOFACT SET
107-1

PUBLISHED
1950

ADMIRAL

MODEL PICTURED
4R11

Three-power operated portable superheterodyne receiver with loop antenna

TUBES
4

POWER SUPPLY
110-120 volts AC/DC or 7.5 volts A & 67.5 volts B supply

TUNING RANGE
540-1620KC

MFR/SUPPLIER
Admiral Corp.

PHOTOFACT SET
108-3

PUBLISHED
1950

ADMIRAL

MODEL PICTURED
5J21

AC/DC operated superheterodyne receiver with loop antenna

TUBES
5

POWER SUPPLY
110-120 volts AC/DC

TUNING RANGE
535-1620KC

MFR/SUPPLIER
Admiral Corp.

PHOTOFACT SET
136-2

PUBLISHED
1951

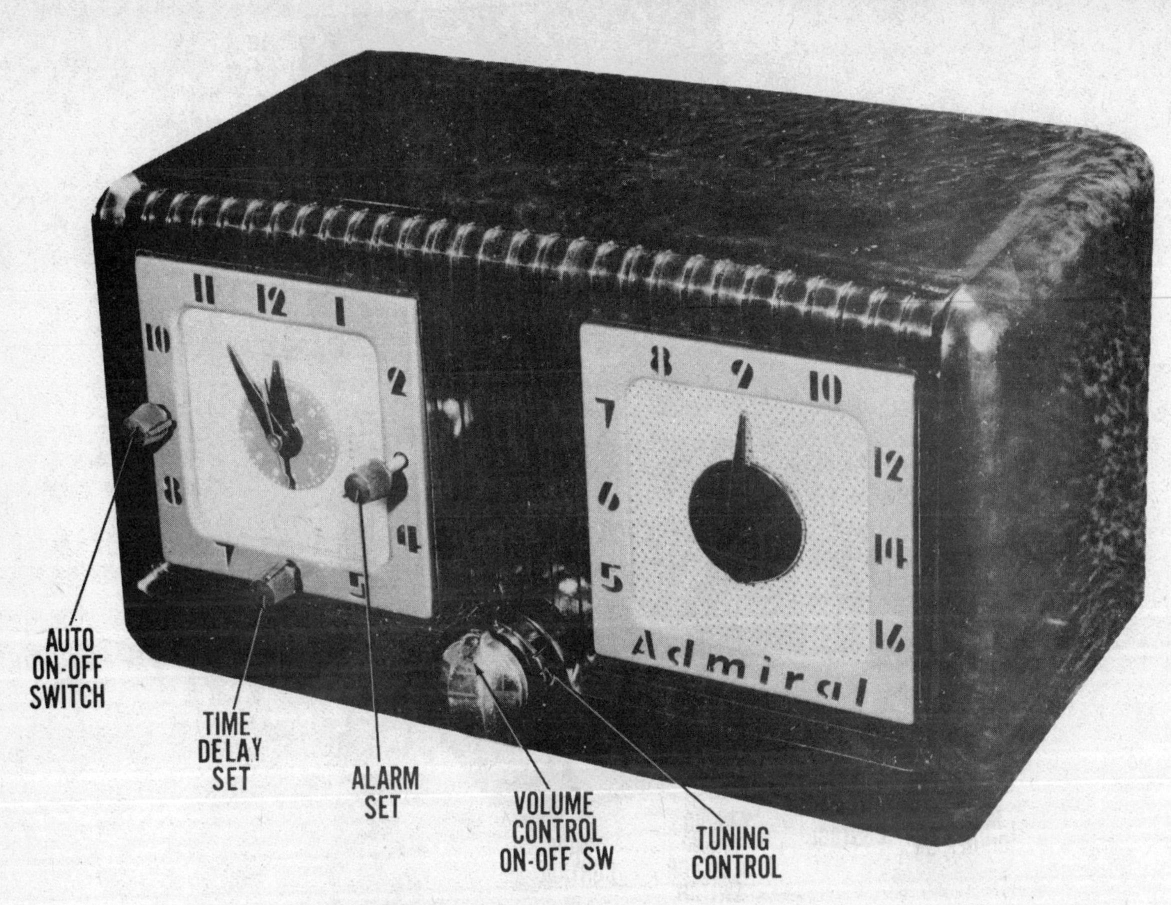

ADMIRAL

MODEL PICTURED
5G22

AC operated superheterodyne receiver with electric clock

TUBES
5

POWER SUPPLY
110-120 volts AC .27 amp @ 117 volts AC

TUNING RANGE
535-1620KC

MFR/SUPPLIER
Admiral Corp.

PHOTOFACT SET
137-2

PUBLISHED
1951

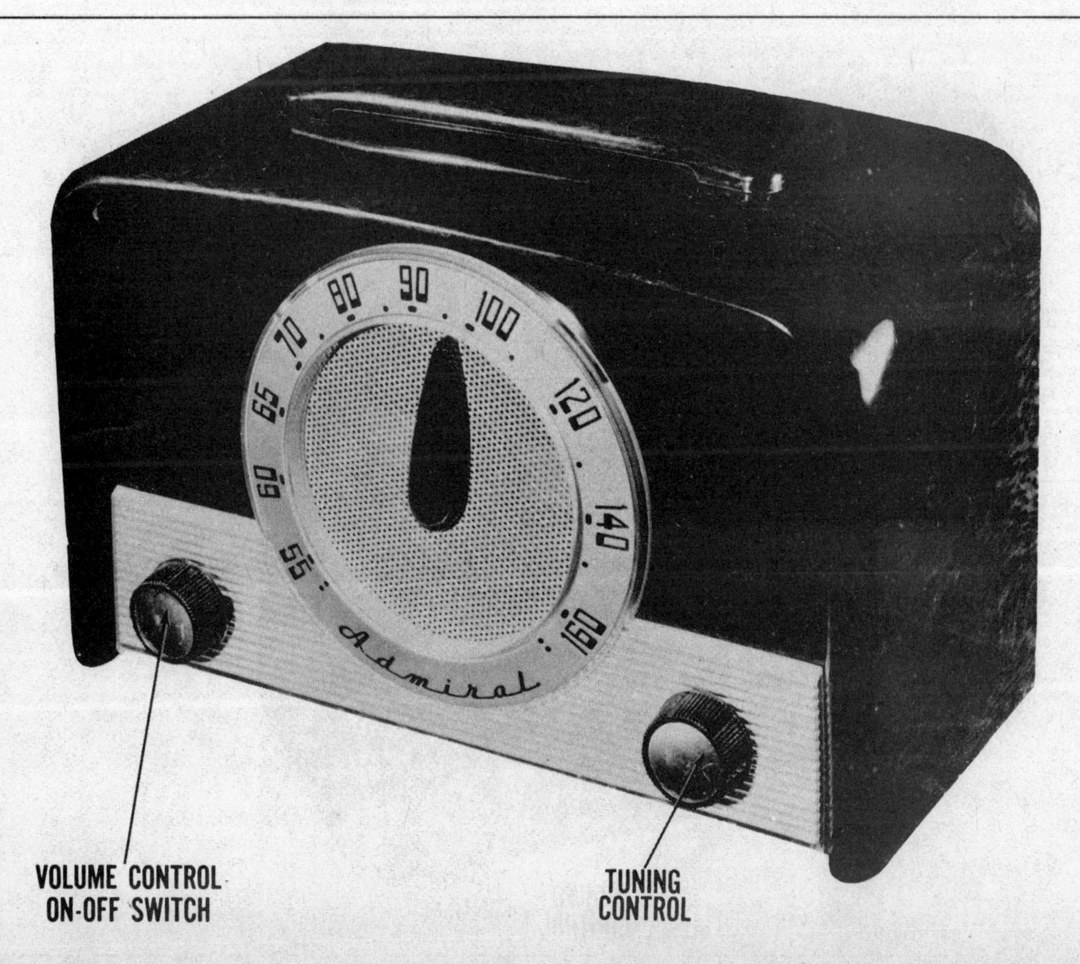

ADMIRAL

MODEL PICTURED
5E22

AC/DC operated superheterodyne receiver with loop antenna

TUBES
5

POWER SUPPLY
110-120 volts AC

TUNING RANGE
540-1620KC

MFR/SUPPLIER
Admiral Corp.

PHOTOFACT SET
139-2

PUBLISHED
1951

ADMIRAL

MODEL PICTURED
6J21

AC operated phono-radio superheterodyne receiver with loop antenna

TUBES
6

POWER SUPPLY
110-120 volts AC

TUNING RANGE
535-1620KC

MFR/SUPPLIER
Admiral Corp.

PHOTOFACT SET
140-2

PUBLISHED
1951

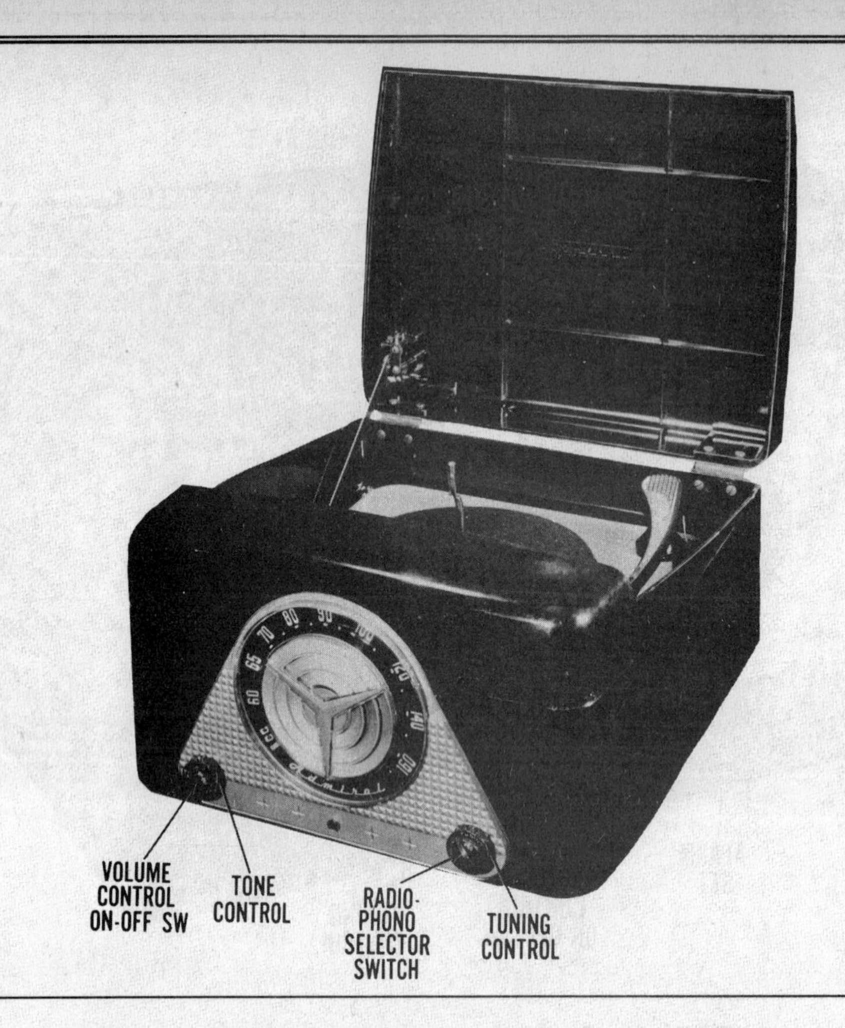

ADMIRAL

MODEL PICTURED
4W19

Three-power operated portable superheterodyne receiver with loop antenna

TUBES
4

POWER SUPPLY
110-120 volts AC/DC or 7.5 volts A supply & 67.5 volts B supply

TUNING RANGE
540-1620KC

MFR/SUPPLIER
Admiral Corp.

PHOTOFACT SET
143-2

PUBLISHED
1951

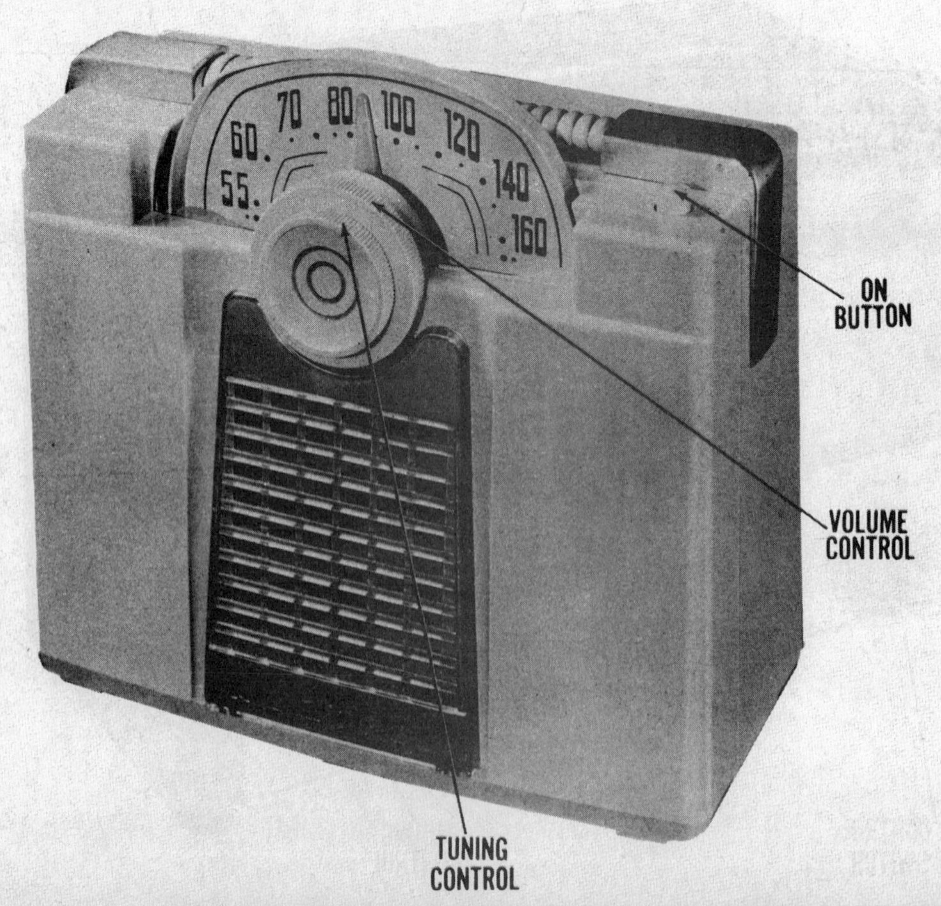

ADMIRAL

MODEL PICTURED
5M21

AC operated radio-phono combo superheterodyne receiver with loop antenna

TUBES
5

POWER SUPPLY
110-120 volts AC, .26 amp @ 117 volts AC

TUNING RANGE
540-1620KC

MFR/SUPPLIER
Admiral Corp.

PHOTOFACT SET
157-2

PUBLISHED
1952

ADMIRAL

MODEL PICTURED
5L21

AC operated superheterodyne receiver with electric clock

TUBES
5

POWER SUPPLY
110-120 volts AC, 60 cycles, .25 amp @ 117 volts AC

TUNING RANGE
540-1620KC

MFR/SUPPLIER
Admiral Corp.

PHOTOFACT SET
160-1

PUBLISHED
1952

ADMIRAL

MODEL PICTURED
6N26

AC operated radio-phono combo superheterodyne receiver with loop antenna

TUBES
6

POWER SUPPLY
110-120 volts AC, .51 amp @ 117 volts AC

TUNING RANGE
540-1620KC

MFR/SUPPLIER
Admiral Corp.

PHOTOFACT SET
165-3

PUBLISHED
1952

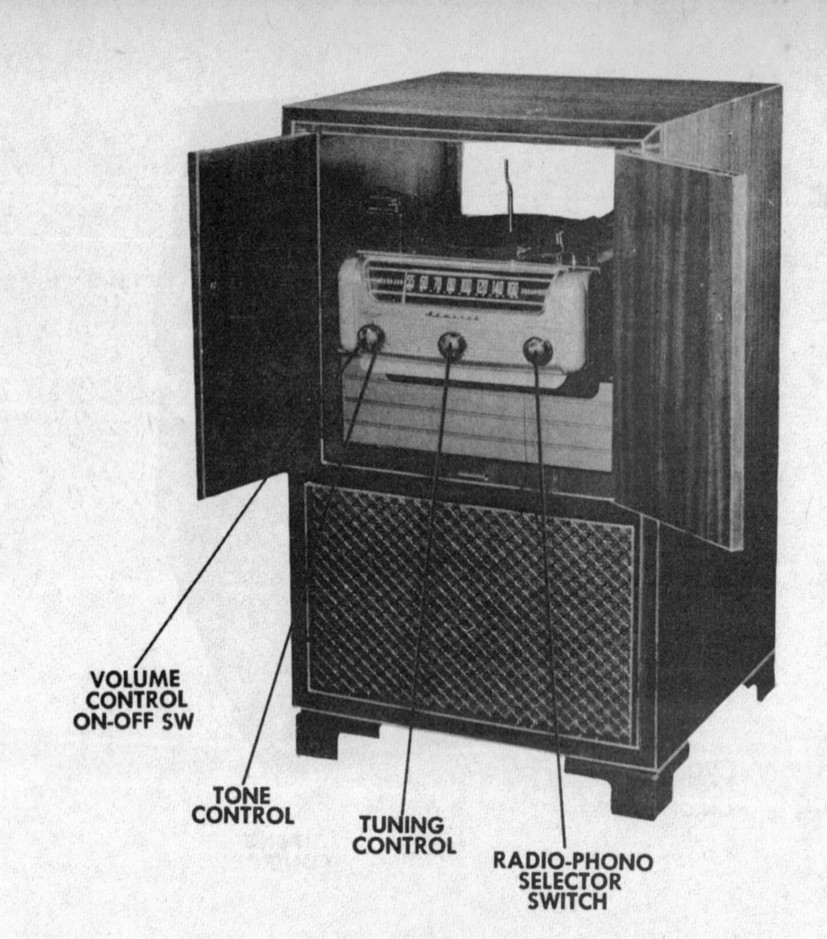

ADMIRAL

MODEL PICTURED
5A32/16

AC superheterodyne receiver with electric clock

TUBES
5

POWER SUPPLY
110-120 volts AC

TUNING RANGE
540-1620KC

MFR/SUPPLIER
Admiral Corp.

PHOTOFACT SET
191-2

PUBLISHED
1953

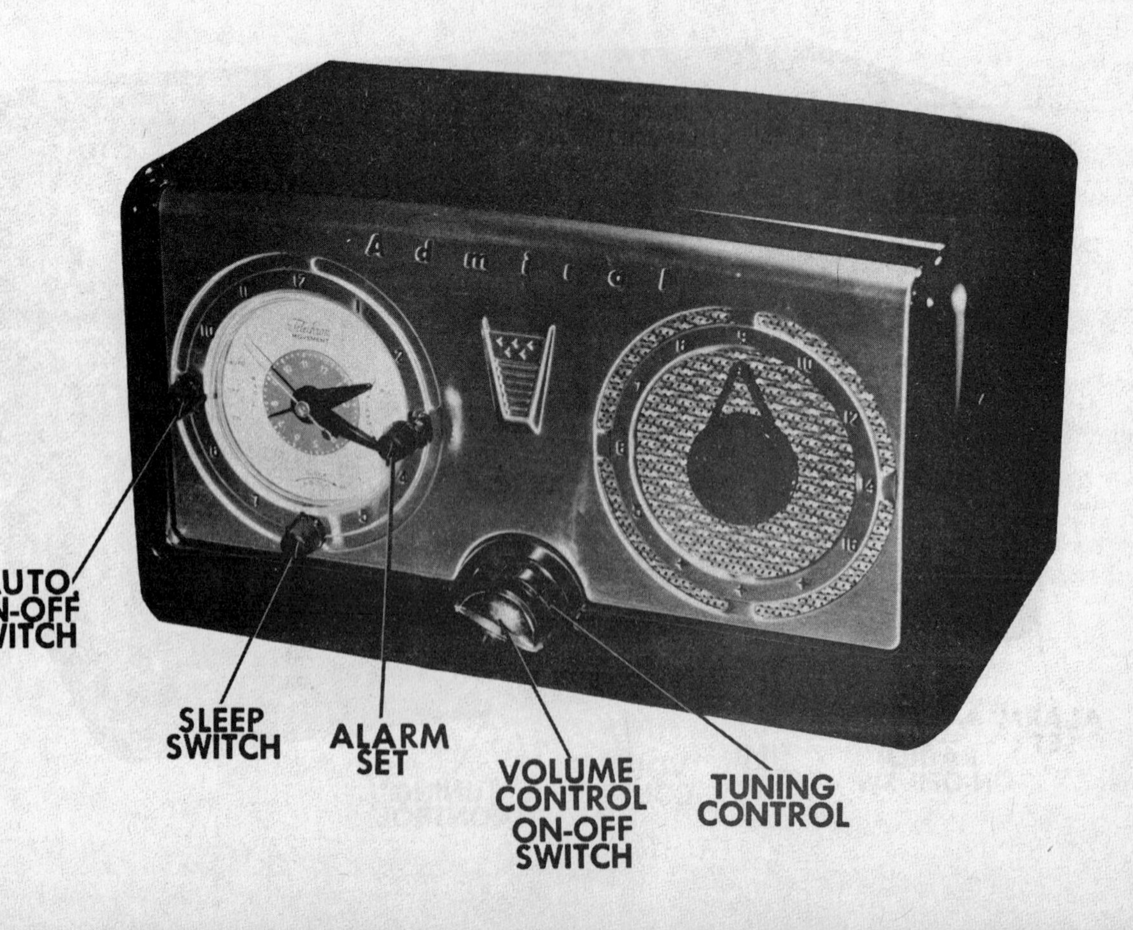

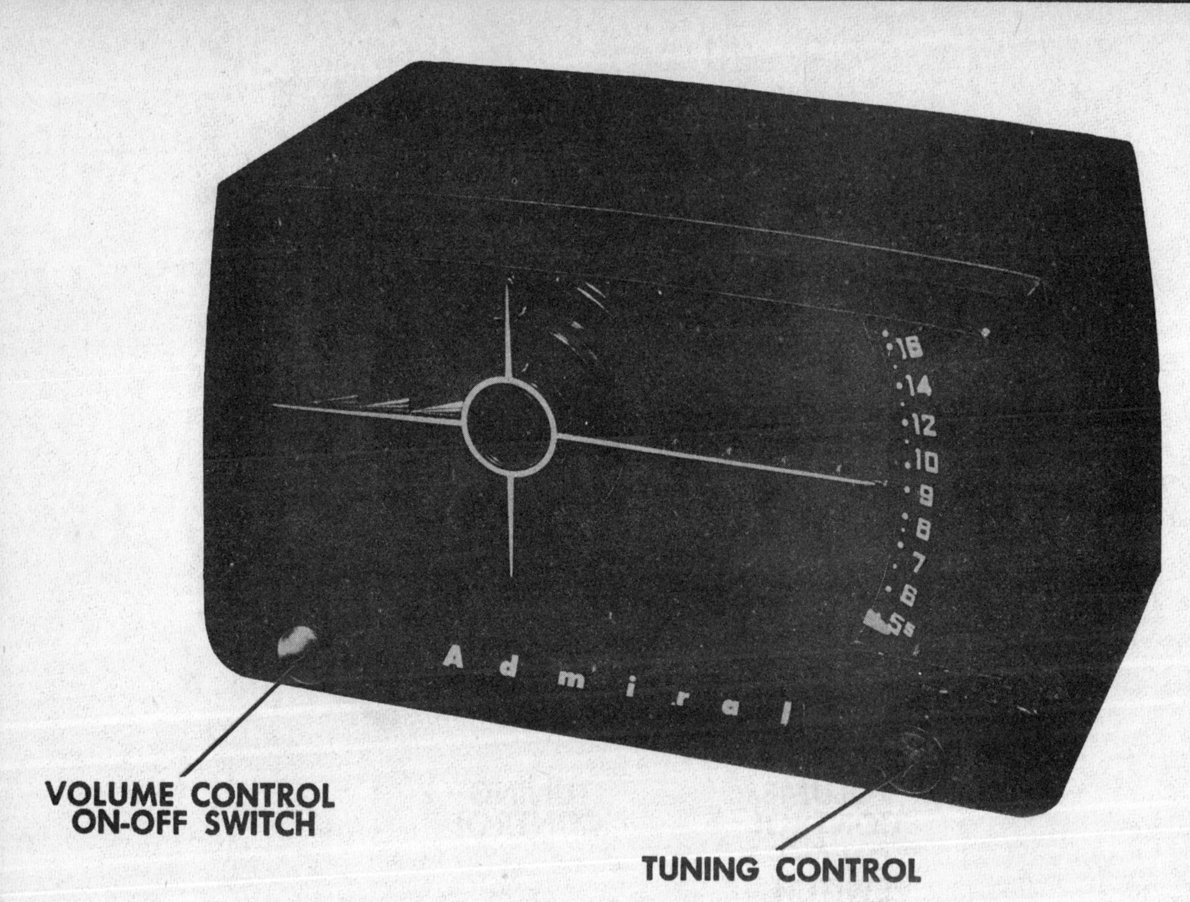

ADMIRAL

MODEL PICTURED
5S22AN

AC/DC operated AM superheterodyne receiver

TUBES
5

POWER SUPPLY
110-120 volts AC/DC, .25 amp @ 117 volts AC

TUNING RANGE
535-1620KC

MFR/SUPPLIER
Admiral Corp.

PHOTOFACT SET
197-2

PUBLISHED
1953

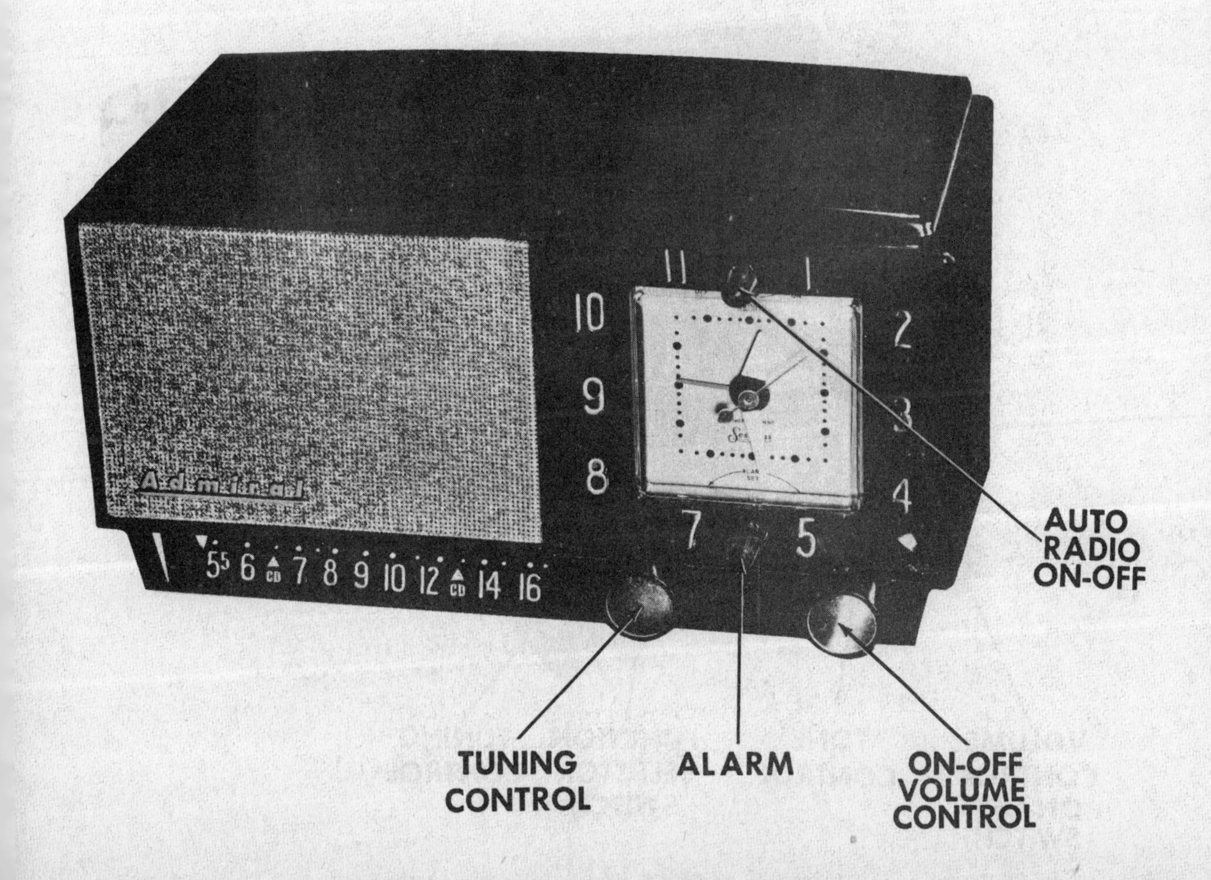

ADMIRAL

MODEL PICTURED
5E31

AC operated AM superheterodyne receiver with electric clock

TUBES
5

POWER SUPPLY
110-120 volts AC, 60 cycles, .25 amp @ 117 volts AC

TUNING RANGE
540-1620KC

MFR/SUPPLIER
Admiral Corp.

PHOTOFACT SET
224-2

PUBLISHED
1953

ADMIRAL

MODEL PICTURED
6C23A

AC/DC AM
superheterodyne receiver

TUBES
6

POWER SUPPLY
110-120 volts AC/DC

TUNING RANGE
540-1620KC

MFR/SUPPLIER
Admiral Corp.

PHOTOFACT SET
252-3

PUBLISHED
1954

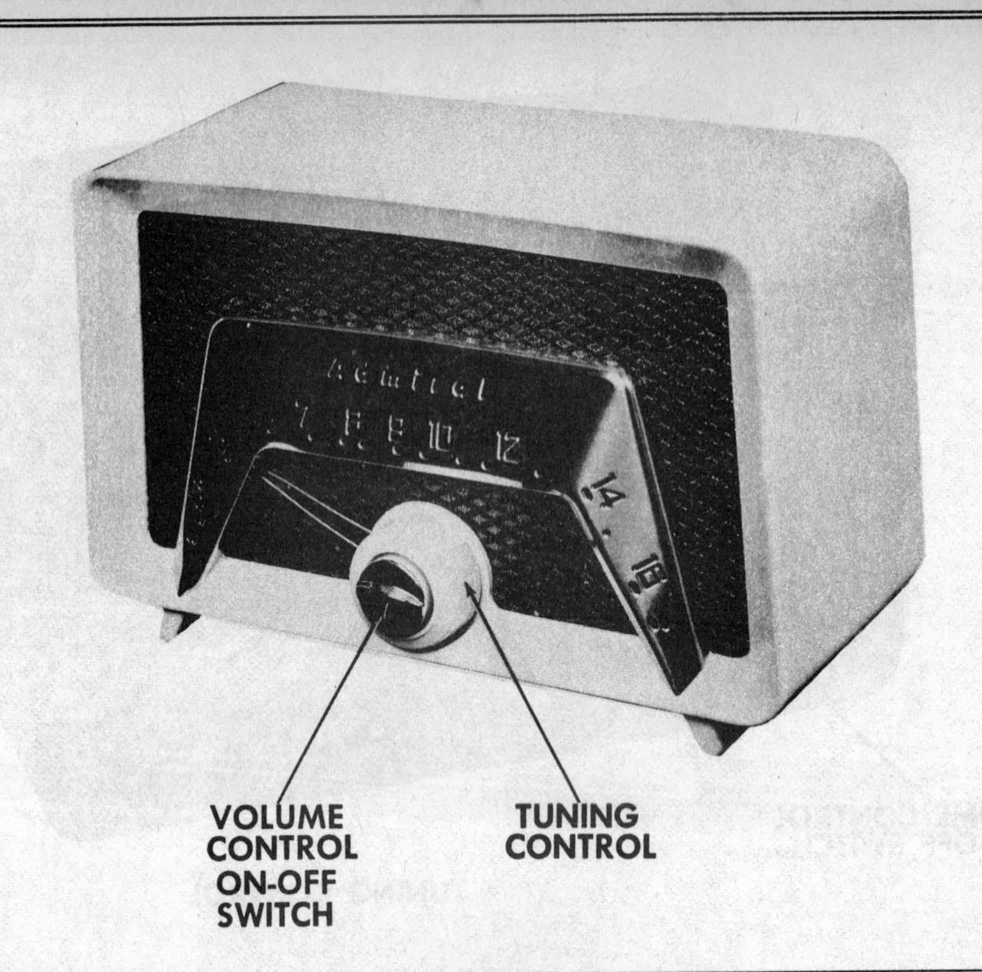

ADMIRAL

MODEL PICTURED
5D31

AC operated radio-phono
superheterodyne receiver

TUBES
5

POWER SUPPLY
110-120 volts AC, 60 cycles

TUNING RANGE
540-1620KC

MFR/SUPPLIER
Admiral Corp.

PHOTOFACT SET
256-3

PUBLISHED
1954

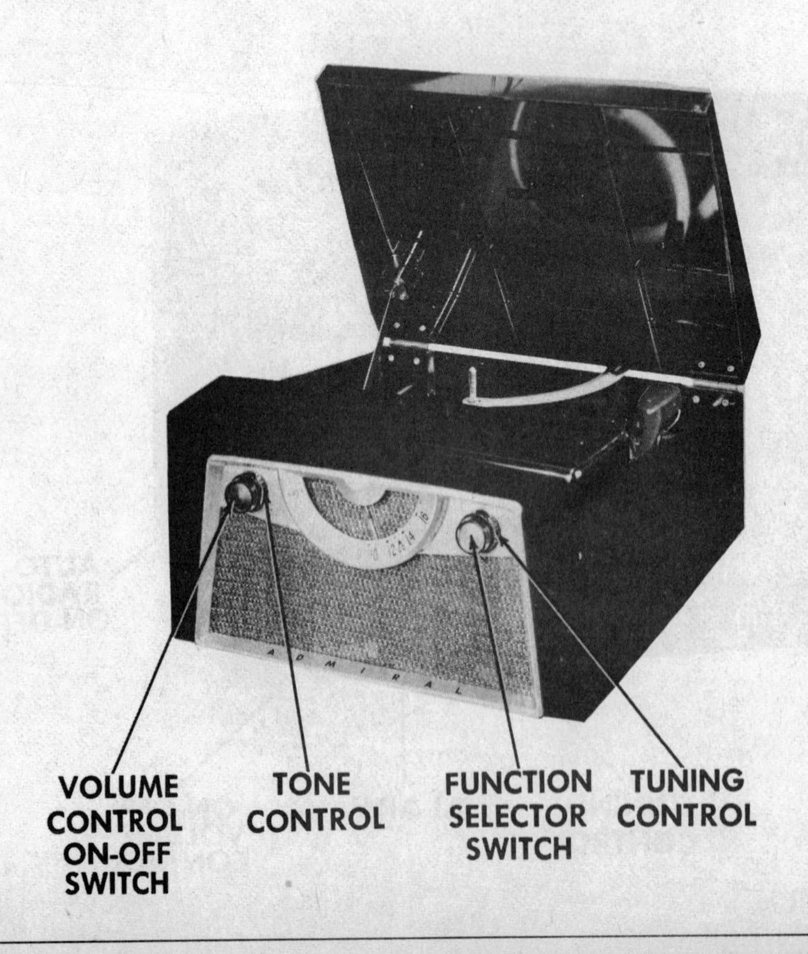

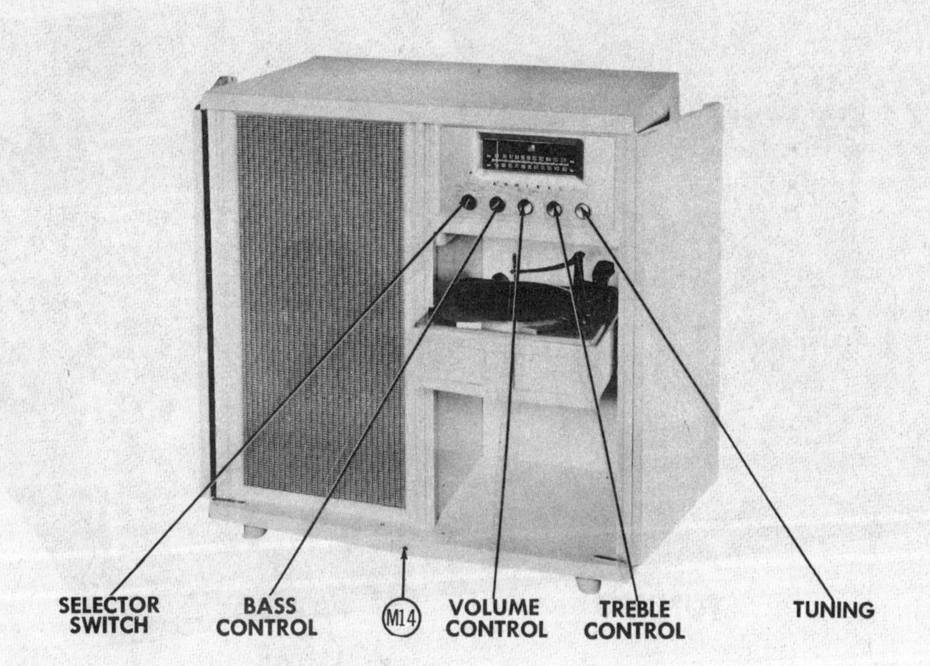

ADMIRAL

MODEL PICTURED
HIFI6

AC operated
AM-FM-phono superhet receiver with three-speed auto record changer

TUBES
20

POWER SUPPLY
110-120 volts AC, 60 cycles

TUNING RANGE
550-1700KC, 88-108MC

MFR/SUPPLIER
Admiral Corp.

PHOTOFACT SET
258-2

PUBLISHED
1954

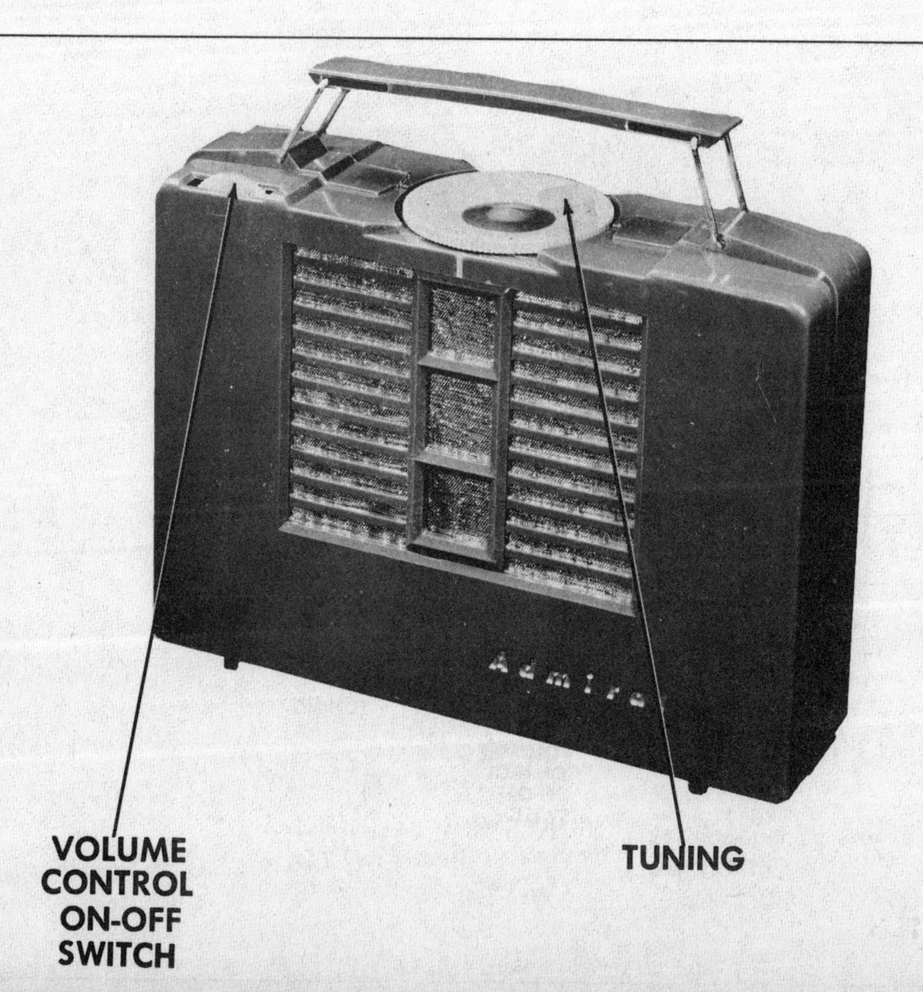

ADMIRAL

MODEL PICTURED
4X11

Battery operated portable AM superheterodyne receiver

TUBES
4

POWER SUPPLY
1.5 volts A & 67.5 volts B supply

TUNING RANGE
535-1620KC

MFR/SUPPLIER
Admiral Corp.

PHOTOFACT SET
261-1

PUBLISHED
1954

ADMIRAL

MODEL PICTURED
5R32

AC/DC operated AM superheterodyne receiver; model 5S3 has electric clock

TUBES
5

POWER SUPPLY
110-120 volts AC/DC

TUNING RANGE
535-1620KC

MFR/SUPPLIER
Admiral Corp.

PHOTOFACT SET
272-1

PUBLISHED
1955

ADMIRAL

MODEL PICTURED
4Z11

Three-power portable AM superheterodyne receiver

TUBES
4

POWER SUPPLY
110-120 volts AC/DC or 7.5 volts A & 67.5 volts B supply

TUNING RANGE
535-1620KC

MFR/SUPPLIER
Admiral Corp.

PHOTOFACT SET
274-2

PUBLISHED
1955

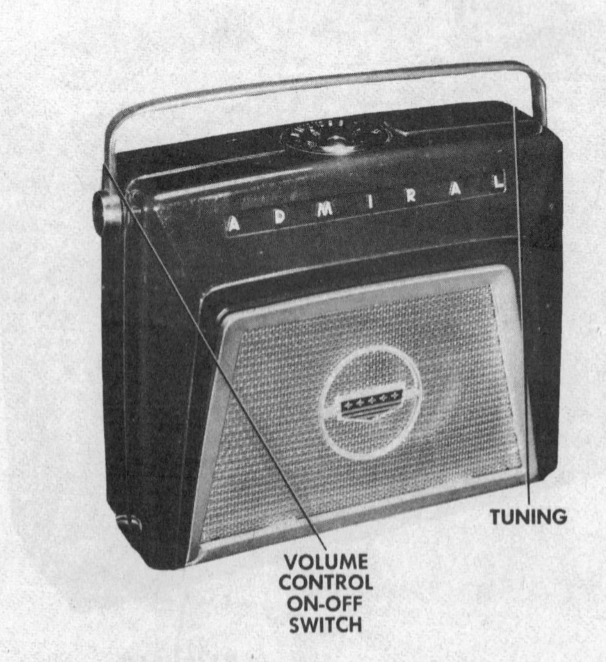

ADMIRAL

MODEL PICTURED
5T31

AC/DC operated AM superheterodyne receiver

TUBES
5

POWER SUPPLY
110-120 volts AC/DC

TUNING RANGE
535-1620KC

MFR/SUPPLIER
Admiral Corp.

PHOTOFACT SET
279-1

PUBLISHED
1955

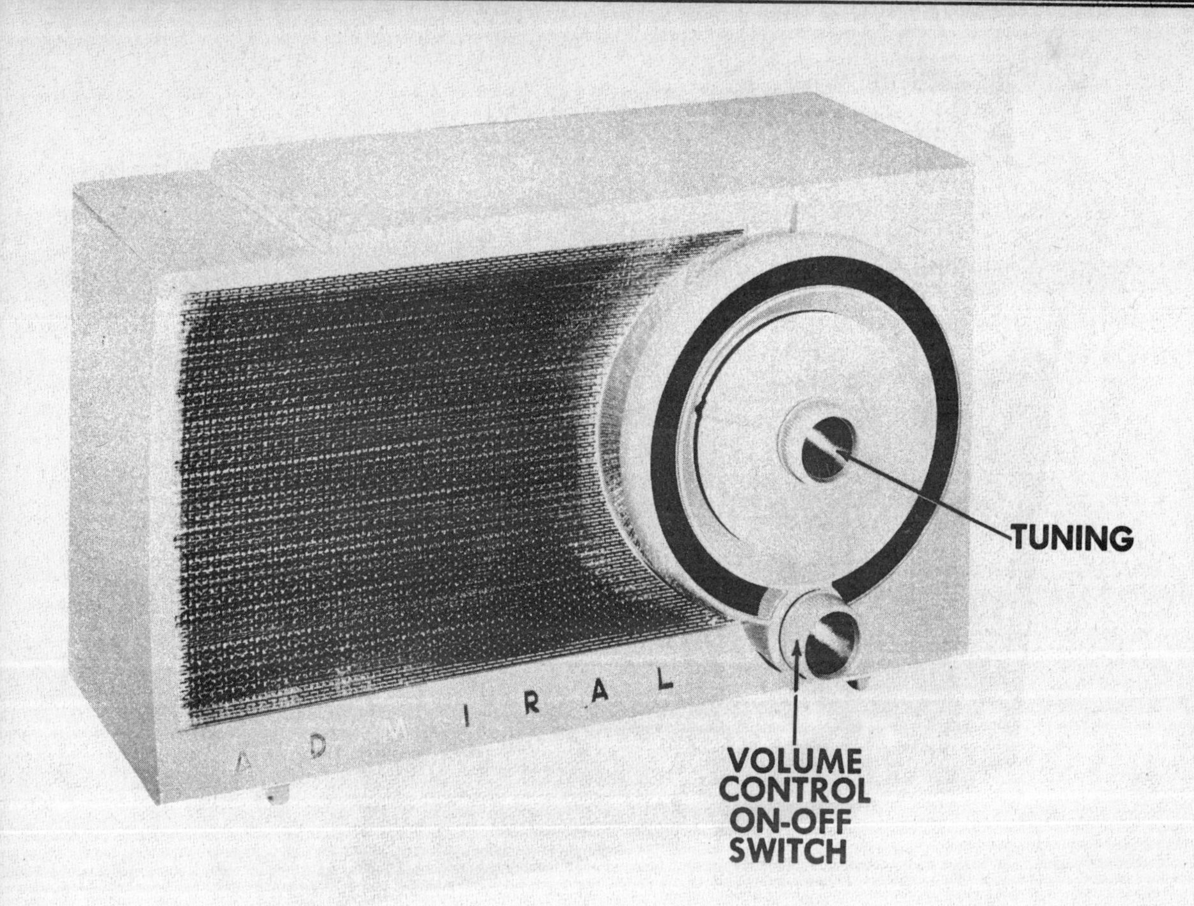

ADMIRAL

MODEL PICTURED
5W32

AC operated AM receiver with electric clock

TUBES
5

POWER SUPPLY
110-120 volts AC, 60 cycles

TUNING RANGE
535-1620KC

MFR/SUPPLIER
Admiral Corp.

PHOTOFACT SET
341-2

PUBLISHED
1956

ADMIRAL

MODEL PICTURED
4E21

Three-power portable AM receiver

TUBES
4

POWER SUPPLY
110-120 volts AC/DC or 7.5 volts A supply & 90 volts B supply

TUNING RANGE
535KC-1620KC

MFR/SUPPLIER
Admiral Corp.

PHOTOFACT SET
354-2

PUBLISHED
1957

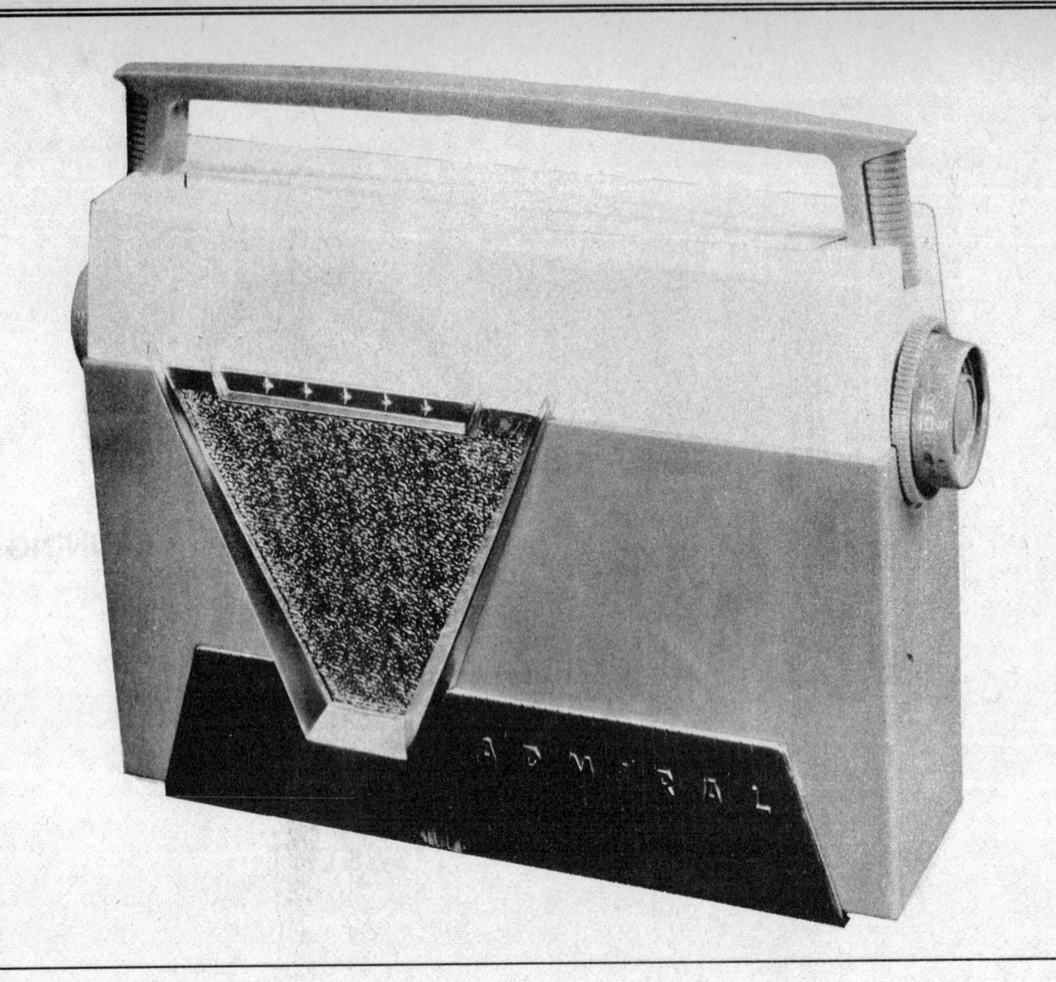

ADMIRAL

MODEL PICTURED
7M14

Battery operated portable AM transistorized receiver

TUBES
0

POWER SUPPLY
6 volts DC

TUNING RANGE
535-1620KC

MFR/SUPPLIER
Admiral Corp.

PHOTOFACT SET
369-1

PUBLISHED
1957

ADMIRAL

MODEL PICTURED
4P24

Battery operated portable AM transistorized receiver

TUBES
0

POWER SUPPLY
6 volts DC

TUNING RANGE
535-1620KC

MFR/SUPPLIER
Admiral Corp.

PHOTOFACT SET
374-1

PUBLISHED
1957

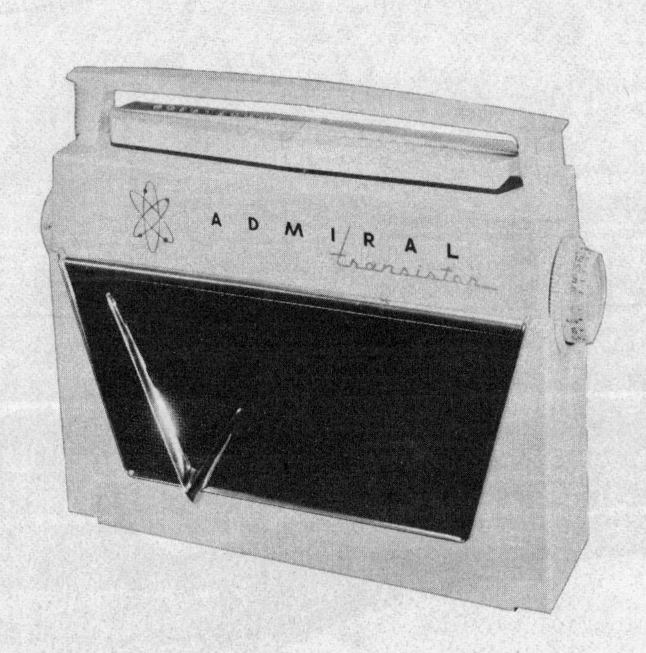

ADMIRAL

MODEL PICTURED
7L16

Battery operated portable AM transistorized receiver

TUBES
0

POWER SUPPLY
9 volts DC

TUNING RANGE
535-1620KC

MFR/SUPPLIER
Admiral Corp.

PHOTOFACT SET
375-6

PUBLISHED
1957

ADMIRAL

MODEL PICTURED
5RP41

AC operated AM receiver with four-speed auto record changer

TUBES
5

POWER SUPPLY
110-120 volts AC, 60 cycles

TUNING RANGE
535-1620KC

MFR/SUPPLIER
Admiral Corp.

PHOTOFACT SET
387-6

PUBLISHED
1958

ADMIRAL

MODEL PICTURED
242

AC/DC operated AM receiver

TUBES
5

POWER SUPPLY
110-120 volts AC/DC

TUNING RANGE
535-1620KC

MFR/SUPPLIER
Admiral Corp.

PHOTOFACT SET
410-4

PUBLISHED
1958

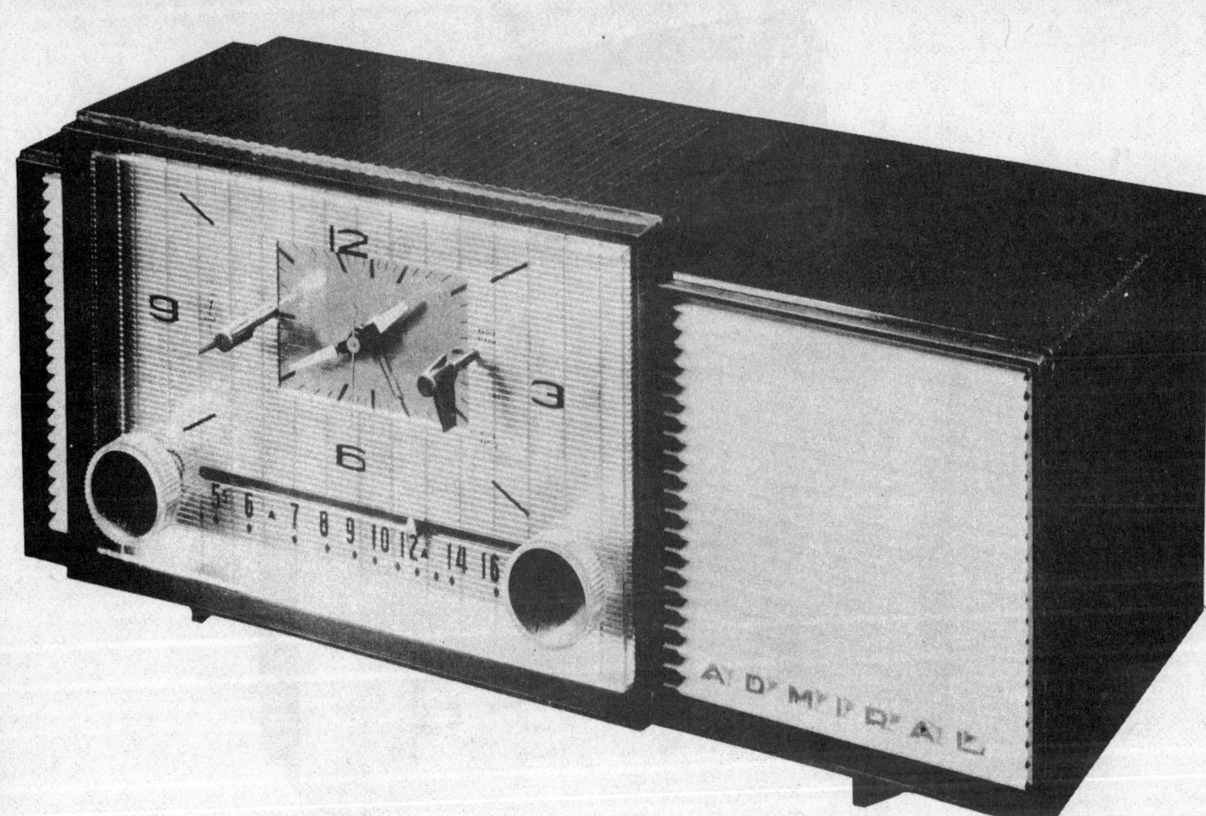

ADMIRAL

MODEL PICTURED
292

AC operated AM receiver with electric clock

TUBES
5

POWER SUPPLY
110-120 volts AC, 60 cycles

TUNING RANGE
535-1620KC

MFR/SUPPLIER
Admiral Corp.

PHOTOFACT SET
410-5

PUBLISHED
1958

ADMIRAL

MODEL PICTURED
202

Three-power portable AM receiver

TUBES
4

POWER SUPPLY
110-120 volts AC/DC or 7.5 volts A & 90 volts B supply

TUNING RANGE
535-1620KC

MFR/SUPPLIER
Admiral Corp.

PHOTOFACT SET
411-4

PUBLISHED
1958

ADMIRAL

MODEL PICTURED
227

Battery operated portable AM transistorized receiver

TUBES
0

POWER SUPPLY
12 volts DC

TUNING RANGE
535-1620KC

MFR/SUPPLIER
Admiral Corp.

PHOTOFACT SET
413-4

PUBLISHED
1958

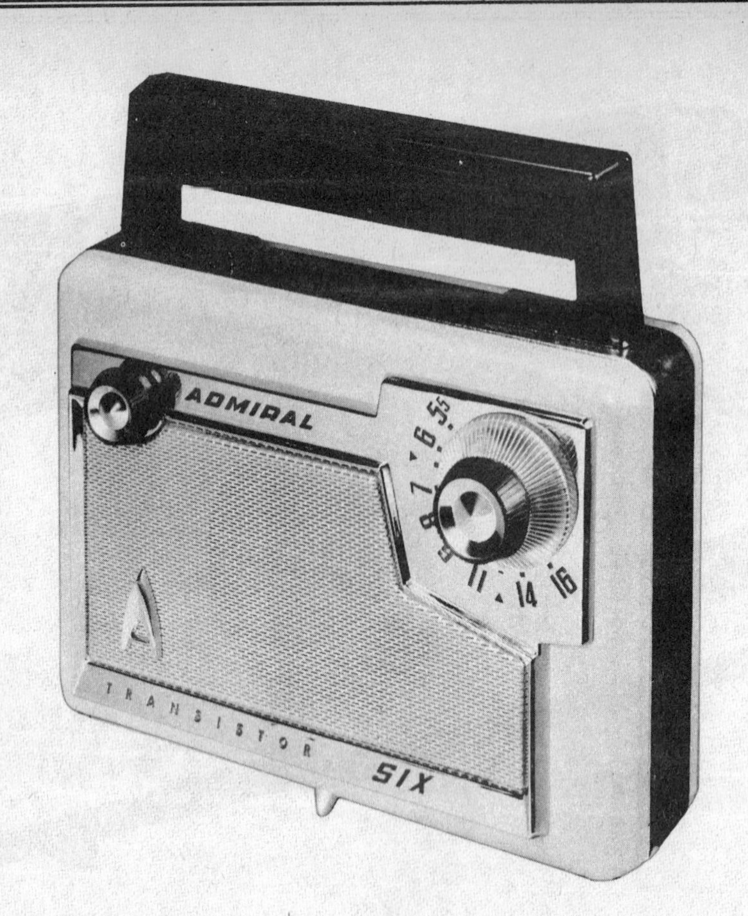

ADMIRAL

MODEL PICTURED
303

AC/DC operated FM-AM receiver

TUBES
6

POWER SUPPLY
110-120 volts AC/DC, .32 amp @ 117 volts AC (29 watts) FM

TUNING RANGE
535-1620KC, 88-108MC

MFR/SUPPLIER
Admiral Corp.

PHOTOFACT SET
426-4

PUBLISHED
1959

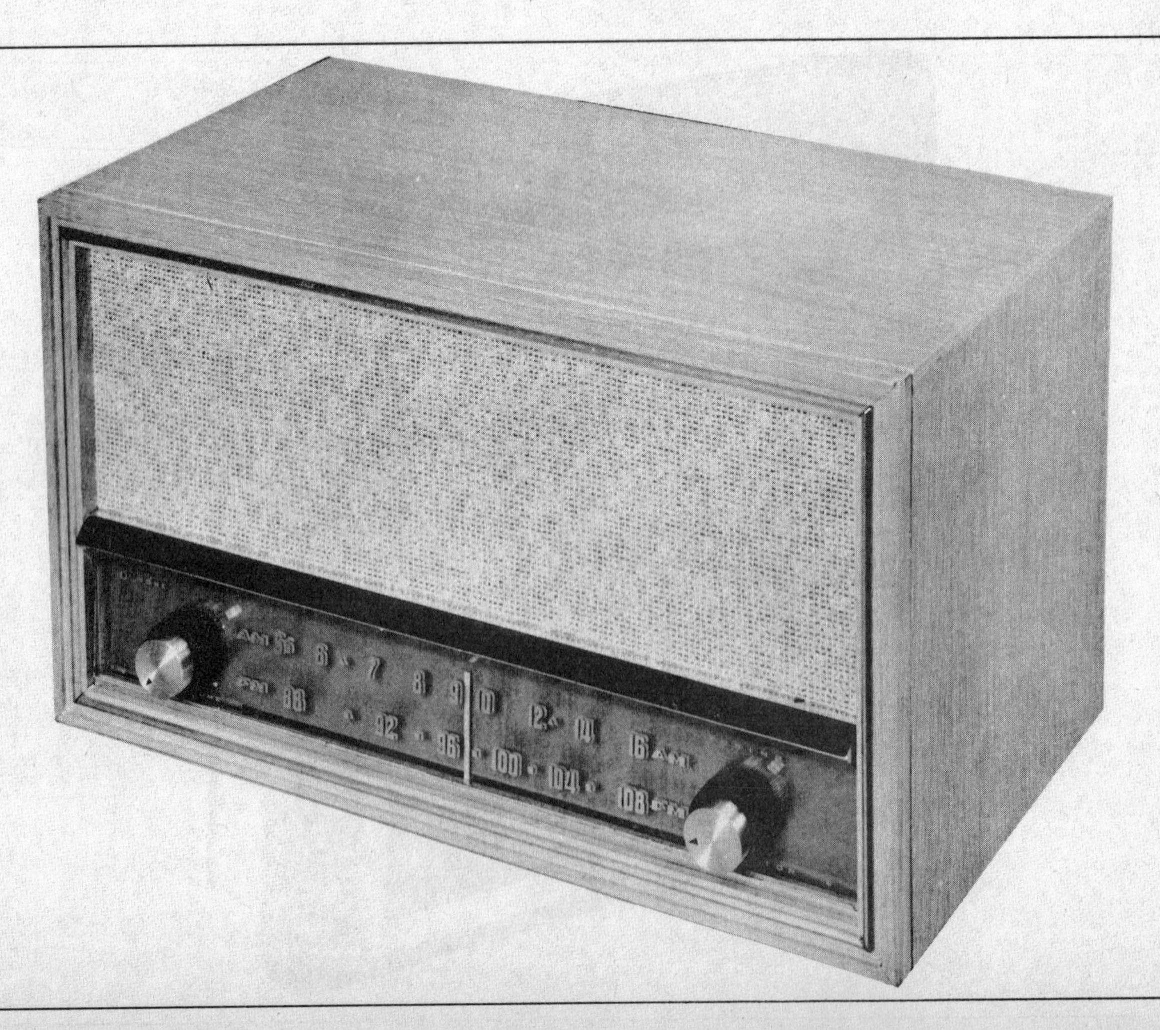

ADMIRAL

MODEL PICTURED
531

Battery operated portable transistorized AM receiver

TUBES
0

POWER SUPPLY
12 volts DC, 9MA @ 12 volts DC (no sig. min. vol.), 11MA @ 12 volts DC(sig. normal vol.)

TUNING RANGE
535-1620KC

MFR/SUPPLIER
Admiral Corp.

PHOTOFACT SET
431-3

PUBLISHED
1959

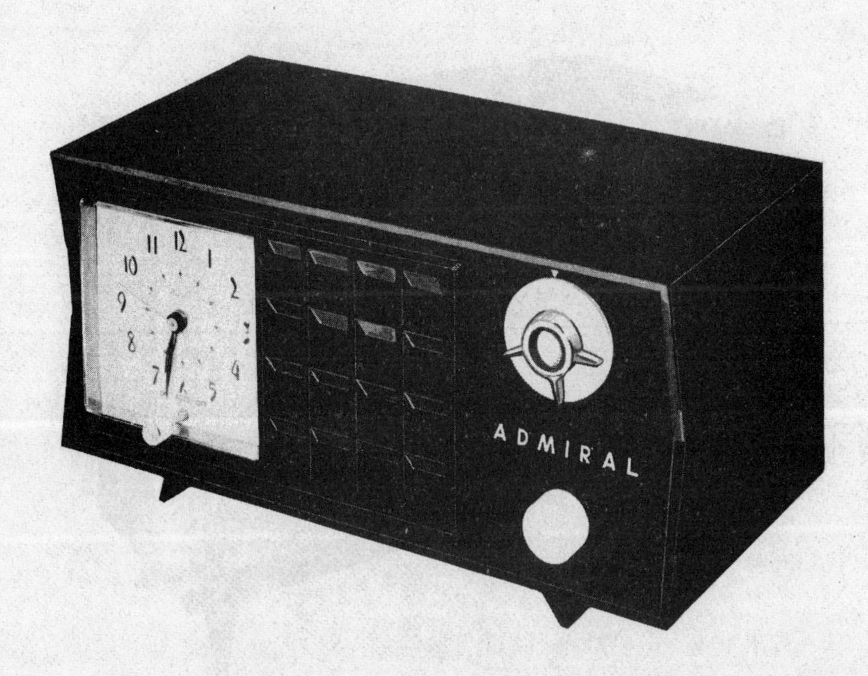

ADMIRAL

MODEL PICTURED
251

AC operated AM receiver, 4L series are not clock versions

TUBES
4

POWER SUPPLY
110-120 volts AC, 60 cycles

TUNING RANGE
535-1620KC

MFR/SUPPLIER
Admiral Corp.

PHOTOFACT SET
446-4

PUBLISHED
1959

ADMIRAL

MODEL PICTURED
581

Battery operated portable transistorized AM receiver

TUBES
0

POWER SUPPLY
6 volts DC

TUNING RANGE
535-1620KC

MFR/SUPPLIER
Admiral Corp.

PHOTOFACT SET
446-5

PUBLISHED
1959

ADMIRAL

MODEL PICTURED
SS642

AC operated AM-FM receiver with four-speed changer

TUBES
17

POWER SUPPLY
110-120 volts AC, 60 cycles

TUNING RANGE
530-1620KC, 88-108MC

MFR/SUPPLIER
Admiral Corp.

PHOTOFACT SET
447-4

PUBLISHED
1959

ADMIRAL

MODEL PICTURED
521

Battery operated portable transistorized AM receiver

TUBES
0

POWER SUPPLY
12 volts DC

TUNING RANGE
535-1600KC

MFR/SUPPLIER
Admiral Corp.

PHOTOFACT SET
448-4

PUBLISHED
1959

ADMIRAL

MODEL PICTURED
566

Battery operated transistorized AM portable receiver

TUBES
0

POWER SUPPLY
9 volts DC (receiver), 1.5 volts DC (clock)

TUNING RANGE
535-1620KC

MFR/SUPPLIER
Admiral Corp.

PHOTOFACT SET
452-3

PUBLISHED
1959

ADMIRAL

MODEL PICTURED
801

Battery operated portable transistorized AM receiver

TUBES
0

POWER SUPPLY
6 volts DC, 28MA @ 6 volts DC

TUNING RANGE
535-1620KC

MFR/SUPPLIER
Admiral Corp.

PHOTOFACT SET
454-4

PUBLISHED
1959

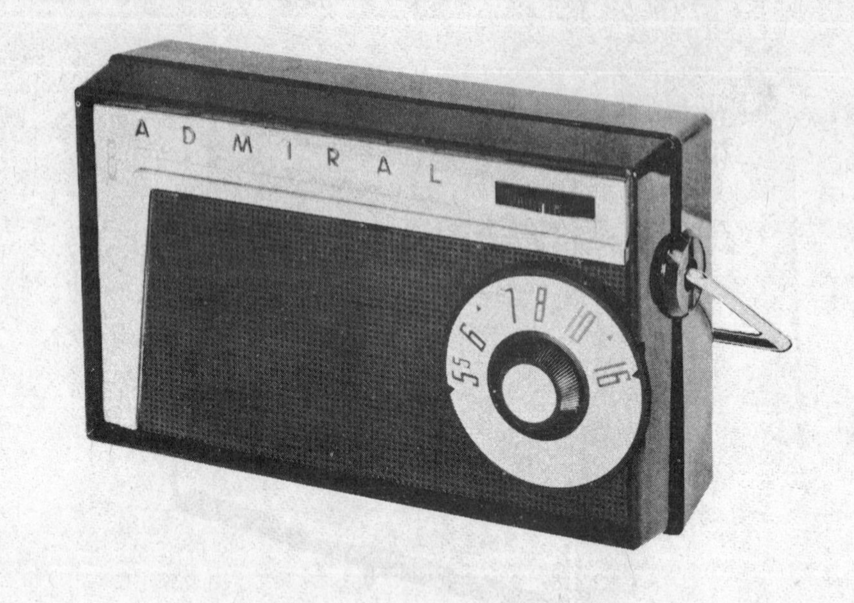

ADMIRAL

MODEL PICTURED
739

Battery operated transistorized portable AM receiver

TUBES
0

POWER SUPPLY
9 volts DC, 8.7MA @ 9 volts DC (no sig. min. vol.), 11MA @ 9 volts DC (sig. normal vol.)

TUNING RANGE
535-1620KC

MFR/SUPPLIER
Admiral Corp.

PHOTOFACT SET
471-3

PUBLISHED
1960

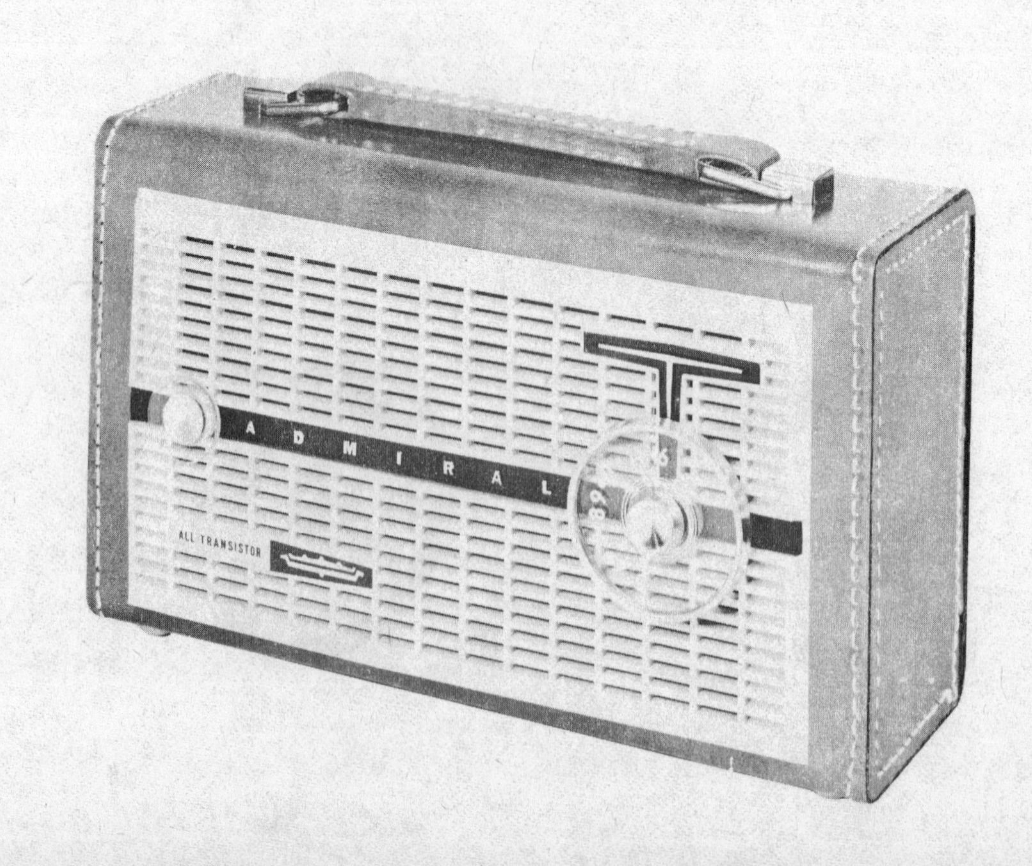

ADMIRAL

MODEL PICTURED
751

Battery operated portable transistorized AM receiver

TUBES
0

POWER SUPPLY
9 volts DC, 12MA @ 9 volts DC (no sig. min. vol.), 15MA @ 9 volts DC (sig. normal vol.)

TUNING RANGE
535-1620KC

MFR/SUPPLIER
Admiral Corp.

PHOTOFACT SET
473-3

PUBLISHED
1960

ADMIRAL

MODEL PICTURED
742

Battery operated transistorized portable AM receiver

TUBES
0

POWER SUPPLY
9 volts DC, 9.5MA @ 9 volts DC (no sig. min. vol.), 25MA @ 9 volts DC (sig. normal vol.)

TUNING RANGE
535-1620KC

MFR/SUPPLIER
Admiral Corp.

PHOTOFACT SET
474-4

PUBLISHED
1960

ADMIRAL

MODEL PICTURED
Y833

AC operated AM receiver (models Y873,Y875,Y878 clock versions)

TUBES
5

POWER SUPPLY
110-120 volts AC, 24 watts, .23 amp @ 117 volts AC (less clock)

TUNING RANGE
535-1620KC

MFR/SUPPLIER
Admiral Corp.

PHOTOFACT SET
476-3

PUBLISHED
1960

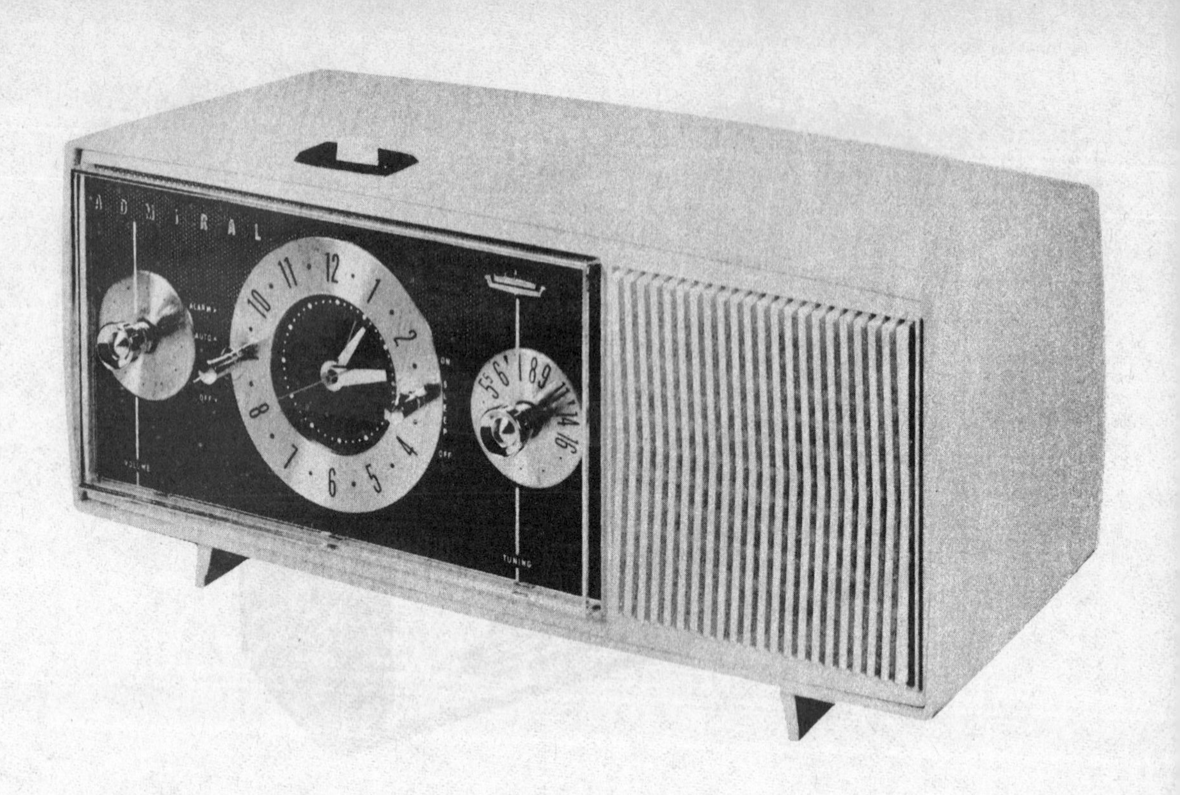

ADMIRAL

MODEL PICTURED
692

Battery operated transistorized AM receiver

TUBES
0

POWER SUPPLY
6 volts DC, 7MA @ 6 volts DC (no sig. min. vol.), 15MA @ 6 volts DC (sig. normal vol.)

TUNING RANGE
535-1620KC

MFR/SUPPLIER
Admiral Corp.

PHOTOFACT SET
478-4

PUBLISHED
1960

ADMIRAL

MODEL PICTURED
703

Battery operated transistorized portable AM receiver

TUBES
0

POWER SUPPLY
6 volts DC, 10MA @ 6 volts DC (no sig. min. vol.), 25MA @ 6 volts DC (sig. normal vol.)

TUNING RANGE
535-1620KC

MFR/SUPPLIER
Admiral Corp.

PHOTOFACT SET
478-5

PUBLISHED
1960

ADMIRAL

MODEL PICTURED
717

Battery operated transistorized portable AM receiver (Y series clock version)

TUBES
0

POWER SUPPLY
6 volts DC (radio), 1.5 volts DC (clock)

TUNING RANGE
532-1620KC

MFR/SUPPLIER
Admiral Corp.

PHOTOFACT SET
479-3

PUBLISHED
1960

ADMIRAL

MODEL PICTURED
Y858

AC operated AM receiver with electric clock

TUBES
5

POWER SUPPLY
110-120 volts AC, 60 cycles, 23 watts, .23 amp @ 117 volts AC (less clock)

TUNING RANGE
535-1620KC

MFR/SUPPLIER
Admiral Corp.

PHOTOFACT SET
483-3

PUBLISHED
1960

ADMIRAL

MODEL PICTURED
816B

Battery operated transistorized portable AM receiver with clock

TUBES
0

POWER SUPPLY
9 volts DC (radio) 1.5 volts DC (clock),

TUNING RANGE
540-1620KC

MFR/SUPPLIER
Admiral Corp.

PHOTOFACT SET
499-3

PUBLISHED
1960

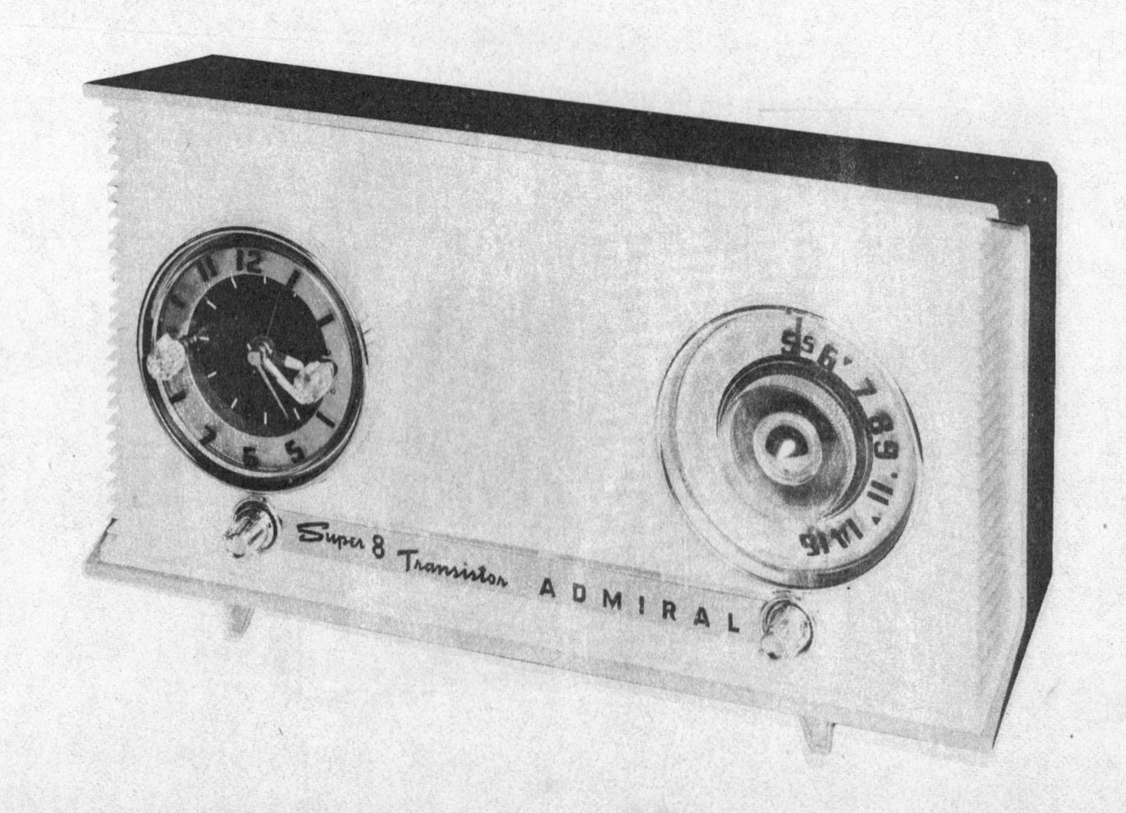

ADMIRAL

MODEL PICTURED
909

Battery operated nine-band transistorized portable receiver

TUBES
0

POWER SUPPLY
12 volts DC (radio), 1.5 volts DC (dial light)

TUNING RANGE
Nine bands

MFR/SUPPLIER
Admiral Corp.

PHOTOFACT SET
502-4

PUBLISHED
1960

ADMIRAL

MODEL PICTURED
Y1149

AC operated FM/AM tuner, stereo amplifier, four-speed auto record changer

TUBES
16

POWER SUPPLY
110-120 volts AC, 60 cycles

TUNING RANGE
535-1620KC, 88-108MC

MFR/SUPPLIER
Admiral Corp.

PHOTOFACT SET
509-3

PUBLISHED
1960

ADMIRAL

MODEL PICTURED
Y2027

Battery operated transistorized portable AM receiver

TUBES
0

POWER SUPPLY
6 volts DC, 12MA @ 6 volts DC (no sig. min. vol.), 20MA @ 6 volts DC (sig. normal vol.)

TUNING RANGE
535-1620KC

MFR/SUPPLIER
Admiral Corp.

PHOTOFACT SET
510-4

PUBLISHED
1960

AERMOTIVE

MODEL PICTURED
181-AD

AC/DC superheterodyne receiver, self-contained loop antenna

TUBES
8

POWER SUPPLY
117 volts AC/DC, .505 amp @ 117 volts AC

TUNING RANGE
540-1700KC

MFR/SUPPLIER
Aermotive Equipment Corp.

PHOTOFACT SET
12-1

PUBLISHED
1947

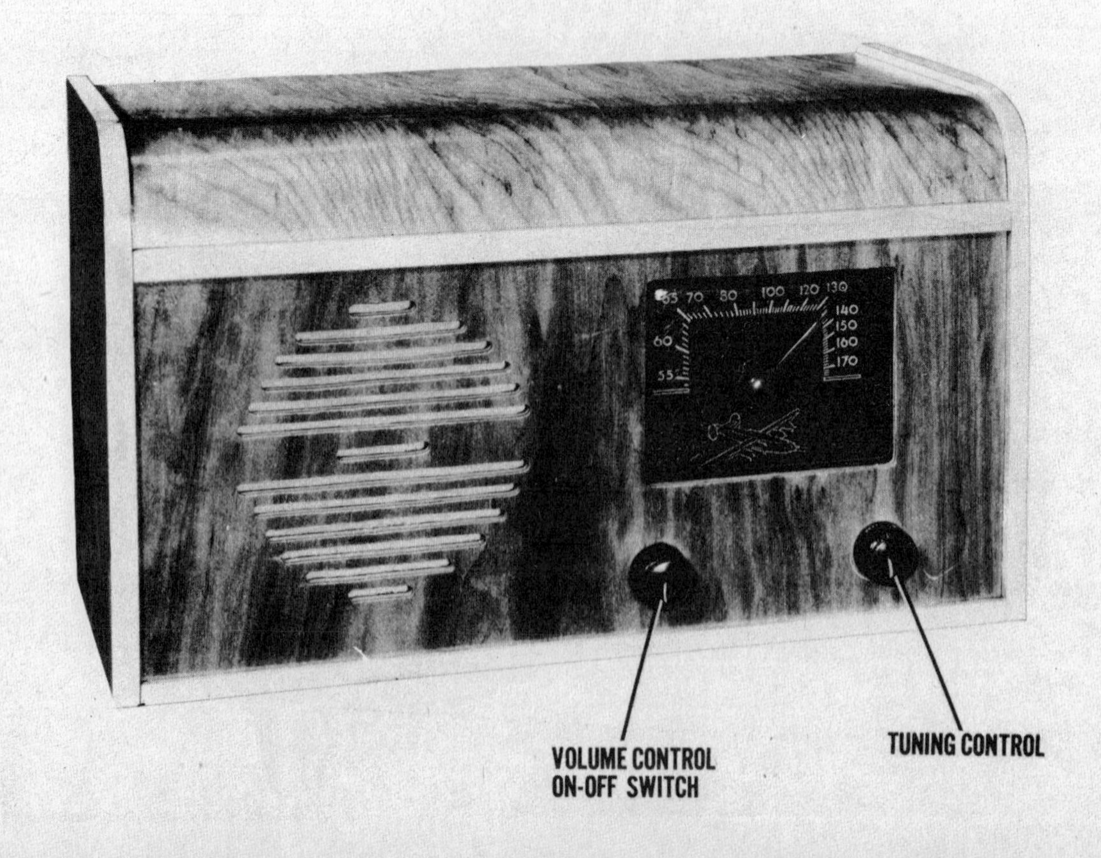

AIR CHIEF

MODEL PICTURED
4-A-24

Battery operated superheterodyne receiver

TUBES
4

POWER SUPPLY
1.5 volts A supply & 90 volts B supply in battery pack form

TUNING RANGE
528-1730KC

MFR/SUPPLIER
Firestone Tire and Rubber Co.

PHOTOFACT SET
13-5

PUBLISHED
1947

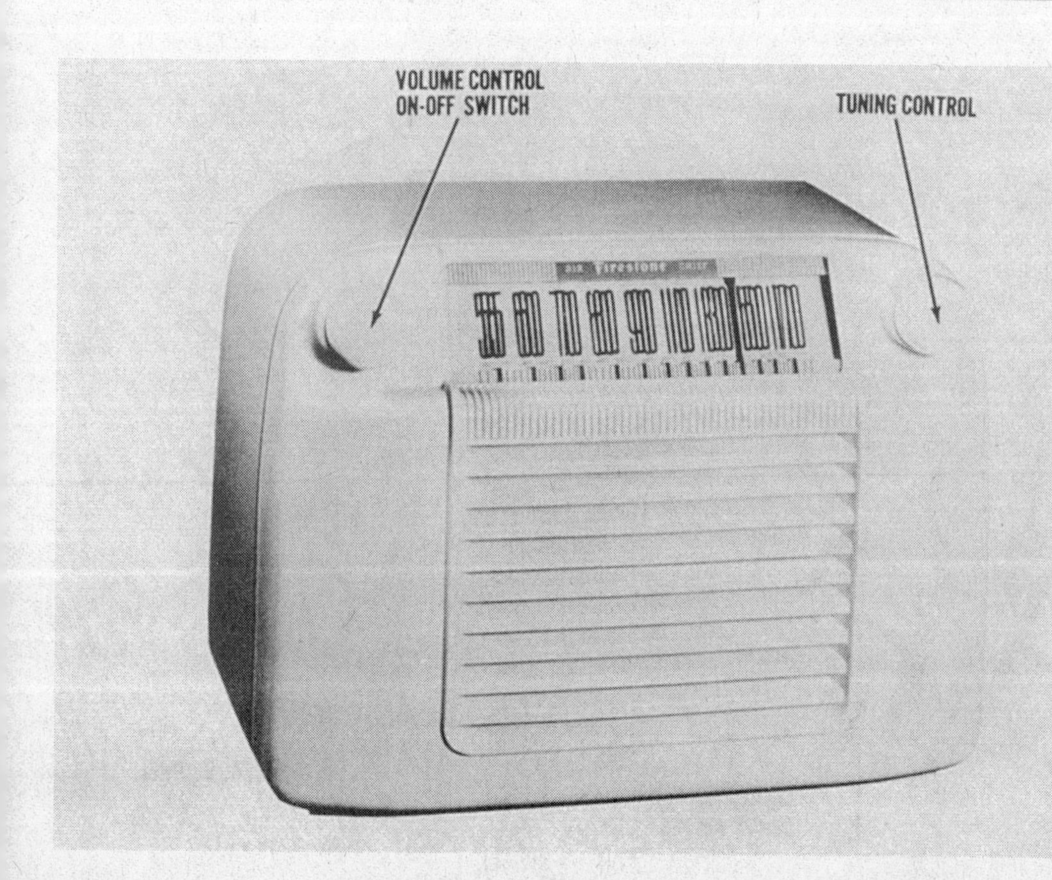

AIR CHIEF

MODEL PICTURED
4-A-25

AC/DC superheterodyne, self-contained loop antenna

TUBES
5

POWER SUPPLY
110-120 volts AC/DC, .235 amp @ 117 volts AC

TUNING RANGE
530-1730KC

MFR/SUPPLIER
Firestone Tire and Rubber Co.

PHOTOFACT SET
13-6

PUBLISHED
1947

AIR CHIEF

MODEL PICTURED
4-A-37

AC operated radio-phono combo superheterodyne, self-contained loop antenna

TUBES
8

POWER SUPPLY
105-120 volts AC

TUNING RANGE
540-1620KC, 9.4-15.4MC

MFR/SUPPLIER
Firestone Tire and Rubber Co.

PHOTOFACT SET
13-7

PUBLISHED
1947

AIR CHIEF

MODEL PICTURED
4-A-2

AC/DC operated superheterodyne receiver, self-contained loop antenna

TUBES
5

POWER SUPPLY
110-120 volts AC/DC

TUNING RANGE
535-1720KC

MFR/SUPPLIER
Firestone Tire and Rubber Co.

PHOTOFACT SET
14-4

PUBLISHED
1947

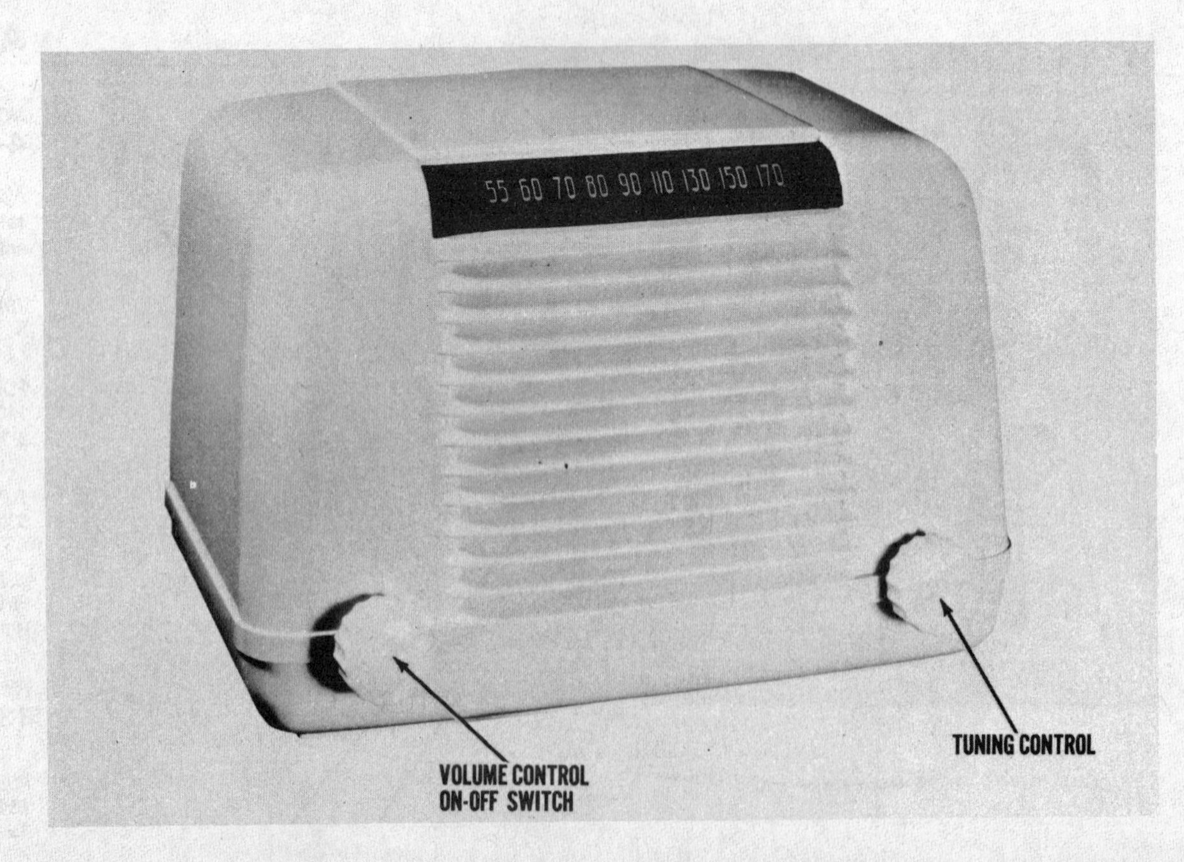

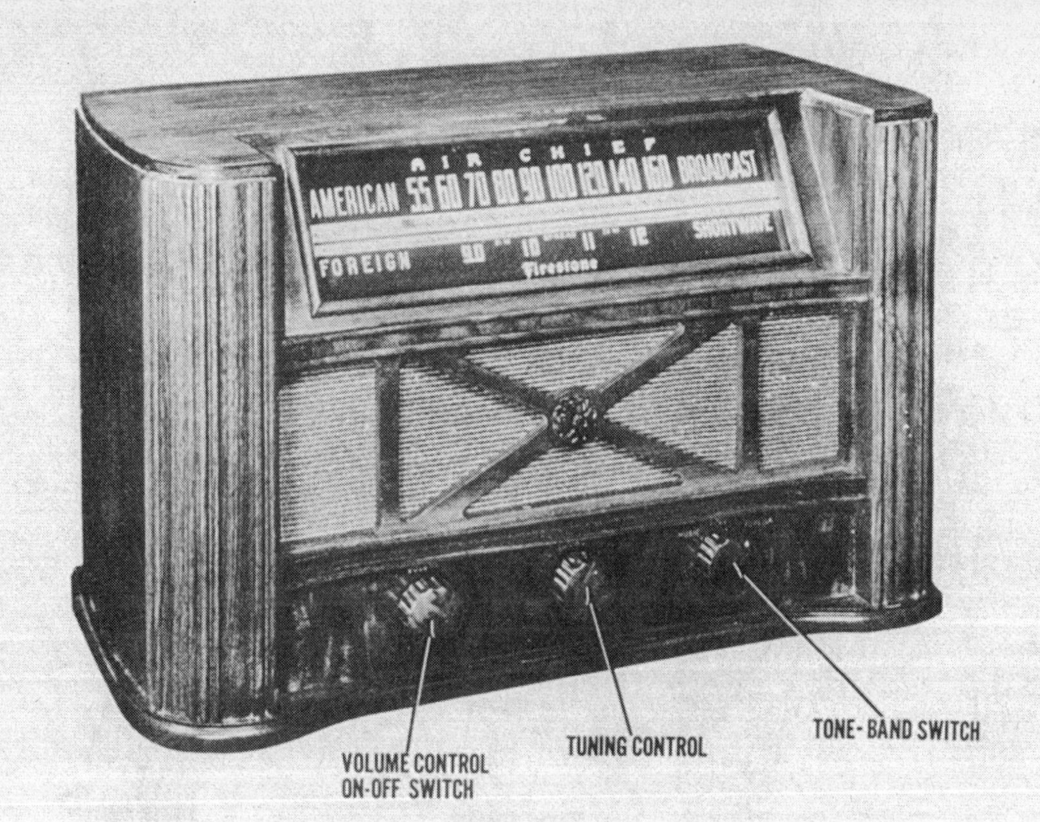

AIR CHIEF

MODEL PICTURED
4-A-20

AC/DC two-band superheterodyne, self-contained loop antenna

TUBES
6

POWER SUPPLY
105-125 volts AC/DC, .240 amp @ 117 volts AC

TUNING RANGE
540-1650KC, 9.0-12.0 MC, 9.0-12.0 MC

MFR/SUPPLIER
Firestone Tire and Rubber Co.

PHOTOFACT SET
15-11

PUBLISHED
1947

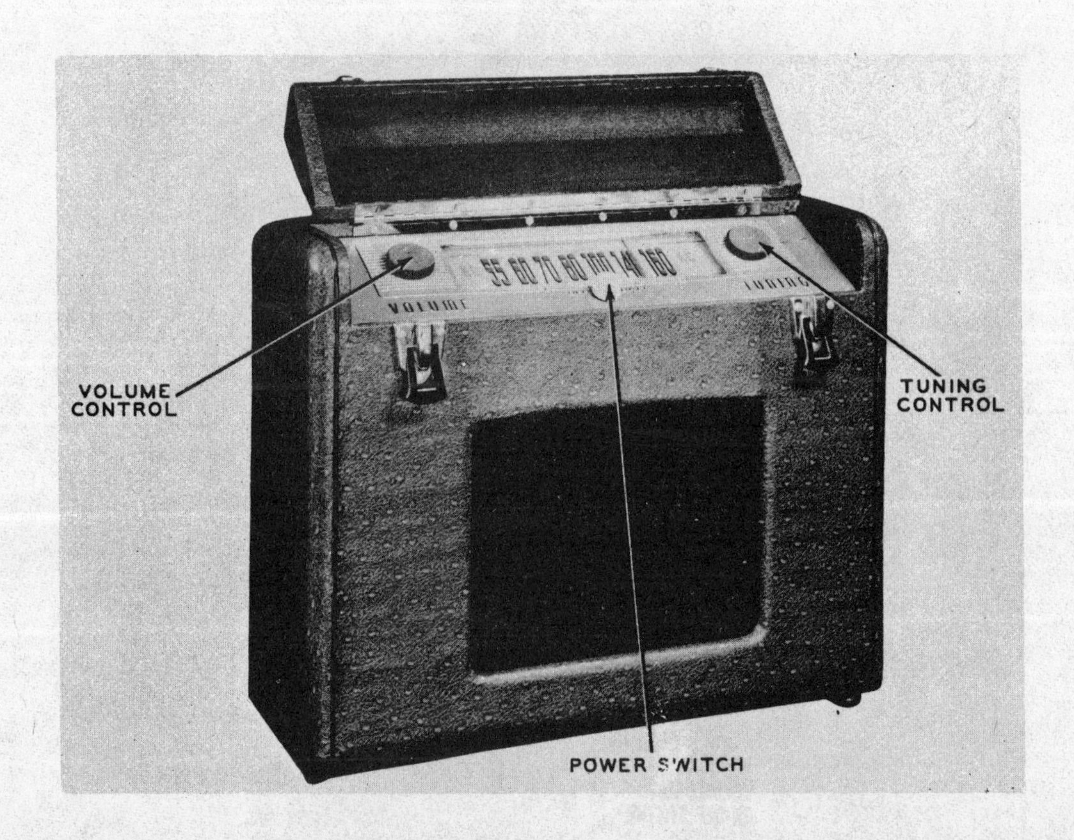

AIR CHIEF

MODEL PICTURED
4-C-3

Three-power portable superheterodyne receiver with loop antenna

TUBES
6

POWER SUPPLY
110-120 volts AC/DC, or 9 volts A supply & 90 volts B supply

TUNING RANGE
530-1620KC

MFR/SUPPLIER
Firestone Tire and Rubber Co.

PHOTOFACT SET
19-17

PUBLISHED
1947

AIR CHIEF

MODEL PICTURED
4-A-10

AC/DC operated superheterodyne receiver with loop antenna

TUBES
5

POWER SUPPLY
110-120 volts AC/DC

TUNING RANGE
535-1620KC

MFR/SUPPLIER
Firestone Tire and Rubber Co.

PHOTOFACT SET
28-11

PUBLISHED
1947

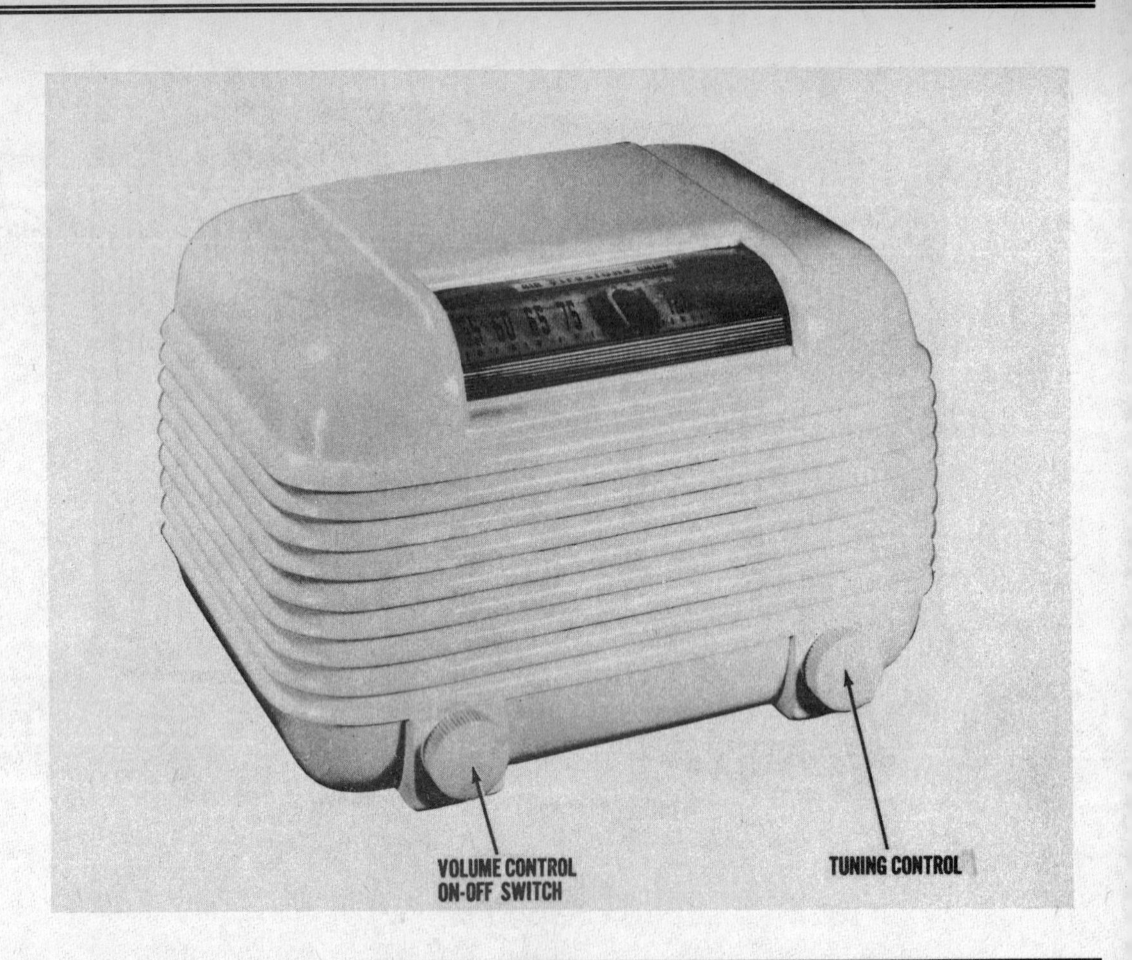

AIR CHIEF

MODEL PICTURED
4-A-27

AC/DC operated superheterodyne receiver with loop antenna

TUBES
5

POWER SUPPLY
110-120 volts AC/DC

TUNING RANGE
540-1630KC

MFR/SUPPLIER
Firestone Tire and Rubber Co.

PHOTOFACT SET
28-12

PUBLISHED
1947

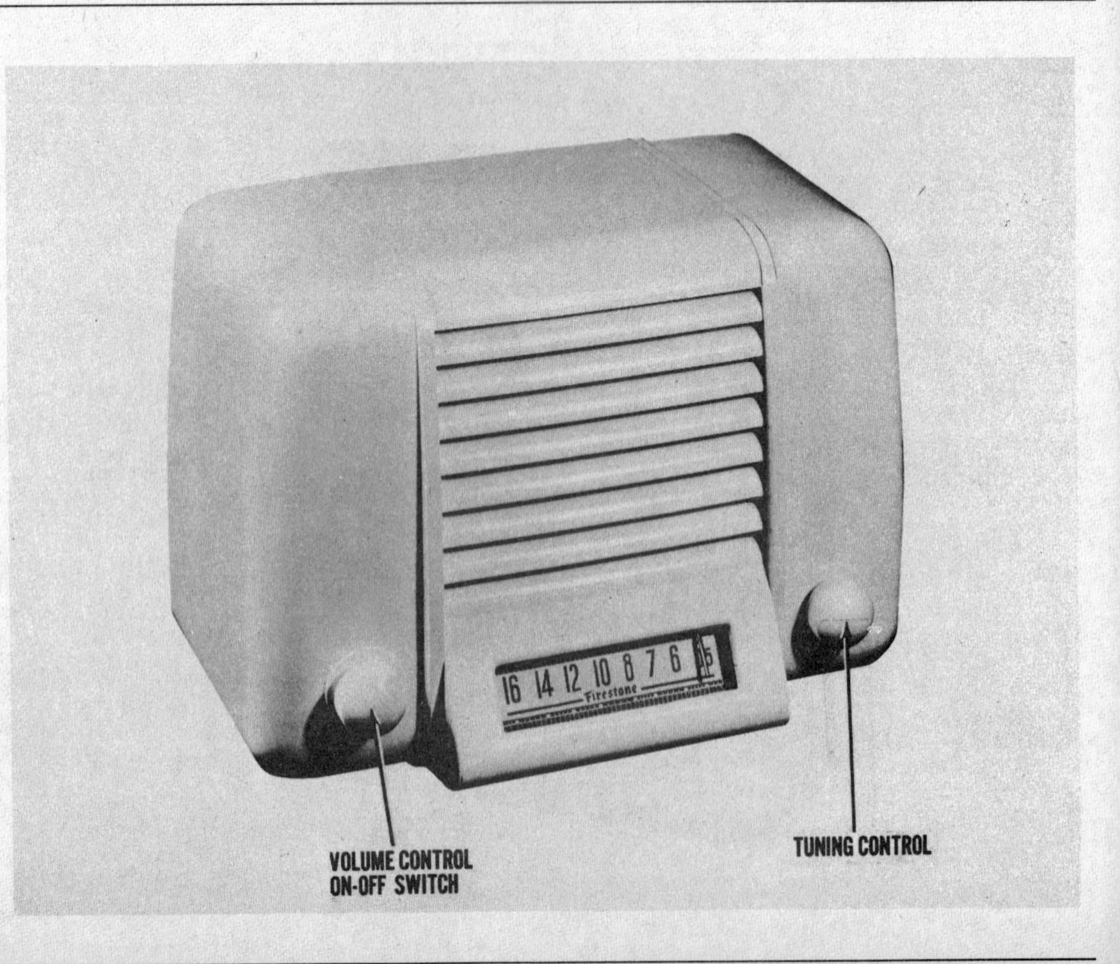

AIR CHIEF

MODEL PICTURED
4-A-42

AC operated phono-radio AM-FM superheterodyne with loop antenna

TUBES
10

POWER SUPPLY
110-120 volts AC

TUNING RANGE
540-1620KC,
87.5-108MC

MFR/SUPPLIER
Firestone Tire and Rubber Co.

PHOTOFACT SET
30-9

PUBLISHED
1947

AIR CHIEF

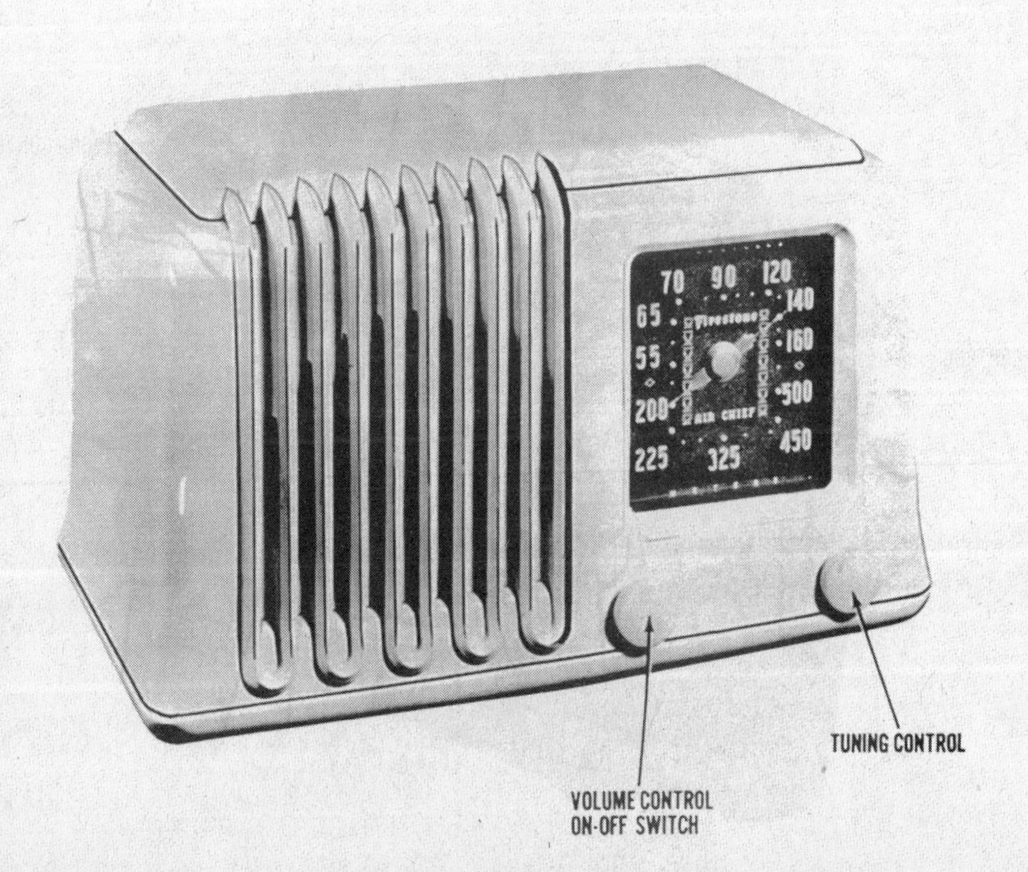

MODEL PICTURED
4-A-3

AC/DC operated superheterodyne receiver with loop antenna

TUBES
6

POWER SUPPLY
110-120 volts AC/DC,
.230 amp @ 117 volts AC

TUNING RANGE
535-1720KC

MFR/SUPPLIER
Firestone Tire and Rubber Co.

PHOTOFACT SET
31-13

PUBLISHED
1948

AIR CHIEF

MODEL PICTURED
4-A-26

AC/DC operated superheterodyne receiver with loop antenna

TUBES
5

POWER SUPPLY
110-120 volts AC/DC, .23 amp @ 117 volts AC

TUNING RANGE
540-1725KC

MFR/SUPPLIER
Firestone Tire and Rubber Co.

PHOTOFACT SET
33-5

PUBLISHED
1948

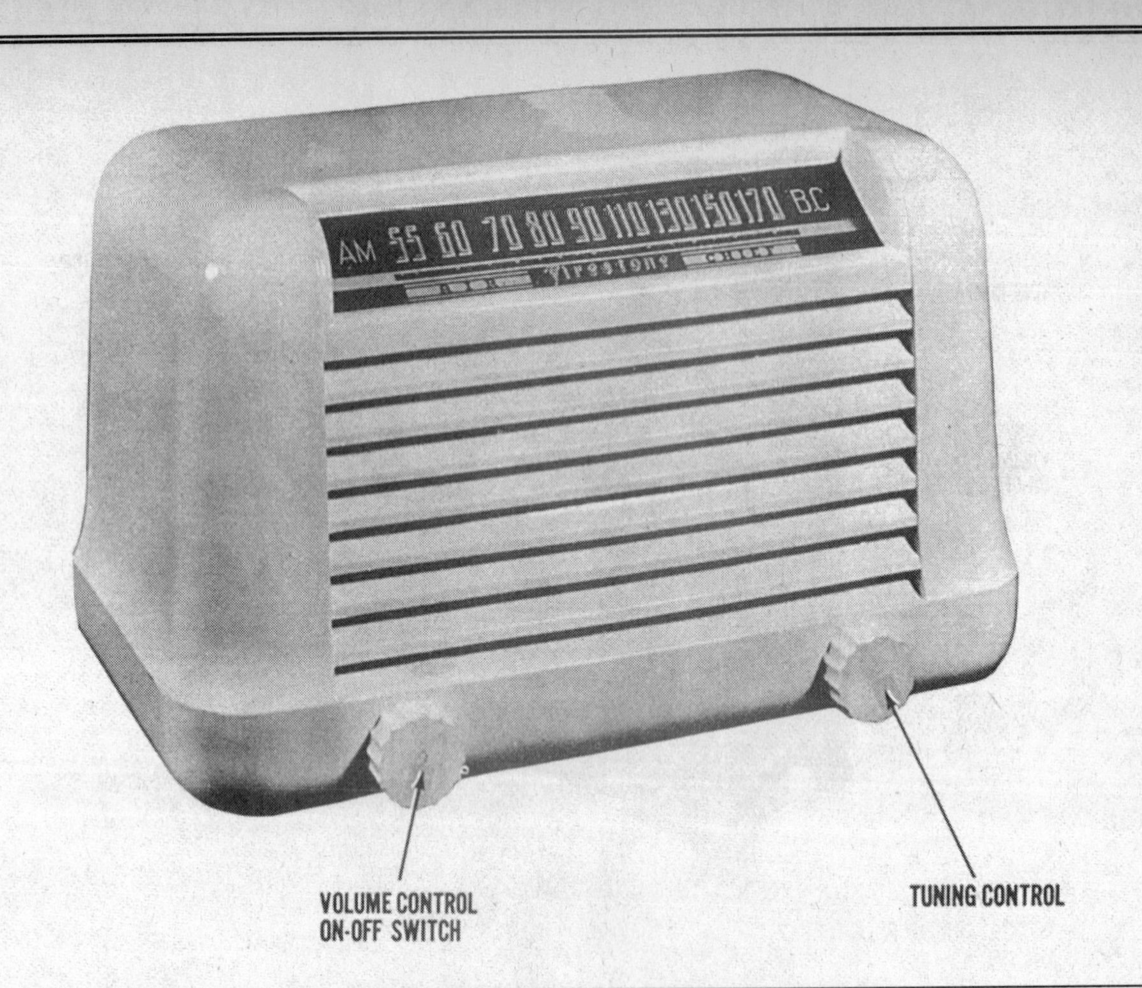

AIR CHIEF

MODEL PICTURED
4-C-5

Three-power operated portable superheterodyne receiver with loop antenna

TUBES
4

POWER SUPPLY
110-120 volts AC/DC or 1.5 volts A supply & 67.5 volts B supply

TUNING RANGE
530-1620KC

MFR/SUPPLIER
Firestone Tire and Rubber Co.

PHOTOFACT SET
33-6

PUBLISHED
1948

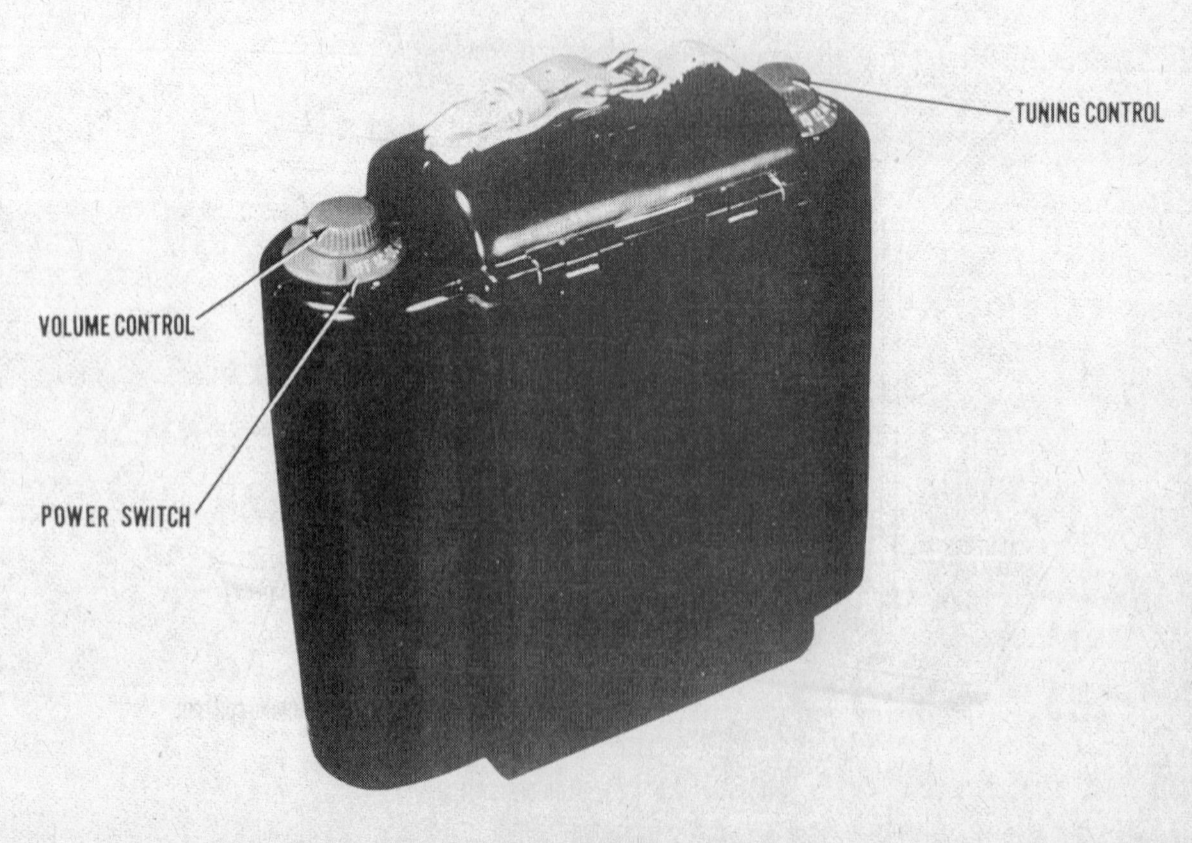

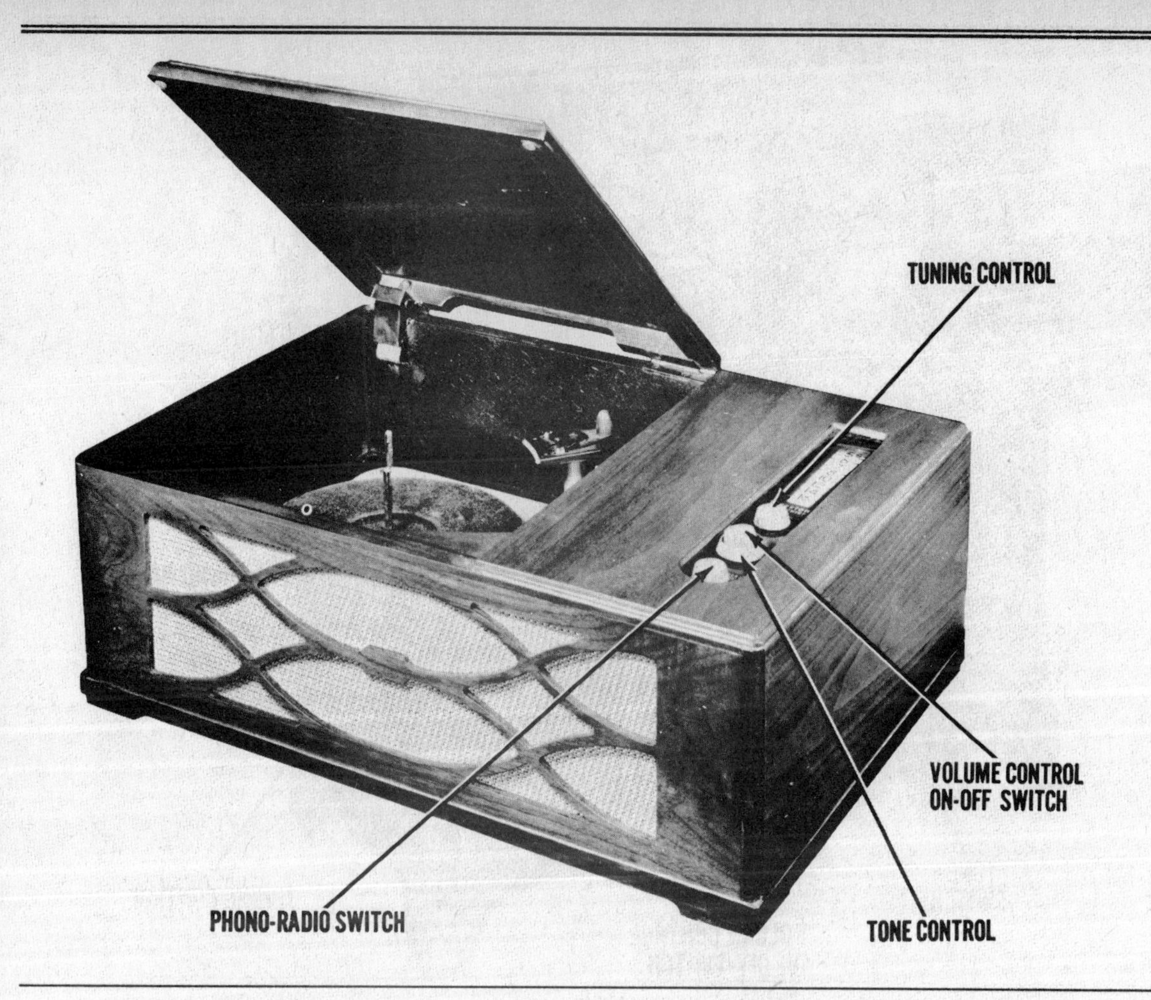

AIR CHIEF

MODEL PICTURED
4-A-17

AC operated phono-radio superheterodyne receiver with loop antenna

TUBES
7

POWER SUPPLY
110-120 volts AC

TUNING RANGE
540-1600KC

MFR/SUPPLIER
Firestone Tire and Rubber Co.

PHOTOFACT SET
35-7

PUBLISHED
1948

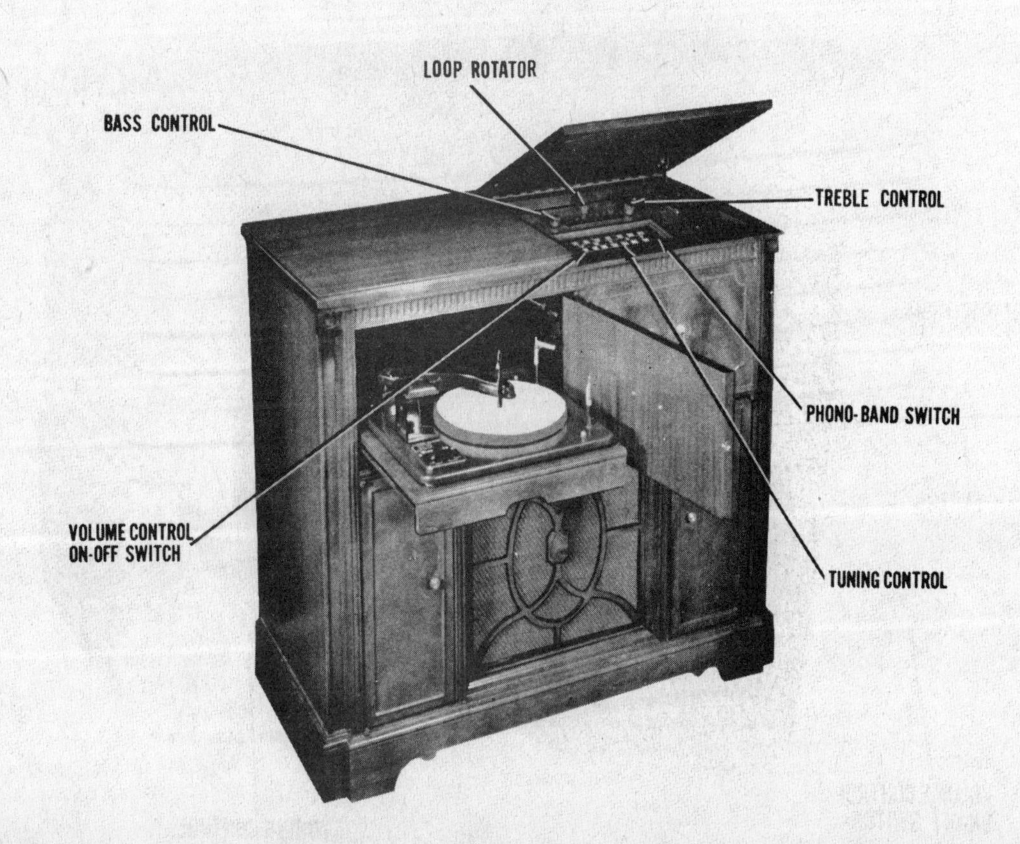

AIR CHIEF

MODEL PICTURED
4-A-15

AC operated combo phono-radio two-band AM-FM super receiver with loop antenna

TUBES
15

POWER SUPPLY
105-125 volts AC

TUNING RANGE
540-1620KC,
87.5-108.5MC,
9.14-15.4MC

MFR/SUPPLIER
Firestone Tire and Rubber Co.

PHOTOFACT SET
36-7

PUBLISHED
1948

AIR CHIEF

MODEL PICTURED
4-A-60

AC operated phono-radio AM-FM superheterodyne receiver with loop antenna

TUBES
9

POWER SUPPLY
110-120 volts AC/DC

TUNING RANGE
88-108MC, 540-1600KC

MFR/SUPPLIER
Firestone Tire and Rubber Co.

PHOTOFACT SET
38-6

PUBLISHED
1948

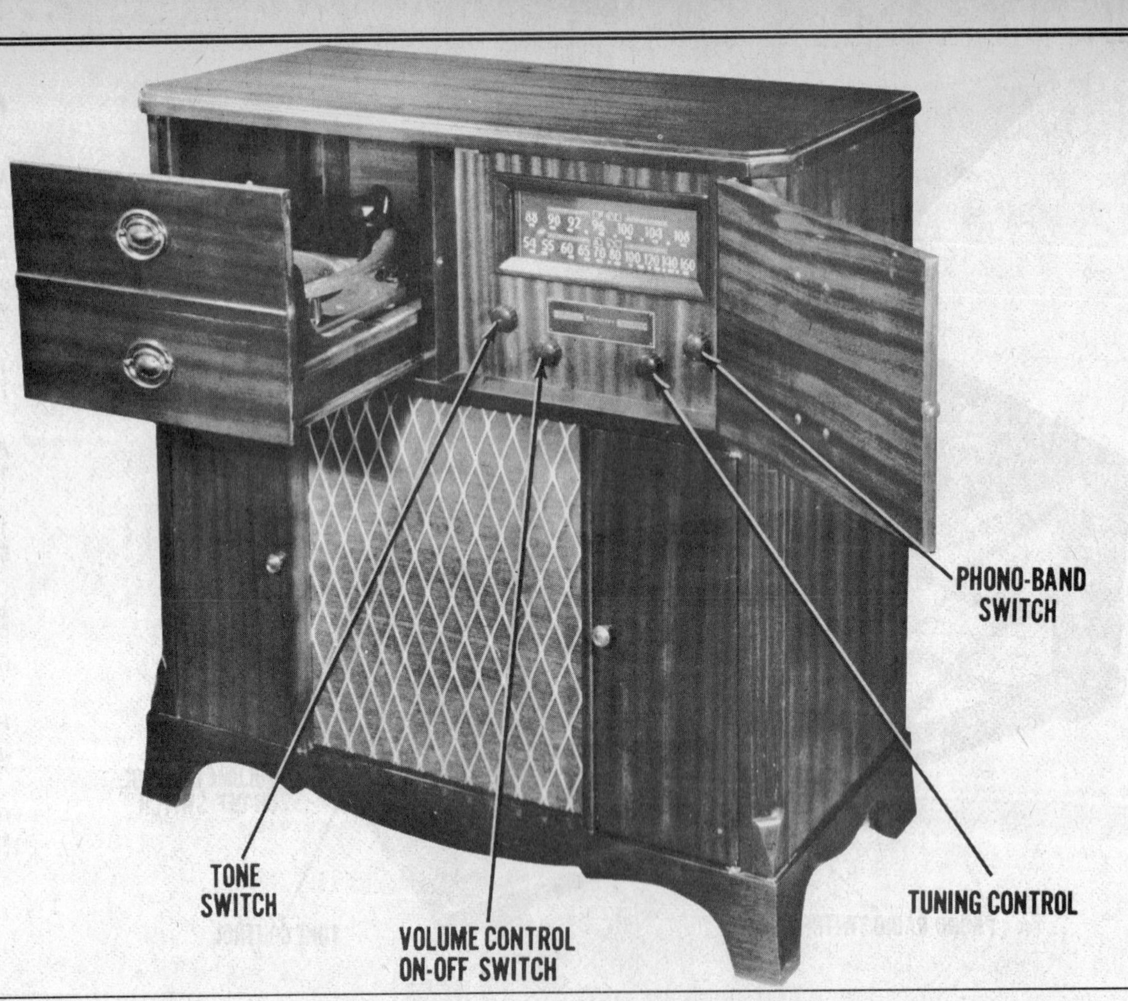

AIR CHIEF

MODEL PICTURED
4-A-11

AC/DC operated superheterodyne receiver with loop antenna

TUBES
6

POWER SUPPLY
105-125 volts AC/DC

TUNING RANGE
540-1600KC

MFR/SUPPLIER
Firestone Tire and Rubber Co.

PHOTOFACT SET
41-7

PUBLISHED
1948

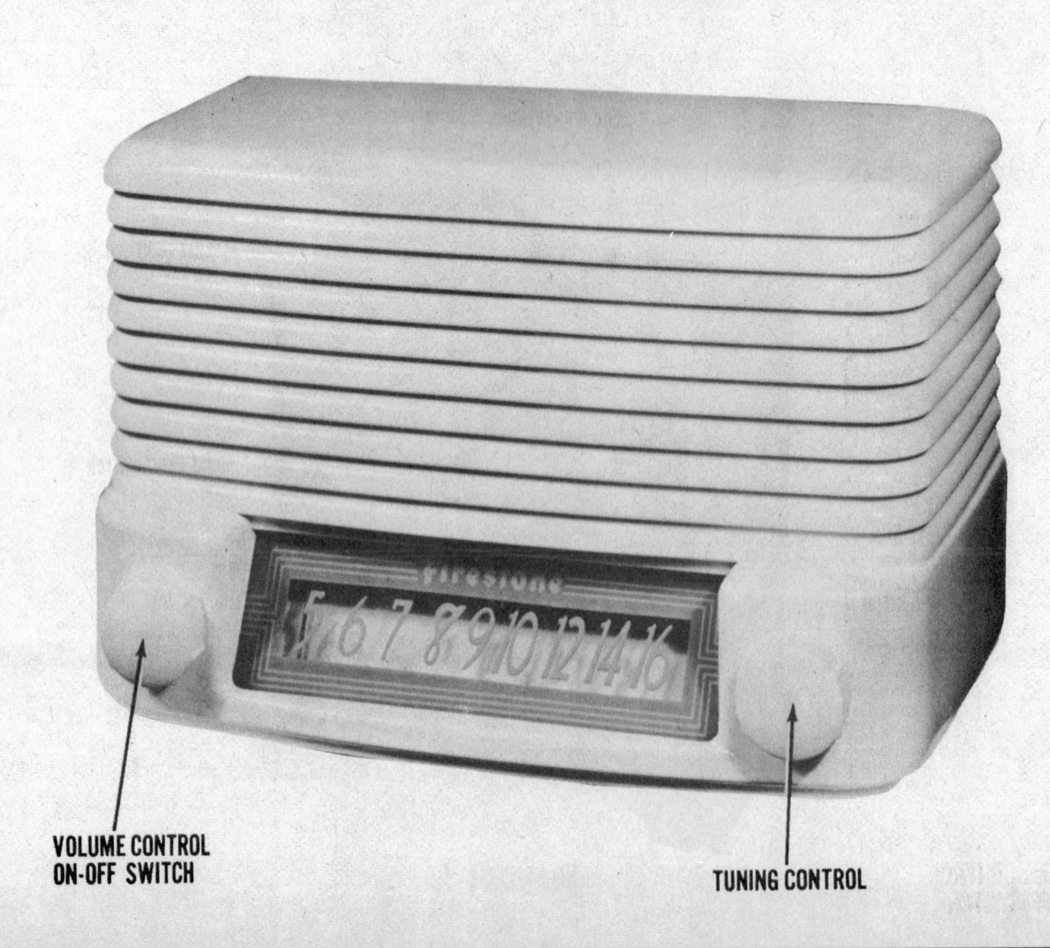

AIR CHIEF

MODEL PICTURED
4-A-61

AC/DC operated superheterodyne receiver with loop antenna

TUBES
5

POWER SUPPLY
105-125 volts AC/DC

TUNING RANGE
540-1630KC

MFR/SUPPLIER
Firestone Tire and Rubber Co.

PHOTOFACT SET
48-7

PUBLISHED
1948

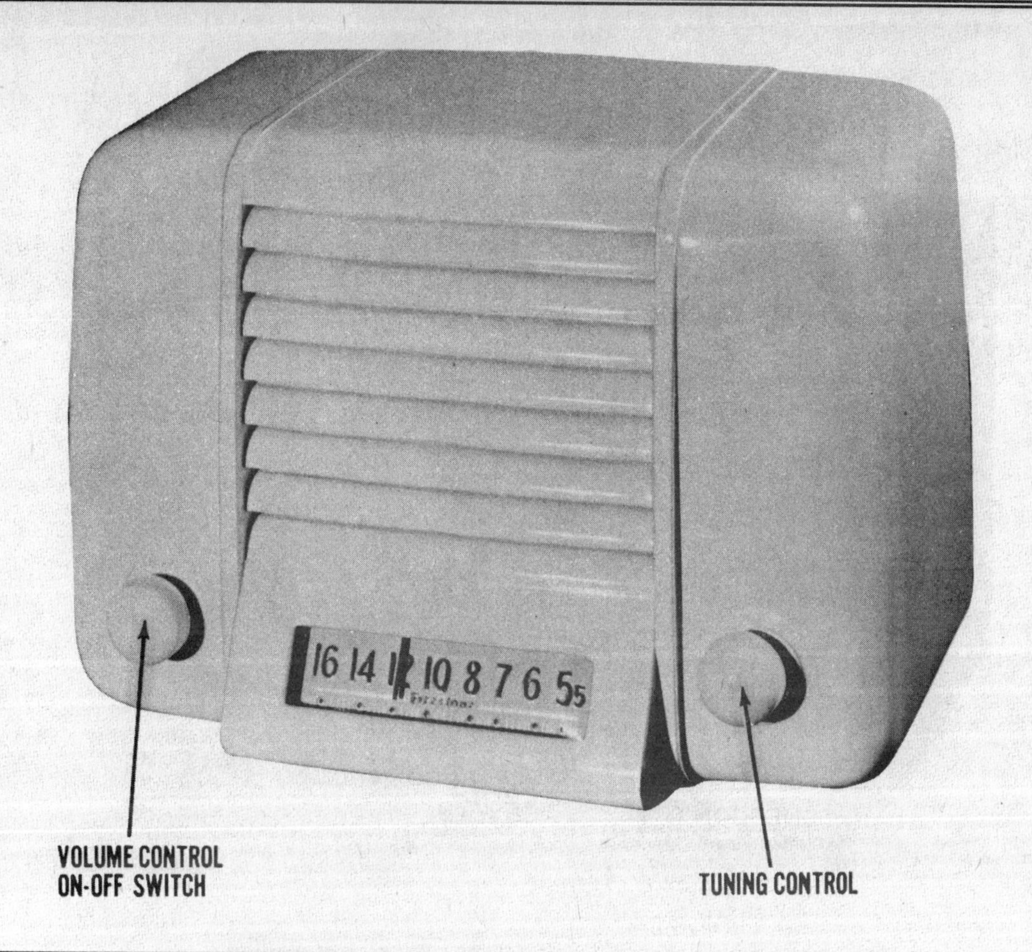

AIR CHIEF

MODEL PICTURED
4-A-12

AC/DC operated AM-FM superheterodyne receiver with loop antenna

TUBES
6

POWER SUPPLY
110-120 volts AC/DC

TUNING RANGE
540-1600KC, 88-108MC

MFR/SUPPLIER
Firestone Tire and Rubber Co.

PHOTOFACT SET
49-8

PUBLISHED
1948

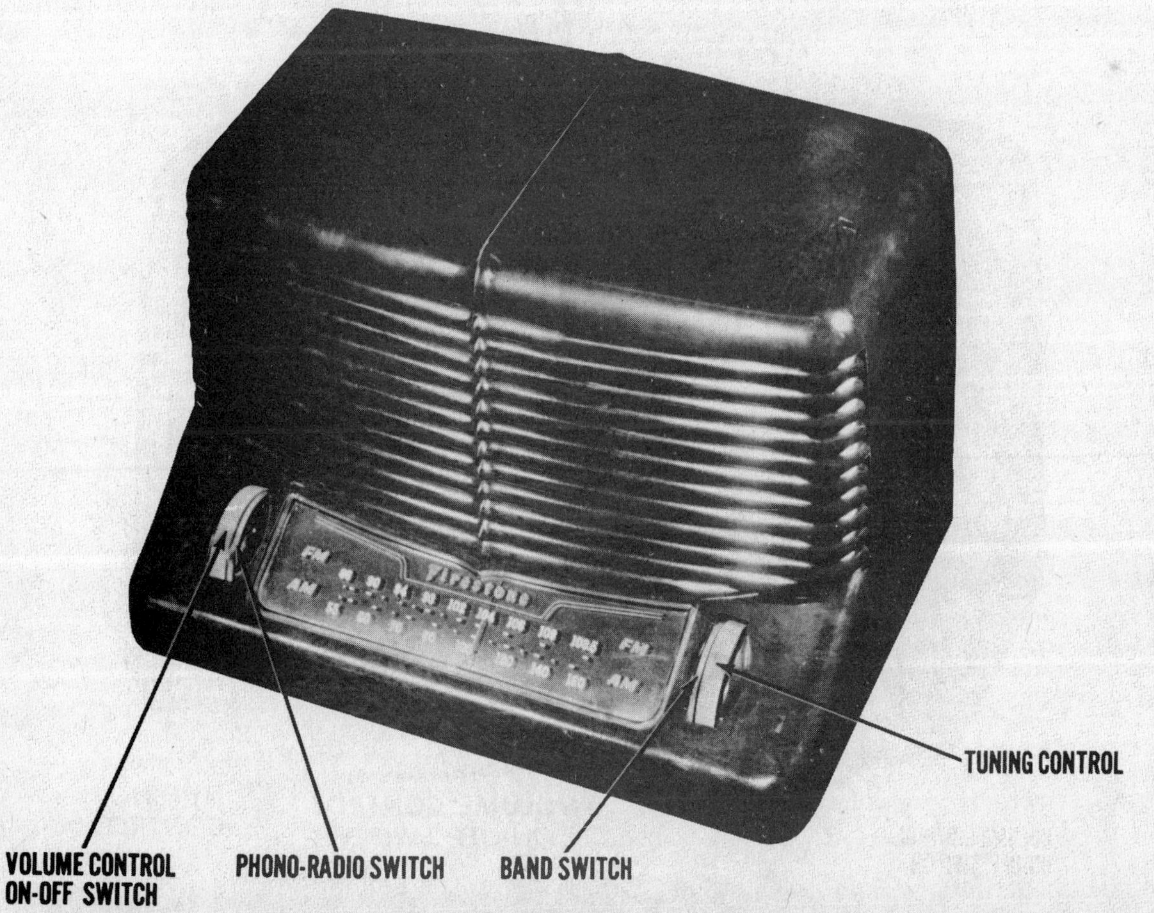

AIR KING

MODEL PICTURED
4604

AC two-band superheterodyne, self-contained loop antenna

TUBES
6

POWER SUPPLY
105-125 volts AC

TUNING RANGE
540-1600KC, 6-18MC

MFR/SUPPLIER
Air King

PHOTOFACT SET
4-25

PUBLISHED
1946

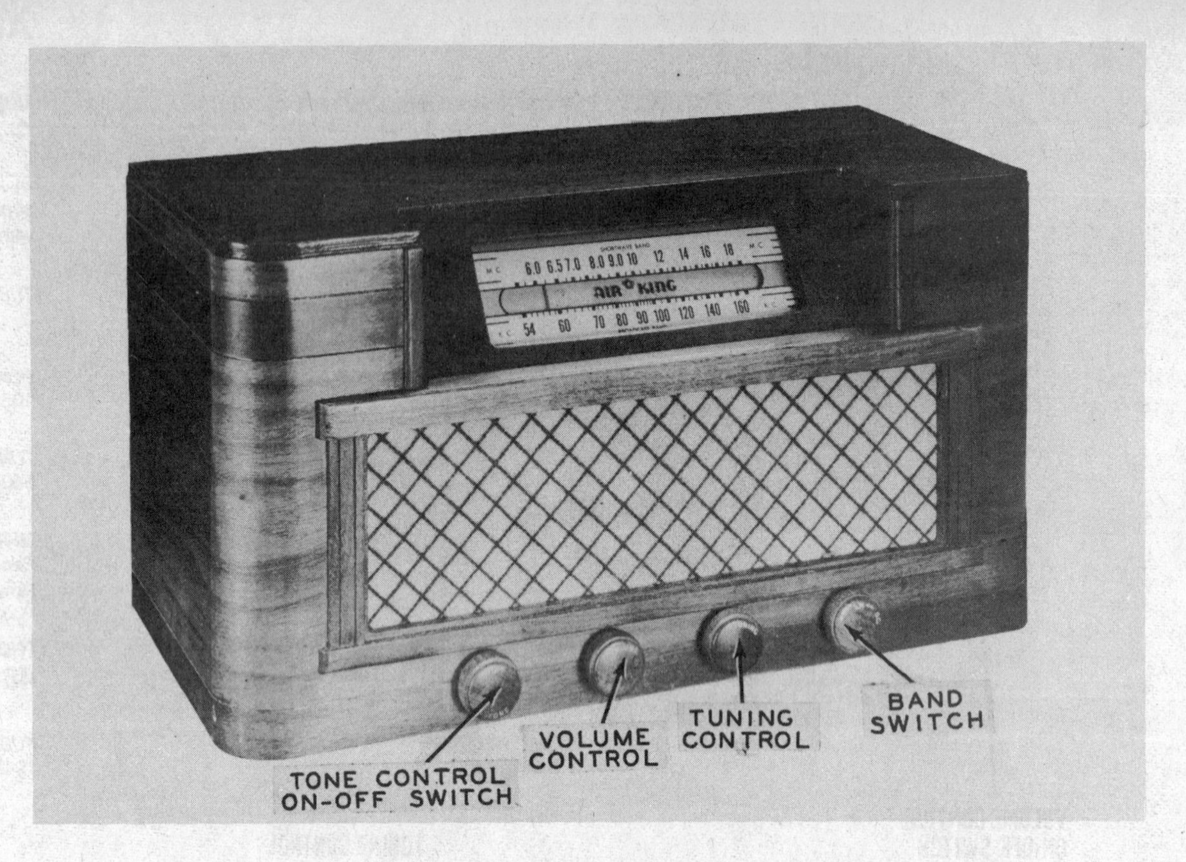

AIR KING

MODEL PICTURED
4705

AC/DC superheterodyne, self-contained loop antenna

TUBES
6

POWER SUPPLY
110-125 volts AC/DC,
.240 amp @ 117 volts AC

TUNING RANGE
540-1700KC

MFR/SUPPLIER
Air King

PHOTOFACT SET
9-1

PUBLISHED
1946

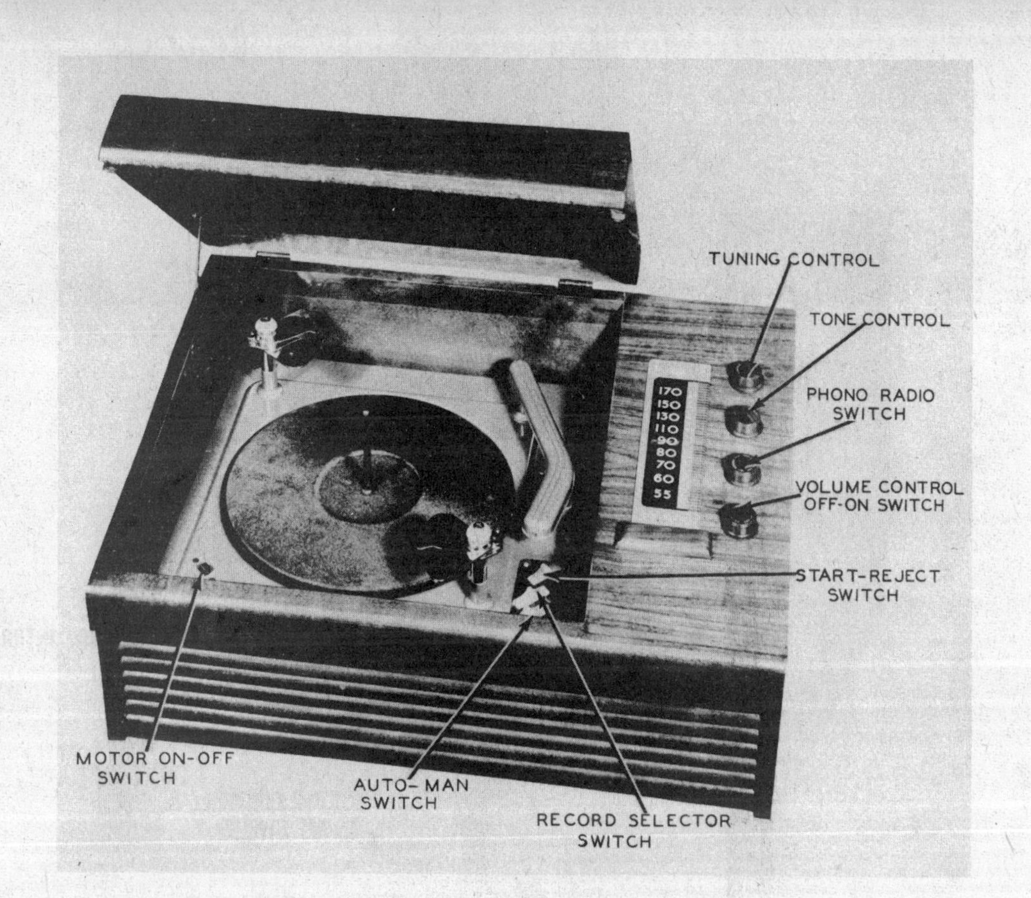

AIR KING

MODEL PICTURED
4704

AC operated auto radio-phono combo, self-contained loop antenna

TUBES
6

POWER SUPPLY
110-125 volts AC

TUNING RANGE
525-1740KC

MFR/SUPPLIER
Air King

PHOTOFACT SET
12-2

PUBLISHED
1947

AIR KING

MODEL PICTURED
A-403

AC operated radio-phono combo superheterodyne with loop antenna

TUBES
4

POWER SUPPLY
110-125 volts AC

TUNING RANGE
535-1600KC

MFR/SUPPLIER
Air King

PHOTOFACT SET
20-2

PUBLISHED
1947

AIR KING

MODEL PICTURED
A-400

AC/DC Superheterodyne receiver

TUBES
4

POWER SUPPLY
110-125 volts AC/DC

TUNING RANGE
535-1600KC

MFR/SUPPLIER
Air King

PHOTOFACT SET
23-1

PUBLISHED
1947

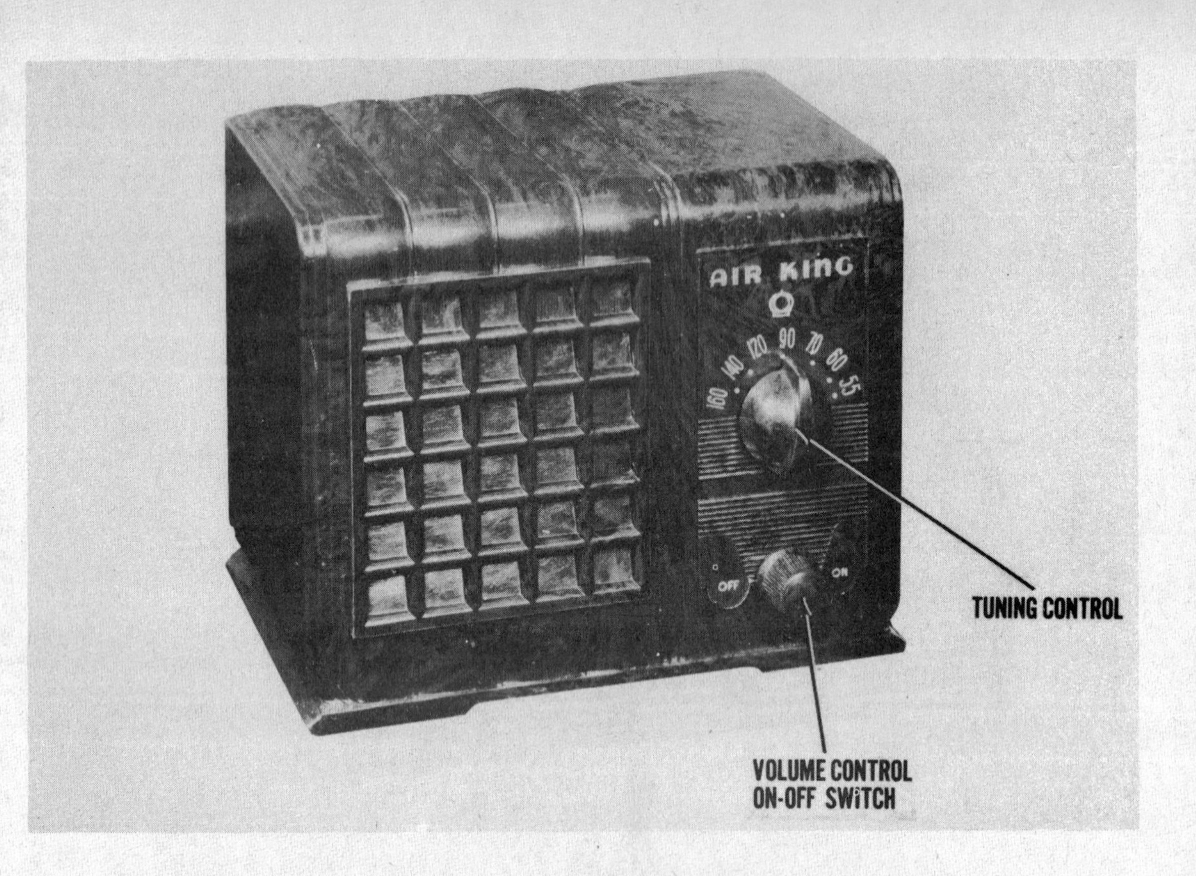

AIR KING

MODEL PICTURED
A-510

Three-power portable superheterodyne receiver with loop antenna

TUBES
4

POWER SUPPLY
110-120 volts AC/DC or 7.5 volts A supply & 90 volts B supply

TUNING RANGE
540-1600KC

MFR/SUPPLIER
Air King

PHOTOFACT SET
24-3

PUBLISHED
1947

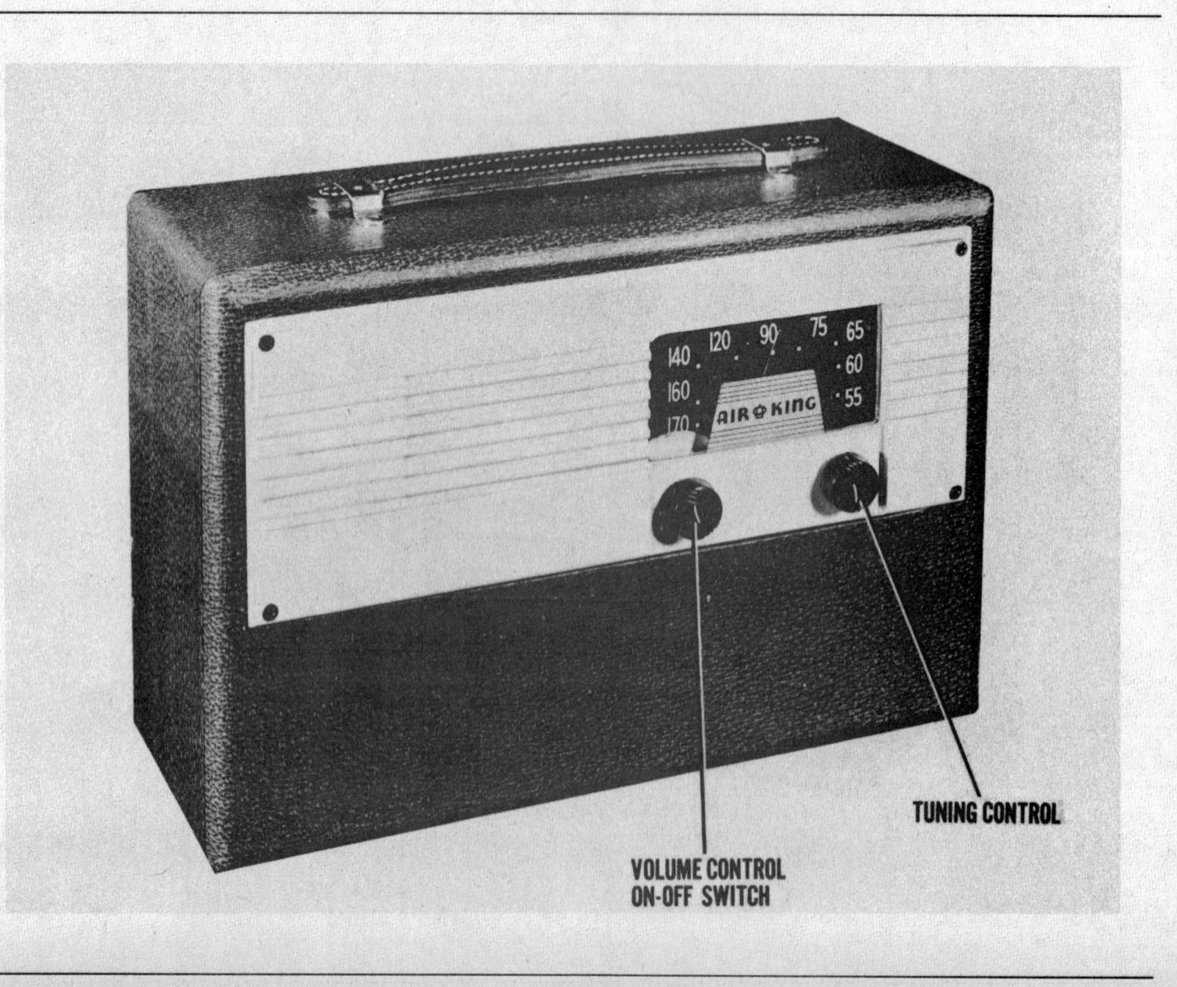

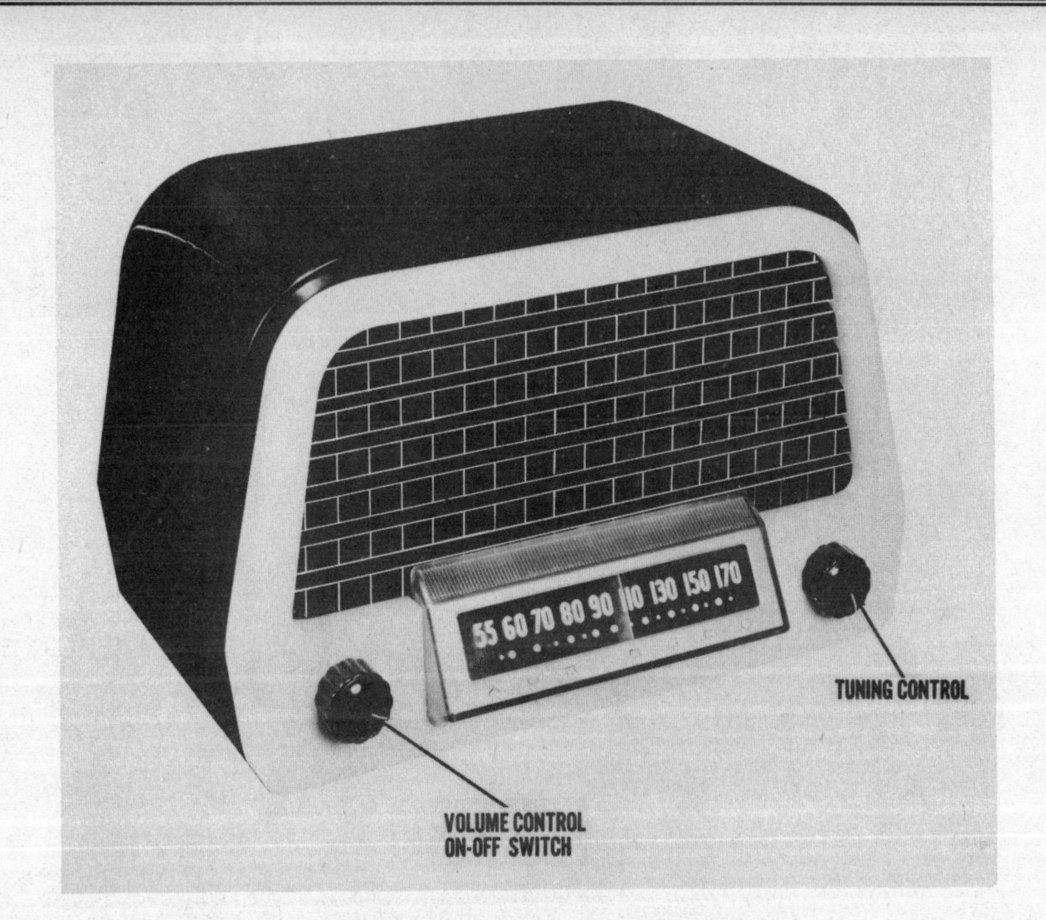

AIR KING

MODEL PICTURED
A-600

AC/DC superheterodyne with loop antenna

TUBES
6

POWER SUPPLY
110-120 volts AC/DC

TUNING RANGE
530-1700KC

MFR/SUPPLIER
Air King

PHOTOFACT SET
26-3

PUBLISHED
1947

AIR KING

MODEL PICTURED
A-511

AC/DC operated superheterodyne receiver with loop antenna

TUBES
5

POWER SUPPLY
110-120 volts AC/DC

TUNING RANGE
535-1750KC

MFR/SUPPLIER
Air King

PHOTOFACT SET
30-2

PUBLISHED
1947

AIR KING

MODEL PICTURED
A-502

AC/DC operated three-band superheterodyne receiver with loop antenna

TUBES
5

POWER SUPPLY
105-125 volts AC/DC, .26 amp @ 117 volts AC

TUNING RANGE
540-1600KC, 6-15MC 13-23MC

MFR/SUPPLIER
Air King

PHOTOFACT SET
31-3

PUBLISHED
1948

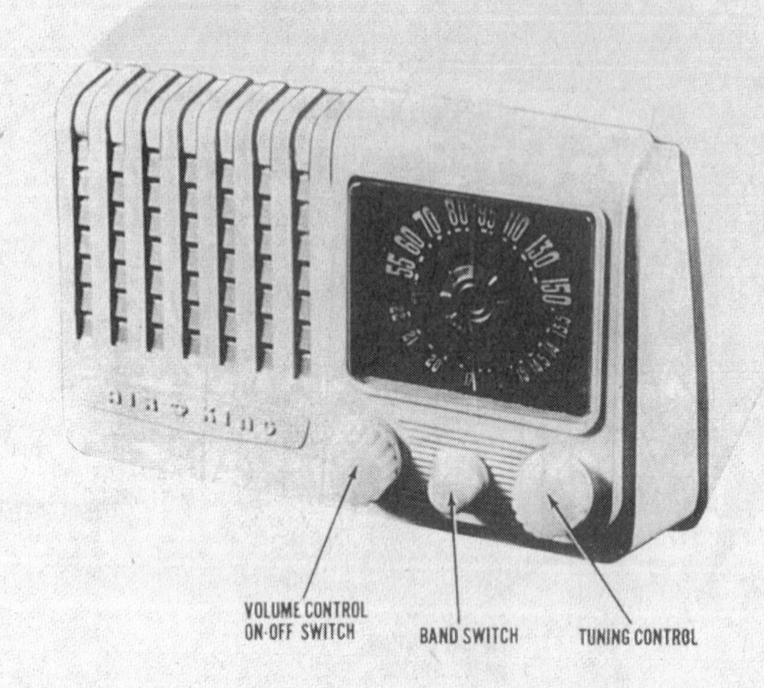

AIR KING

MODEL PICTURED
A-410

Battery operated portable superheterodyne receiver with loop antenna and camera

TUBES
4

POWER SUPPLY
1.5 volts A & 67.5 volts B supply

TUNING RANGE
540-1600KC

MFR/SUPPLIER
Air King

PHOTOFACT SET
34-1

PUBLISHED
1947

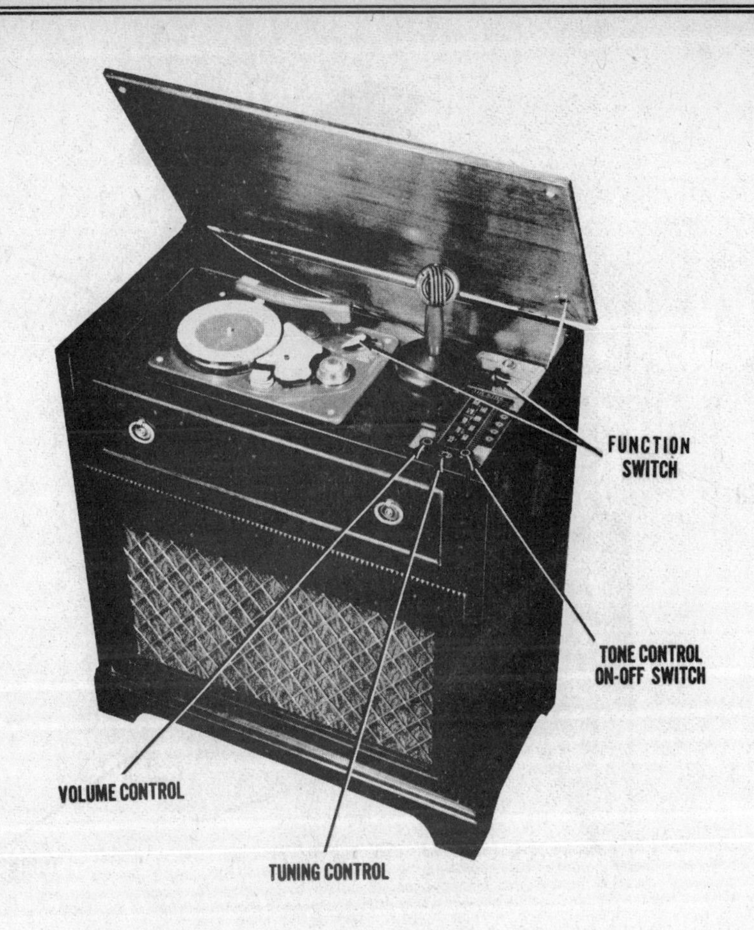

AIR KING

MODEL PICTURED
4700

AC operated superheterodyne receiver and phono with wire recorder

TUBES
7

POWER SUPPLY
110-120 volts AC

TUNING RANGE
540-1670KC

MFR/SUPPLIER
Air King

PHOTOFACT SET
39-1

PUBLISHED
1948

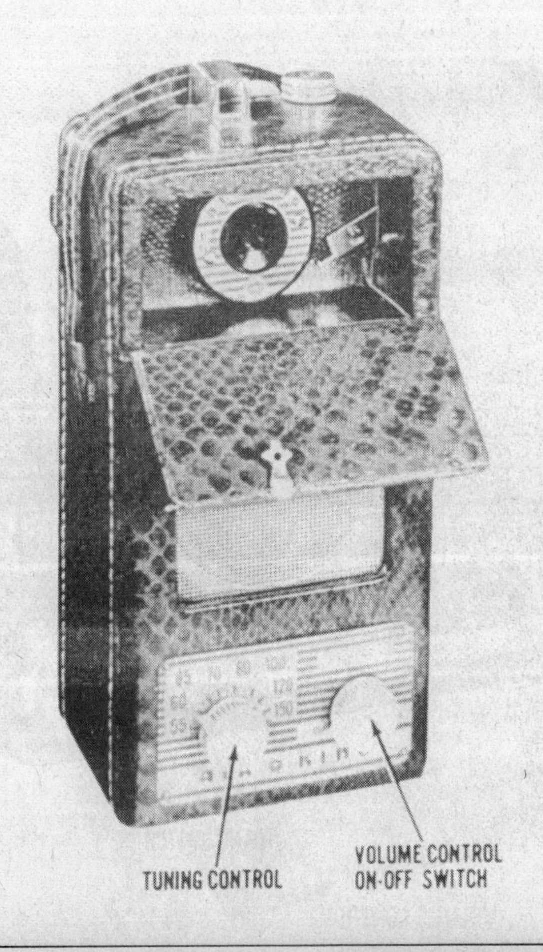

AIR KING

MODEL PICTURED
A-410 (REVISED)

Battery operated portable superheterodyne receiver with camera

TUBES
4

POWER SUPPLY
1.5 volts A supply & 67.5 volts B supply

TUNING RANGE
540-1600KC

MFR/SUPPLIER
Air King

PHOTOFACT SET
40-1

PUBLISHED
1948

AIR KING

MODEL PICTURED
A-426

Battery operated portable superheterodyne receiver with loop antenna

TUBES
4

POWER SUPPLY
1.5 volts A & 45 volts B supply

TUNING RANGE
540-1650KC

MFR/SUPPLIER
Air King

PHOTOFACT SET
43-1

PUBLISHED
1948

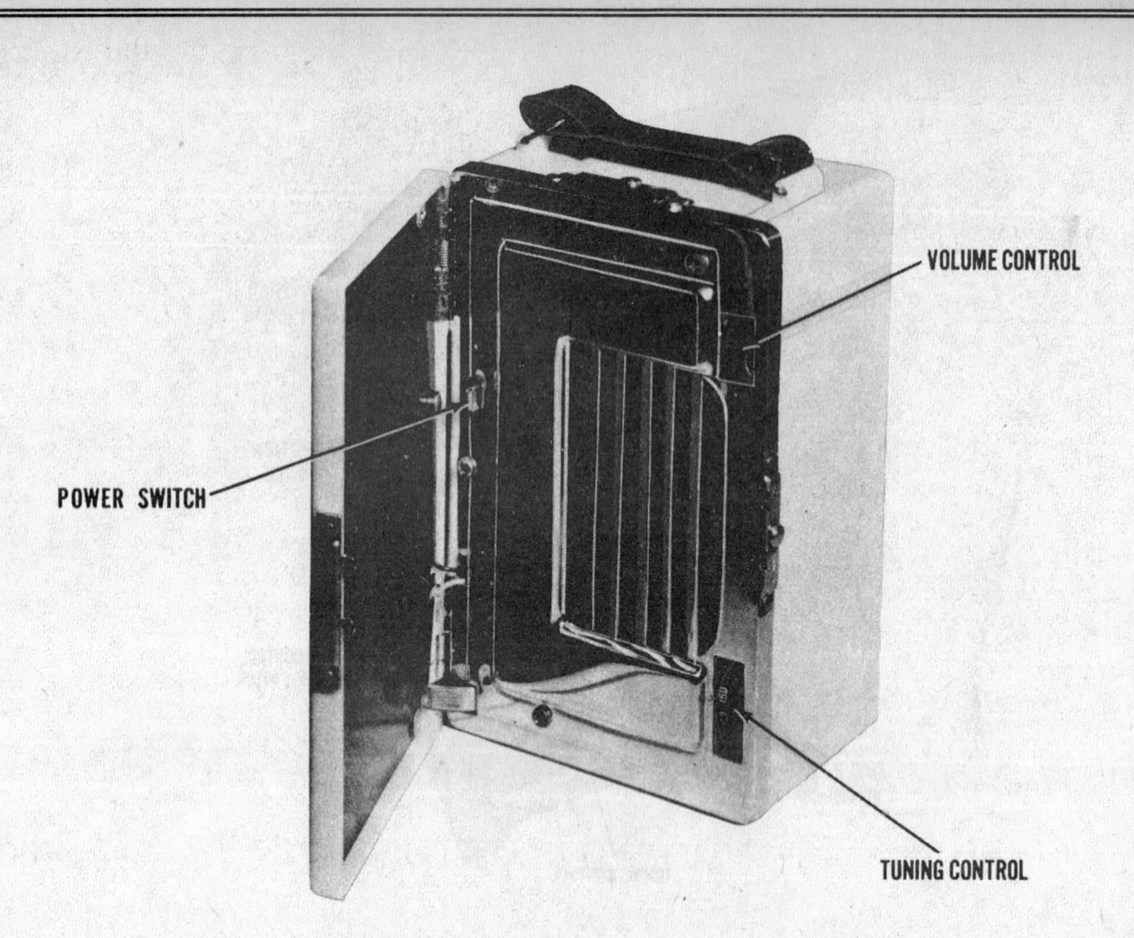

AIR KING

MODEL PICTURED
A-650

AC/DC operated AM-FM superheterodyne receiver with loop antenna

TUBES
6

POWER SUPPLY
110-120 volts AC/DC

TUNING RANGE
535-1650KC

MFR/SUPPLIER
Air King

PHOTOFACT SET
45-4

PUBLISHED
1948

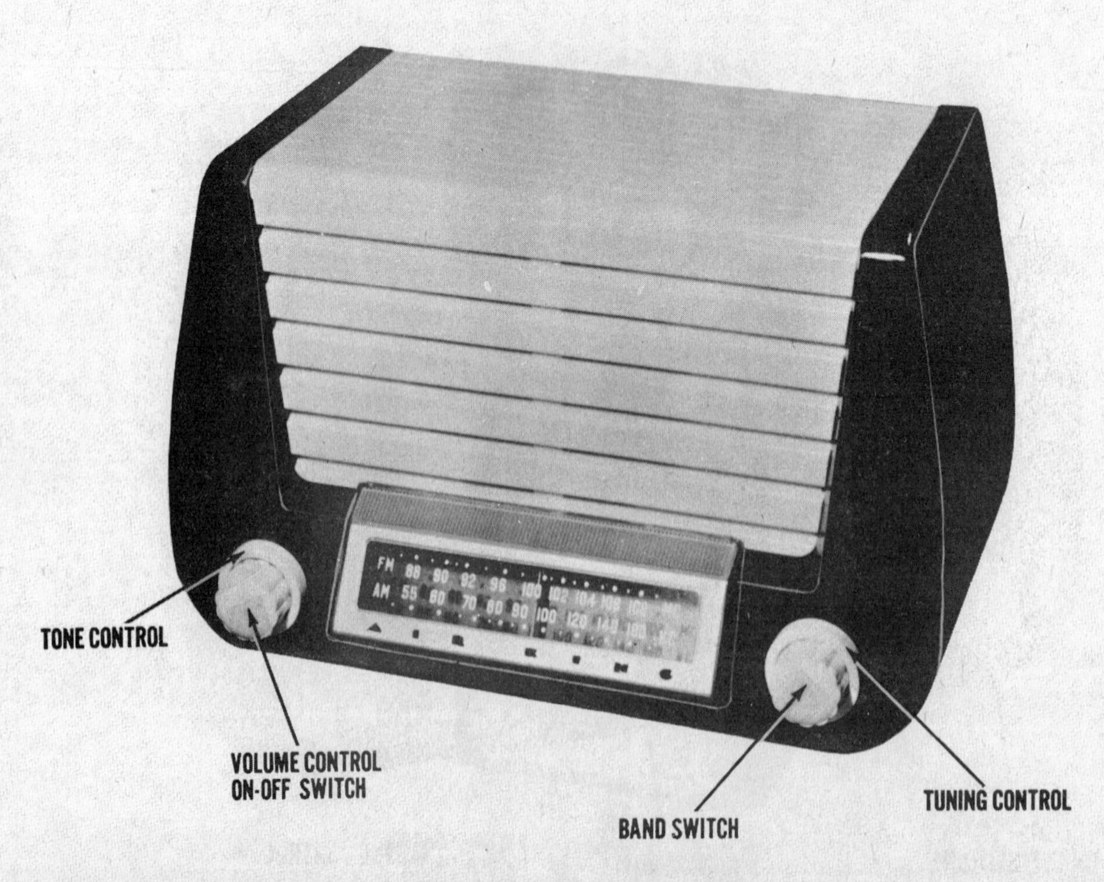

AIR KING

MODEL PICTURED
A-520

Three-power operated portable superheterodyne receiver with loop antenna

TUBES
4

POWER SUPPLY
110-120 volts AC/DC or 4.5 volts A supply and 67.5 volts B supply

TUNING RANGE
540-1640KC

MFR/SUPPLIER
Air King

PHOTOFACT SET
49-4

PUBLISHED
1948

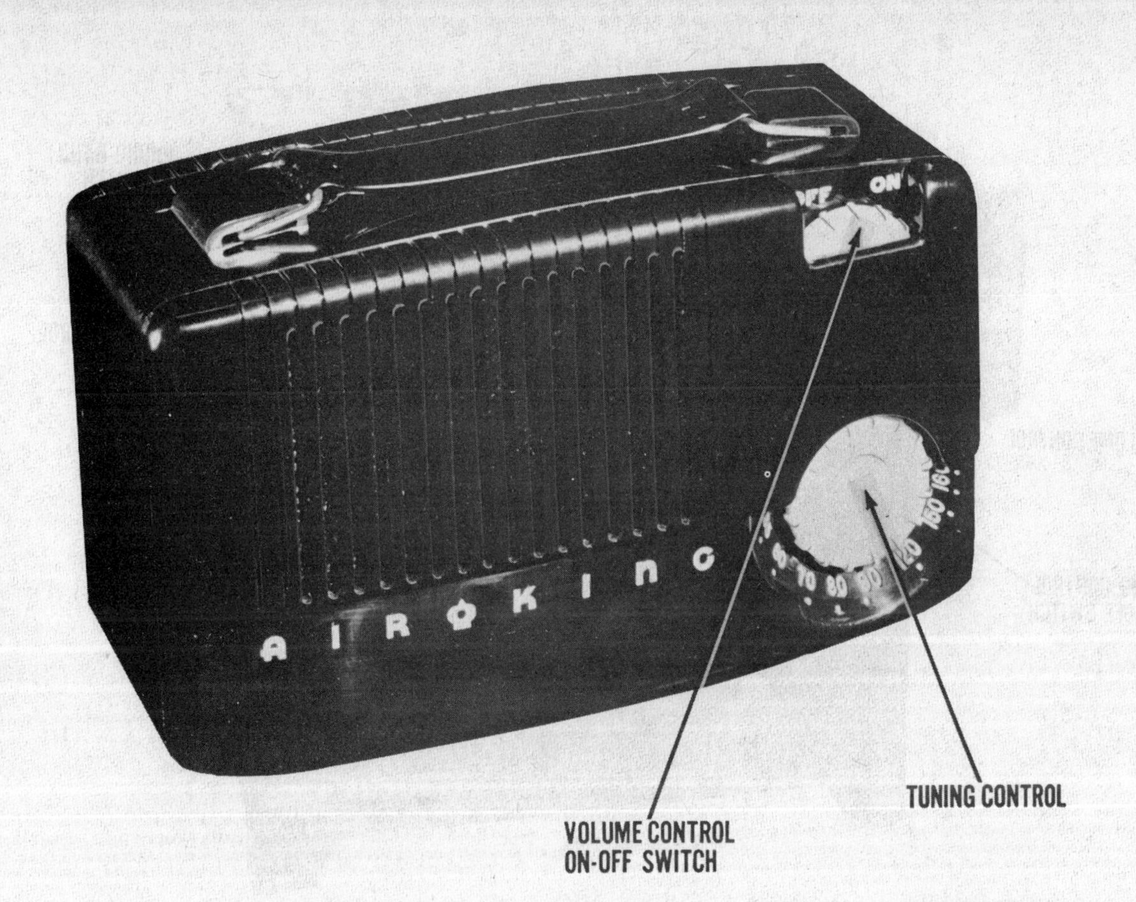

AIR KING

MODEL PICTURED
A-625

AC/DC operated superheterodyne receiver with loop antenna

TUBES
6

POWER SUPPLY
110-120 volts AC/DC

TUNING RANGE
540-1600KC

MFR/SUPPLIER
Air King

PHOTOFACT SET
50-3

PUBLISHED
1948

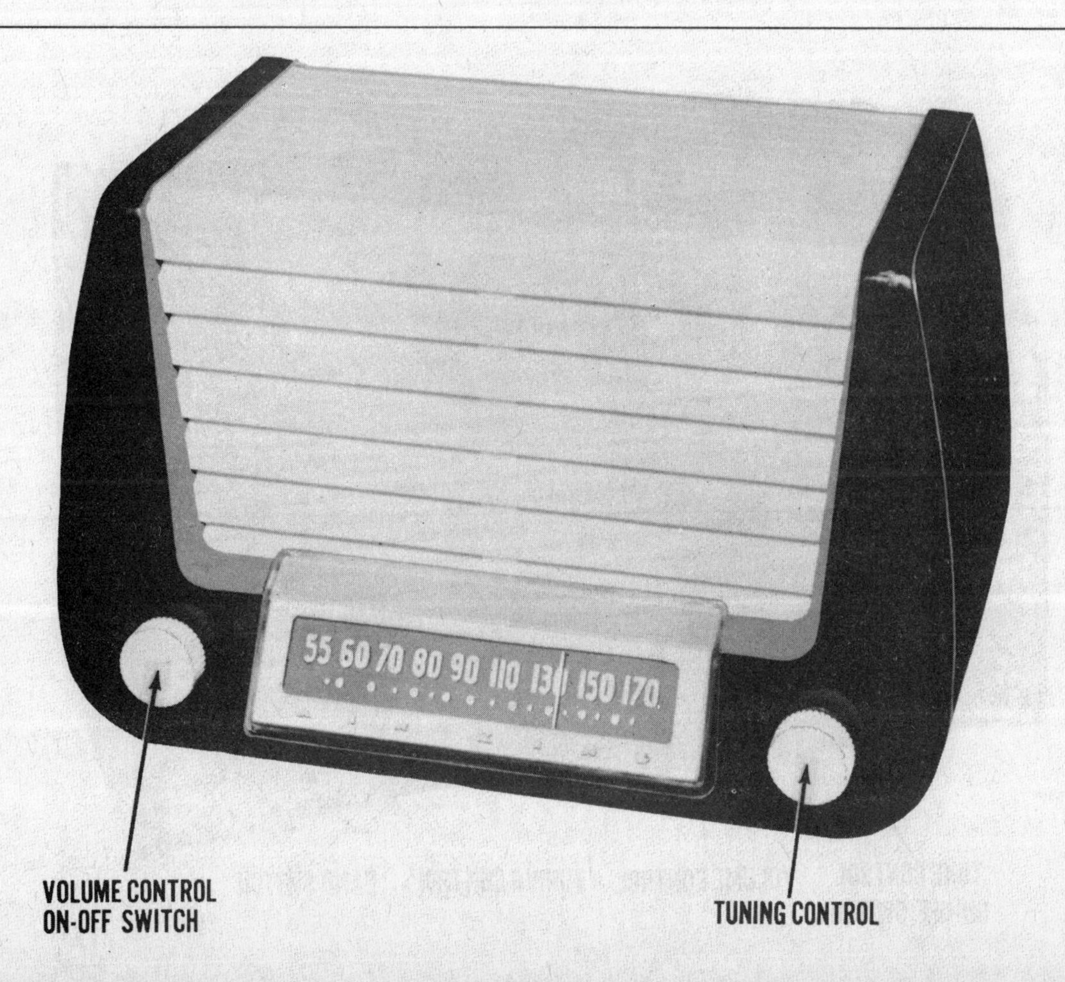

AIR KING

MODEL PICTURED
800

AC operated combo phono-radio, AM/FM superheterodyne receiver with loop antenna

TUBES
8

POWER SUPPLY
110-120 volts AC, .56 amp @ 117 volts AC

TUNING RANGE
535-1620KC, 87.25-108.75MC

MFR/SUPPLIER
Air King

PHOTOFACT SET
66-1

PUBLISHED
1949

AIR KING

MODEL PICTURED
A-604

AC operated two-band superheterodyne receiver with loop antenna

TUBES
6

POWER SUPPLY
110-120 volts AC, .39 amp @ 117 volts AC

TUNING RANGE
540-1600KC

MFR/SUPPLIER
Air King

PHOTOFACT SET
81-2

PUBLISHED
1950

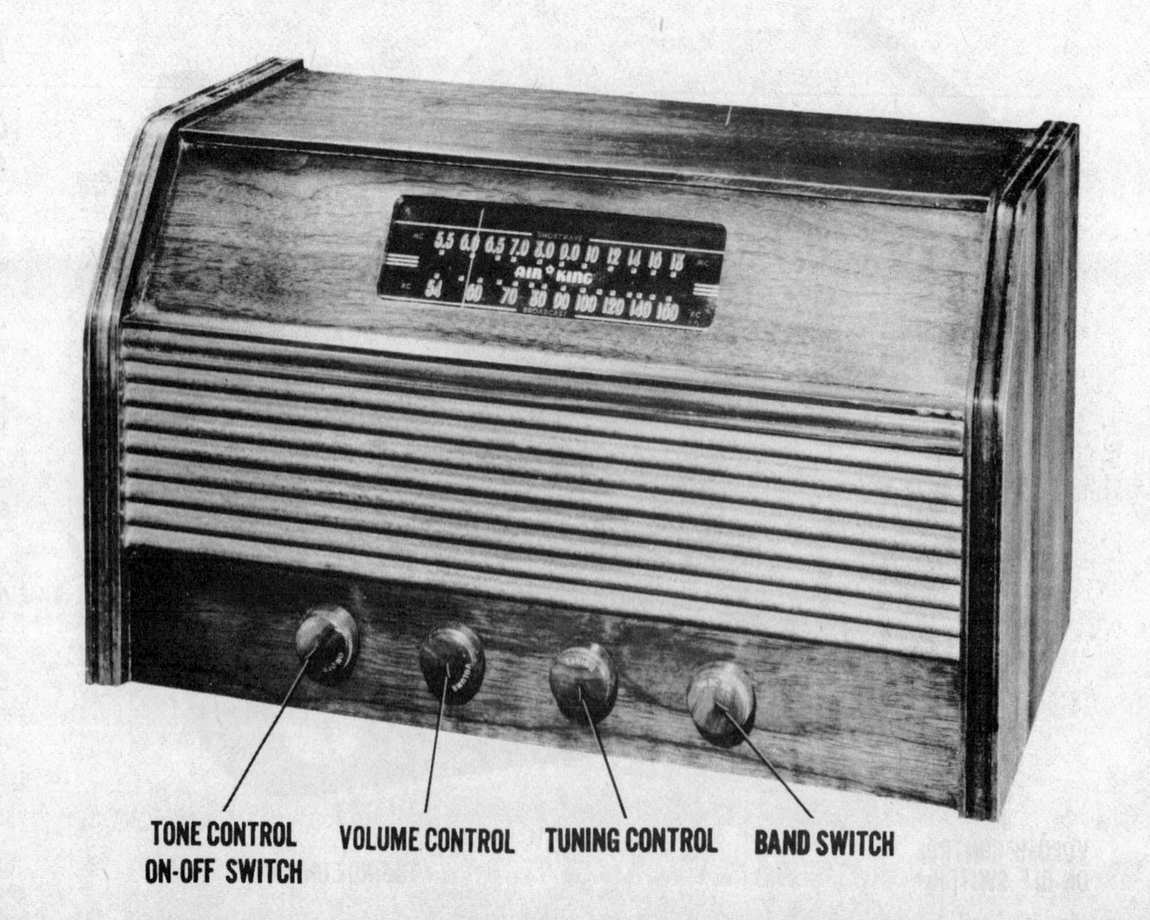

AIR KNIGHT

MODEL PICTURED
N5-RD291

AC operated phono-radio combo superheterodyne with loop antenna

TUBES
5

POWER SUPPLY
110-120 volts AC

TUNING RANGE
540-1720KC

MFR/SUPPLIER
Butler Bro. Randolph and Canal

PHOTOFACT SET
17-3

PUBLISHED
1947

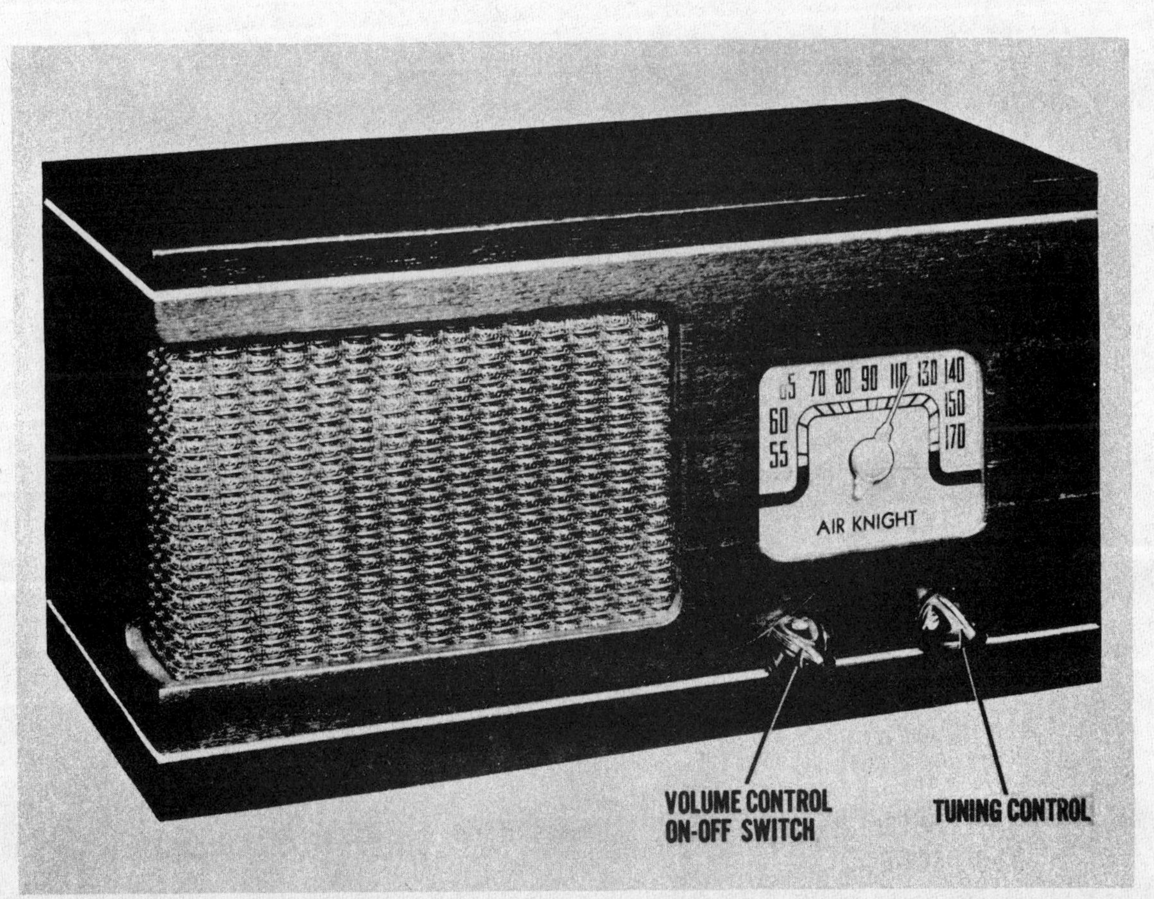

AIR KNIGHT

MODEL PICTURED
CA-500

AC/DC operated superheterodyne receiver, self-contained loop antenna

TUBES
5

POWER SUPPLY
110-120 volts AC/DC

TUNING RANGE
540-1720KC

MFR/SUPPLIER
Butler Bro. Randolph and Canal

PHOTOFACT SET
17-4

PUBLISHED
1947

AIRADIO

MODEL PICTURED
3100

AC/DC operated FM receiver

TUBES
8

POWER SUPPLY
110-120 volts AC/DC

TUNING RANGE
88-108MC

MFR/SUPPLIER
Airadio Inc.

PHOTOFACT SET
37-1

PUBLISHED
1948

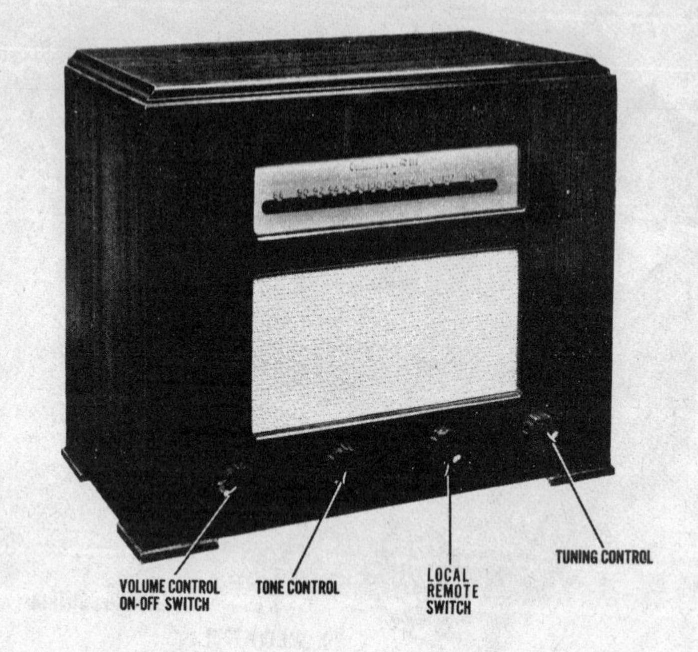

AIRCASTLE

MODEL PICTURED
106B

AC/DC operated superheterodyne receiver, self-contained loop antenna

TUBES
5

POWER SUPPLY
117 volts AC/DC, .260 amp @ 117 volts AC

TUNING RANGE
540-1630KC

MFR/SUPPLIER
Spiegel Inc.

PHOTOFACT SET
13-3

PUBLISHED
1947

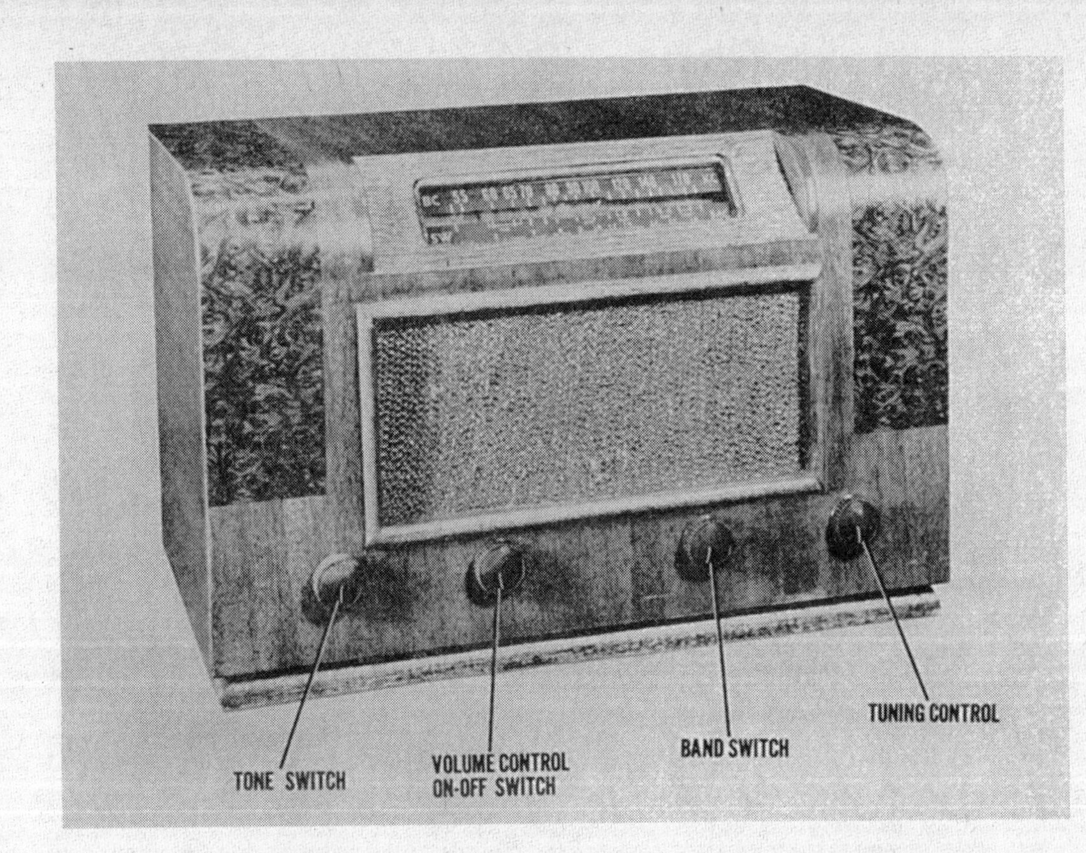

AIRCASTLE

MODEL PICTURED
5011

AC/DC two-band superheterodyne, self-contained loop antenna

TUBES
6

POWER SUPPLY
110-125 volts AC/DC, .240 amp @ 117 volts AC

TUNING RANGE
520-1730KC

MFR/SUPPLIER
Spiegel Inc.

PHOTOFACT SET
13-4

PUBLISHED
1947

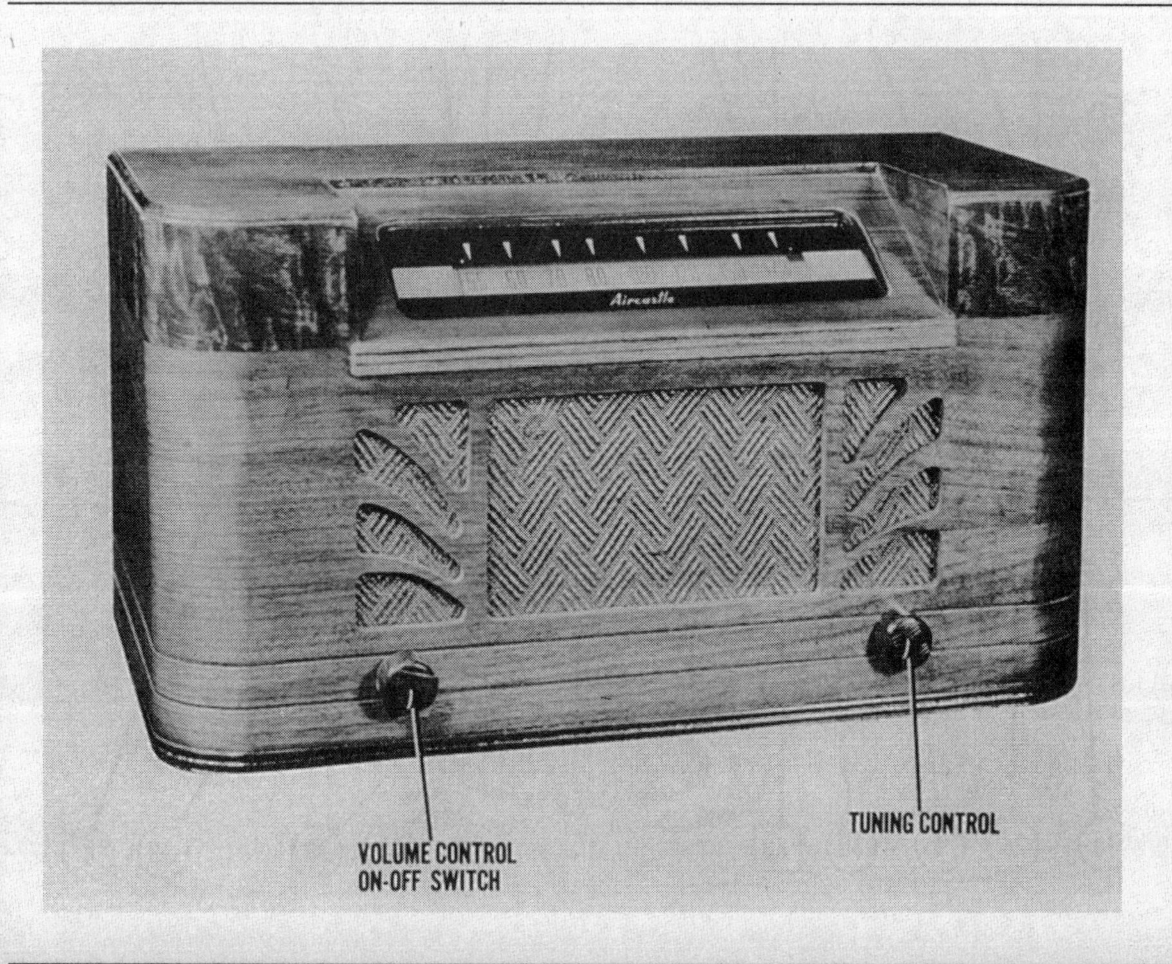

AIRCASTLE

MODEL PICTURED
PX

Battery operated superheterodyne receiver

TUBES
4

POWER SUPPLY
90 volts B battery & 1.5 volts A battery, Burgess type 17G-D60 or equivalent.

TUNING RANGE
535-1725KC

MFR/SUPPLIER
Spiegel Inc.

PHOTOFACT SET
13-35

PUBLISHED
1947

AIRCASTLE

MODEL PICTURED
568

AC/DC two-band superheterodyne receiver

TUBES
5

POWER SUPPLY
105-125 volts AC/DC

TUNING RANGE
540-1600KC, 6-18MC

MFR/SUPPLIER
Spiegel Inc.

PHOTOFACT SET
14-1

PUBLISHED
1947

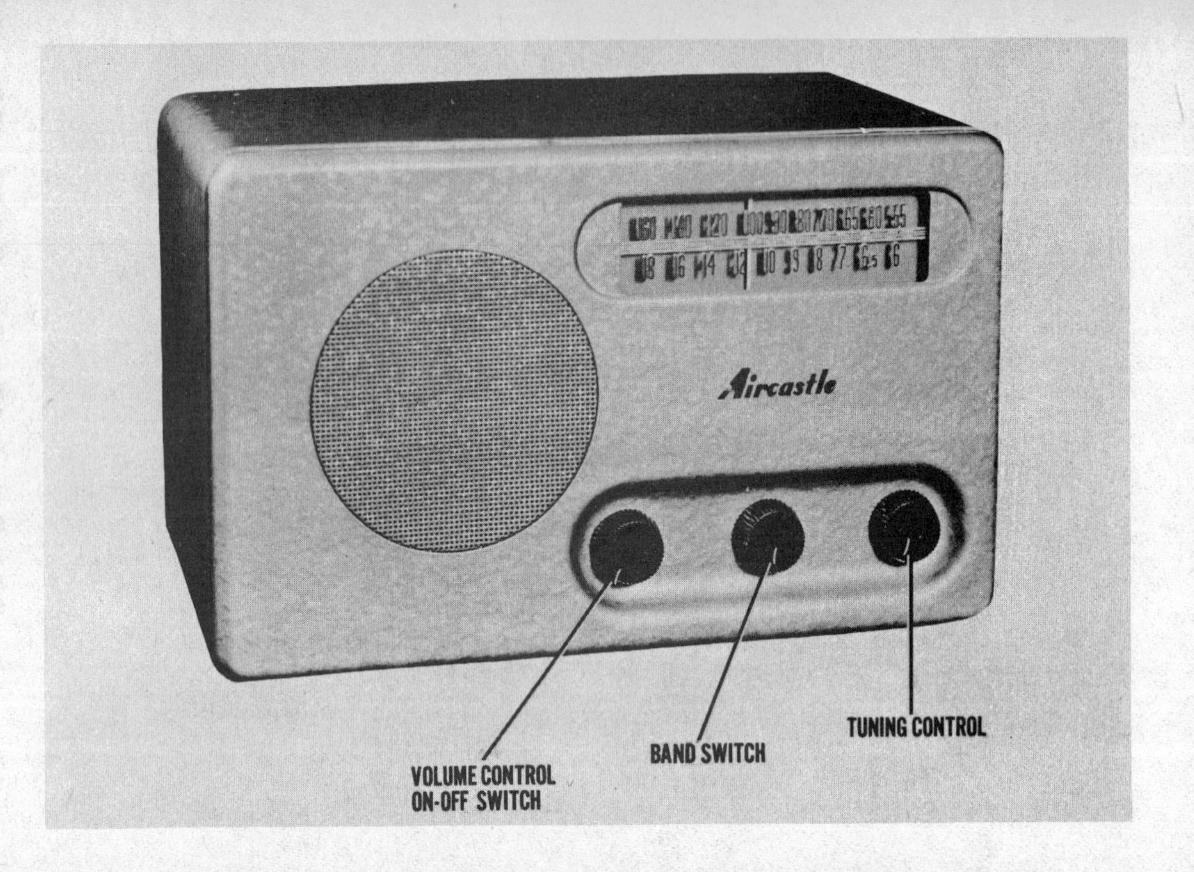

AIRCASTLE

MODEL PICTURED
651

AC/DC superheterodyne, self-contained loop antenna

TUBES
5

POWER SUPPLY
105-125 volts AC/DC,
.235 amp @ 117 volts AC

TUNING RANGE
540-1700KC

MFR/SUPPLIER
Spiegel Inc.

PHOTOFACT SET
15-1

PUBLISHED
1947

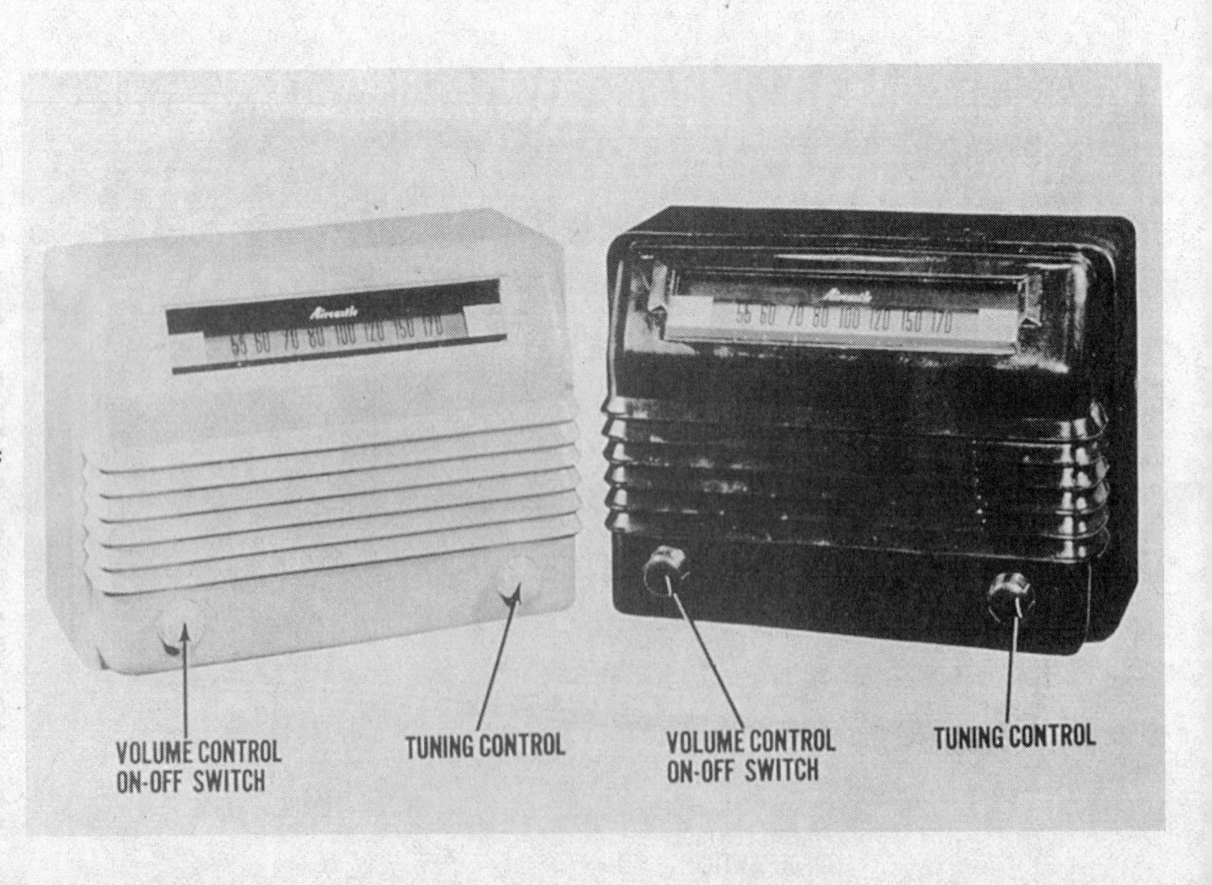

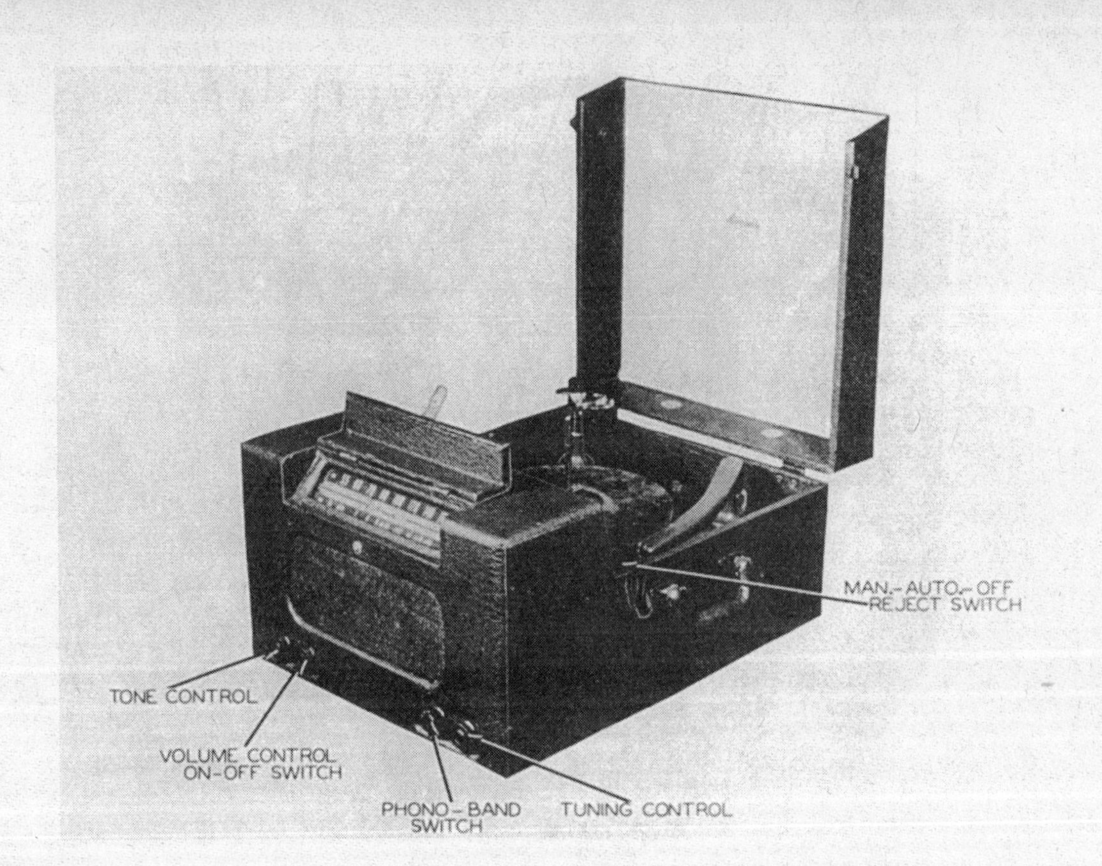

AIRCASTLE

MODEL PICTURED
6634

AC operated combo radio-phono superheterodyne with loop antenna

TUBES
6

POWER SUPPLY
105-125 volts AC

TUNING RANGE
540-1660KC, 5.5-18.6MC

MFR/SUPPLIER
Spiegel Inc.

PHOTOFACT SET
15-2

PUBLISHED
1947

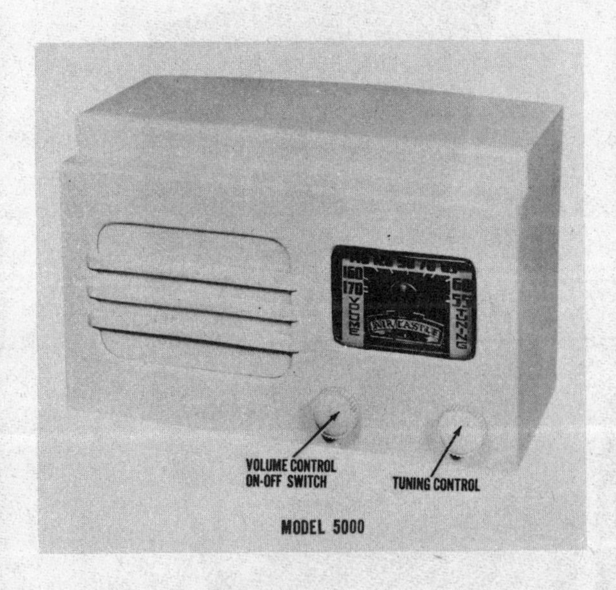

MODEL 5000

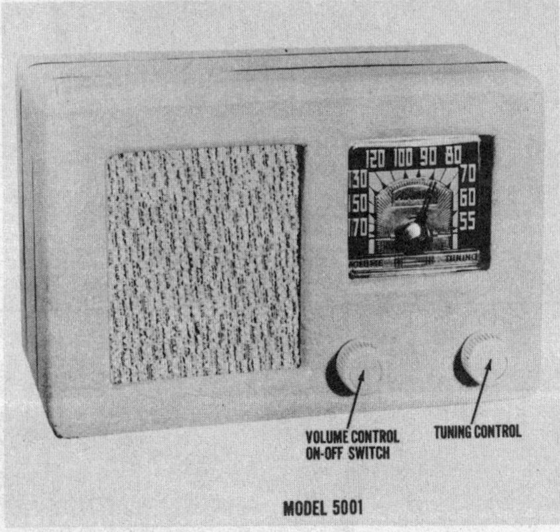

MODEL 5001

AIRCASTLE

MODEL PICTURED
5000 and 5001

AC/DC superheterodyne receiver, self-contained loop antenna

TUBES
5

POWER SUPPLY
110-120 volts AC, .230 amp @ 117 volts AC

TUNING RANGE
540-1720KC

MFR/SUPPLIER
Spiegel Inc.

PHOTOFACT SET
16-2

PUBLISHED
1947

AIRCASTLE

MODEL PICTURED
5020

Three-power portable, AC/DC battery superheterodyne receiver

TUBES
4

POWER SUPPLY
110-125 volts AC/DC or two-4.5 volt batteries A supply & two-45 volt batteries B supply

TUNING RANGE
540-1720KC

MFR/SUPPLIER
Spiegel Inc.

PHOTOFACT SET
16-3

PUBLISHED
1947

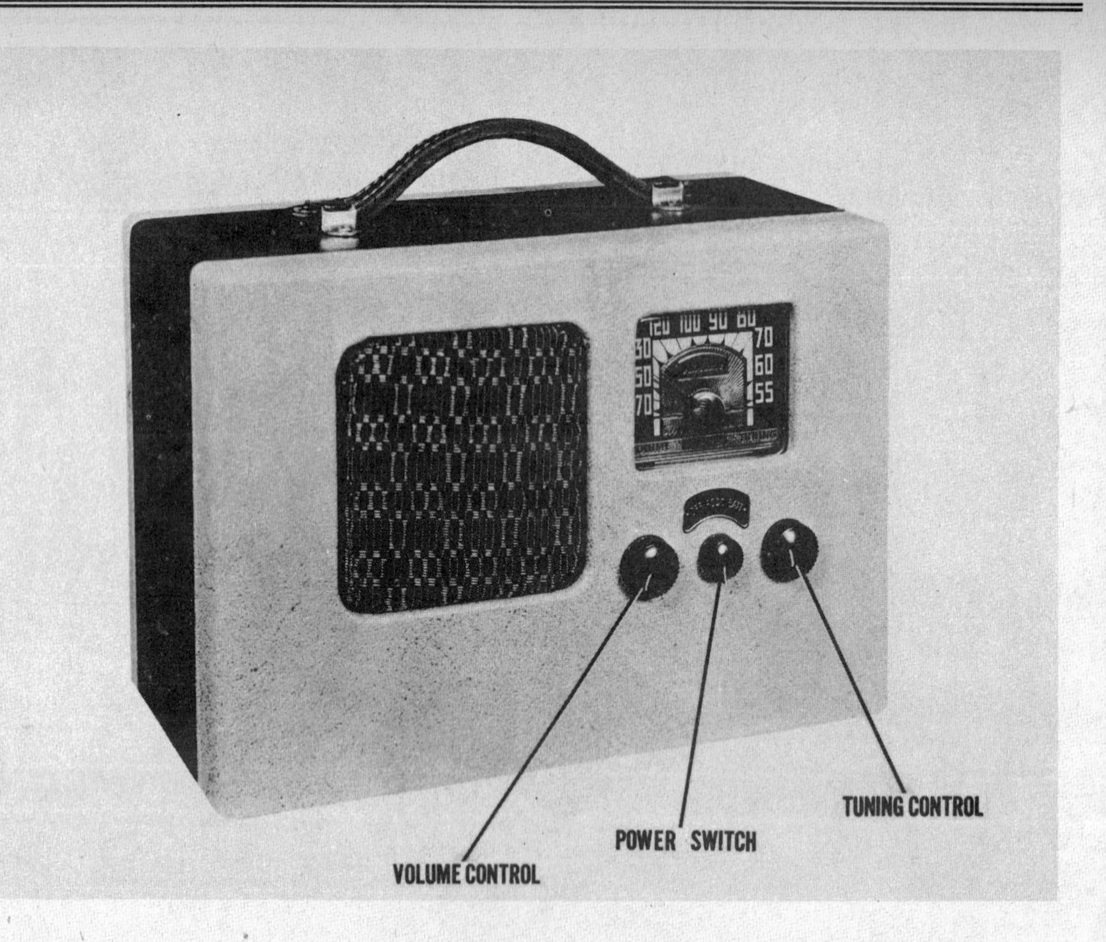

AIRCASTLE

MODEL PICTURED
6541, 6547

AC operated auto-phono combo superheterodyne receiver, self-contained loop antenna

TUBES
5

POWER SUPPLY
105-125 volts AC

TUNING RANGE
540-1700KC

MFR/SUPPLIER
Spiegel Inc.

PHOTOFACT SET
17-2

PUBLISHED
1947

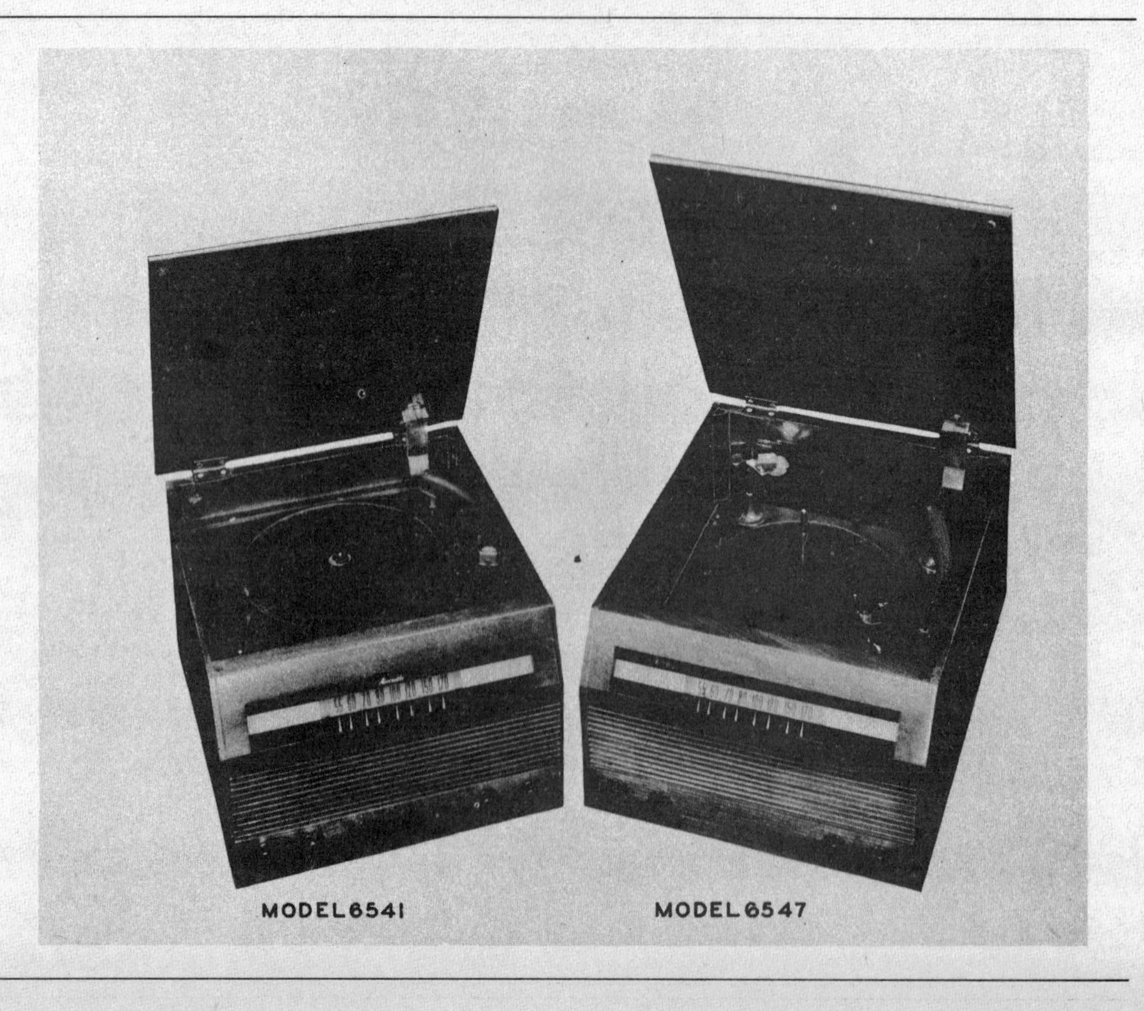

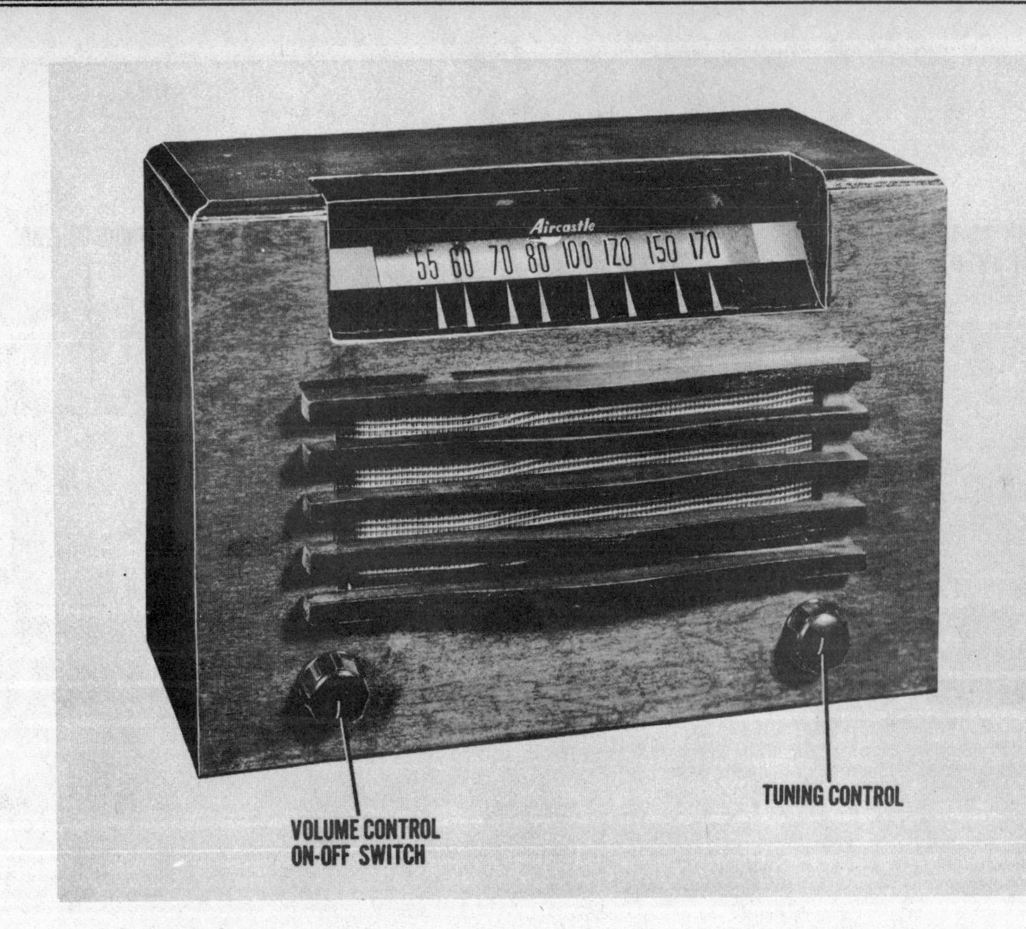

AIRCASTLE

MODEL PICTURED
6514

AC/DC superheterodyne receiver, self-contained loop antenna

TUBES
5

POWER SUPPLY
105-125 volts AC/DC

TUNING RANGE
540-1700KC

MFR/SUPPLIER
Spiegel Inc.

PHOTOFACT SET
18-4

PUBLISHED
1947

AIRCASTLE

MODEL PICTURED
5002

AC/DC operated superheterodyne receiver, self-contained loop antenna

TUBES
6

POWER SUPPLY
110-125 volts AC/DC, .220 amp @ 117 volts AC

TUNING RANGE
540-1720KC

MFR/SUPPLIER
Spiegel Inc.

PHOTOFACT SET
19-1

PUBLISHED
1947

AIRCASTLE

MODEL PICTURED
5003

AC/DC superheterodyne receiver with loop antenna

TUBES
5

POWER SUPPLY
110-120 volts AC/DC

TUNING RANGE
530-1720KC

MFR/SUPPLIER
Spiegel Inc.

PHOTOFACT SET
20-1

PUBLISHED
1947

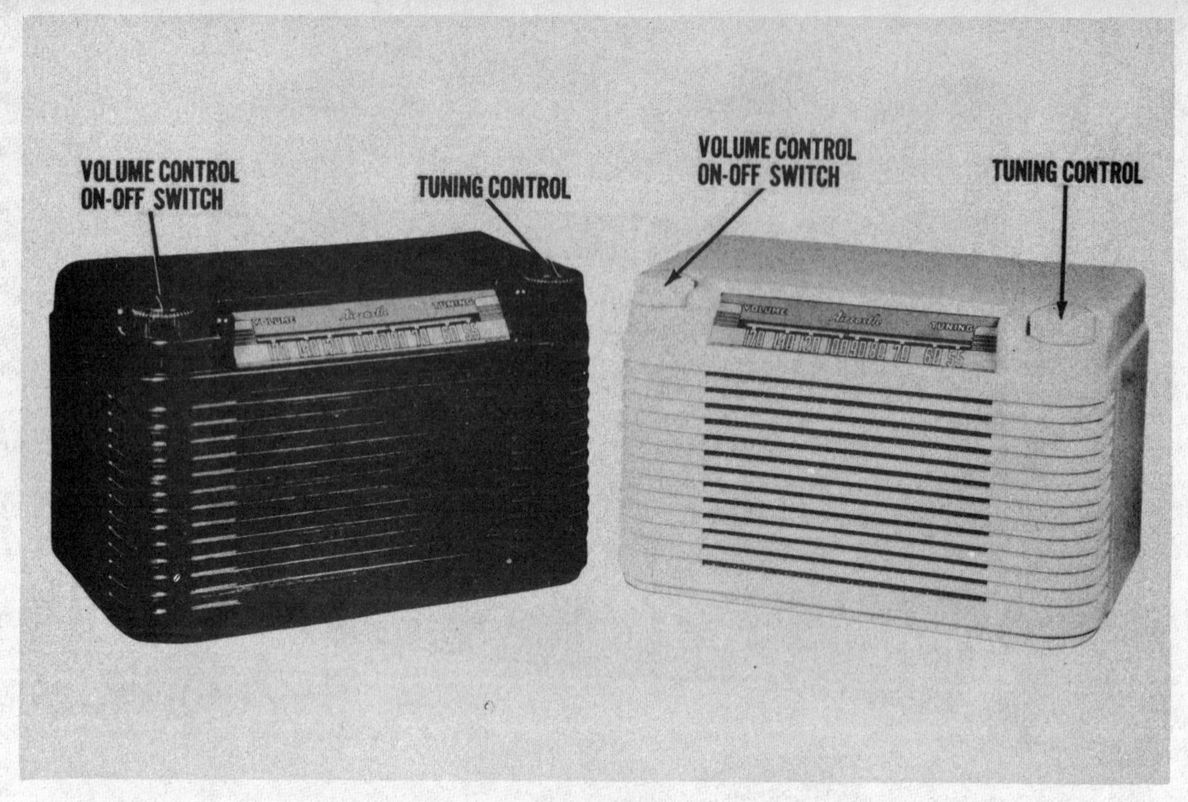

AIRCASTLE

MODEL PICTURED
5025

Three-power portable superheterodyne receiver with loop antenna

TUBES
4

POWER SUPPLY
110-120 volts AC/DC or 9 volts A & 90 volts B supply

TUNING RANGE
540-1720KC

MFR/SUPPLIER
Spiegel Inc.

PHOTOFACT SET
24-2

PUBLISHED
1947

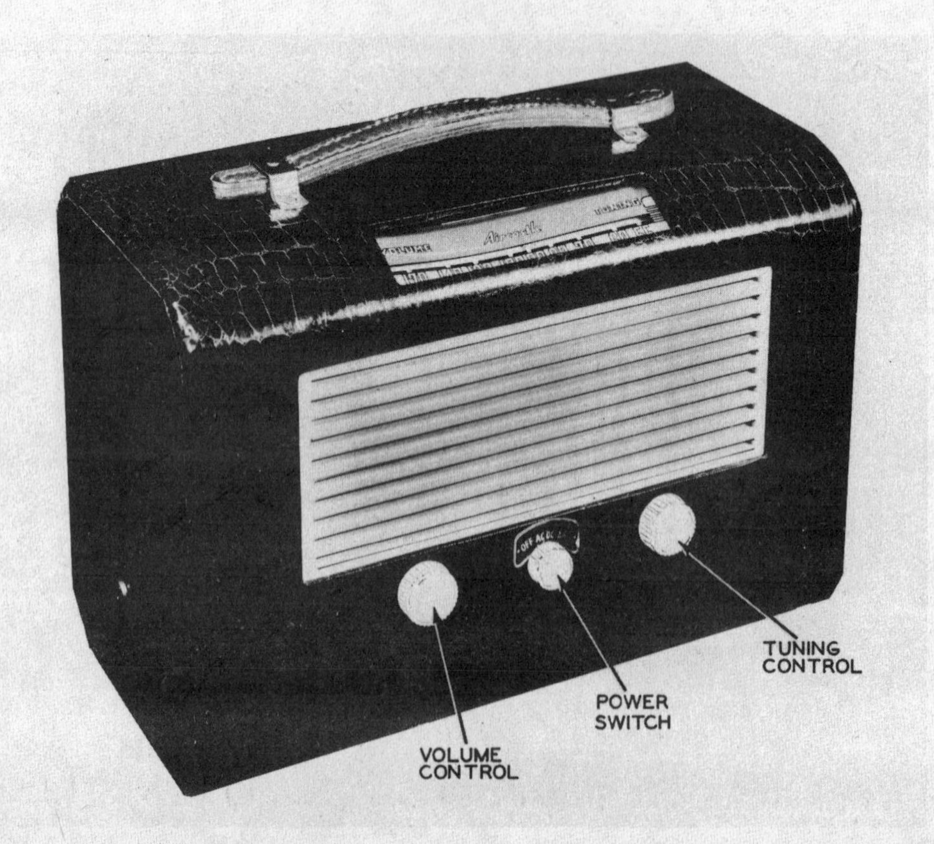

AIRCASTLE

MODEL PICTURED
5024

Three-power operated portable superheterodyne receiver with loop antenna

TUBES
4

POWER SUPPLY
110-125 volts AC/DC or 9 volts A & 90 volts B supply in pack form

TUNING RANGE
540-1650KC

MFR/SUPPLIER
Spiegel Inc.

PHOTOFACT SET
45-1

PUBLISHED
1948

AIRCASTLE

MODEL PICTURED
5052

AC/DC operated superheterodyne receiver with loop antenna

TUBES
5

POWER SUPPLY
110-125 volts AC/DC

TUNING RANGE
540-1650KC

MFR/SUPPLIER
Spiegel Inc.

PHOTOFACT SET
45-2

PUBLISHED
1948

AIRCASTLE

MODEL PICTURED
7553

AC/DC operated superheterodyne receiver with loop antenna

TUBES
5

POWER SUPPLY
110-120 volts AC/DC

TUNING RANGE
540-1700KC

MFR/SUPPLIER
Spiegel Inc.

PHOTOFACT SET
45-3

PUBLISHED
1948

AIRCASTLE

MODEL PICTURED
5008

AC/DC operated superheterodyne receiver with loop antenna

TUBES
6

POWER SUPPLY
110-125 volts AC/DC, .20 amp @ 117 volts AC

TUNING RANGE
540-1730KC

MFR/SUPPLIER
Spiegel Inc.

PHOTOFACT SET
46-1

PUBLISHED
1948

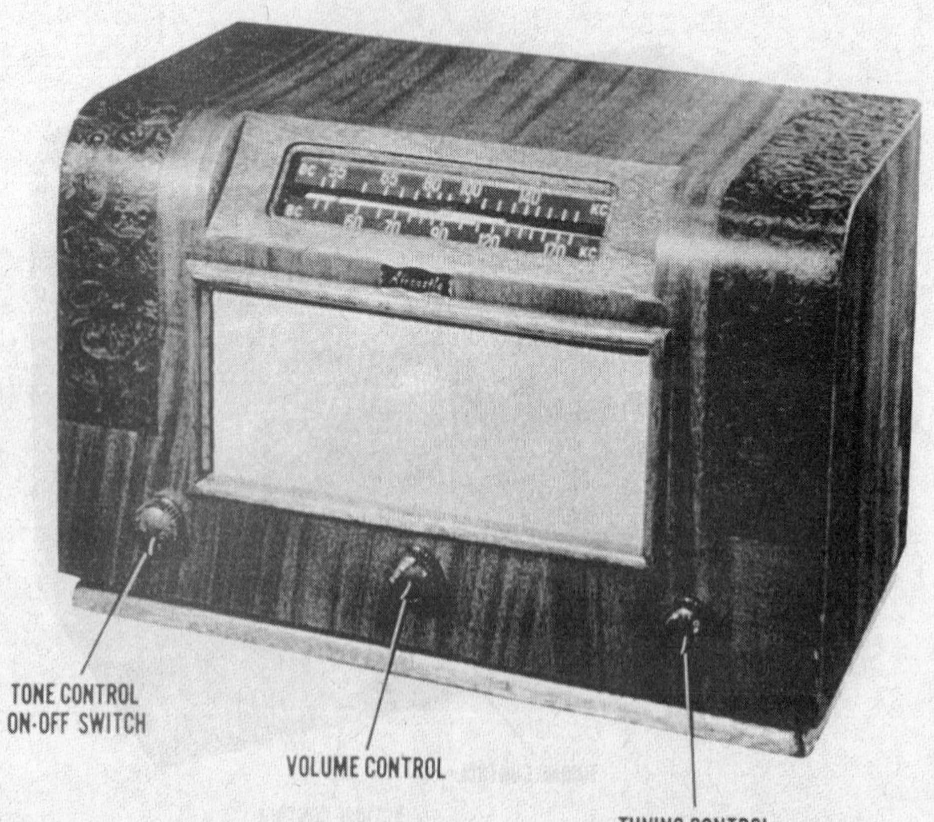

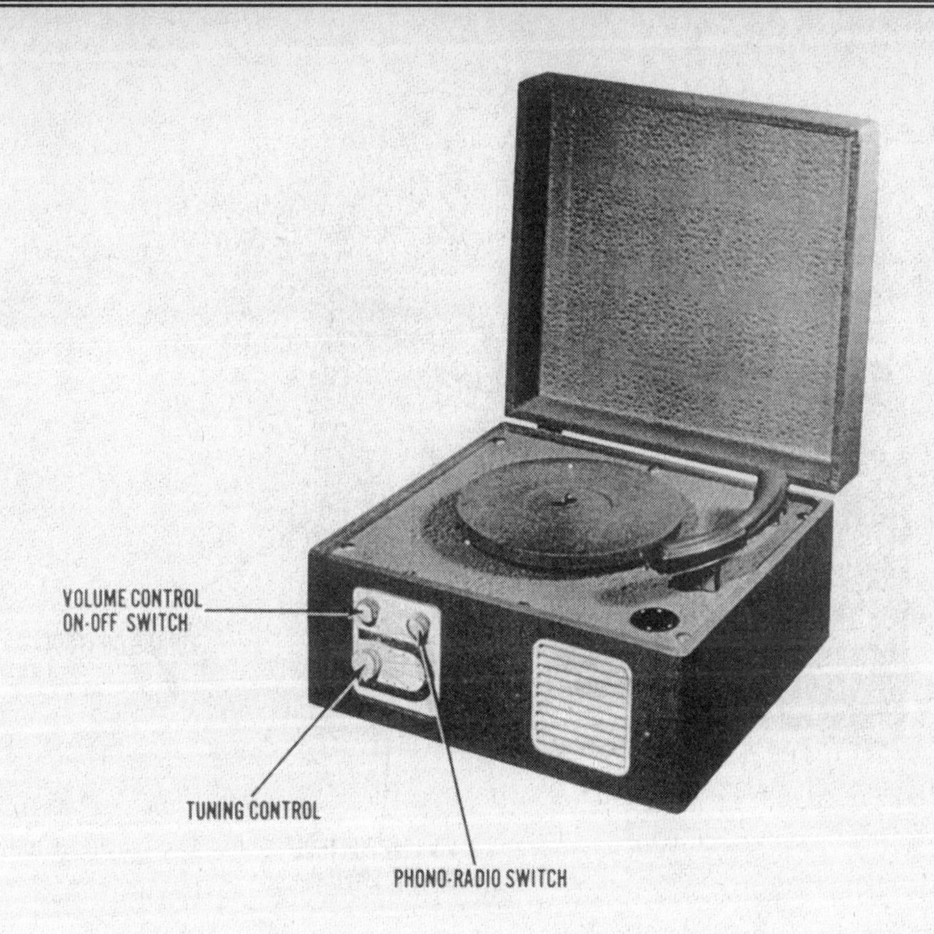

AIRCASTLE

MODEL PICTURED
5035

AC combination phono-radio superheterodyne receiver with loop antenna

TUBES
5

POWER SUPPLY
110-125 volts AC

TUNING RANGE
540-1650KC

MFR/SUPPLIER
Spiegel Inc.

PHOTOFACT SET
46-2

PUBLISHED
1948

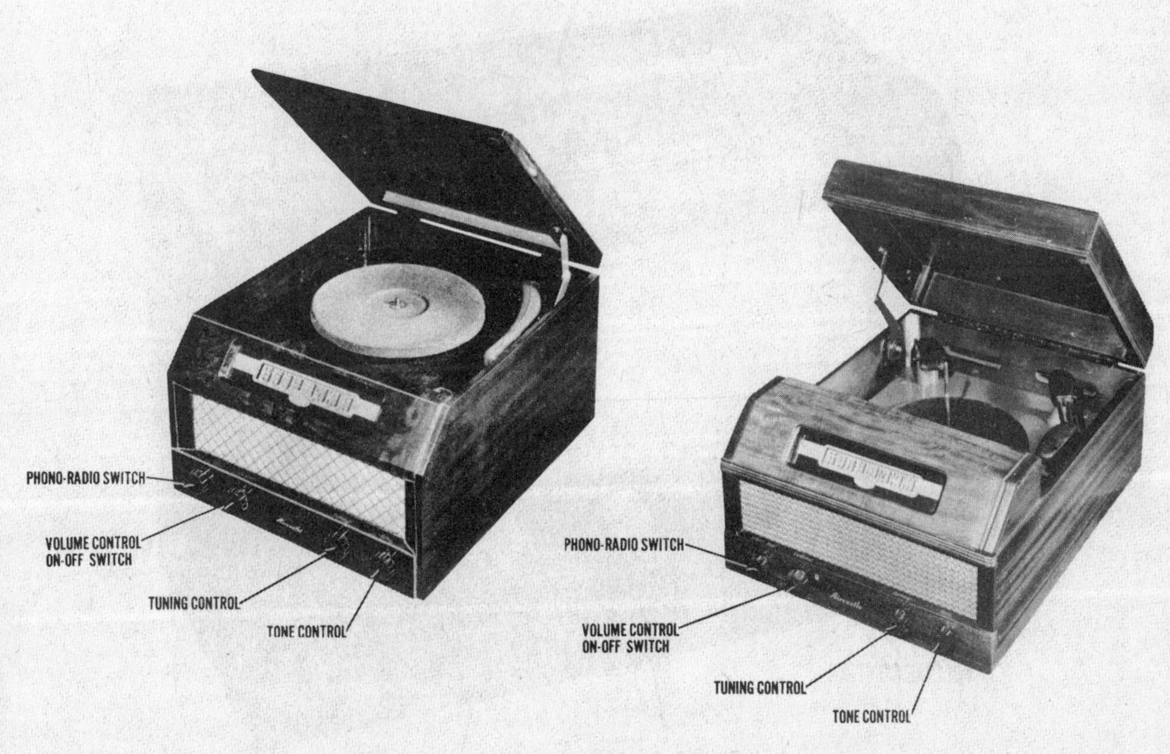

AIRCASTLE

MODEL PICTURED
G-516, G-518

AC operated phono-radio superheterodyne receiver, self-contained loop antenna

TUBES
5

POWER SUPPLY
110-120 volts AC

TUNING RANGE
532-1700KC

MFR/SUPPLIER
Spiegel Inc.

PHOTOFACT SET
48-3

PUBLISHED
1948

AIRCASTLE

MODEL PICTURED
5050

AC/DC operated superheterodyne receiver with loop antenna

TUBES
5

POWER SUPPLY
110-125 volts AC/DC

TUNING RANGE
540-1650KC

MFR/SUPPLIER
Spiegel Inc.

PHOTOFACT SET
48-4

PUBLISHED
1948

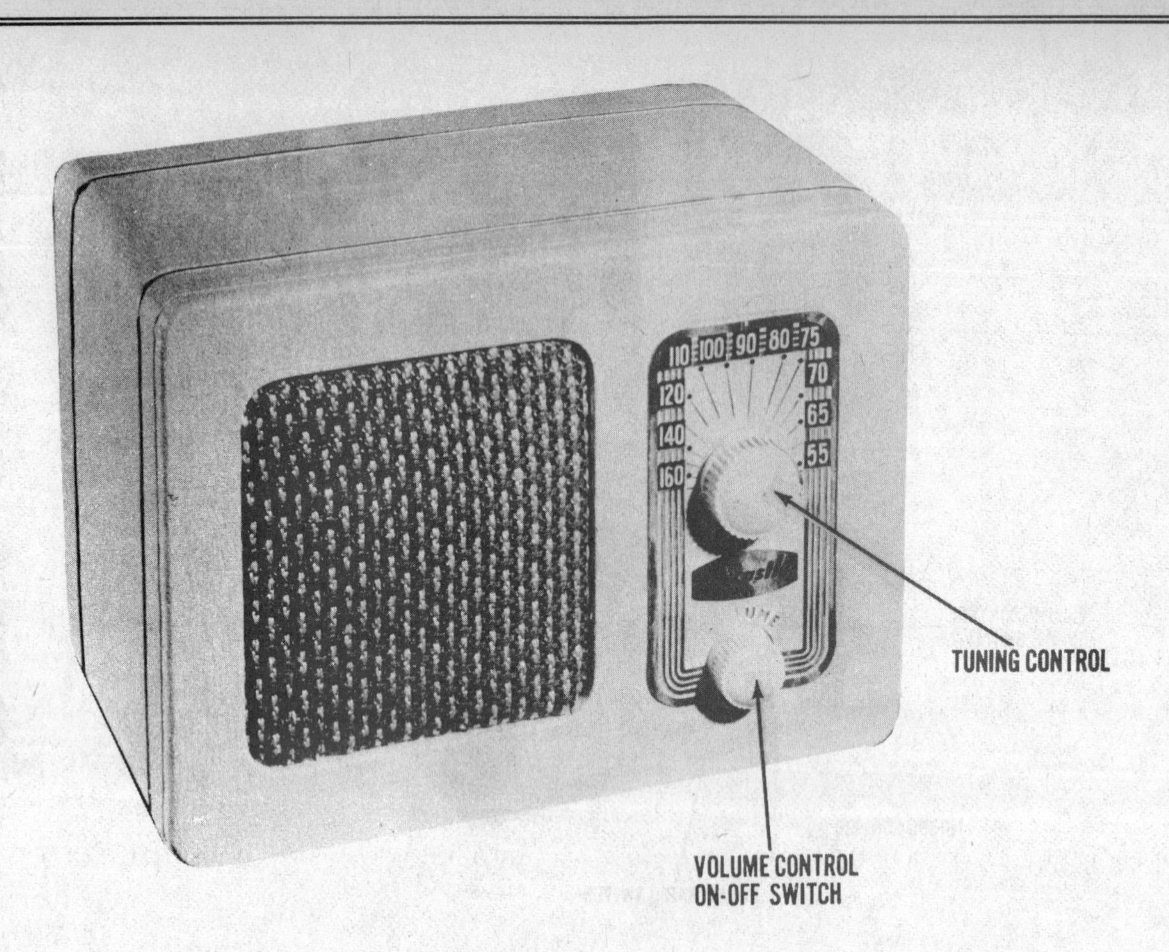

AIRCASTLE

MODEL PICTURED
5027

Three-power portable superheterodyne with loop antenna

TUBES
4

POWER SUPPLY
110-125 volts AC/DC or 9 volts A & 90 volts B supply

TUNING RANGE
540-1650KC

MFR/SUPPLIER
Spiegel Inc.

PHOTOFACT SET
49-3

PUBLISHED
1948

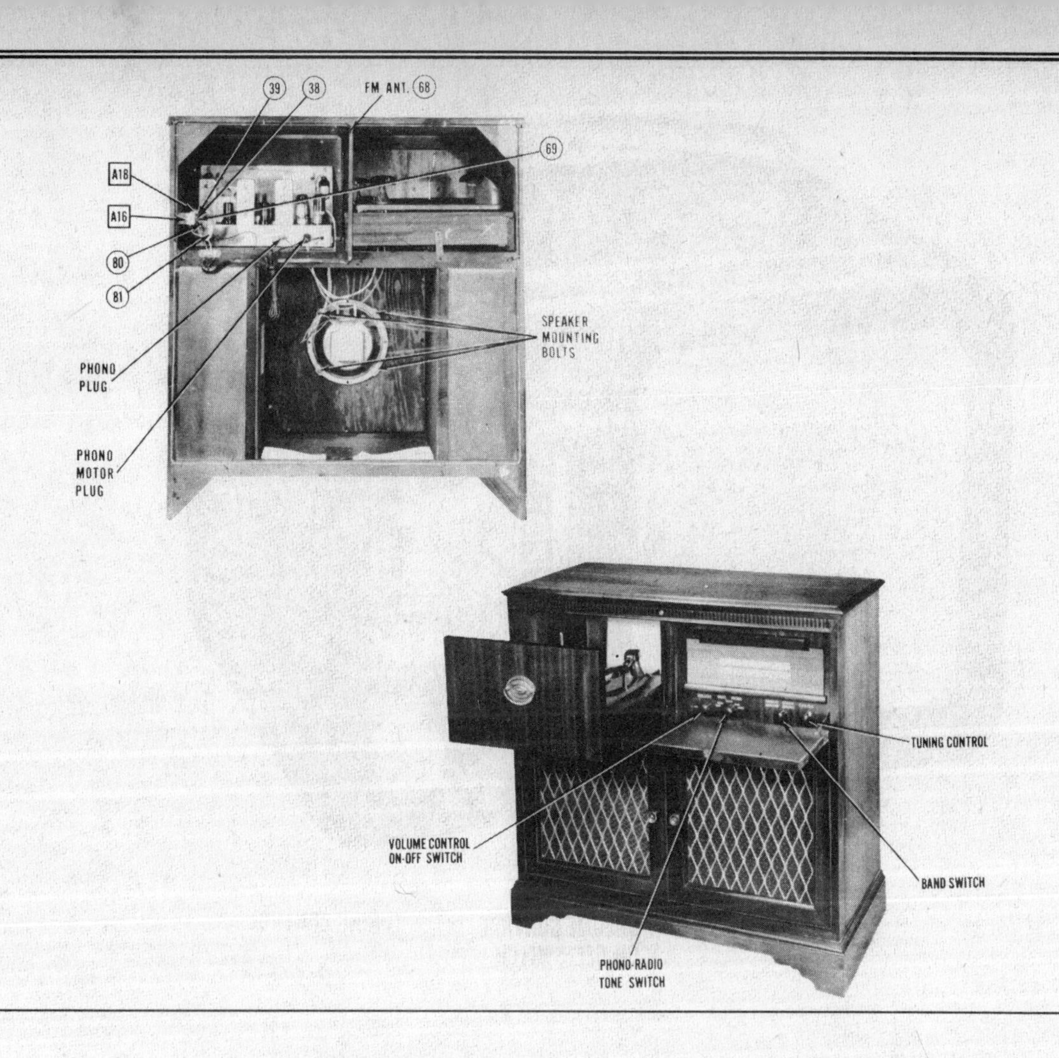

AIRCASTLE

MODEL PICTURED
G-725

AC/DC operated AM-FM superheterodyne receiver with loop antenna

TUBES
7

POWER SUPPLY
105-125 volts AC or 120 volts DC

TUNING RANGE
535-1620KC, 87.6-108.4MC

MFR/SUPPLIER
Spiegel Inc.

PHOTOFACT SET
50-1

PUBLISHED
1948

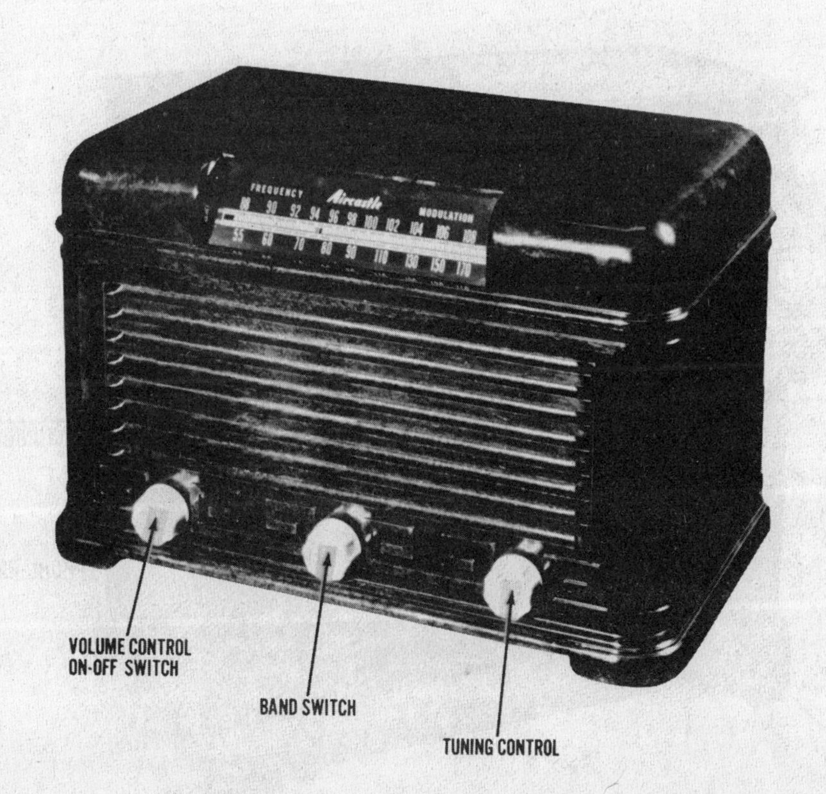

AIRCASTLE

MODEL PICTURED
9

AC/DC operated AM-FM superheterodyne receiver with loop antenna

TUBES
6

POWER SUPPLY
105-125 volts AC

TUNING RANGE
535-1700KC, 88-108MC

MFR/SUPPLIER
Spiegel Inc.

PHOTOFACT SET
50-2

PUBLISHED
1948

AIRCASTLE

MODEL PICTURED
5029

Battery operated superheterodyne receiver with loop antenna

TUBES
4

POWER SUPPLY
1.5 volts A supply & 67.5 volts B supply

TUNING RANGE
540-1620KC

MFR/SUPPLIER
Spiegel Inc.

PHOTOFACT SET
51-1

PUBLISHED
1948

AIRCASTLE

MODEL PICTURED
7B

AC operated phono-radio AM-FM superheterodyne receiver with loop antenna

TUBES
11

POWER SUPPLY
105-125 volts AC

TUNING RANGE
540-1700KC, 88-108MC

MFR/SUPPLIER
Spiegel Inc.

PHOTOFACT SET
52-1

PUBLISHED
1948

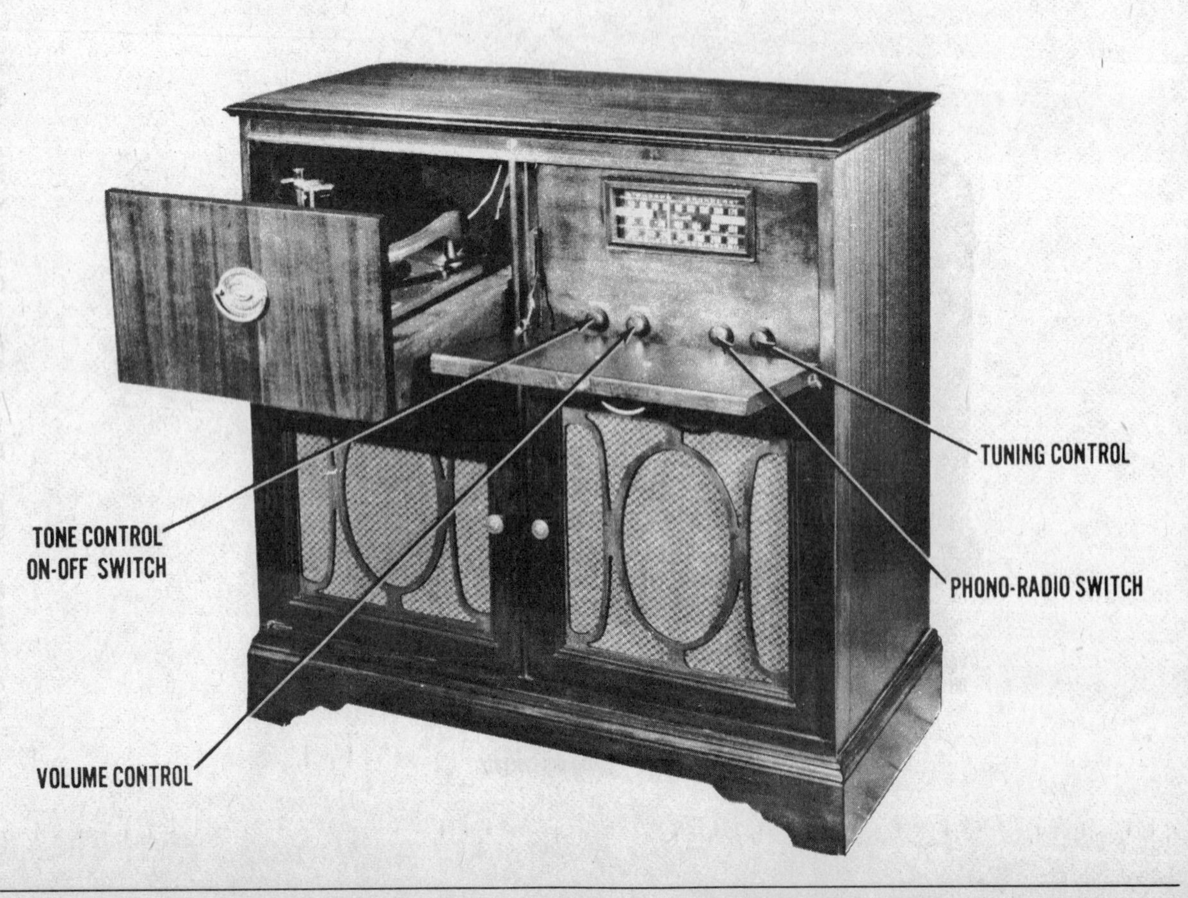

AIRCASTLE

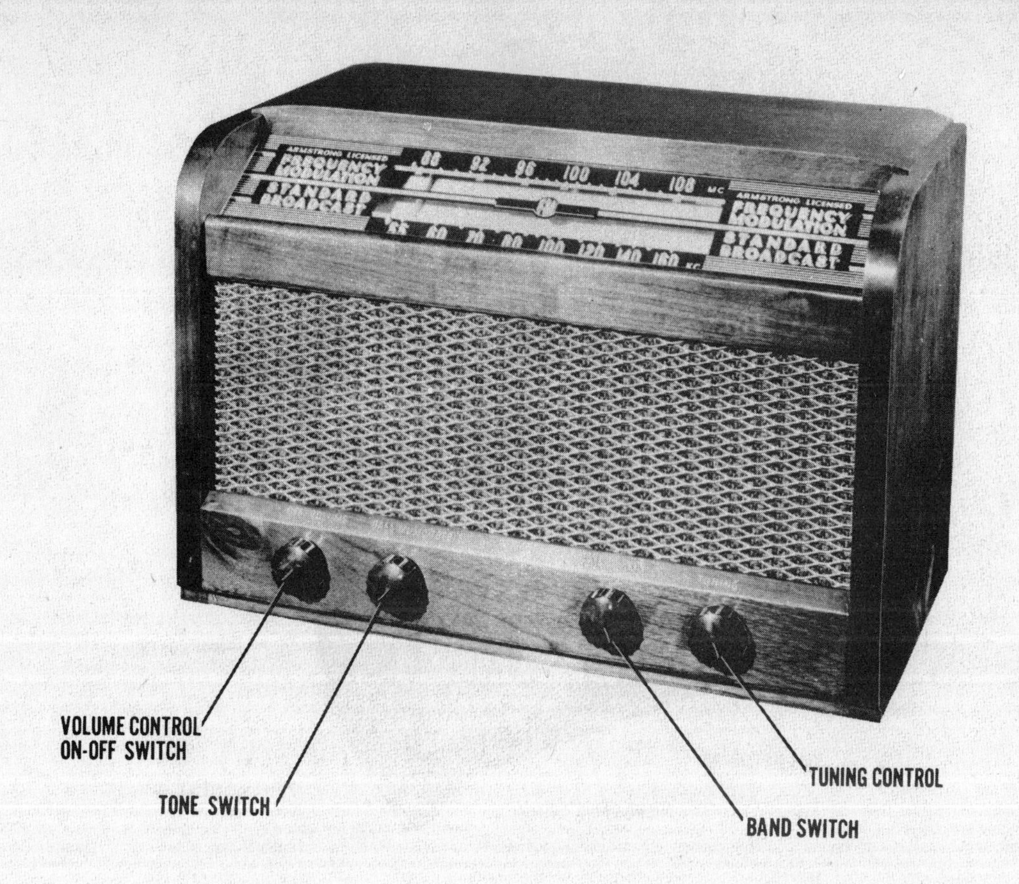

MODEL PICTURED
G-724

AC/DC operated AM-FM superheterodyne receiver with loop antenna

TUBES
7

POWER SUPPLY
105-125 volts AC or 120 volts DC

TUNING RANGE
535-1620KC,
87.6-108.4MC

MFR/SUPPLIER
Spiegel Inc.

PHOTOFACT SET
52-25

PUBLISHED
1948

AIRCASTLE

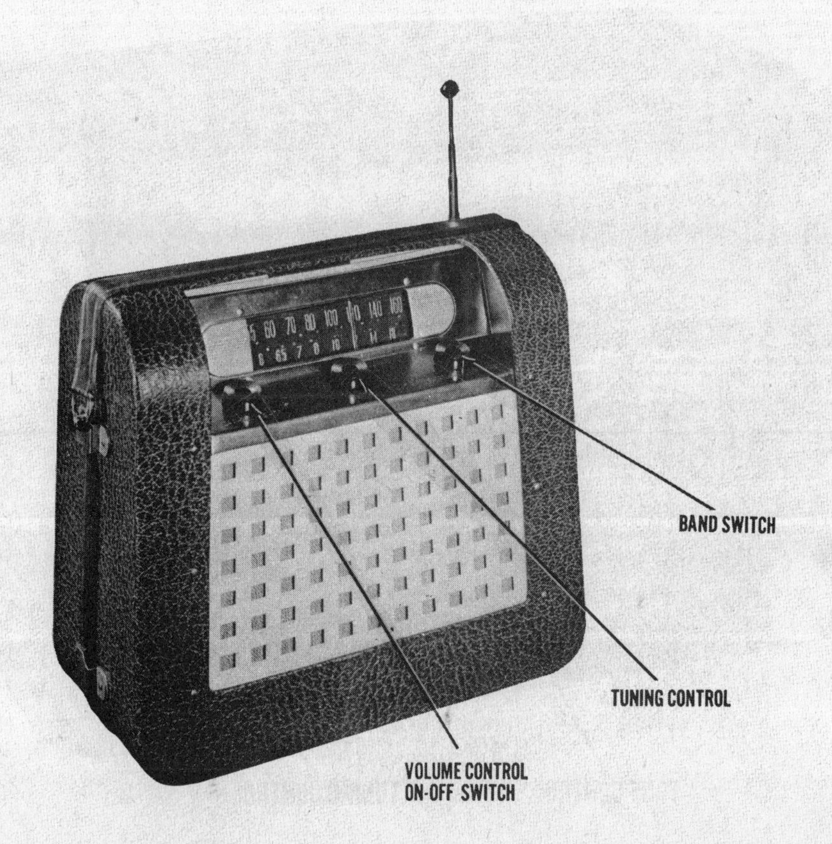

MODEL PICTURED
G521

Three-power portable superheterodyne receiver with loop antenna

TUBES
5

POWER SUPPLY
110-120 volts AC/DC or 9 volts A supply & 90 volts B supply

TUNING RANGE
535-1620KC, 5.6-18.5MC

MFR/SUPPLIER
Speigel, Inc.

PHOTOFACT SET
54-3

PUBLISHED
1949

AIRCASTLE

MODEL PICTURED
138104

AC operated combo phono-radio, superheterodyne receiver with loop antenna

TUBES
6

POWER SUPPLY
110-120 volts AC, .44 amp @ 117 volts AC

TUNING RANGE
535-1725KC

MFR/SUPPLIER
Spiegel Inc.

PHOTOFACT SET
54-23

PUBLISHED
1949

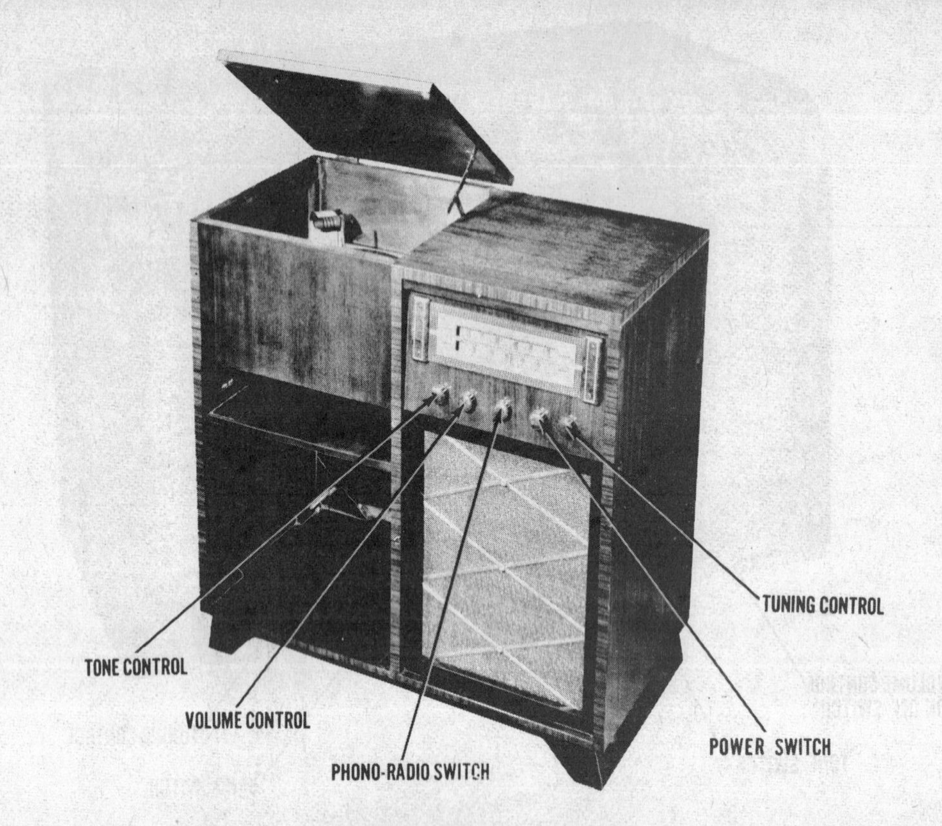

AIRCASTLE

MODEL PICTURED
10002

AC/DC operated superheterodyne receiver with loop antenna

TUBES
6

POWER SUPPLY
105-125 volts AC/DC

TUNING RANGE
550-1600KC

MFR/SUPPLIER
Spiegel Inc.

PHOTOFACT SET
56- 1

PUBLISHED
1949

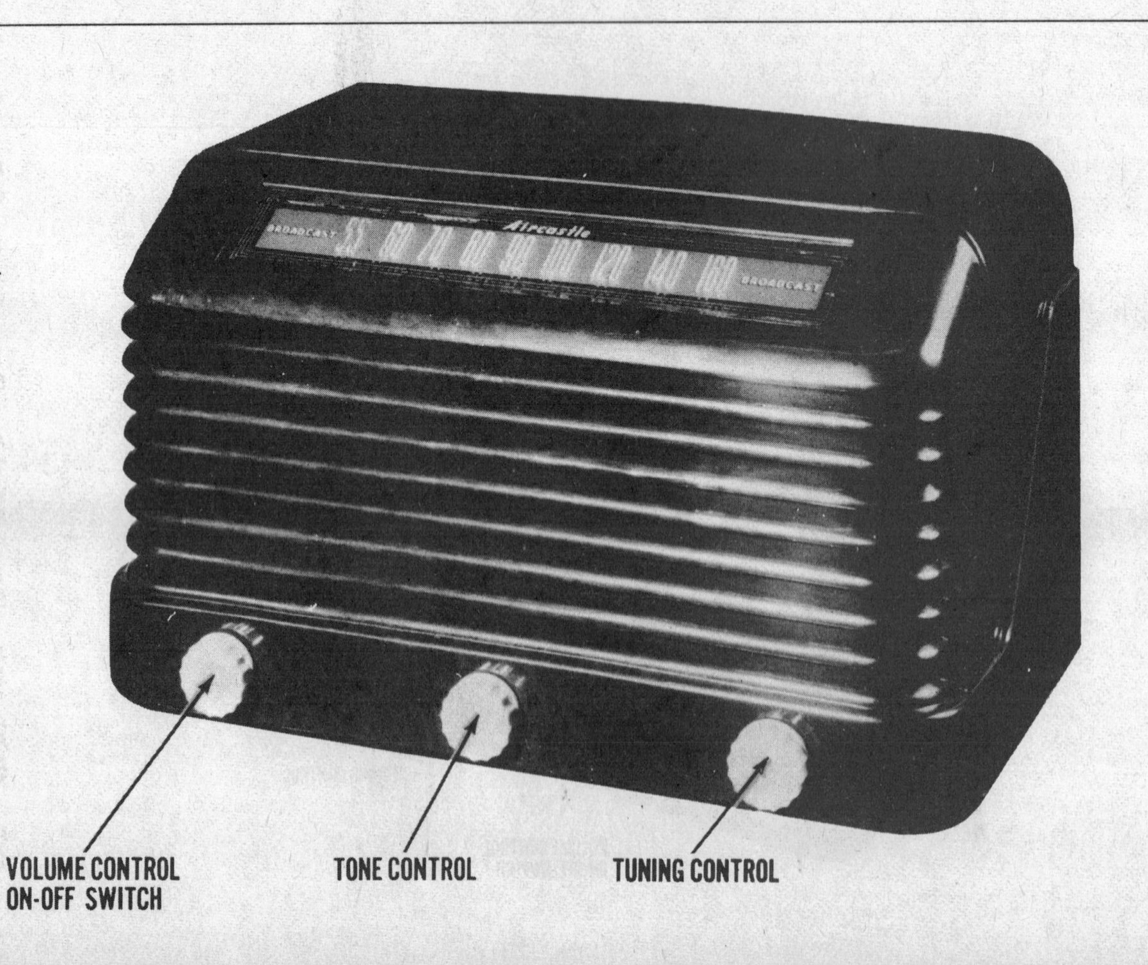

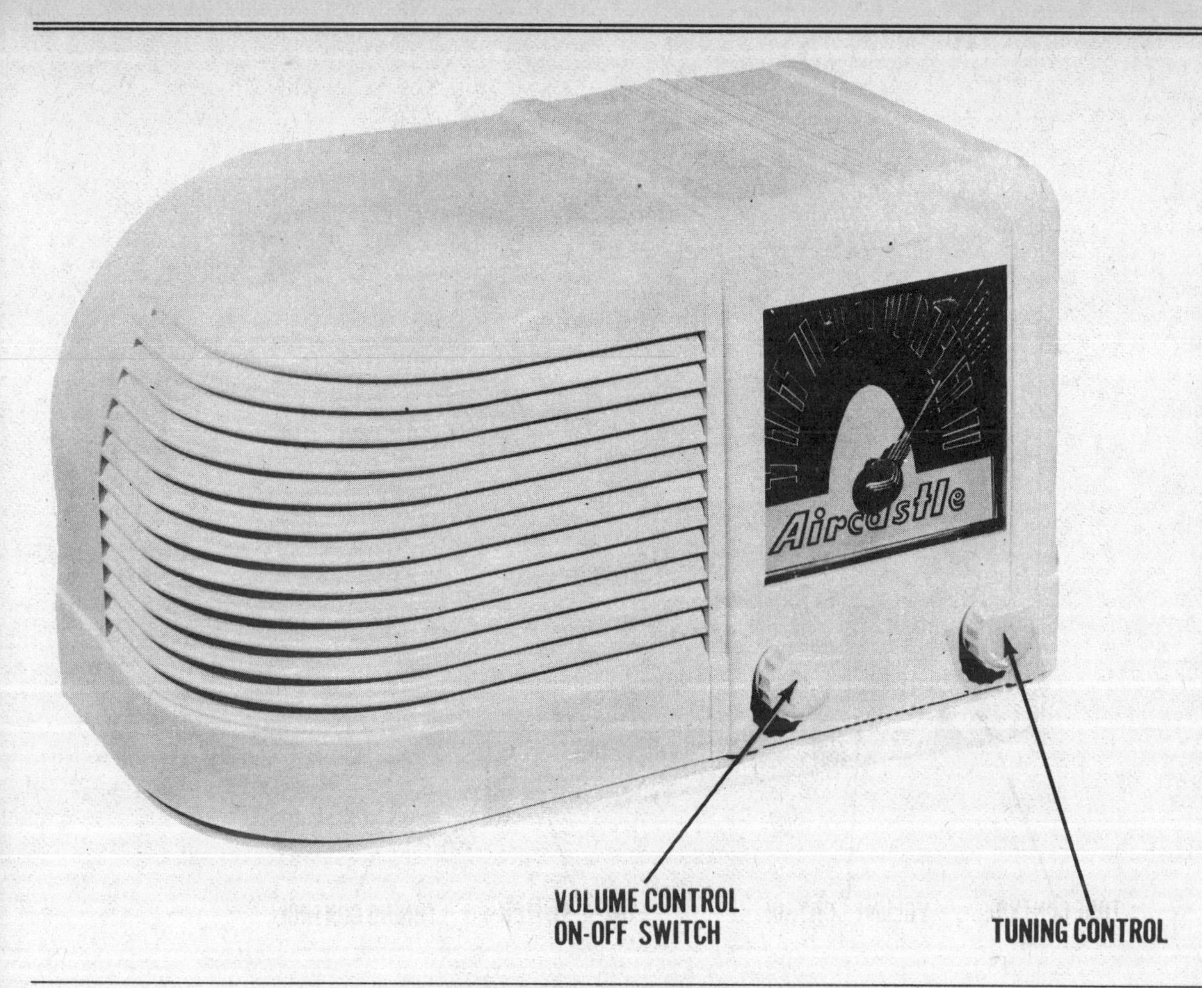

AIRCASTLE

MODEL PICTURED
10003-I

AC/DC operated superheterodyne receiver with loop antenna

TUBES
5

POWER SUPPLY
110-120 volts AC/DC

TUNING RANGE
540-1700KC

MFR/SUPPLIER
Spiegel Inc.

PHOTOFACT SET
56-2

PUBLISHED
1949

AIRCASTLE

MODEL PICTURED
147114

Three-power operated portable superheterodyne receiver with loop antenna

TUBES
4

POWER SUPPLY
105-125 volts AC/DC or 7.5 volts A & 67.5 volts B battery supply

TUNING RANGE
540-1600KC

MFR/SUPPLIER
Spiegel Inc.

PHOTOFACT SET
56-3

PUBLISHED
1949

AIRCASTLE

MODEL PICTURED
108014

AC operated superheterodyne receiver with loop antenna

TUBES
6

POWER SUPPLY
110-120 volts AC

TUNING RANGE
540-1725KC

MFR/SUPPLIER
Spiegel, INC.

PHOTOFACT SET
57-4

PUBLISHED
1949

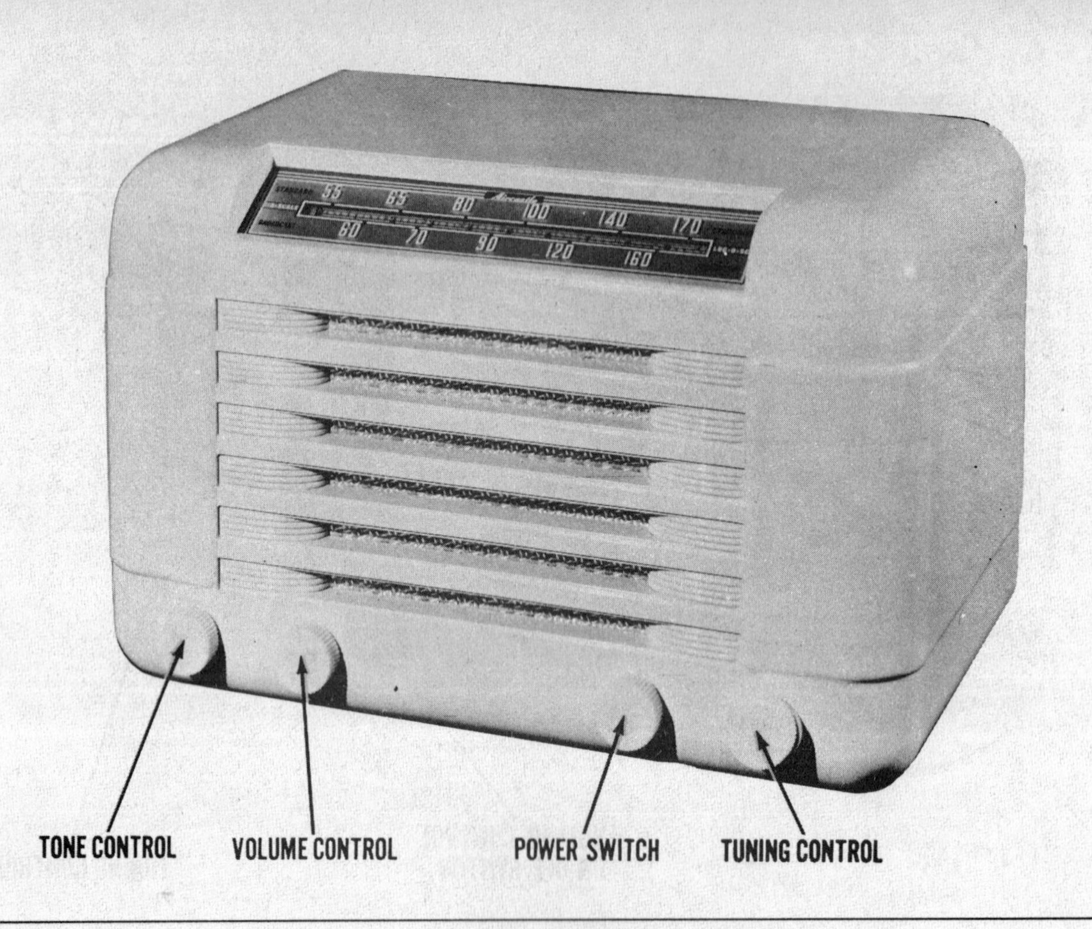

AIRCASTLE

MODEL PICTURED
10023

AC operated phono-radio, superheterodyne receiver with loop antenna

TUBES
5

POWER SUPPLY
105-125 volts AC

TUNING RANGE
540-1620KC

MFR/SUPPLIER
Spiegel Inc.

PHOTOFACT SET
58-1

PUBLISHED
1949

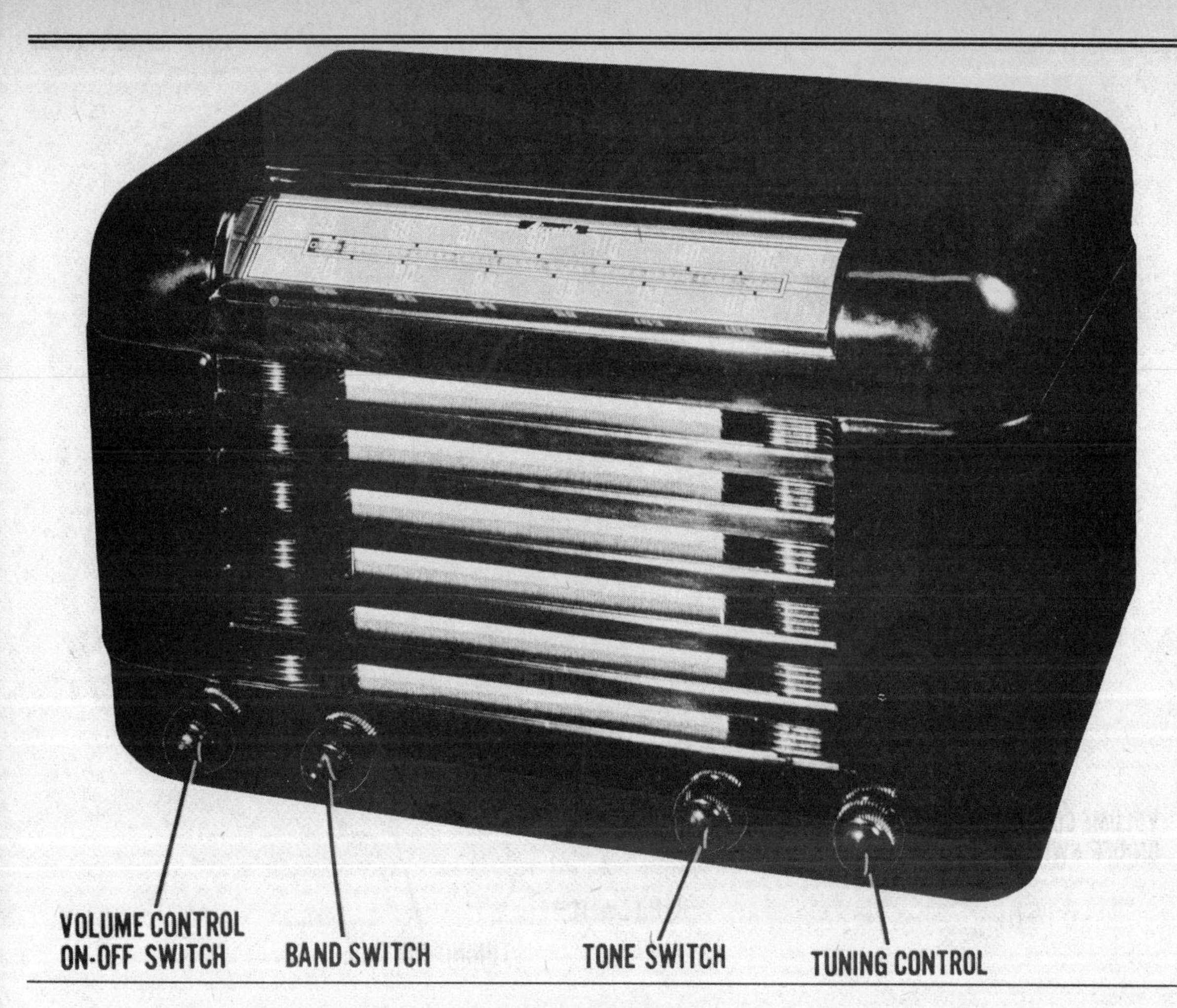

AIRCASTLE

MODEL PICTURED
131504

AC/DC operated AM-FM superheterodyne receiver

TUBES
10

POWER SUPPLY
110-120 volts AC/DC

TUNING RANGE
540-1600KC, 88-108MC

MFR/SUPPLIER
Spiegel Inc.

PHOTOFACT SET
60-2

PUBLISHED
1949

AIRCASTLE

MODEL PICTURED
SC-448

AC operated FM superheterodyne tuner

TUBES
4

POWER SUPPLY
105-125 volts AC

TUNING RANGE
87.5-108.5MC

MFR/SUPPLIER
Spiegel Inc.

PHOTOFACT SET
62-2

PUBLISHED
1949

AIRCASTLE

MODEL PICTURED
10005

AC/DC operated AM-FM superheterodyne receiver with loop antenna

TUBES
8

POWER SUPPLY
105-125 volts AC/DC

TUNING RANGE
540-1720KC, 88-108MC

MFR/SUPPLIER
Spiegel Inc.

PHOTOFACT SET
62-3

PUBLISHED
1949

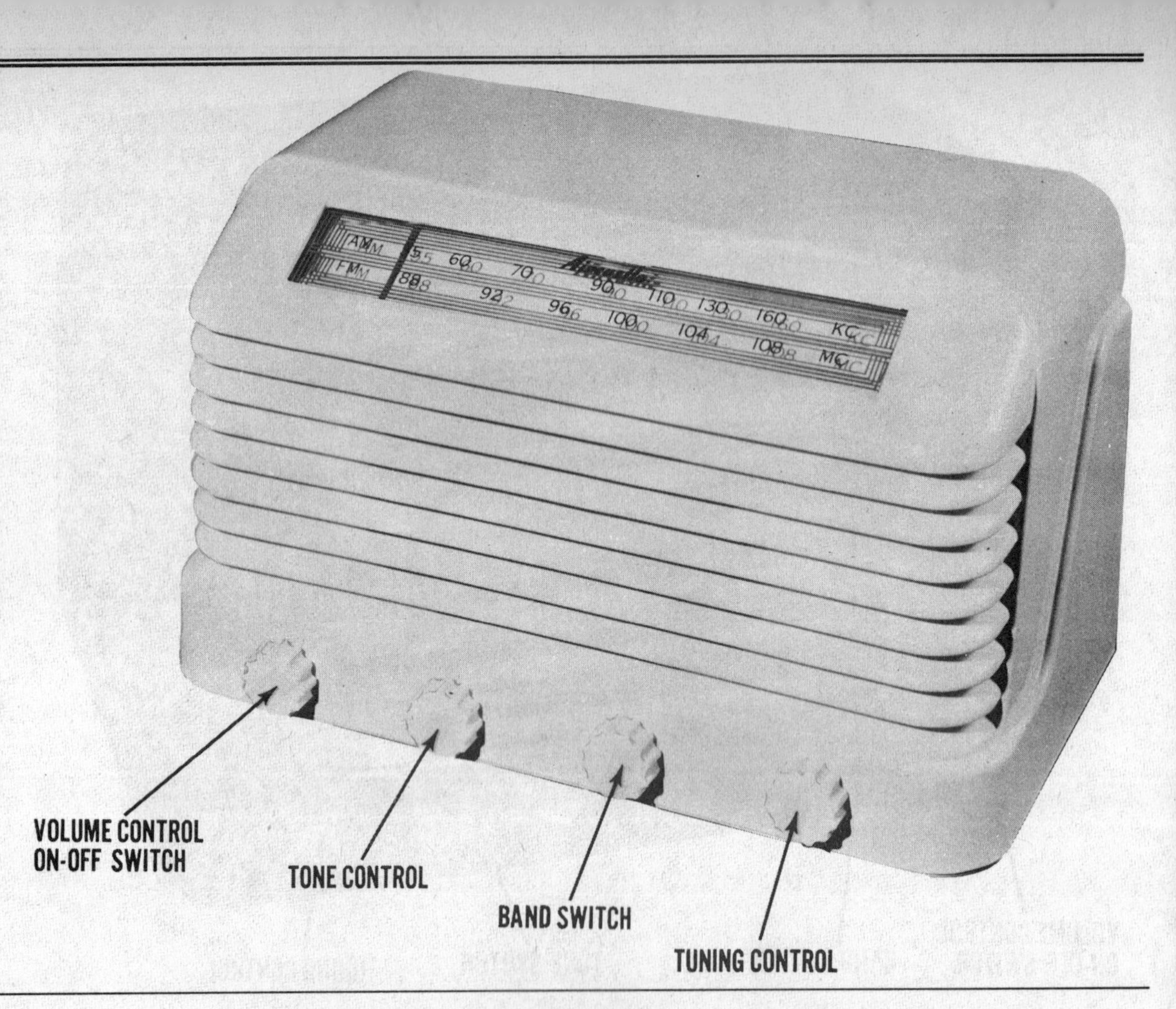

AIRCASTLE

MODEL PICTURED
213

Three-power operated portable superheterodyne receiver with loop antenna

TUBES
5

POWER SUPPLY
105-125 volts AC/DC or 4.5 volts A & 67.5 volts B supply

TUNING RANGE
535-1620KC

MFR/SUPPLIER
Spiegel Inc.

PHOTOFACT SET
63-1

PUBLISHED
1949

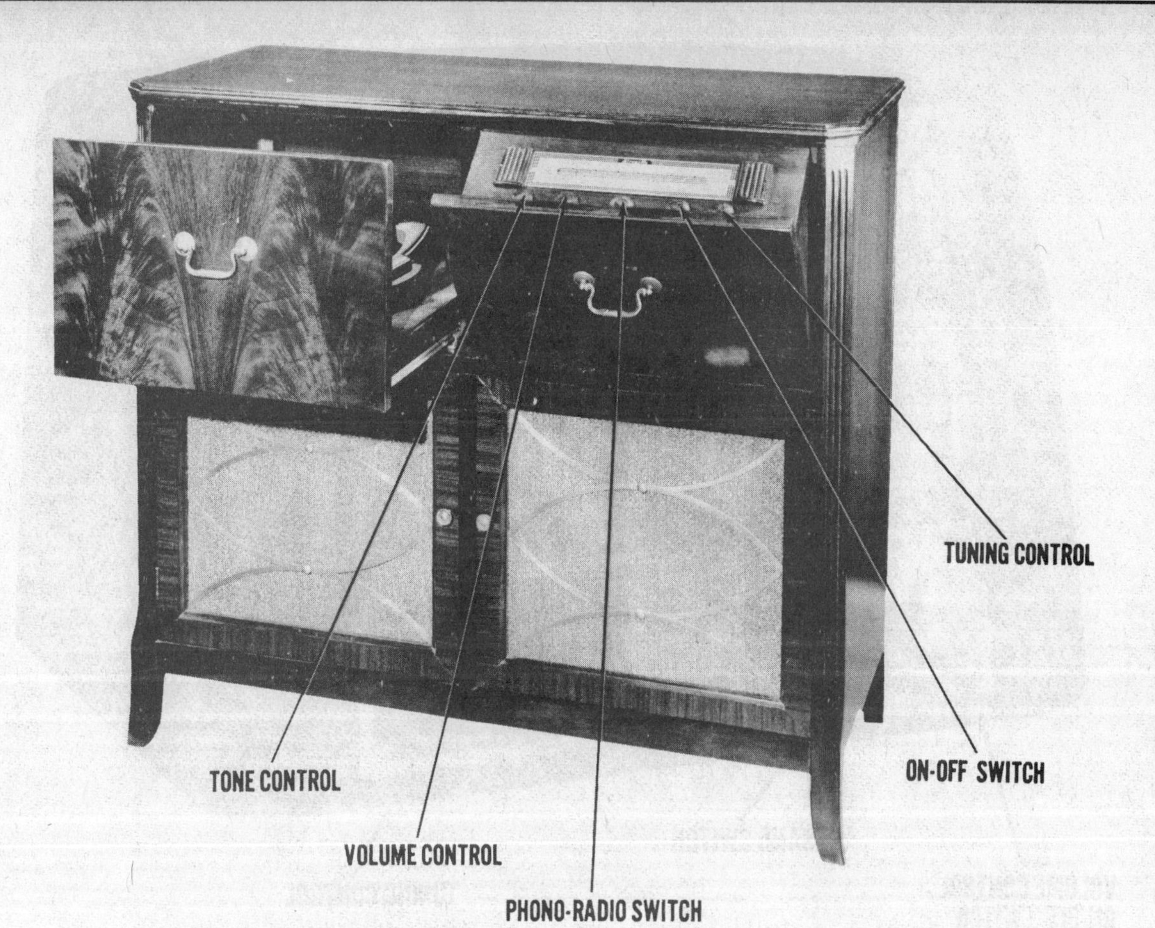

AIRCASTLE

MODEL PICTURED
138124

AC operated combo phono-radio, superheterodyne receiver with loop antenna

TUBES
6

POWER SUPPLY
110-120 volts AC, .49 amp @ 117 volts AC

TUNING RANGE
530-1725KC

MFR/SUPPLIER
Spiegel Inc.

PHOTOFACT SET
64-1

PUBLISHED
1949

AIRCASTLE

MODEL PICTURED
211

AC/DC operated superheterodyne receiver

TUBES
4

POWER SUPPLY
105-125 volts AC/DC, .26 amp @ 117 volts AC

TUNING RANGE
530-1620KC

MFR/SUPPLIER
Spiegel Inc.

PHOTOFACT SET
65-1

PUBLISHED
1949

AIRCASTLE

MODEL PICTURED
212

AC operated AM-FM superheterodyne receiver with loop antenna

TUBES
8

POWER SUPPLY
105-125 volts AC, .38 amp @ 117 volts AC

TUNING RANGE
535-1600KC, 88-108MC

MFR/SUPPLIER
Spiegel Inc.

PHOTOFACT SET
68-3

PUBLISHED
1949

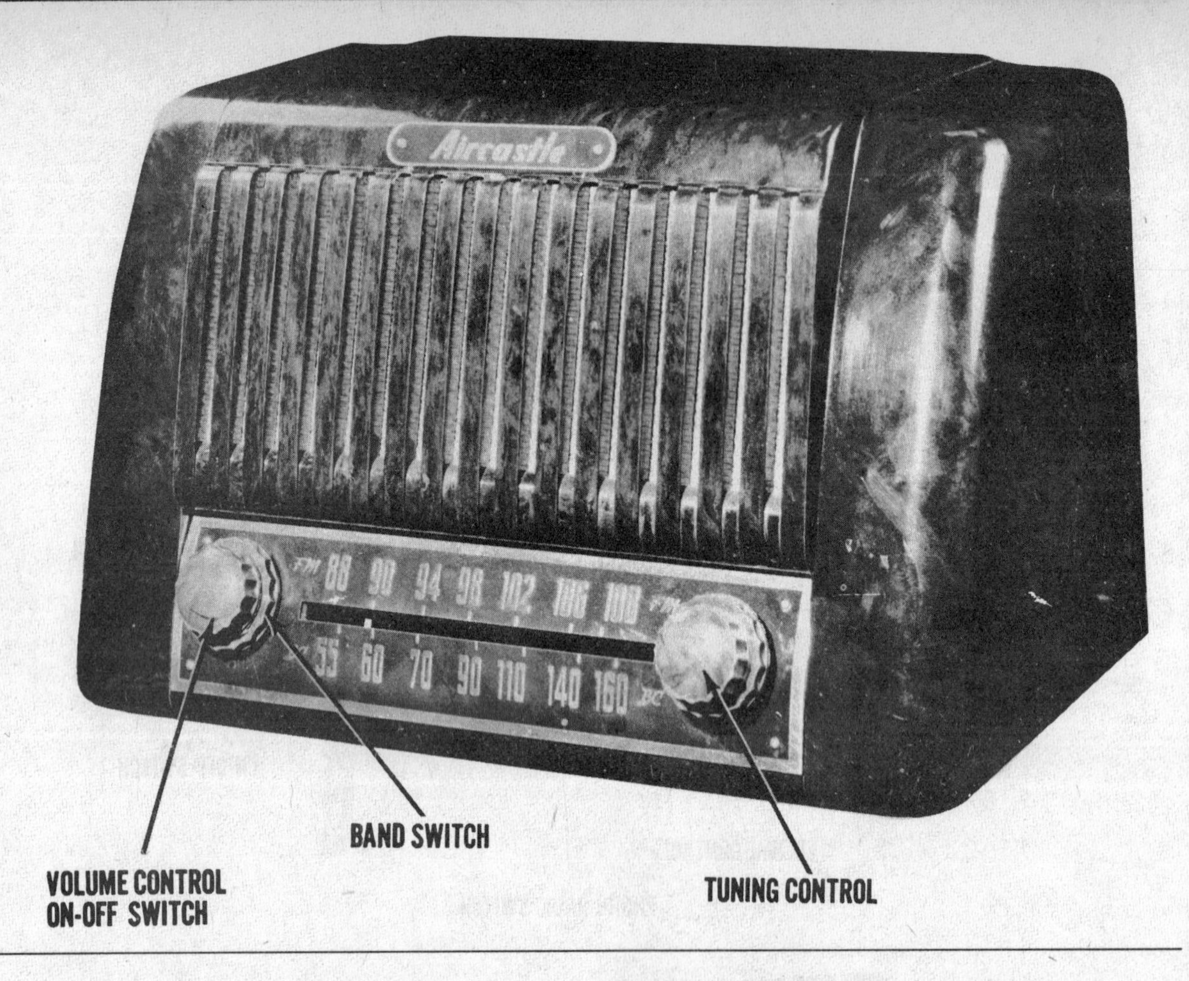

AIRCASTLE

MODEL PICTURED
132564

Battery operated superheterodyne receiver

TUBES
4

POWER SUPPLY
1.5 volts A supply and 90 volts B supply in pack form

TUNING RANGE
540-1725KC

MFR/SUPPLIER
Spiegel Inc.

PHOTOFACT SET
69-1

PUBLISHED
1949

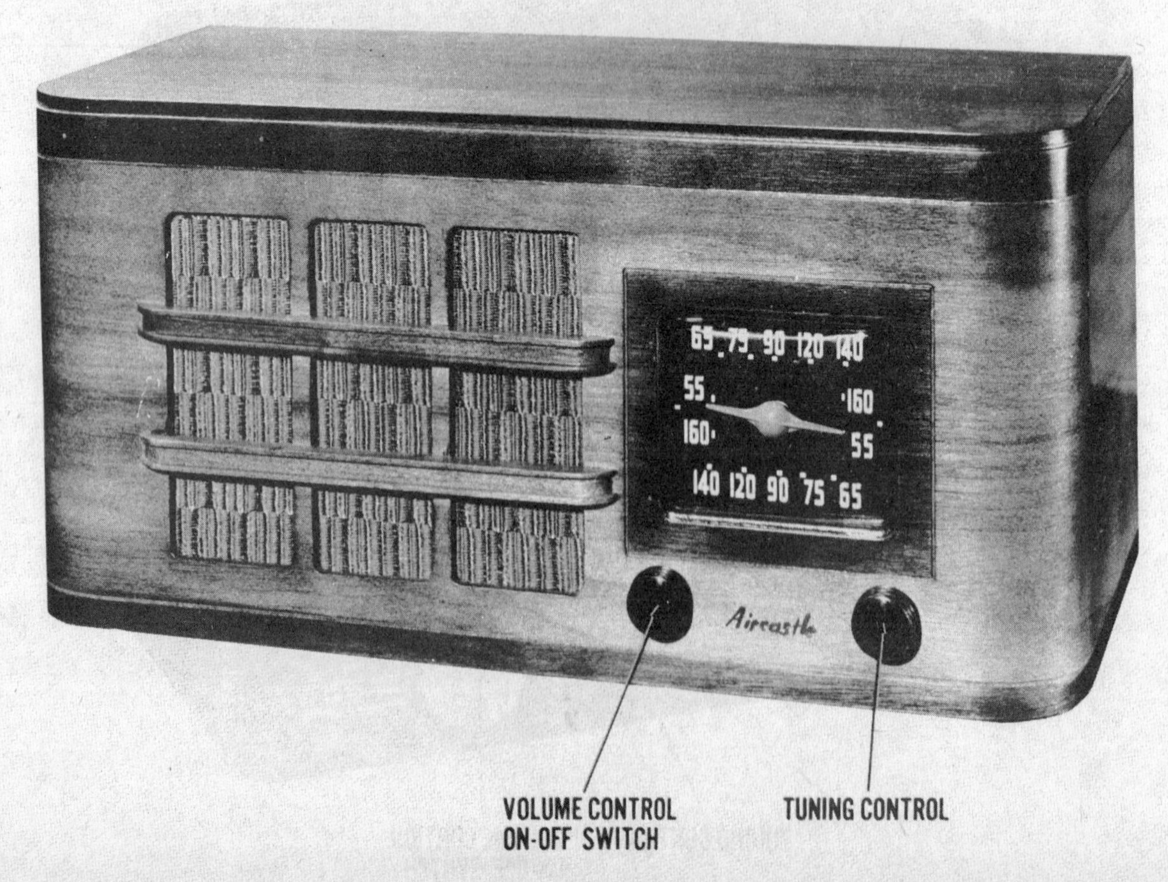

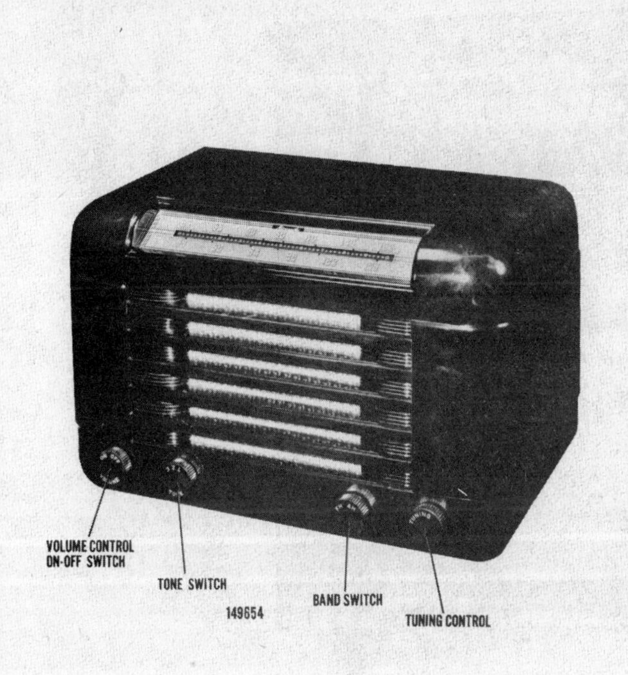

AIRCASTLE

MODEL PICTURED
150084

AC operated combo phono-radio, AM/FM superheterodyne receiver with loop antenna

TUBES
8

POWER SUPPLY
105-125 volts AC

TUNING RANGE
540-1600KC, 88-108MC

MFR/SUPPLIER
Spiegel Inc.

PHOTOFACT SET
71-4

PUBLISHED
1949

AIRCASTLE

MODEL PICTURED
5036

AC operated combo phono-radio, superheterodyne receiver

TUBES
4

POWER SUPPLY
110-125 volts AC

TUNING RANGE
540-1620KC

MFR/SUPPLIER
Spiegel Inc.

PHOTOFACT SET
72-2

PUBLISHED
1949

AIRCASTLE

MODEL PICTURED
121104

AC operated combo phono-radio AM-FM superheterodyne receiver with loop antenna

TUBES
10

POWER SUPPLY
110-120 volts AC

TUNING RANGE
540-1600KC, 88-108MC

MFR/SUPPLIER
Spiegel Inc.

PHOTOFACT SET
73-1

PUBLISHED
1949

AIRCASTLE

MODEL PICTURED
6050

AC operated combo phono-radio, superheterodyne receiver with loop antenna

TUBES
5

POWER SUPPLY
110-125 volts AC

TUNING RANGE
540-1720KC

MFR/SUPPLIER
Spiegel Inc.

PHOTOFACT SET
74-1

PUBLISHED
1949

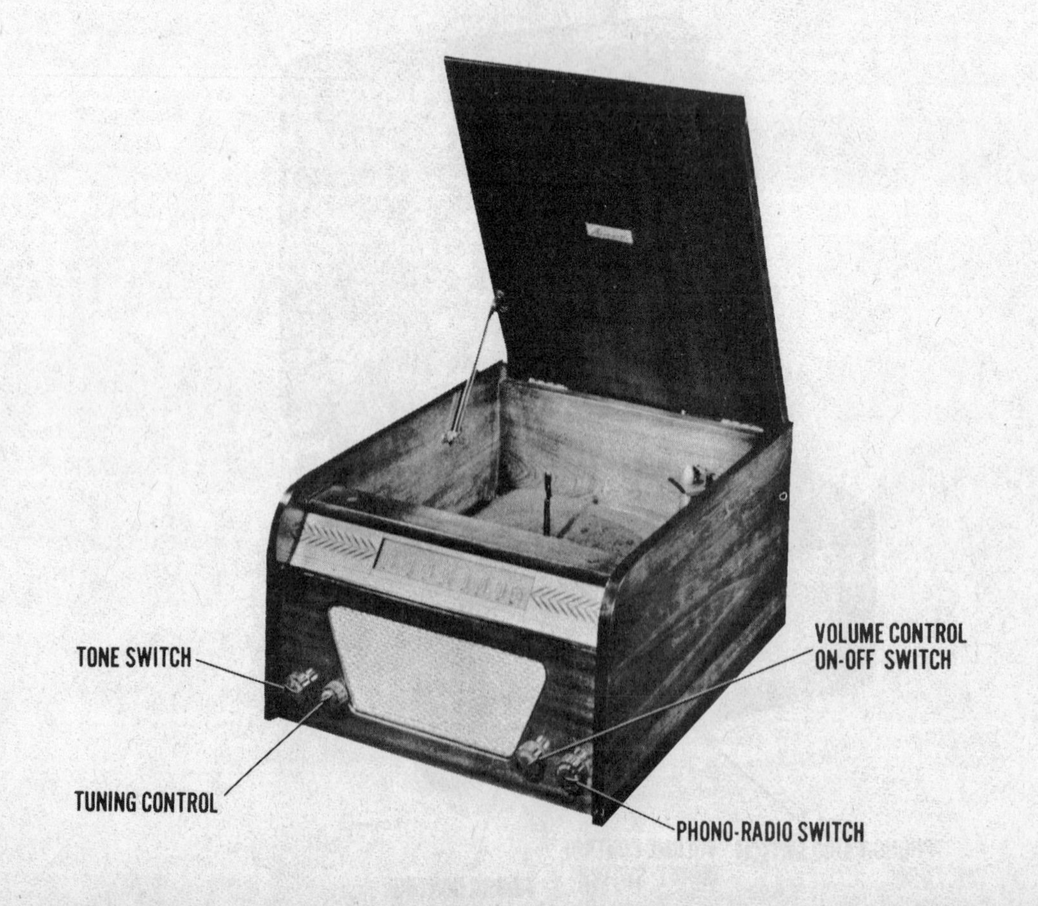

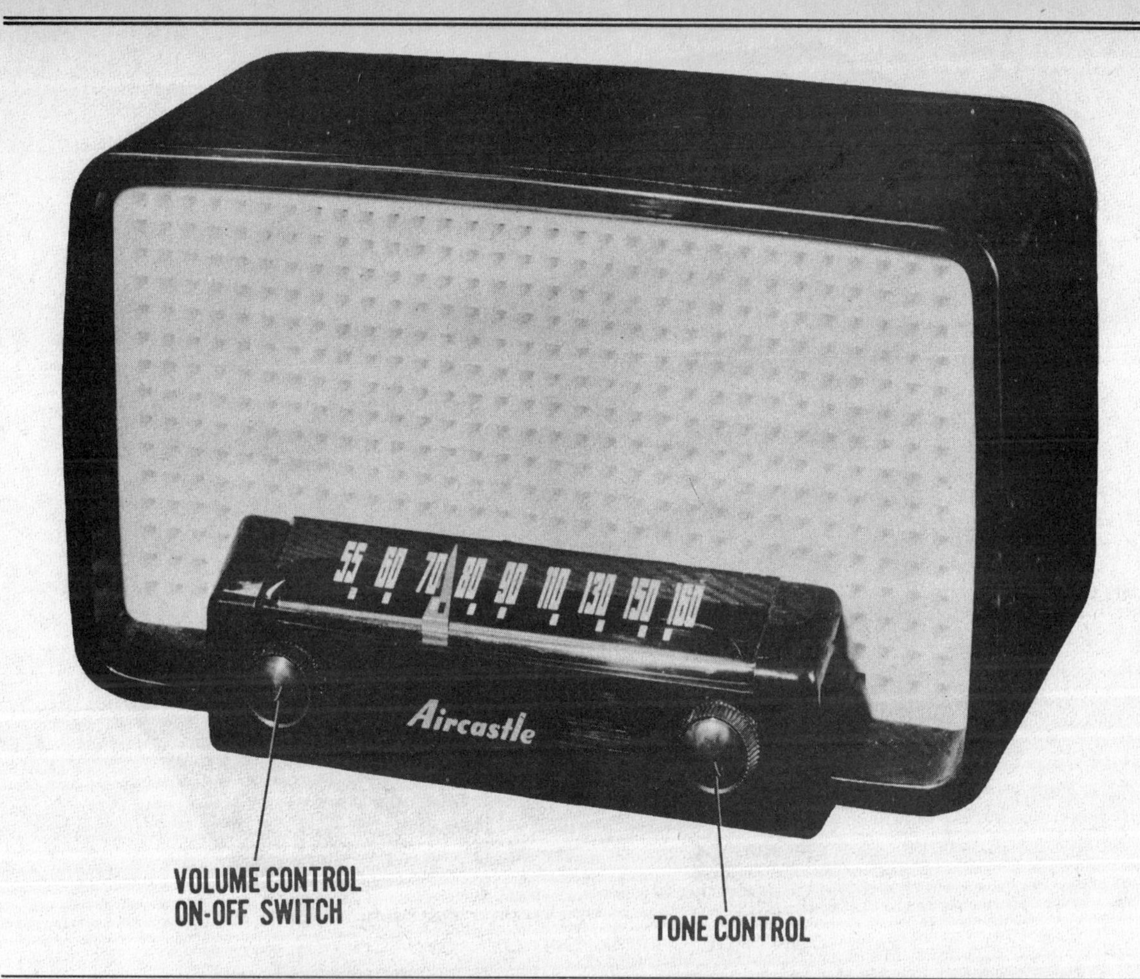

AIRCASTLE

MODEL PICTURED
201

AC/DC operated superheterodyne receiver with loop antenna

TUBES
6

POWER SUPPLY
105-125 volts AC/DC, .24 amp @ 117 volts AC

TUNING RANGE
550-1600KC

MFR/SUPPLIER
Spiegel Inc.

PHOTOFACT SET
81-1

PUBLISHED
1950

AIRCASTLE

MODEL PICTURED
198

AC operated AM-FM superheterodyne receiver with loop antenna

TUBES
8

POWER SUPPLY
105-125 volts AC .39 amp @ 117 volts AC

TUNING RANGE
535-1620KC

MFR/SUPPLIER
Spiegel Inc.

PHOTOFACT SET
83-1

PUBLISHED
1950

AIRCASTLE

MODEL PICTURED
227I

AC/DC operated superheterodyne receiver

TUBES
4

POWER SUPPLY
105-125 volts AC/DC

TUNING RANGE
540-1700KC

MFR/SUPPLIER
Spiegel Inc.

PHOTOFACT SET
84-1

PUBLISHED
1950

AIRCASTLE

MODEL PICTURED
DM700

Three-power operated portable superheterodyne receiver

TUBES
4

POWER SUPPLY
110-120 volts AC/DC or 6 volts A & 67.5 volts B supply

TUNING RANGE
540-1720KC

MFR/SUPPLIER
Spiegel Inc.

PHOTOFACT SET
85-1

PUBLISHED
1950

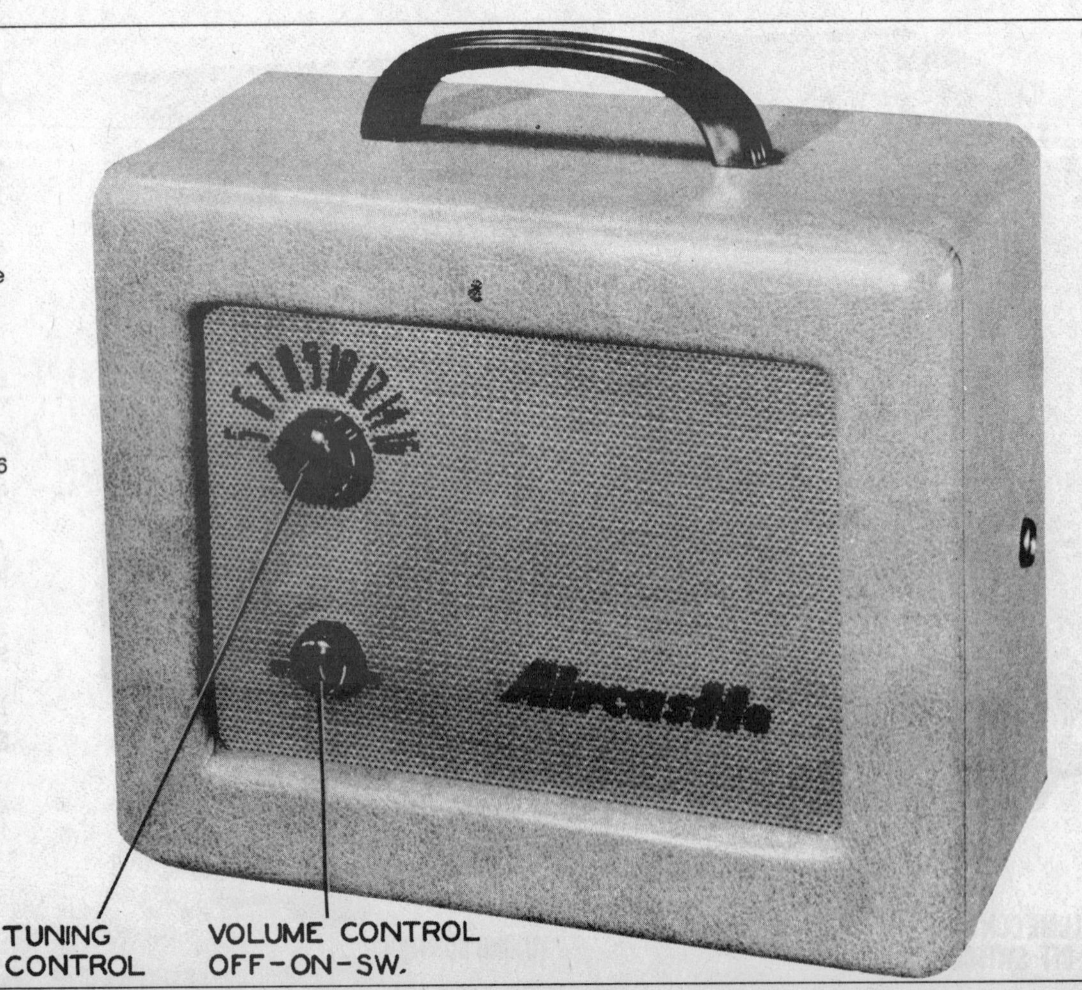

AIRCASTLE

MODEL PICTURED
WEU-262

AC/DC operated AM-FM superheterodyne receiver with loop antenna

TUBES
8

POWER SUPPLY
110-120 volts AC .44 amp @ 117 volts AC

TUNING RANGE
535-1620KC, 88-108MC

MFR/SUPPLIER
Spiegel Inc.

PHOTOFACT SET
91-1

PUBLISHED
1950

AIRCASTLE

MODEL PICTURED
9012W

AC/DC operated superheterodyne receiver with loop antenna

TUBES
4

POWER SUPPLY
101-125 volts AC/DC

TUNING RANGE
550-1720KC

MFR/SUPPLIER
Spiegel Inc.

PHOTOFACT SET
94-1

PUBLISHED
1950

AIRCASTLE

MODEL PICTURED
171

AC/DC operated superheterodyne receiver with loop antenna

TUBES
5

POWER SUPPLY
105-120 volts AC/DC .25 amp @ 117 volts AC

TUNING RANGE
535-1620KC

MFR/SUPPLIER
Spiegel Inc.

PHOTOFACT SET
96-1

PUBLISHED
1950

AIRCASTLE

MODEL PICTURED
6053

AC operated phono-radio superheterodyne receiver with loop antenna

TUBES
5

POWER SUPPLY
110-125 volts AC

TUNING RANGE
540-1720KC

MFR/SUPPLIER
Spiegel Inc.

PHOTOFACT SET
97-1

PUBLISHED
1950

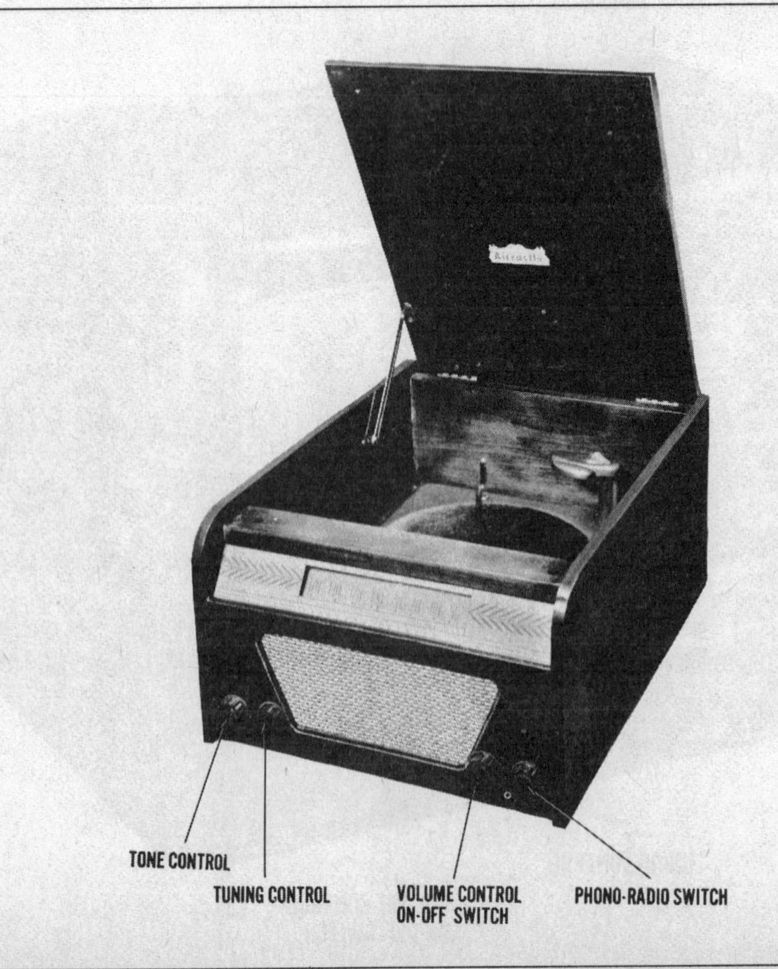

AIRCASTLE

MODEL PICTURED
9009W

AC/DC operated superheterodyne receiver with loop antenna

TUBES
5

POWER SUPPLY
110-120 volts AC/DC

TUNING RANGE
55-1700KC

MFR/SUPPLIER
Spiegel Inc.

PHOTOFACT SET
97-2

PUBLISHED
1950

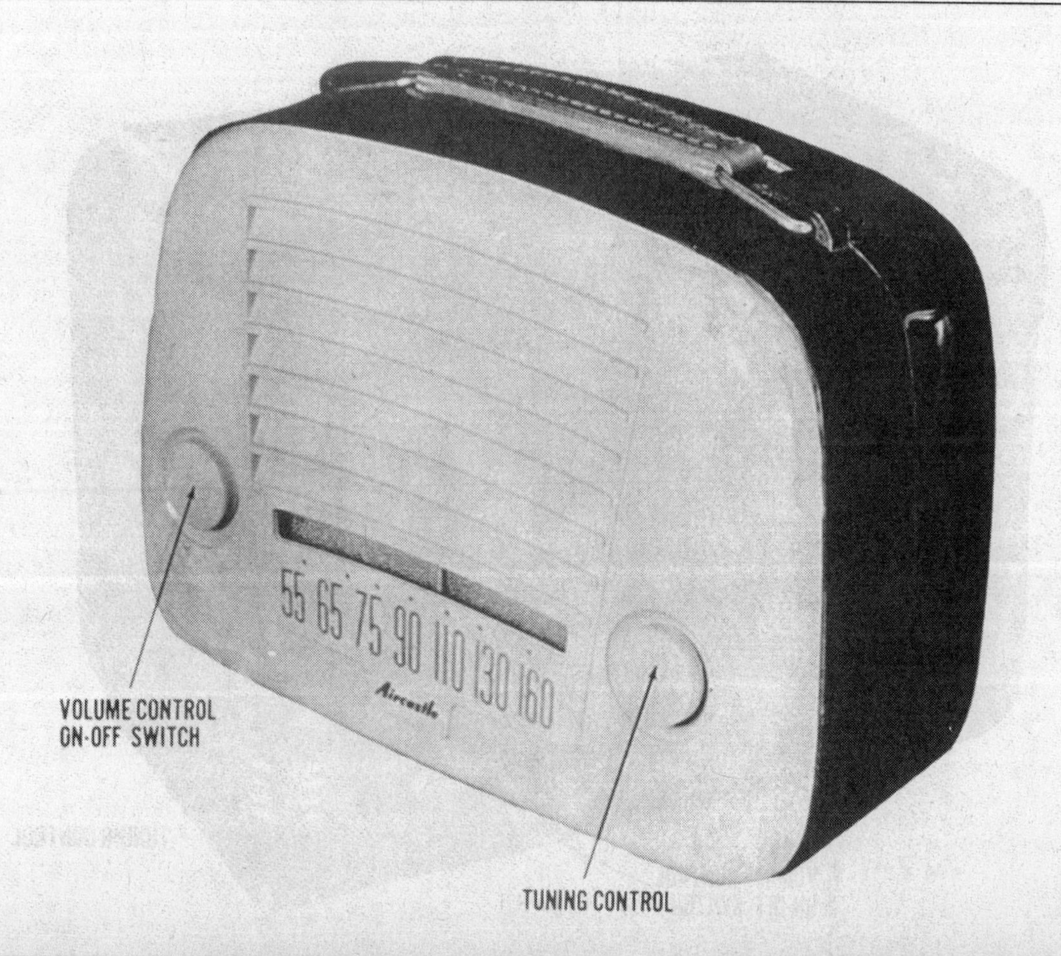

AIRCASTLE

MODEL PICTURED
102B

Three-power operated portable superheterodyne receiver with loop antenna

TUBES
5

POWER SUPPLY
105-120 volts AC/DC or 4.5 volts A supply & 90 volts B supply

TUNING RANGE
540-1620KC

MFR/SUPPLIER
Spiegel Inc.

PHOTOFACT SET
98-2

PUBLISHED
1950

AIRCASTLE

MODEL PICTURED
PC-8

AC operated combo phono-radio superheterodyne receiver with loop antenna

TUBES
5

POWER SUPPLY
110-120 volts AC, .22 amp @ 117 volts AC

TUNING RANGE
540-1620KC

MFR/SUPPLIER
Spiegel Inc.

PHOTOFACT SET
99-1

PUBLISHED
1950

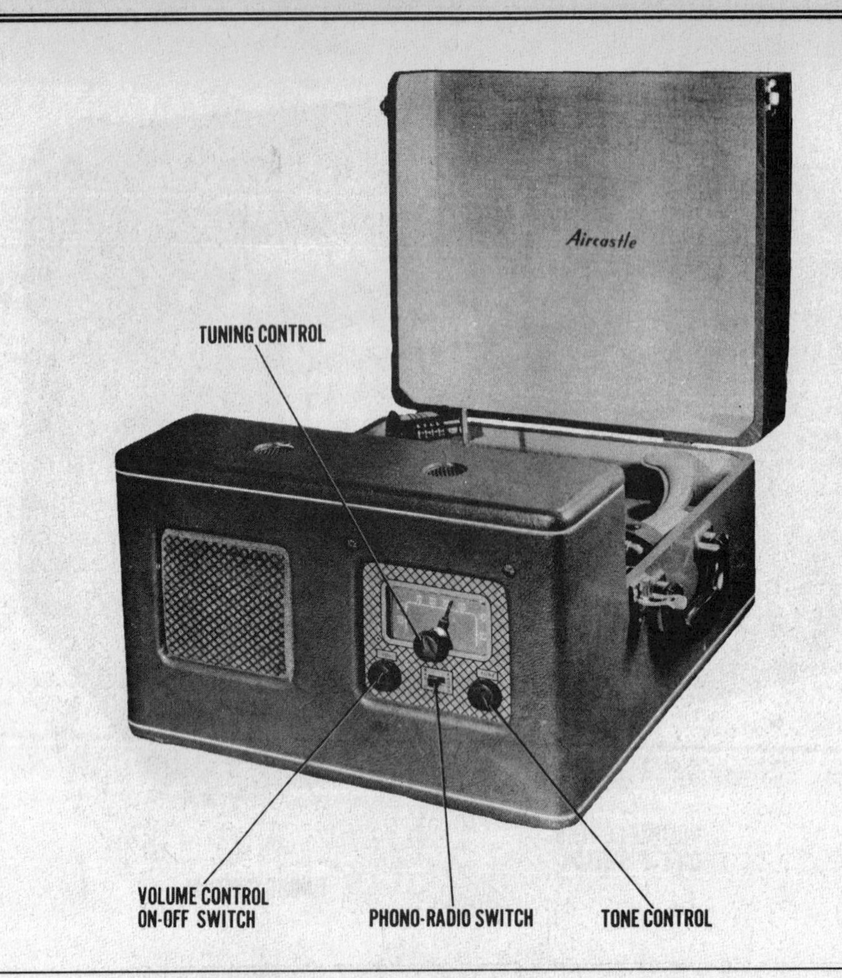

AIRCASTLE

MODEL PICTURED
9008W

AC/DC operated superheterodyne receiver with loop antenna

TUBES
4

POWER SUPPLY
105-125 volts AC/DC, .25 amp @ 117 volts AC

TUNING RANGE
550-1700KC

MFR/SUPPLIER
Spiegel Inc.

PHOTOFACT SET
99-2

PUBLISHED
1950

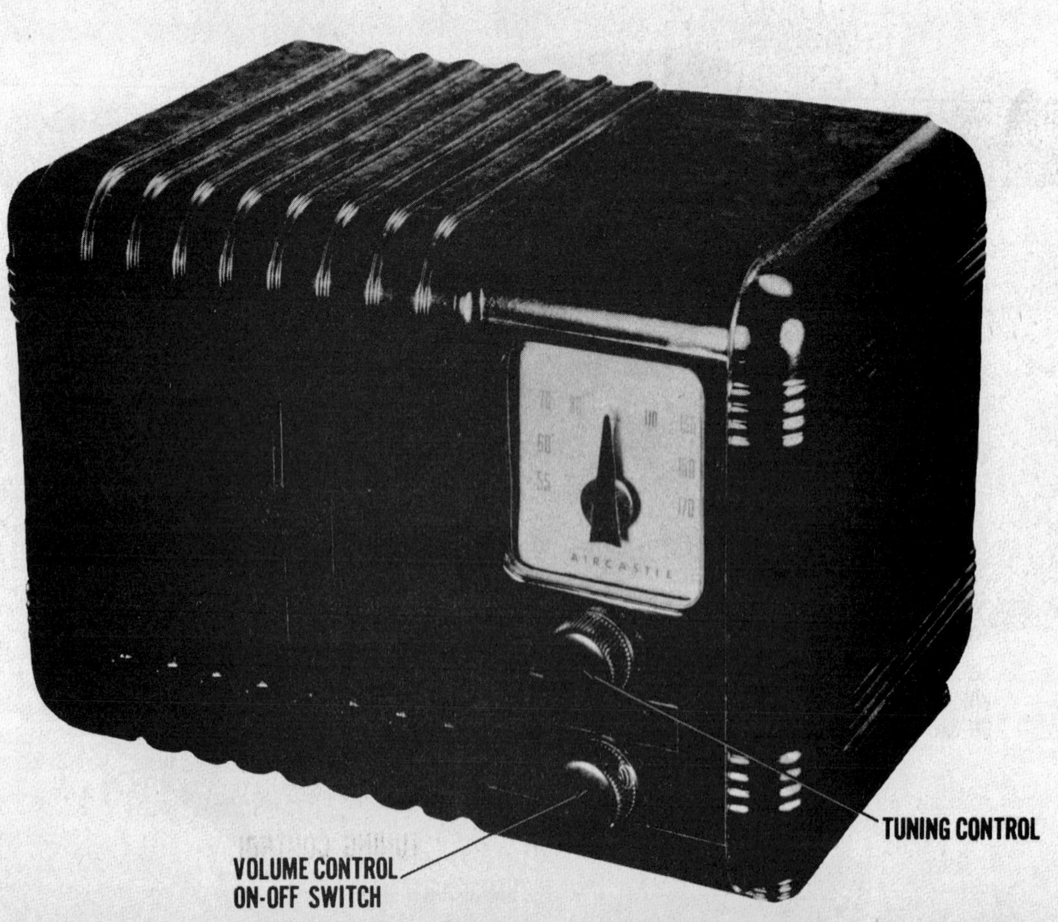

AIRCASTLE

MODEL PICTURED
5015.1

AC/DC operated superheterodyne receiver with loop antenna

TUBES
5

POWER SUPPLY
110-125 volts AC/DC

TUNING RANGE
540-1720KC

MFR/SUPPLIER
Spiegel Inc.

PHOTOFACT SET
118-3

PUBLISHED
1950

AIRCASTLE

MODEL PICTURED
606-400WB

Battery operated superheterodyne receiver

TUBES
4

POWER SUPPLY
1.5 volts A & 90 volts B battery in pack form

TUNING RANGE
540-1725KC

MFR/SUPPLIER
Spiegel Inc.

PHOTOFACT SET
119-2

PUBLISHED
1951

AIRCASTLE

MODEL PICTURED
5056-A

AC/DC operated superheterodyne receiver with loop antenna

TUBES
4

POWER SUPPLY
110-120 volts AC/DC

TUNING RANGE
540-1650KC

MFR/SUPPLIER
Spiegel Inc.

PHOTOFACT SET
120-2

PUBLISHED
1951

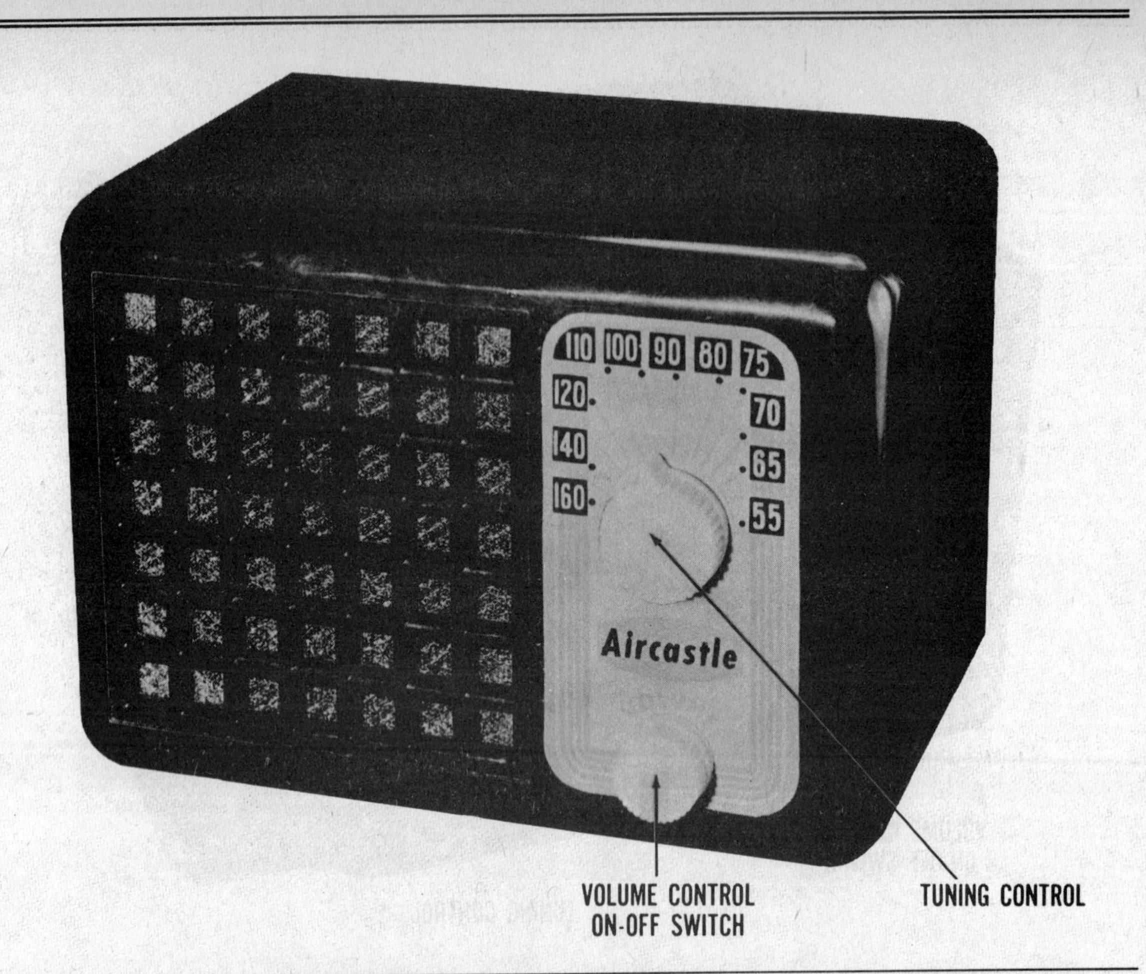

AIRCASTLE

MODEL PICTURED
607-314

AC/DC operated superheterodyne receiver with loop antenna

TUBES
6

POWER SUPPLY
105-120 volts AC/DC ,24 amp @ 117 volts AC

TUNING RANGE
535-1620KC

MFR/SUPPLIER
Spiegel Inc.

PHOTOFACT SET
122-2

PUBLISHED
1951

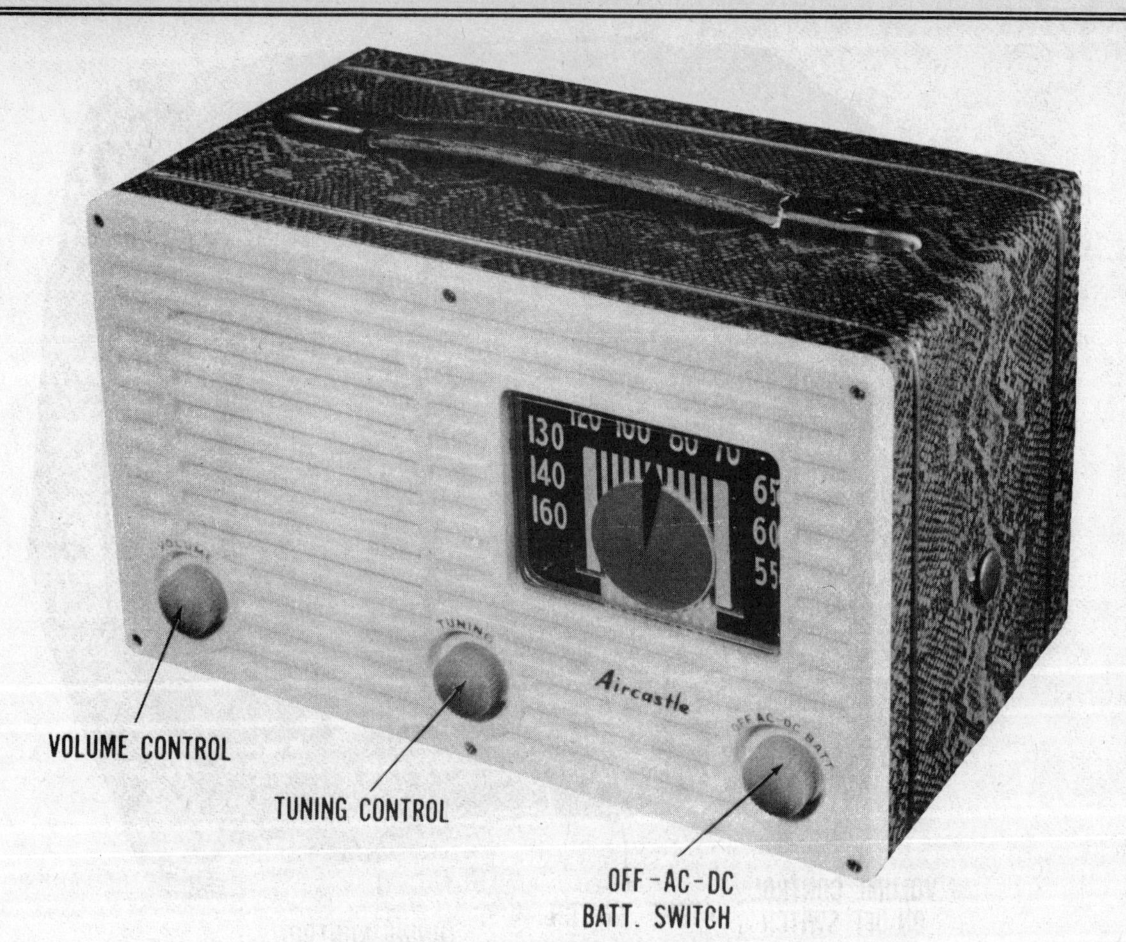

AIRCASTLE

MODEL PICTURED
5022

Three-power operated portable superheterodyne receiver with loop antenna

TUBES
4

POWER SUPPLY
110-125 volts AC/DC or 3 volts A supply & 67.5 volts B supply

TUNING RANGE
540-1650KC

MFR/SUPPLIER
Spiegel Inc.

PHOTOFACT SET
123-2

PUBLISHED
1951

AIRCASTLE

MODEL PICTURED
153

AC operated radio-phono superheterodyne receiver with loop antenna

TUBES
7

POWER SUPPLY
110-120 volts AC

TUNING RANGE
535-1620KC

MFR/SUPPLIER
Spiegel Inc.

PHOTOFACT SET
126-2

PUBLISHED
1951

AIRCASTLE

MODEL PICTURED
REV248

AC/DC operated superheterodyne receiver with loop antenna

TUBES
6

POWER SUPPLY
110-120 volts AC/DC

TUNING RANGE
540-1620KC

MFR/SUPPLIER
Spiegel Inc.

PHOTOFACT SET
127-2

PUBLISHED
1951

AIRCASTLE

MODEL PICTURED
935

AC operated superheterodyne receiver with electric clock

TUBES
5

POWER SUPPLY
110-120 volts AC, .23 amp @ 117 volts AC

TUNING RANGE
540-1620KC

MFR/SUPPLIER
Spiegel Inc.

PHOTOFACT SET
128-2

PUBLISHED
1951

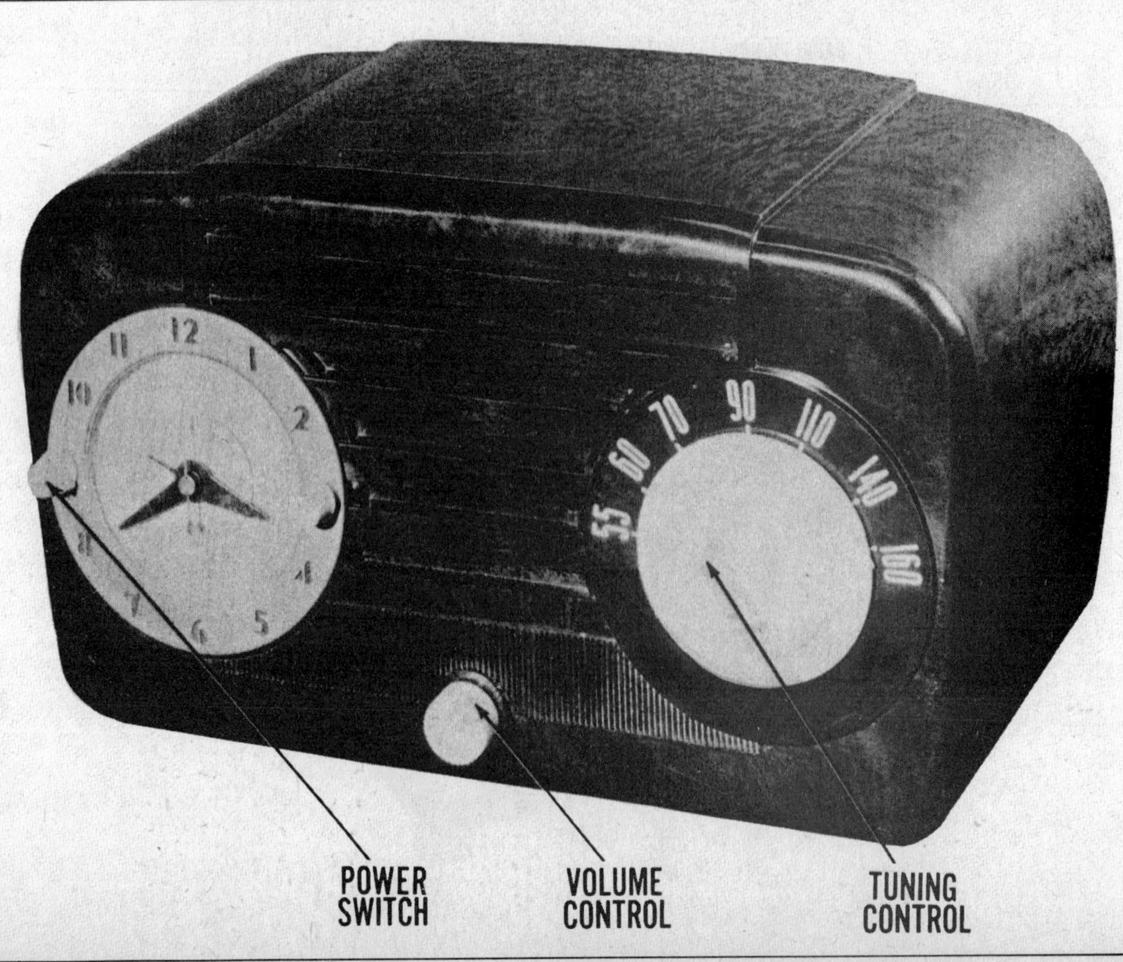

AIRCASTLE

MODEL PICTURED
915I,W

AC operated RF receiver with electric clock

TUBES
4

POWER SUPPLY
105-125 volts AC

TUNING RANGE
540-1650KC

MFR/SUPPLIER
Spiegel Inc.

PHOTOFACT SET
129-2

PUBLISHED
1951

AIRCASTLE

MODEL PICTURED
350

AC operated phono-radio AM-FM superheterodyne receiver

TUBES
8

POWER SUPPLY
105-125 volts AC

TUNING RANGE
540-1600KC, 88-108MC

MFR/SUPPLIER
Spiegel Inc.

PHOTOFACT SET
136-4

PUBLISHED
1951

AIRCASTLE

MODEL PICTURED
607-316-1

AC/DC operated superheterodyne receiver with loop antenna

TUBES
5

POWER SUPPLY
110-120 volts AC/DC

TUNING RANGE
535-1620KC

MFR/SUPPLIER
Spiegel Inc.

PHOTOFACT SET
138-2

PUBLISHED
1951

AIRCASTLE

MODEL PICTURED
472-053VM

AC operated AM-phono combo superheterodyne receiver with loop antenna

TUBES
5

POWER SUPPLY
110-120 volts AC

TUNING RANGE
540-1650KC

MFR/SUPPLIER
Spiegel Inc.

PHOTOFACT SET
163-2

PUBLISHED
1952

AIRCASTLE

MODEL PICTURED
659.511

AC operated superheterodyne receiver with electric clock

TUBES
4

POWER SUPPLY
110-120 volts AC, 60 cycle, .23 amp @ 117 volts AC

TUNING RANGE
520-1650KC

MFR/SUPPLIER
Spiegel Inc.

PHOTOFACT SET
167-2

PUBLISHED
1952

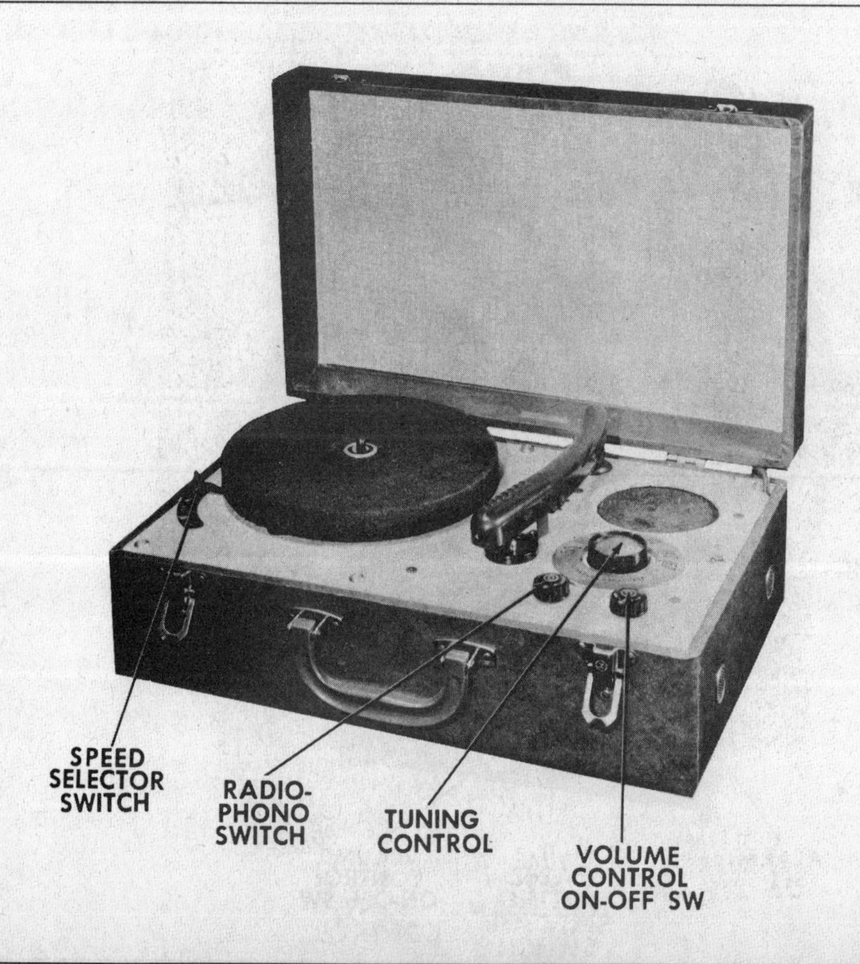

AIRCASTLE

MODEL PICTURED
652.505

AC operated AM-phono superheterodyne receiver with three-speed manual record player

TUBES
5

POWER SUPPLY
110-120 volts AC 60 CYCLE

TUNING RANGE
530-1650KC

MFR/SUPPLIER
Spiegel Inc.

PHOTOFACT SET
168-2

PUBLISHED
1952

AIRCASTLE

MODEL PICTURED
607.299

AC/DC operated AM superheterodyne receiver

TUBES
4

POWER SUPPLY
110-120 volts AC/DC

TUNING RANGE
530-1629KC

MFR/SUPPLIER
Spiegel Inc.

PHOTOFACT SET
177-3

PUBLISHED
1952

AIRCASTLE

MODEL PICTURED
659.520E

AC operated AM superheterodyne receiver with electric clock and lamp

TUBES
4

POWER SUPPLY
110-120 volts AC, 60 cycle

TUNING RANGE
540-1620KC

MFR/SUPPLIER
Spiegel, Inc.

PHOTOFACT SET
185-4

PUBLISHED
1952

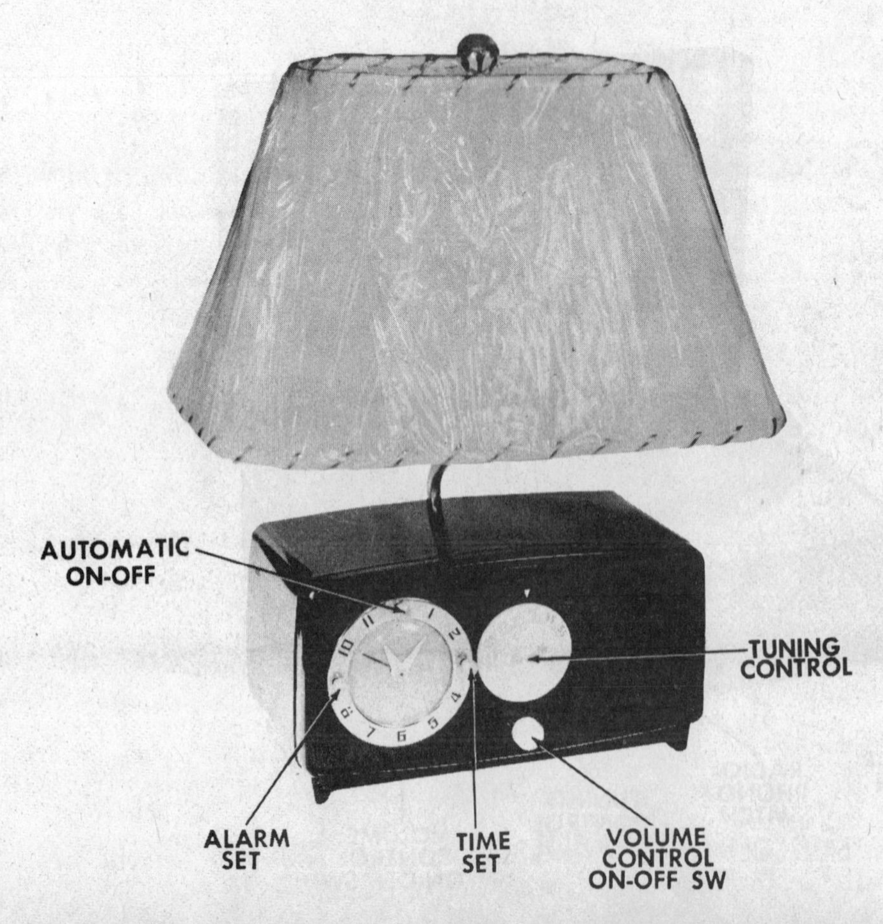

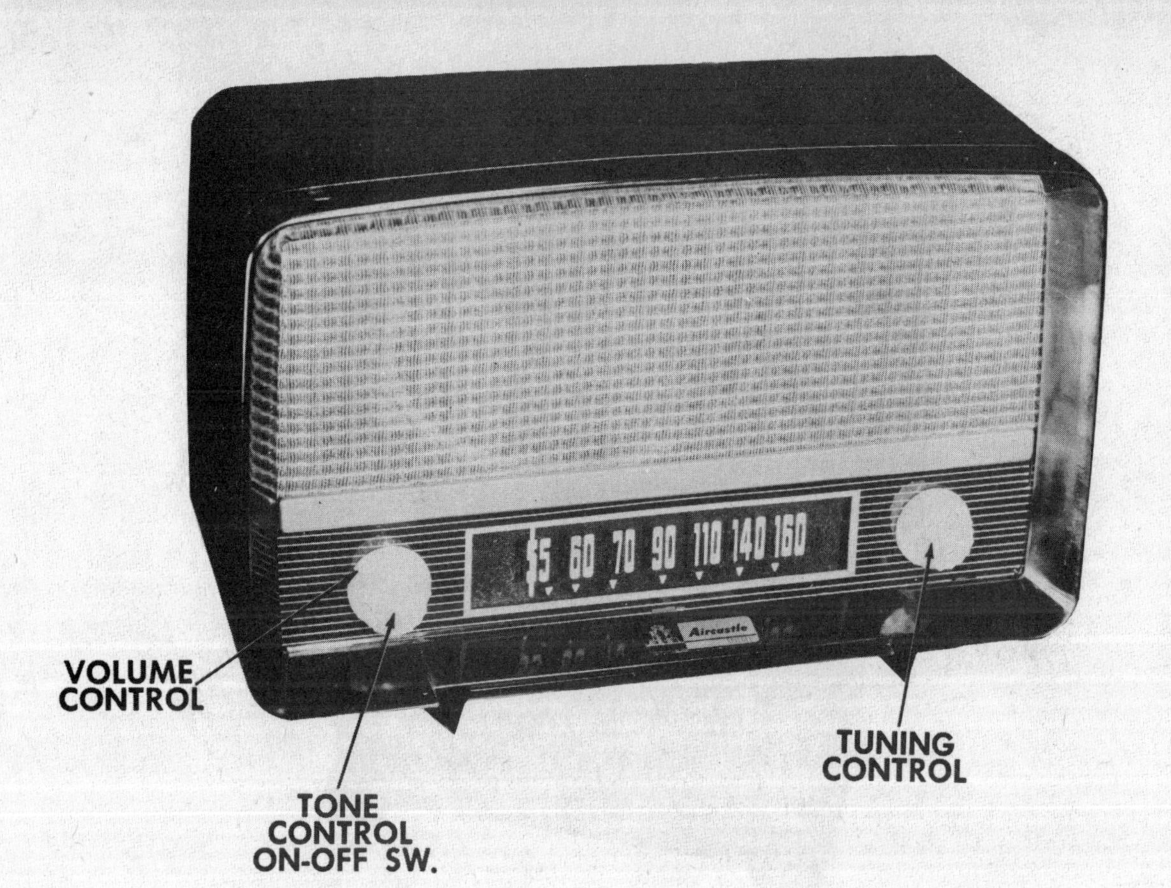

AIRCASTLE

MODEL PICTURED
652.6T1E

AC/DC operated AM superheterodyne receiver

TUBES
6

POWER SUPPLY
110-120 volts AC/DC, .23 amp. @ 117 volts AC

TUNING RANGE
540-1610KC

MFR/SUPPLIER
Spiegel Inc.

PHOTOFACT SET
205-2

PUBLISHED
1953

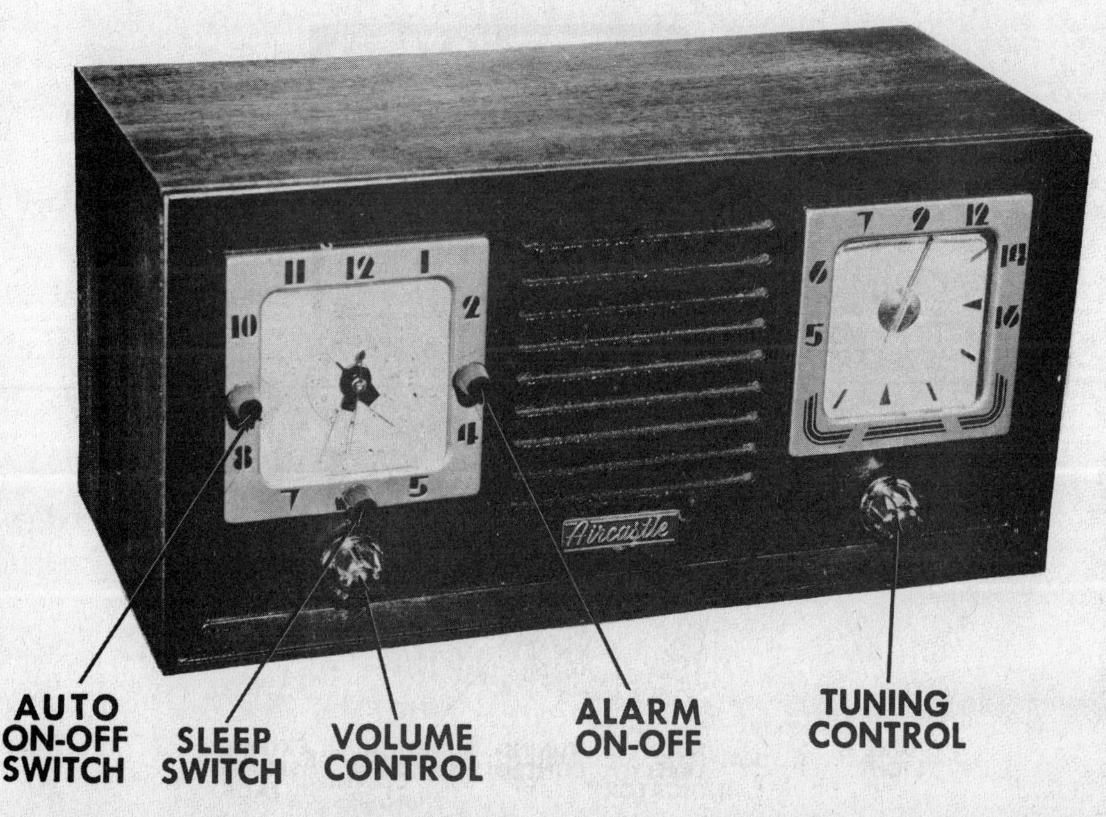

AIRCASTLE

MODEL PICTURED
610.CL152B

AC operated AM superheterodyne receiver with electric clock

TUBES
5

POWER SUPPLY
105-125 volts AC, 60 cycles .23 amp. @ 117 volts AC

TUNING RANGE
538-1650 KC

MFR/SUPPLIER
Spiegel Inc.

PHOTOFACT SET
208-1

PUBLISHED
1953

AIRCASTLE

MODEL PICTURED
472.254

AC operated radio-phono combo AM superheterodyne receiver

TUBES
7

POWER SUPPLY
110-120 volts AC, 60 cycle

TUNING RANGE
540-1600KC

MFR/SUPPLIER
Spiegel Inc.

PHOTOFACT SET
215-2

PUBLISHED
1953

AIRCASTLE

MODEL PICTURED
603.880

AC operated AM superheterodyne receiver with disc recorder and playback unit

TUBES
6

POWER SUPPLY
105-125 volts AC, 60 cycle

TUNING RANGE
535-1620KC

MFR/SUPPLIER
Spiegel Inc.

PHOTOFACT SET
230-2

PUBLISHED
1954

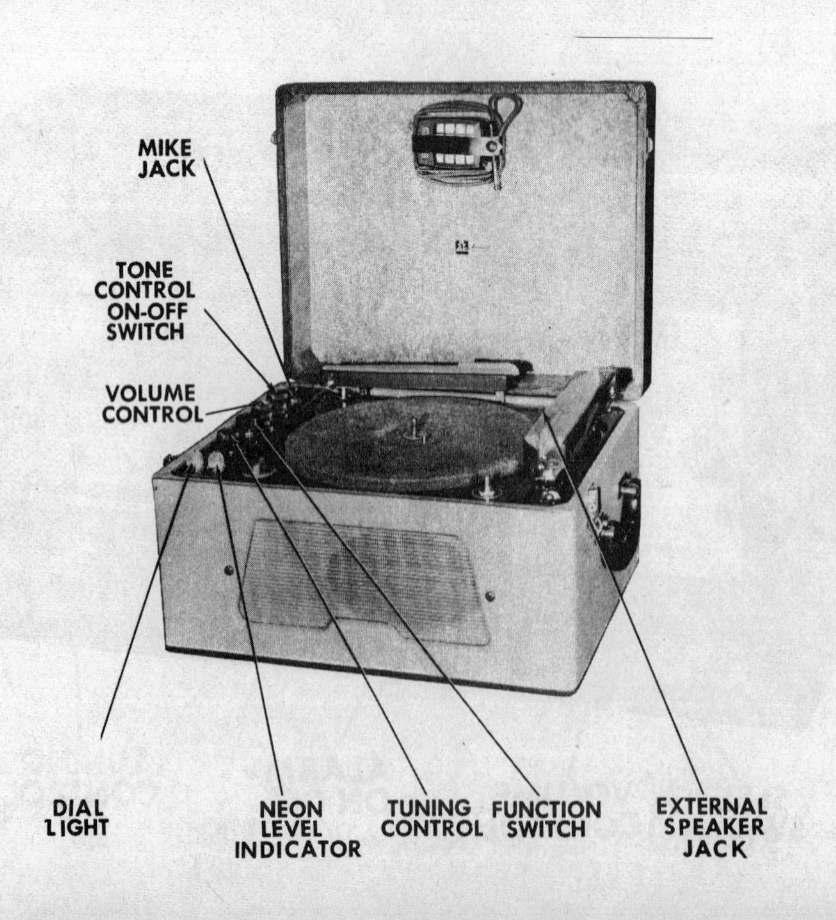

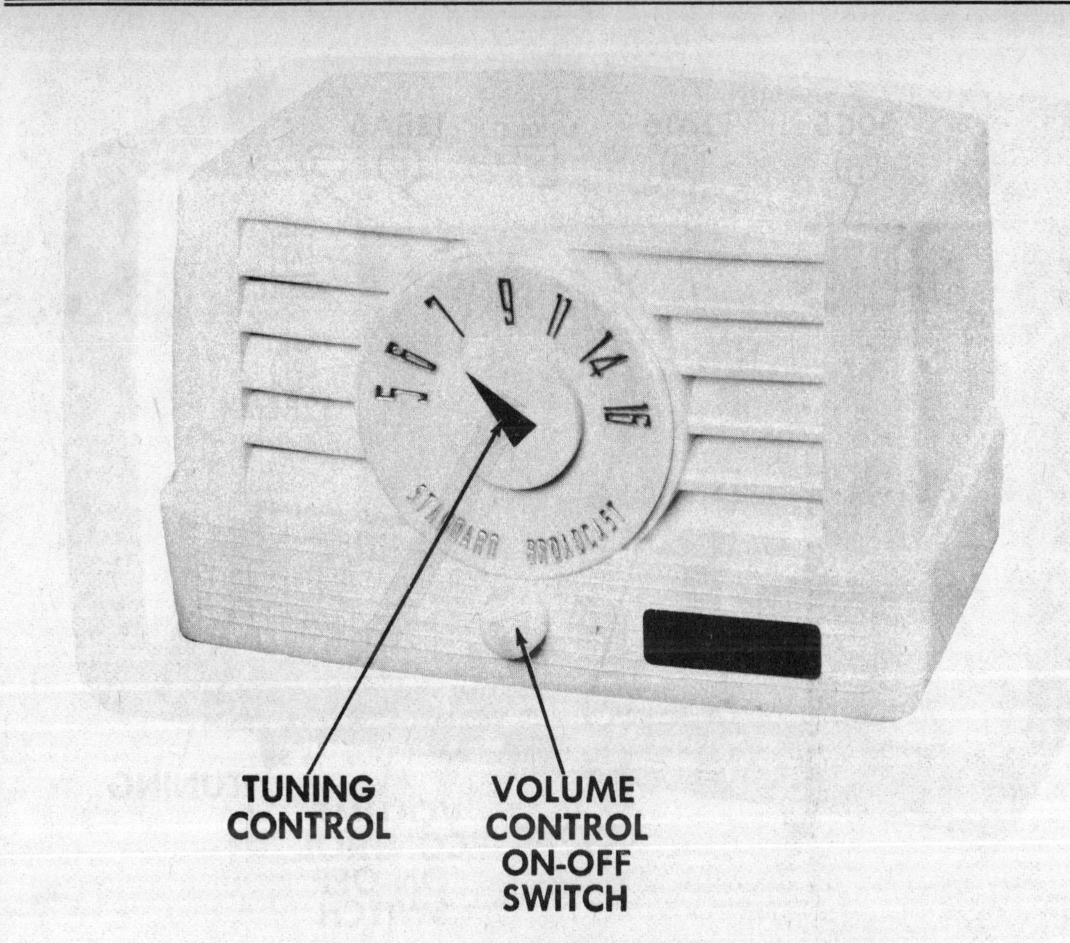

AIRCASTLE

MODEL PICTURED
652.5C1M

AC/DC operated AM superheterodyne receiver

TUBES
5

POWER SUPPLY
110-120 volts AC/DC (model 652.5C1 AC only) .24 amp @ 117 volts AC

TUNING RANGE
535-1620KC

MFR/SUPPLIER
Spiegel Inc.

PHOTOFACT SET
246-1

PUBLISHED
1954

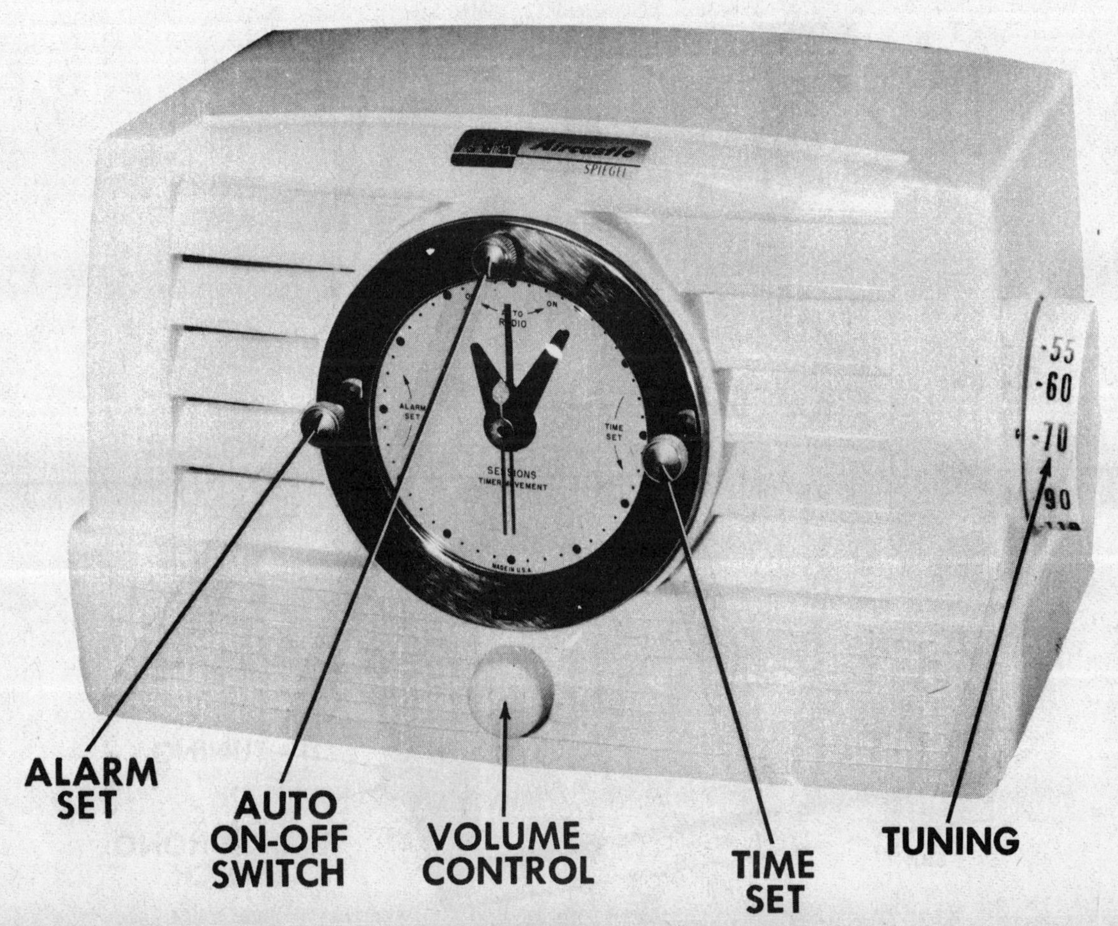

AIRCASTLE

MODEL PICTURED
652.5C1M

AC operated AM superheterodyne receiver with electric clock

TUBES
5

POWER SUPPLY
110-120 volts AC, 60 cycle

TUNING RANGE
535-1620KC

MFR/SUPPLIER
Spiegel Inc.

PHOTOFACT SET
260-3

PUBLISHED
1954

AIRCASTLE

MODEL PICTURED
652.5T5E

AC/DC operated AM superheterodyne receiver

TUBES
5

POWER SUPPLY
105-1250 volts AC/DC

TUNING RANGE
535-1620KC

MFR/SUPPLIER
Spiegel Inc.

PHOTOFACT SET
260-4

PUBLISHED
1954

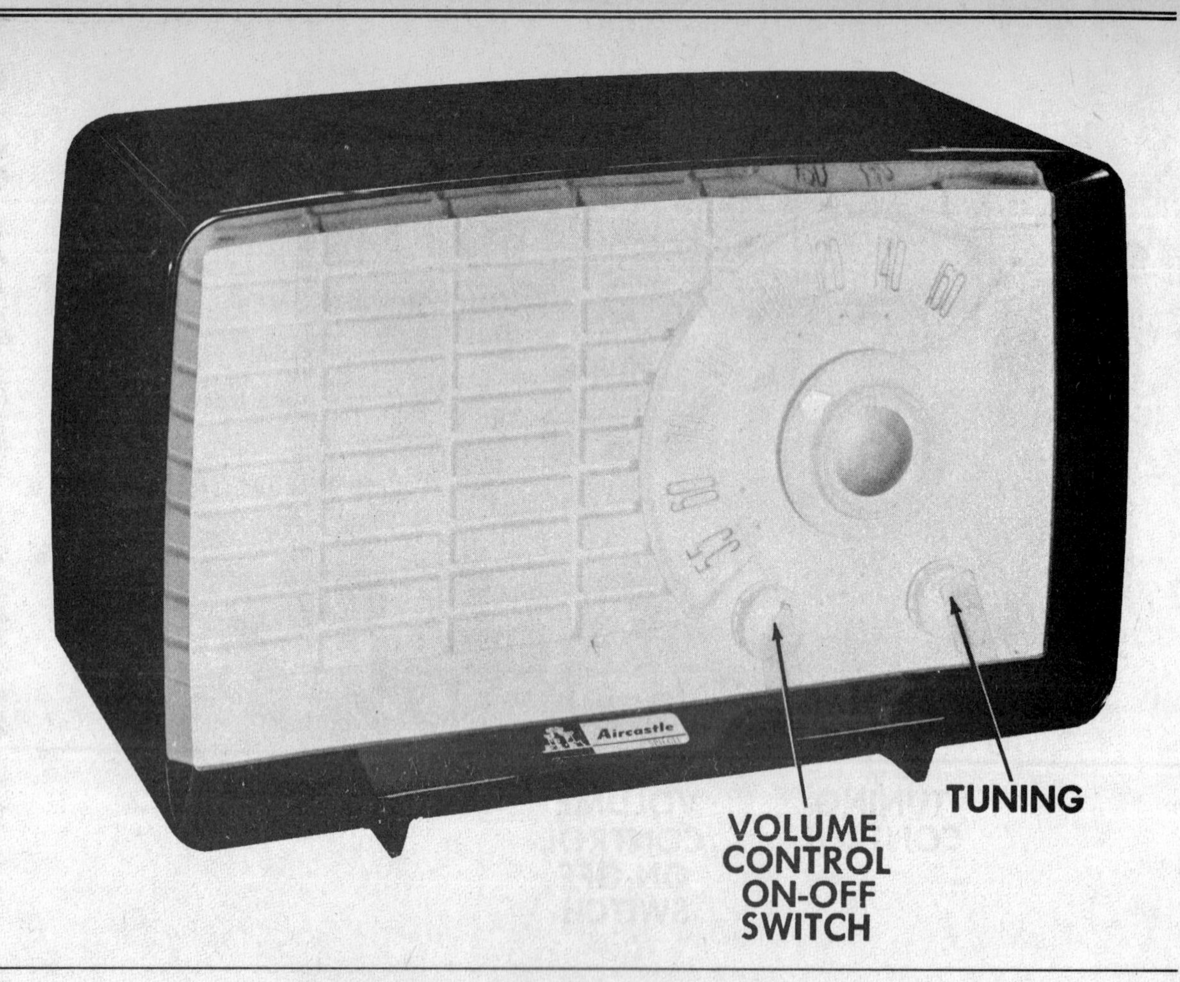

AIRCASTLE

MODEL PICTURED
652.5X5

AC operated radio-phono superheterodyne receiver with three-speed auto record changer

TUBES
6

POWER SUPPLY
110-120 volts AC, 60 cycle

TUNING RANGE
535-1620KC

MFR/SUPPLIER
Spiegel Inc.

PHOTOFACT SET
286-2

PUBLISHED
1955

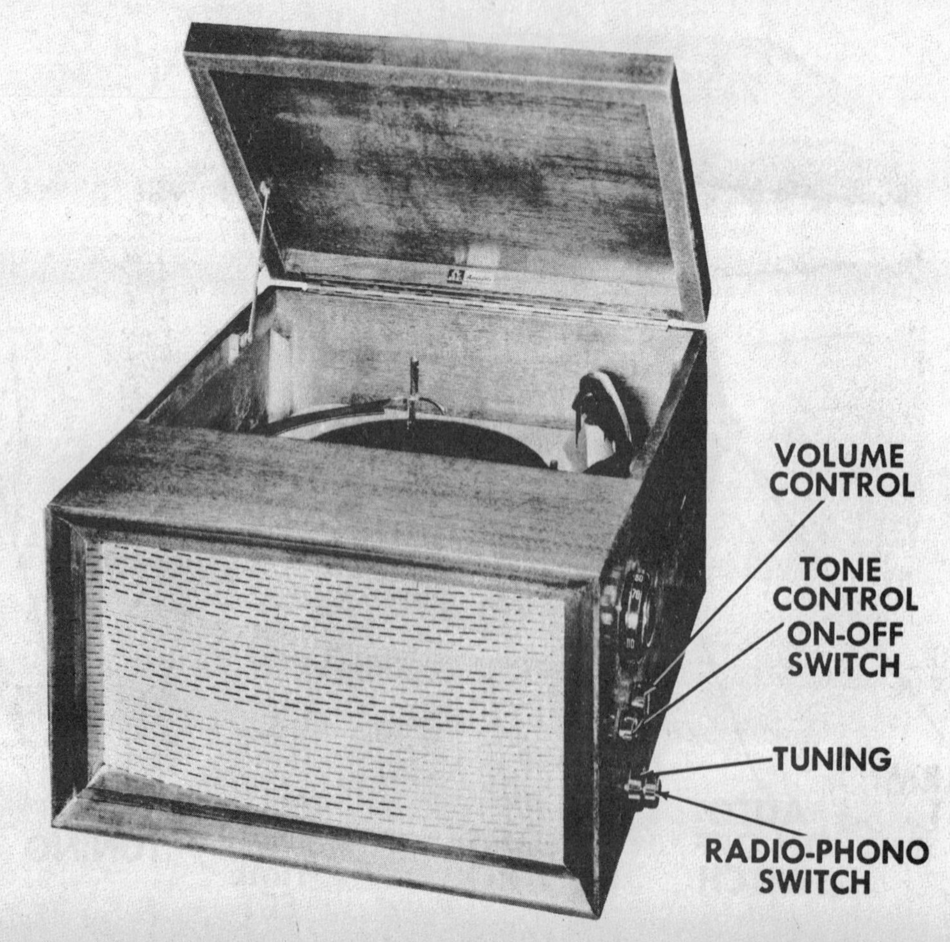

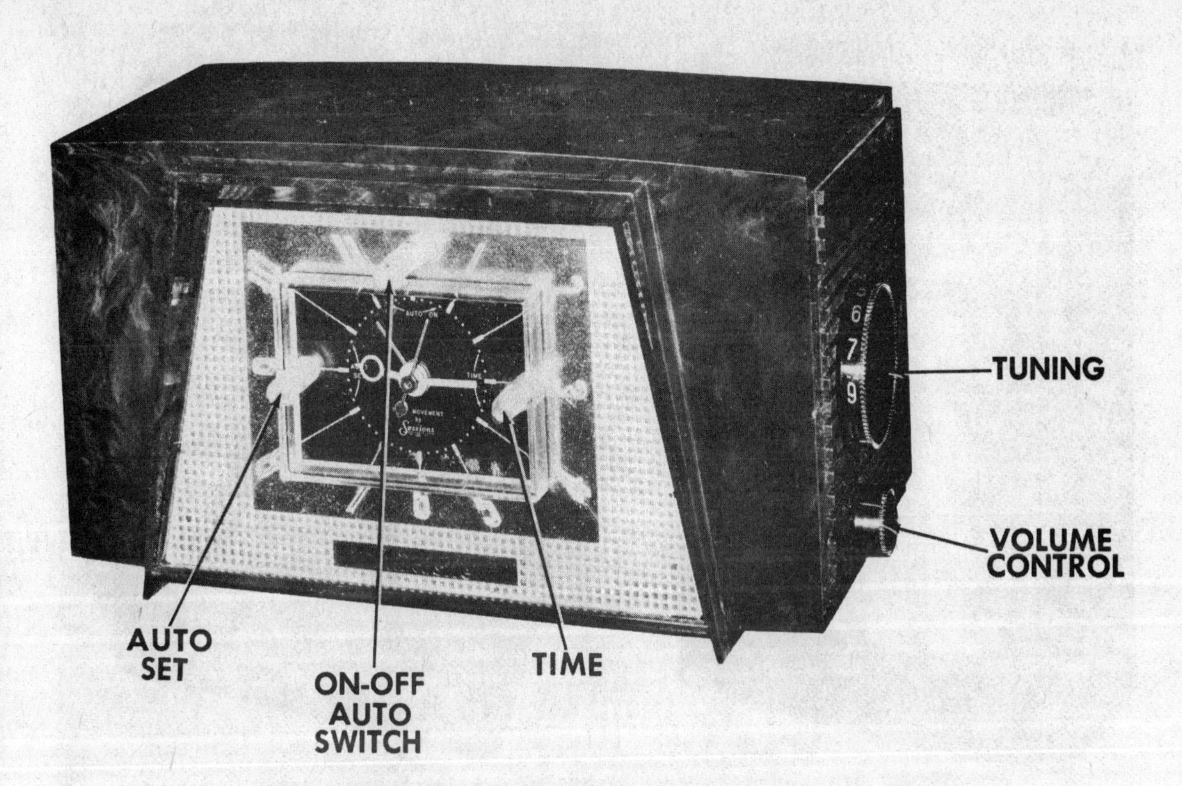

AIRCASTLE

MODEL PICTURED
782.5C1

AC operated AM superheterodyne receiver with electric clock

TUBES
5

POWER SUPPLY
110-120 volts AC, 60 cycle

TUNING RANGE
538-1650KC

MFR/SUPPLIER
Spiegel Inc.

PHOTOFACT SET
287-2

PUBLISHED
1955

AIRCASTLE

MODEL PICTURED
782.FM-99-AC

AC operated AM-FM superheterodyne receiver

TUBES
9

POWER SUPPLY
105-120 volts AC, 60 cycle

TUNING RANGE
535-1650KC, 88-108MC

MFR/SUPPLIER
Spiegel Inc.

PHOTOFACT SET
290-2

PUBLISHED
1955

AIRLINE

MODEL PICTURED
54WG-2500A and 54WG-2700A

AC operated two-band superheterodyne receiver, self-contained loop antenna

TUBES
7

POWER SUPPLY
105-125 volts AC

TUNING RANGE
528-1600KC, 5.75-18.3MC

MFR/SUPPLIER
Montgomery Ward and Co.

PHOTOFACT SET
4-15

PUBLISHED
1946

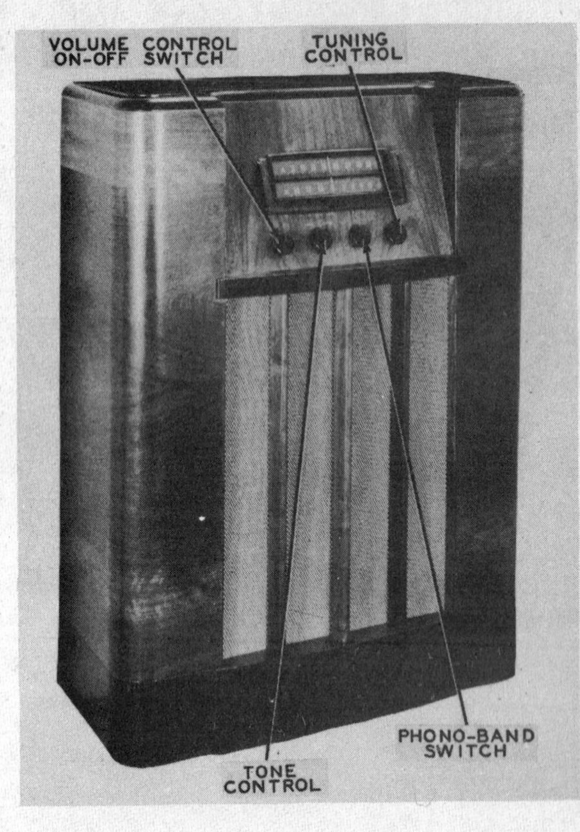

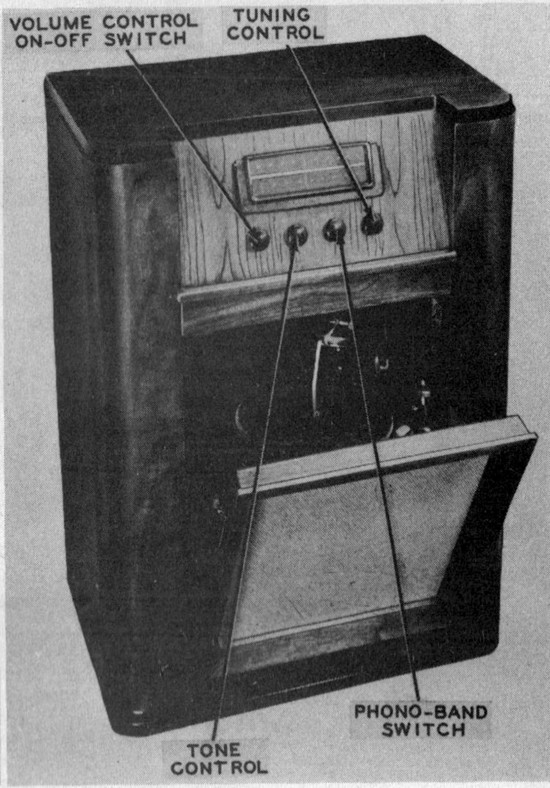

AIRLINE

MODEL PICTURED
64WG-1804B

AC/DC superheterodyne receiver, self-contained loop antenna

TUBES
6

POWER SUPPLY
105-125 volts AC/DC

TUNING RANGE
535-1620KC

MFR/SUPPLIER
Montgomery Ward and Co.

PHOTOFACT SET
4-27

PUBLISHED
1946

AIRLINE

MODEL PICTURED
64WG-1801C

AC/DC superheterodyne receiver, self-contained loop antenna

TUBES
5

POWER SUPPLY
105-125 volts AC/DC

TUNING RANGE
540-1600KC

MFR/SUPPLIER
Montgomery Ward and Co.

PHOTOFACT SET
4-33

PUBLISHED
1946

AIRLINE

MODEL PICTURED
64WG-1807A

AC two-band superheterodyne receiver, self-contained loop antenna

TUBES
6

POWER SUPPLY
105-125 volts AC

TUNING RANGE
540-1600KC, 9-15.6 MC

MFR/SUPPLIER
Montgomery Ward and Co.

PHOTOFACT SET
5-4

PUBLISHED
1946

AIRLINE

MODEL PICTURED
64WG-1511A
64WG-1809A

AC/DC superheterodyne receiver with self-contained loop antenna

TUBES
6

POWER SUPPLY
105-125 volts AC/DC

TUNING RANGE
540-1600KC

MFR/SUPPLIER
Montgomery Ward and Co.

PHOTOFACT SET
5-5

PUBLISHED
1946

MODEL 1511A

MODEL 1809A

AIRLINE

MODEL PICTURED
64WG-2007B

AC operated combo phono-superheterodyne receiver with self-contained loop antenna

TUBES
5

POWER SUPPLY
105-125 volts AC

TUNING RANGE
540-1600KC

MFR/SUPPLIER
Montgomery Ward and Co.

PHOTOFACT SET
5-6

PUBLISHED
1946

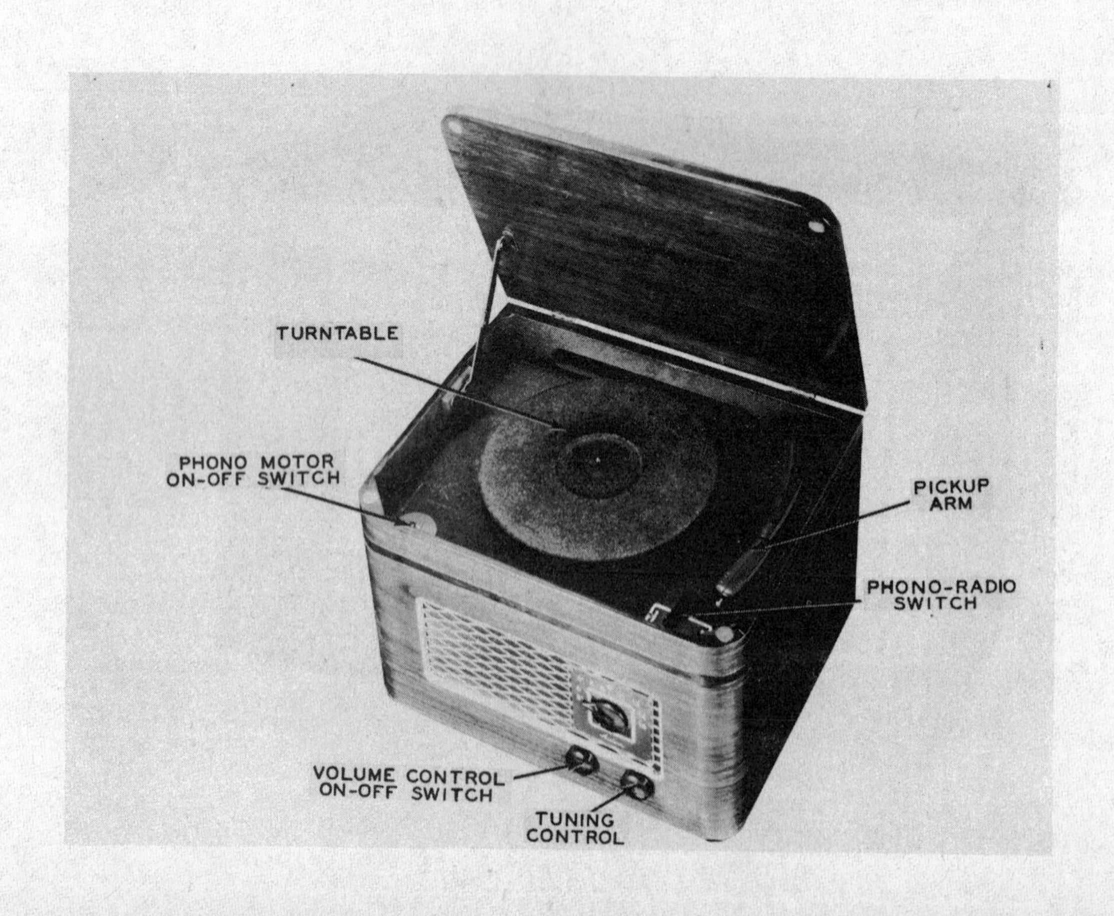

AIRLINE

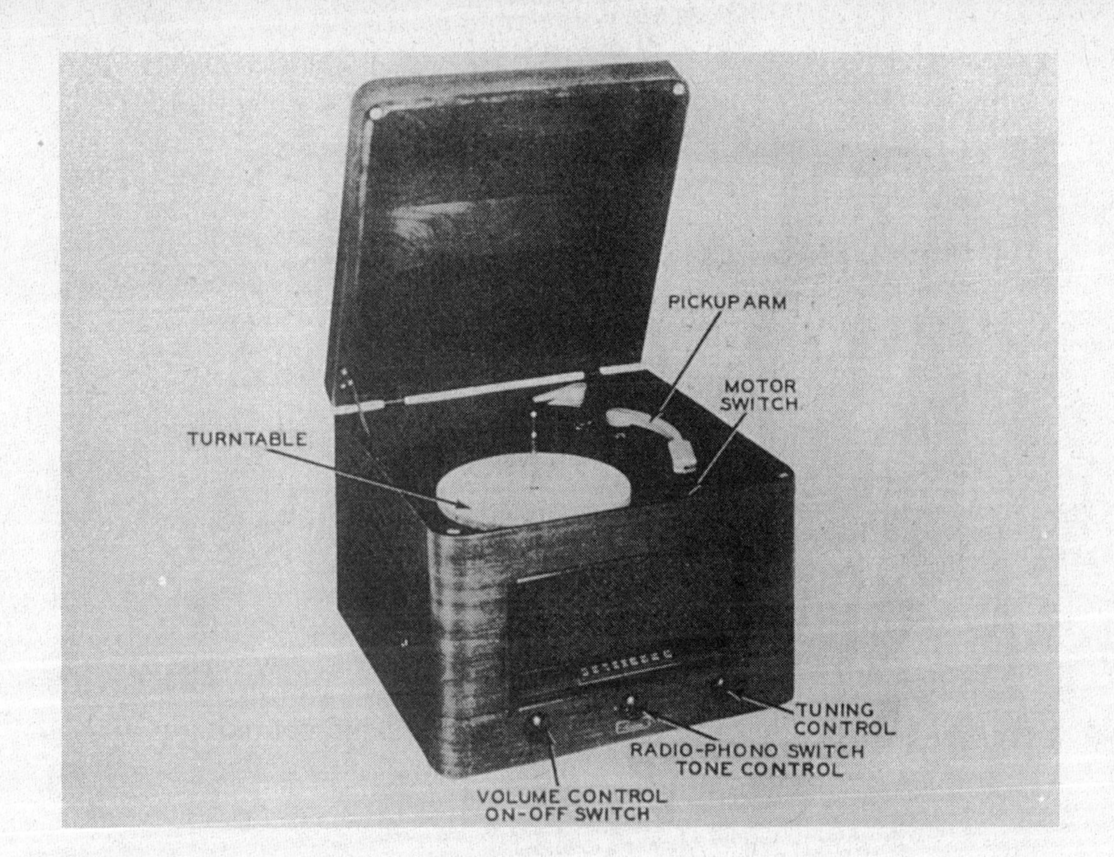

MODEL PICTURED
64WG-2009A

AC operated combo automatic phono-superheterodyne receiver with self-contained loop antenna

TUBES
6

POWER SUPPLY
105-125 volts AC

TUNING RANGE
540-1600KC

MFR/SUPPLIER
Montgomery Ward and Co.

PHOTOFACT SET
6-2

PUBLISHED
1946

AIRLINE

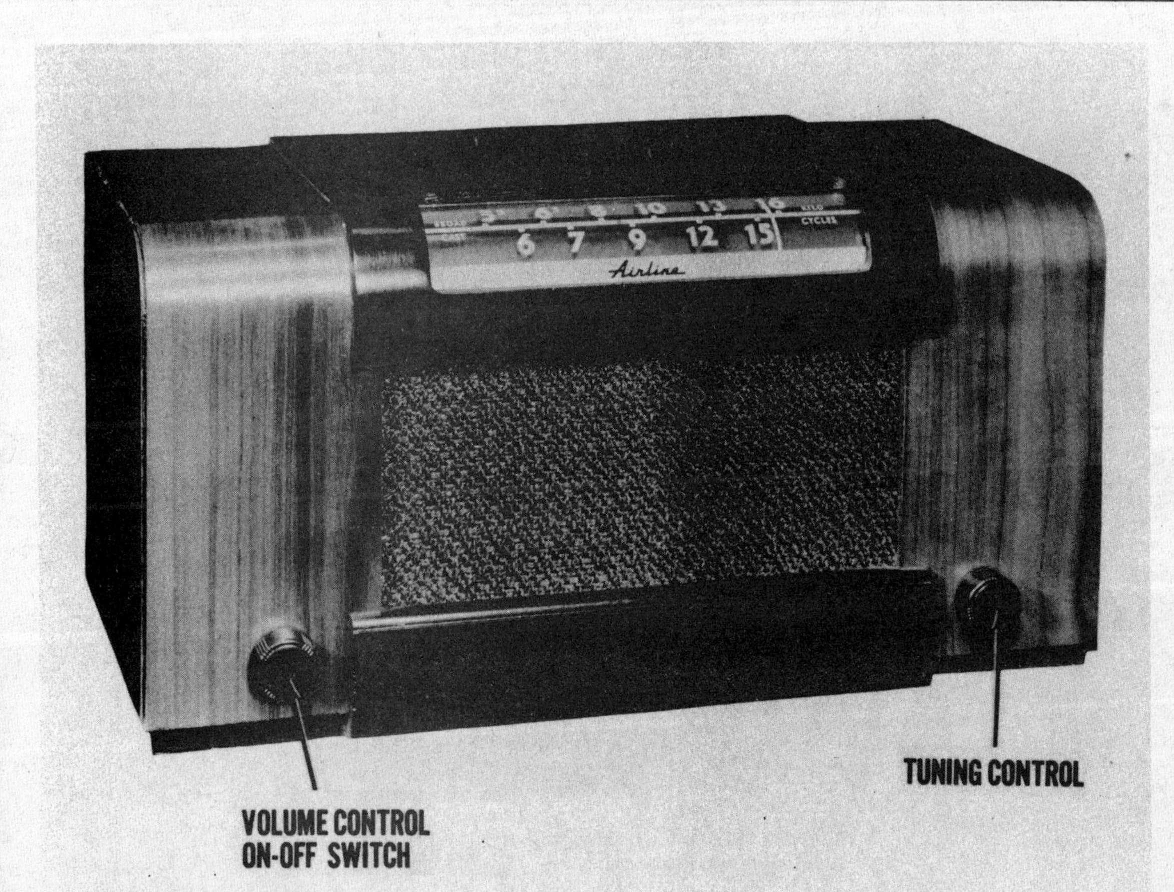

MODEL PICTURED
54KP-1209B

Battery operated superheterodyne receiver, self-contained loop antenna

TUBES
5

POWER SUPPLY
1.5 volts A & 90 volts B supply in pack form

TUNING RANGE
535-1620KC

MFR/SUPPLIER
Montgomery Ward and Co.

PHOTOFACT SET
8-1

PUBLISHED
1946

AIRLINE

MODEL PICTURED
64WG-1052A

Three-power portable AD/DC battery superheterodyne receiver

TUBES
5

POWER SUPPLY
105-125 volts AC/DC or 9 volts A battery & 90 volts B supply in battery pack form

TUNING RANGE
540-1600KC

MFR/SUPPLIER
Montgomery Ward and Co.

PHOTOFACT SET
9-2

PUBLISHED
1946

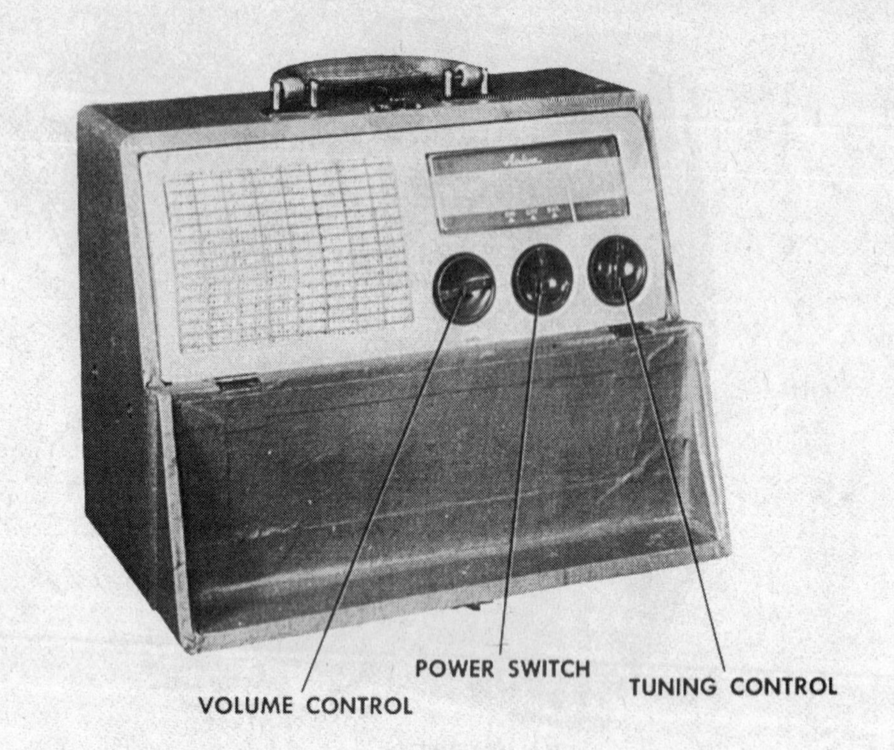

AIRLINE

MODEL PICTURED
64WG-1050A

Three-power portable AC/DC battery superheterodyne receiver, self-contained loop antenna

TUBES
4

POWER SUPPLY
105-125 volts AC/DC or 1.5 volts A & 60 volts B battery

TUNING RANGE
540-1600KC

MFR/SUPPLIER
Montgomery Ward and Co.

PHOTOFACT SET
10-2

PUBLISHED
1946

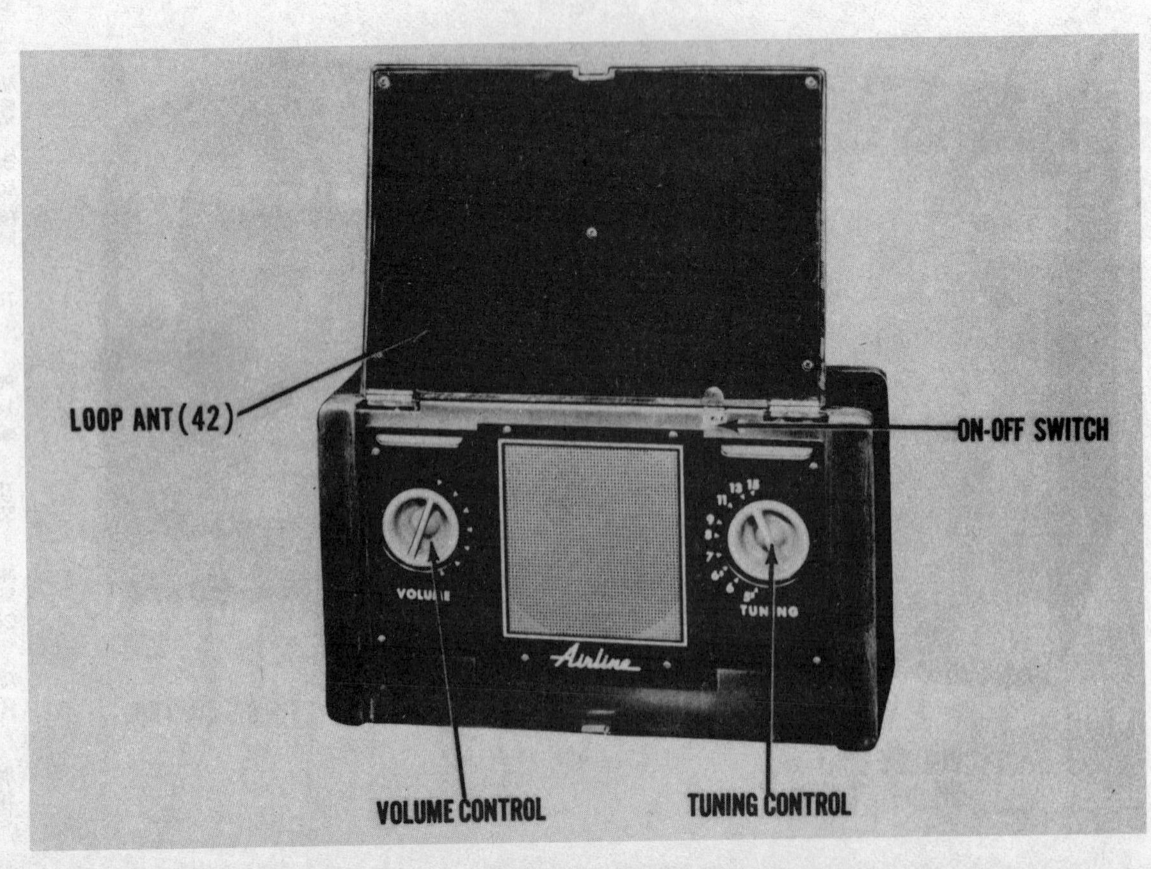

AIRLINE

MODEL PICTURED
64BR-1205A

Battery operated superheterodyne receiver

TUBES
4

POWER SUPPLY
1.5 volts A battery & 90 volts B battery 270mA @ 1.6volts DC & 12.5mA @ 97volts DC

TUNING RANGE
540-1700KC

MFR/SUPPLIER
Montgomery Ward and Co.

PHOTOFACT SET
10-3

PUBLISHED
1946

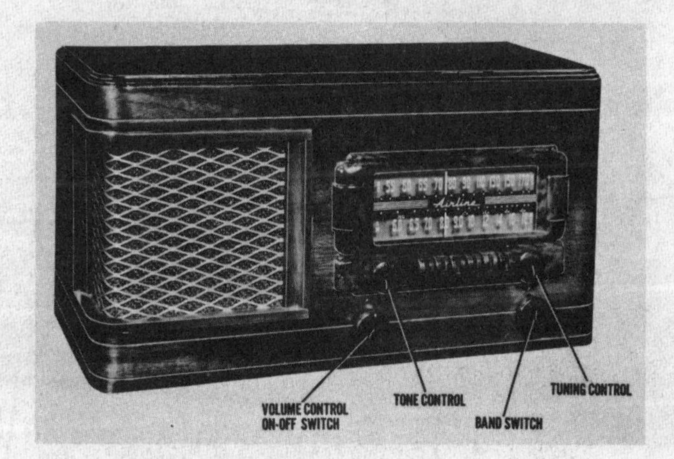

AIRLINE

MODEL PICTURED
64BR-1208A

Battery operated superheterodyne receiver with pushbutton tuning

TUBES
5

POWER SUPPLY
1.5 volts A battery & 90 volts B battery in pack form

TUNING RANGE
535-1720KC, 5.6-18.1MC

MFR/SUPPLIER
Montgomery Ward and Co.

PHOTOFACT SET
16-4

PUBLISHED
1947

AIRLINE

MODEL PICTURED
64BR-1808A

AC operated multi-band superheterodyne receiver with phono provisions and loop antenna

TUBES
8

POWER SUPPLY
105-125 volts AC, .700 amp @ 117 volts AC

TUNING RANGE
540-1600KC, plus four short-wave bands

MFR/SUPPLIER
Montgomery Ward and Co.

PHOTOFACT SET
16-5

PUBLISHED
1947

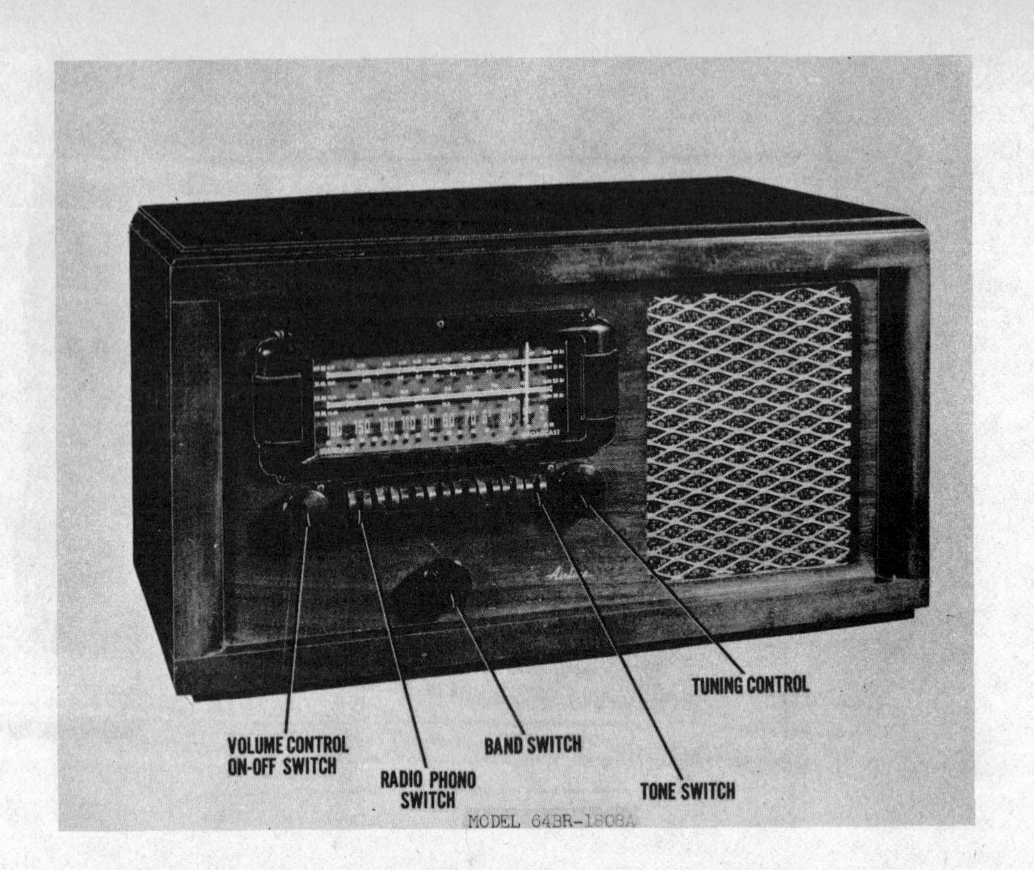

AIRLINE

MODEL PICTURED
64WG-1207B

Battery operated superheterodyne receiver

TUBES
4

POWER SUPPLY
1.5 volts A battery & 90 volts B battery in pack form

TUNING RANGE
540-1600KC

MFR/SUPPLIER
Montgomery Ward and Co.

PHOTOFACT SET
18-5

PUBLISHED
1947

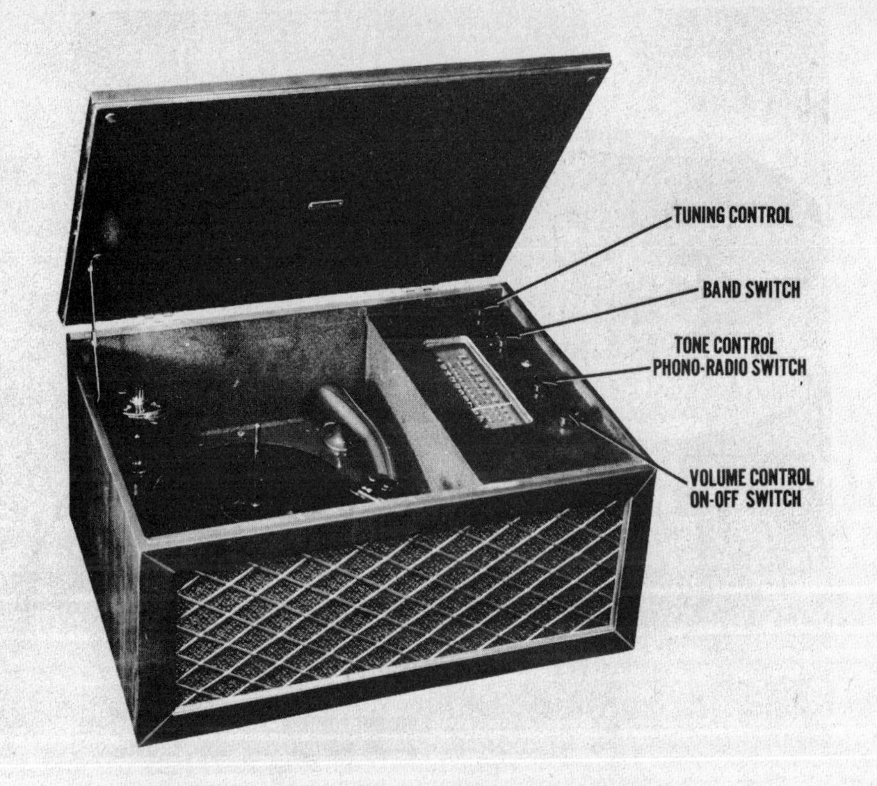

AIRLINE

MODEL PICTURED
74WG-2010B

AC operated phono-radio combo superheterodyne receiver with loop antenna

TUBES
6

POWER SUPPLY
105-125 volts AC

TUNING RANGE
540-1600KC, 9-15.5MC

MFR/SUPPLIER
Montgomery Ward and Co.

PHOTOFACT SET
18-6

PUBLISHED
1947

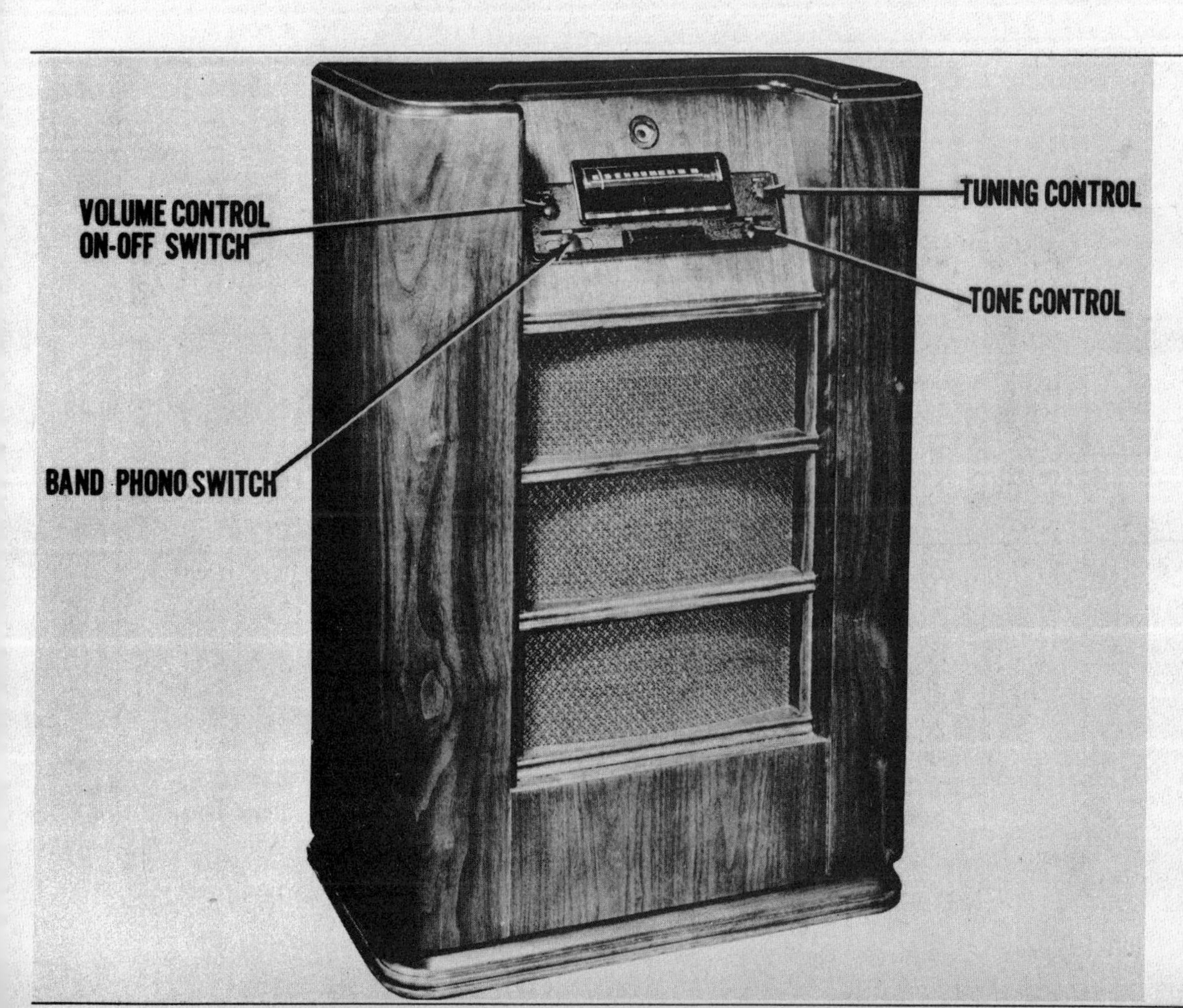

AIRLINE

MODEL PICTURED
74WG-2505A

AC operated FM-AM superheterodyne receiver with loop antenna and phono provisions

TUBES
10

POWER SUPPLY
105-125 volts AC

TUNING RANGE
540-1600KC, 88-108MC, 9-15.5MC

MFR/SUPPLIER
Montgomery Ward and Co.

PHOTOFACT SET
18-7

PUBLISHED
1947

AIRLINE

MODEL PICTURED
74BR-2001B

AC operated radio-phono combination

TUBES
5

POWER SUPPLY
105-125 volts AC

TUNING RANGE
535-1690KC

MFR/SUPPLIER
Montgomery Ward and Co.

PHOTOFACT SET
23-2

PUBLISHED
1947

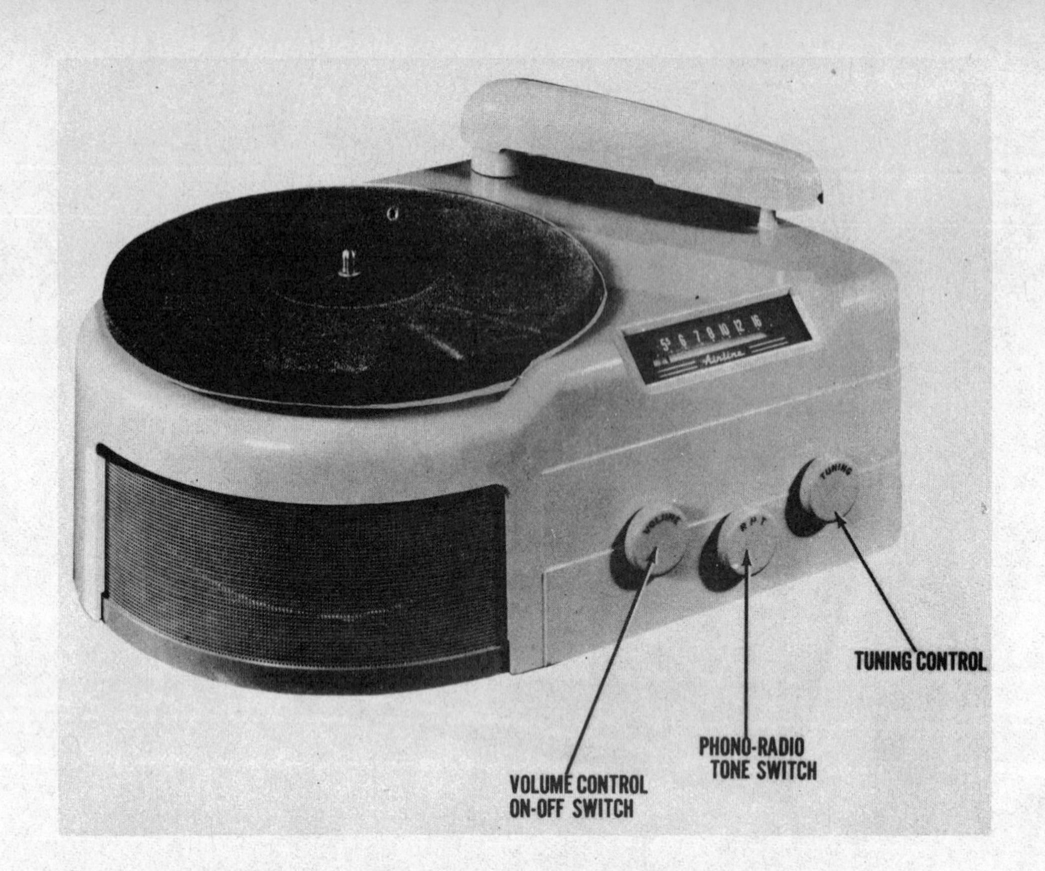

AIRLINE

MODEL PICTURED
74BR-1514B

AC/DC two-band superheterodyne receiver with loop antenna

TUBES
6

POWER SUPPLY
105-125 volts AC/DC

TUNING RANGE
530-1600KC, 9-12MC

MFR/SUPPLIER
Montgomery Ward and Co.

PHOTOFACT SET
24-4

PUBLISHED
1947

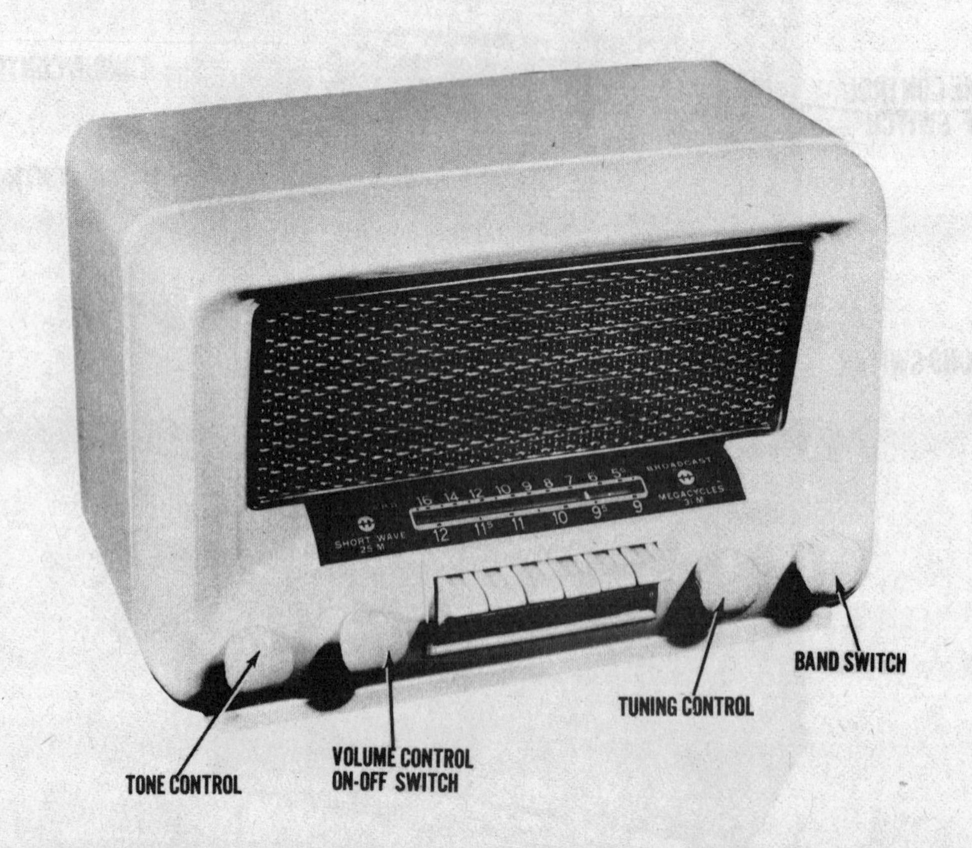

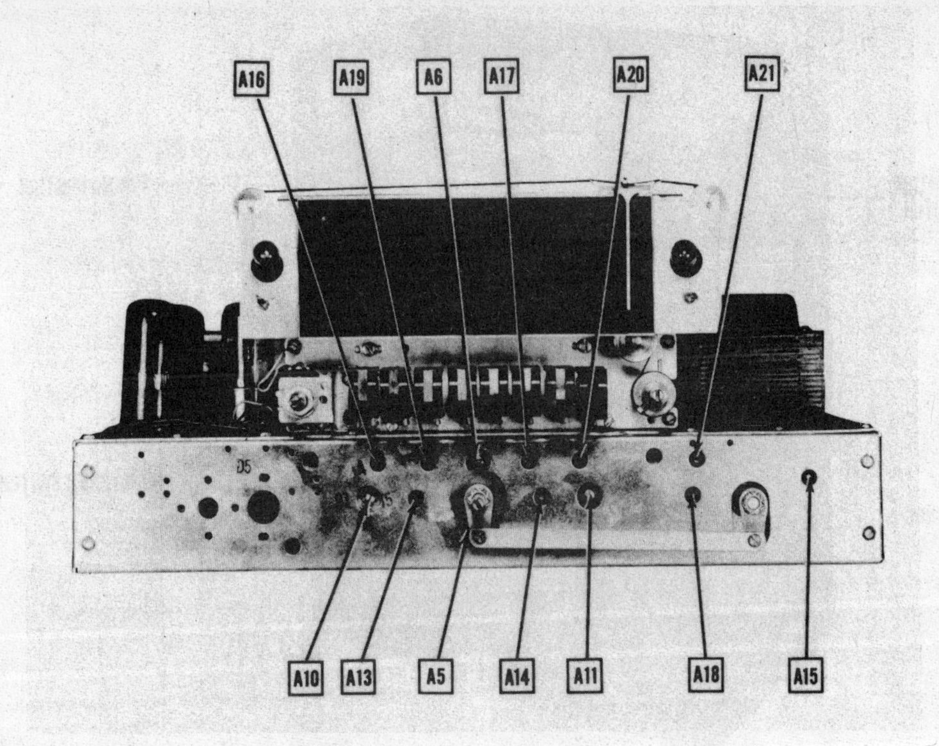

AIRLINE

MODEL PICTURED
74BR-2701A

AC operated combo radio-phono five-band superheterodyne receiver

TUBES
8

POWER SUPPLY
105-125 volts AC

TUNING RANGE
540-1600KC, plus four short-wave bands

MFR/SUPPLIER
Montgomery Ward and Co.

PHOTOFACT SET
24-5

PUBLISHED
1947

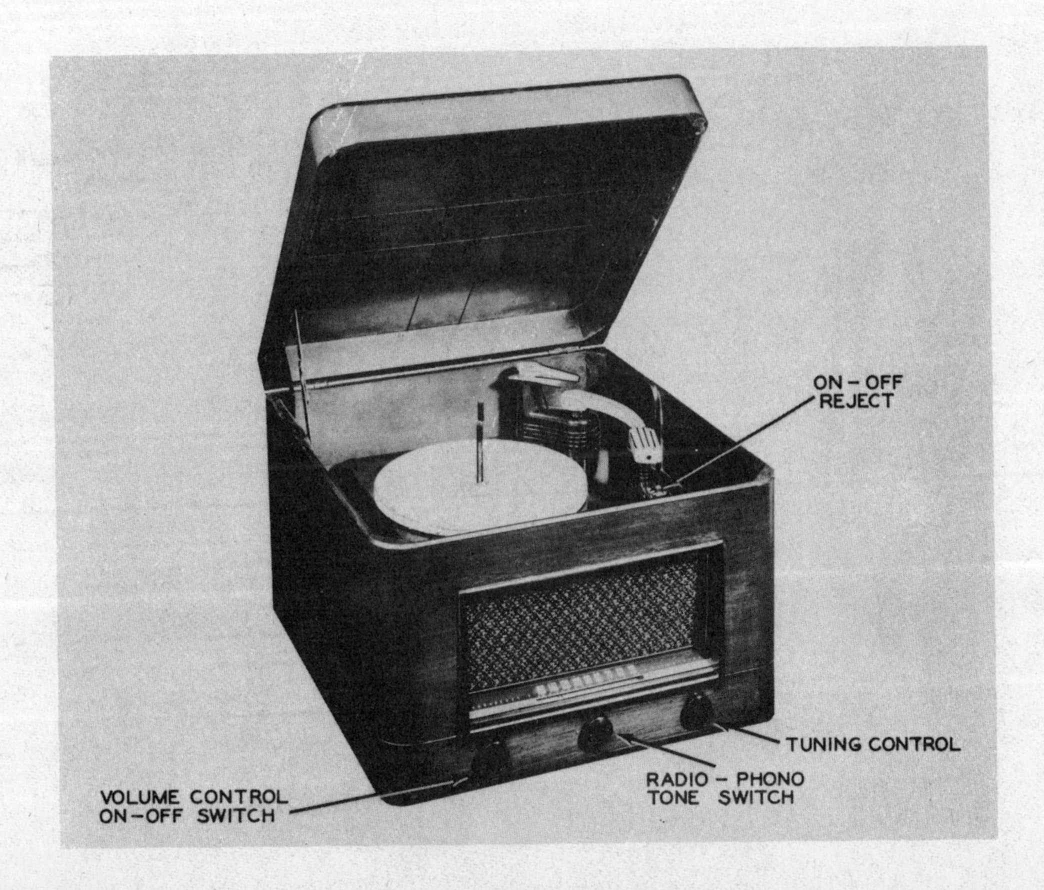

AIRLINE

MODEL PICTURED
74WG-2002A

AC operated radio-phono combo superheterodyne receiver with loop antenna

TUBES
6

POWER SUPPLY
105-125 volts AC

TUNING RANGE
540-1600KC

MFR/SUPPLIER
Montgomery Ward and Co.

PHOTOFACT SET
26-4

PUBLISHED
1947

AIRLINE

MODEL PICTURED
74WG-2709A

AC operated radio-phono combo superheterodyne receiver with loop antenna

TUBES
7

POWER SUPPLY
105-125 volts AC

TUNING RANGE
540-1600KC, 5.75-18.3MC

MFR/SUPPLIER
Montgomery Ward and Co.

PHOTOFACT SET
26-5

PUBLISHED
1947

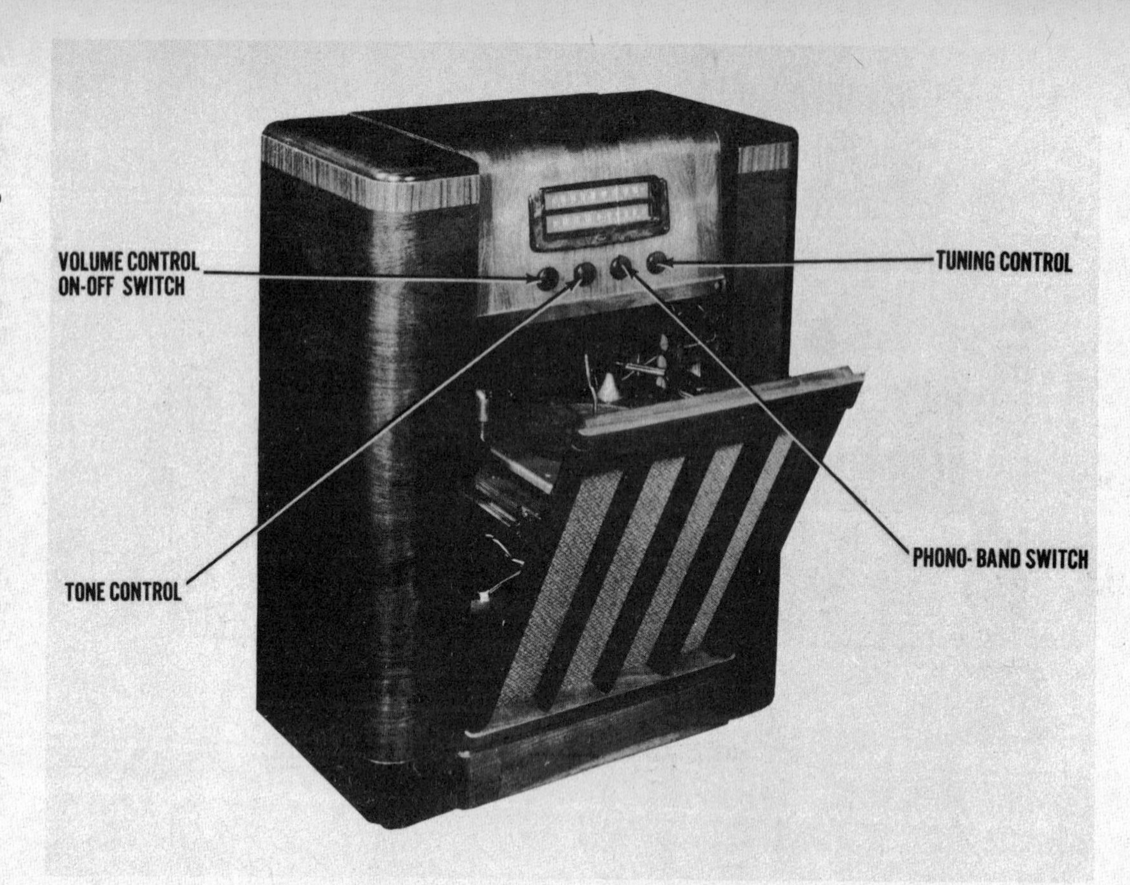

AIRLINE

MODEL PICTURED
74WG-1510A

AC/DC operated superheterodyne receiver with loop antenna

TUBES
6

POWER SUPPLY
105-125 volts AC/DC, .200 amp @ 117 volts AC

TUNING RANGE
540-1600KC

MFR/SUPPLIER
Montgomery Ward and Co.

PHOTOFACT SET
27-1

PUBLISHED
1947

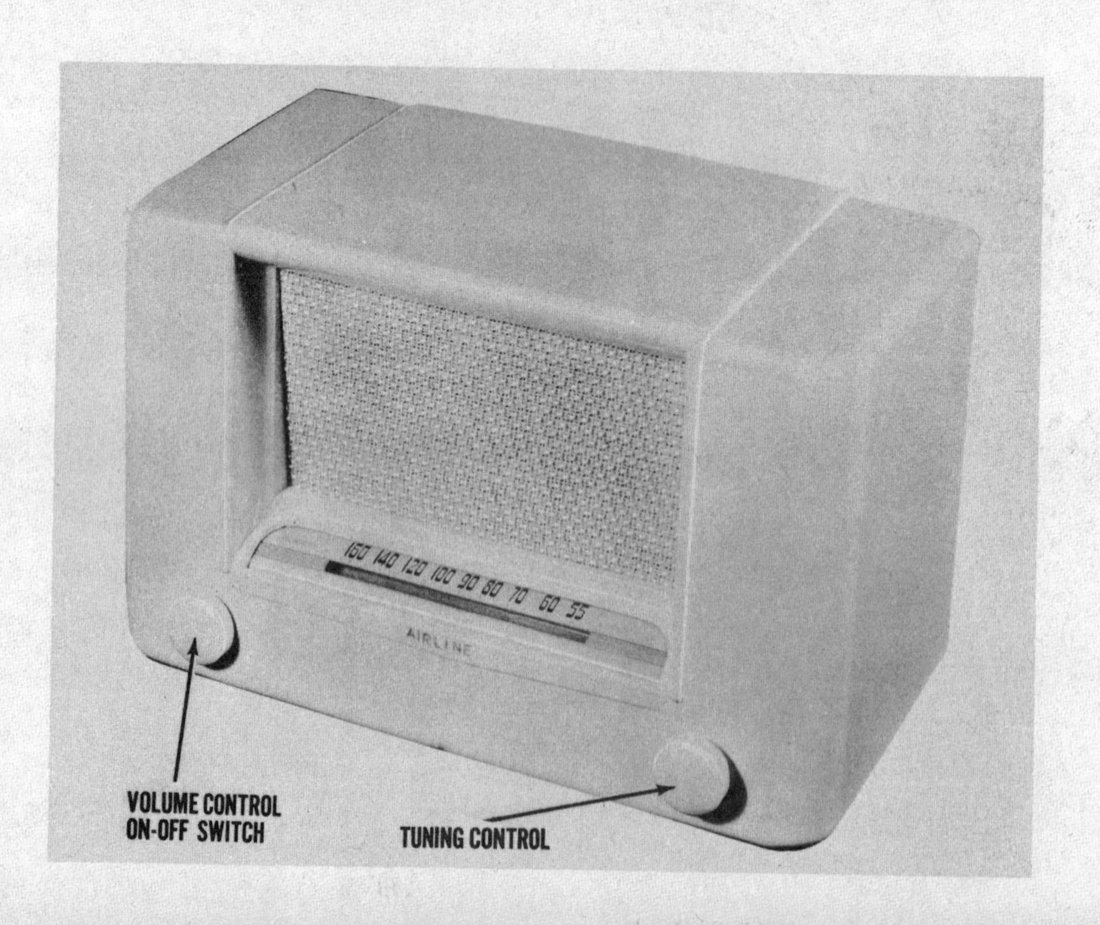

VOLUME CONTROL
ON-OFF SWITCH

TONE CONTROL
PHONO-RADIO SWITCH

TUNING CONTROL

BAND SWITCH

MODEL 74WG-2504A

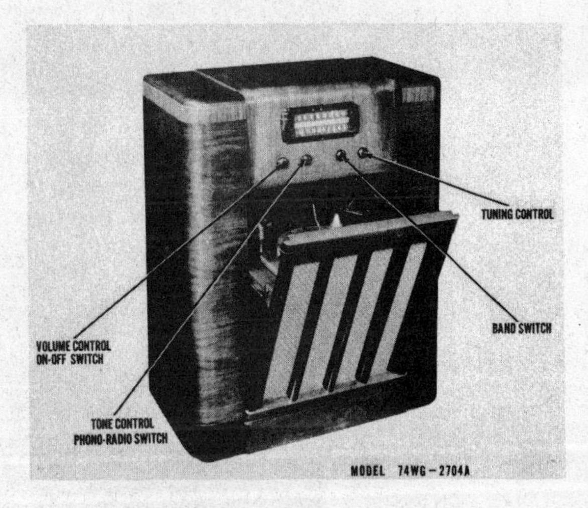

MODEL 74WG-2704A

AIRLINE

MODEL PICTURED
74WG-2504A, 74WG-2704A

AC operated two-band superheterodyne receiver with loop antenna

TUBES
6

POWER SUPPLY
105-125 volts AC

TUNING RANGE
540-1600KC, 5.75-18.3MC

MFR/SUPPLIER
Montgomery Ward and Co.

PHOTOFACT SET
28-1

PUBLISHED
1947

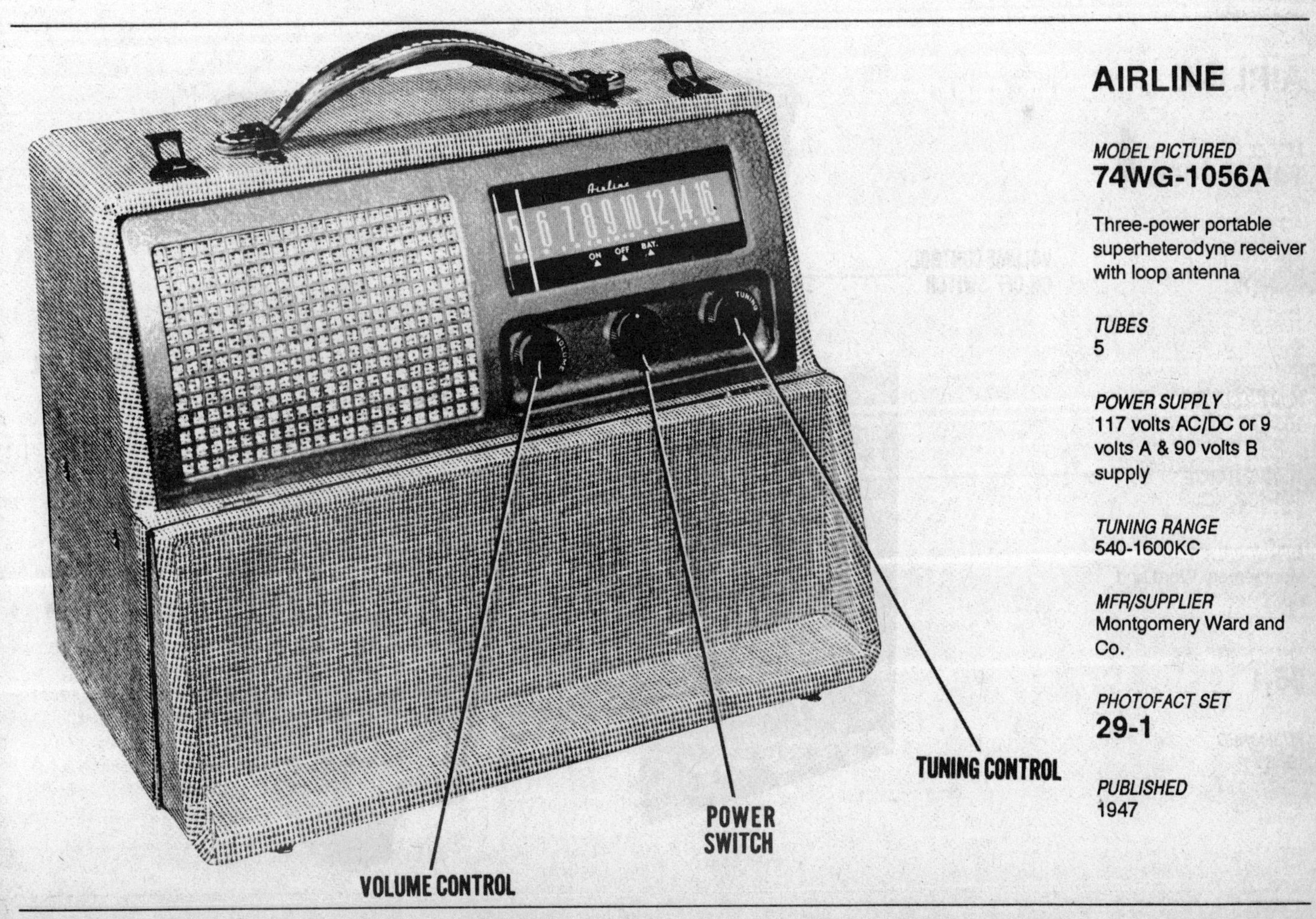

AIRLINE

MODEL PICTURED
74WG-1056A

Three-power portable superheterodyne receiver with loop antenna

TUBES
5

POWER SUPPLY
117 volts AC/DC or 9 volts A & 90 volts B supply

TUNING RANGE
540-1600KC

MFR/SUPPLIER
Montgomery Ward and Co.

PHOTOFACT SET
29-1

PUBLISHED
1947

AIRLINE

MODEL PICTURED
74WG-1057A

Three-power portable superheterodyne receiver with loop antenna

TUBES
4

POWER SUPPLY
105-125 volts AC/DC or 1.5 volts A supply & 60 volts B supply in pack form

TUNING RANGE
540-1600KC

MFR/SUPPLIER
Montgomery Ward and Co.

PHOTOFACT SET
32-2

PUBLISHED
1948

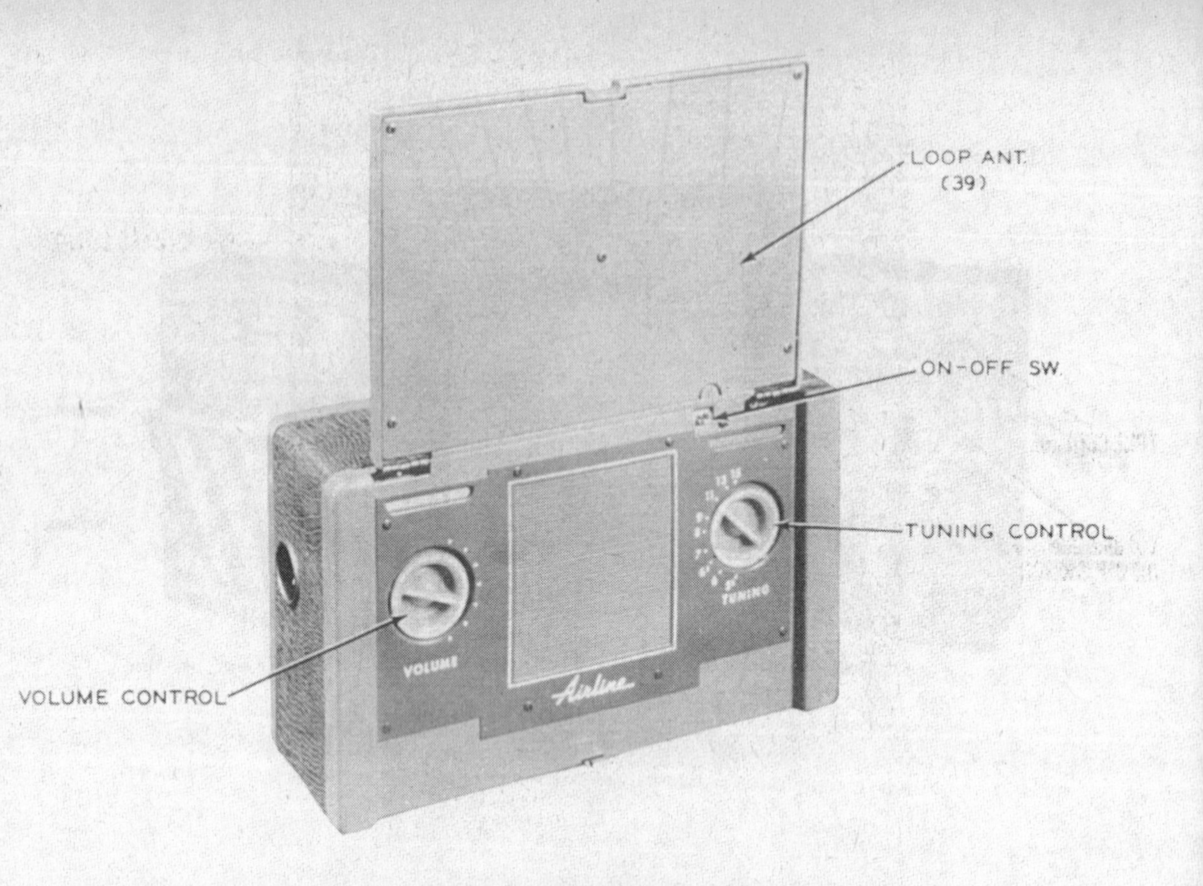

AIRLINE

MODEL PICTURED
74KR2706B

AC operated phono-radio superheterodyne receiver with loop antenna

TUBES
6

POWER SUPPLY
105-125 volts AC

TUNING RANGE
540-1620KC

MFR/SUPPLIER
Montgomery Ward and Co.

PHOTOFACT SET
35-1

PUBLISHED
1948

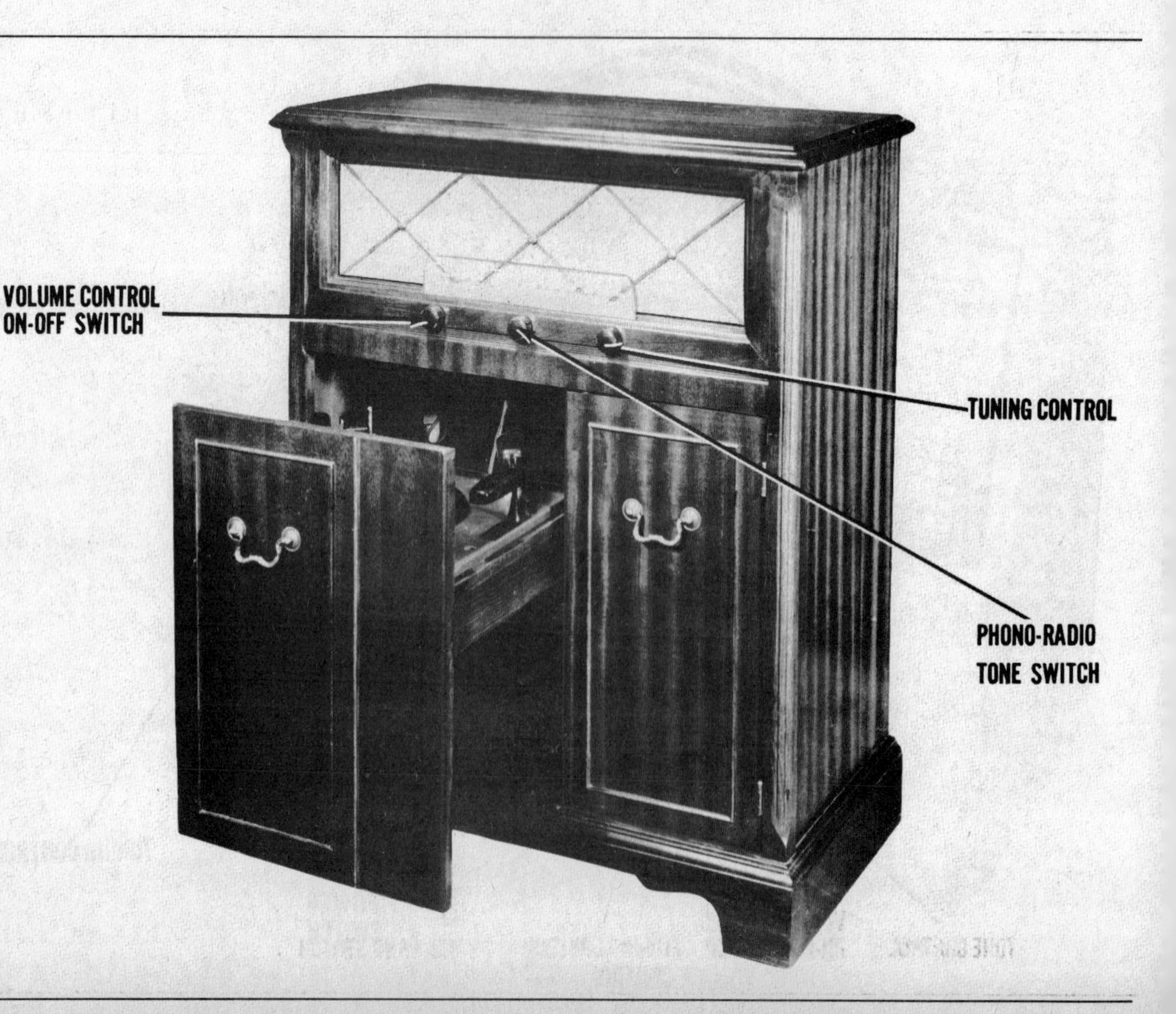

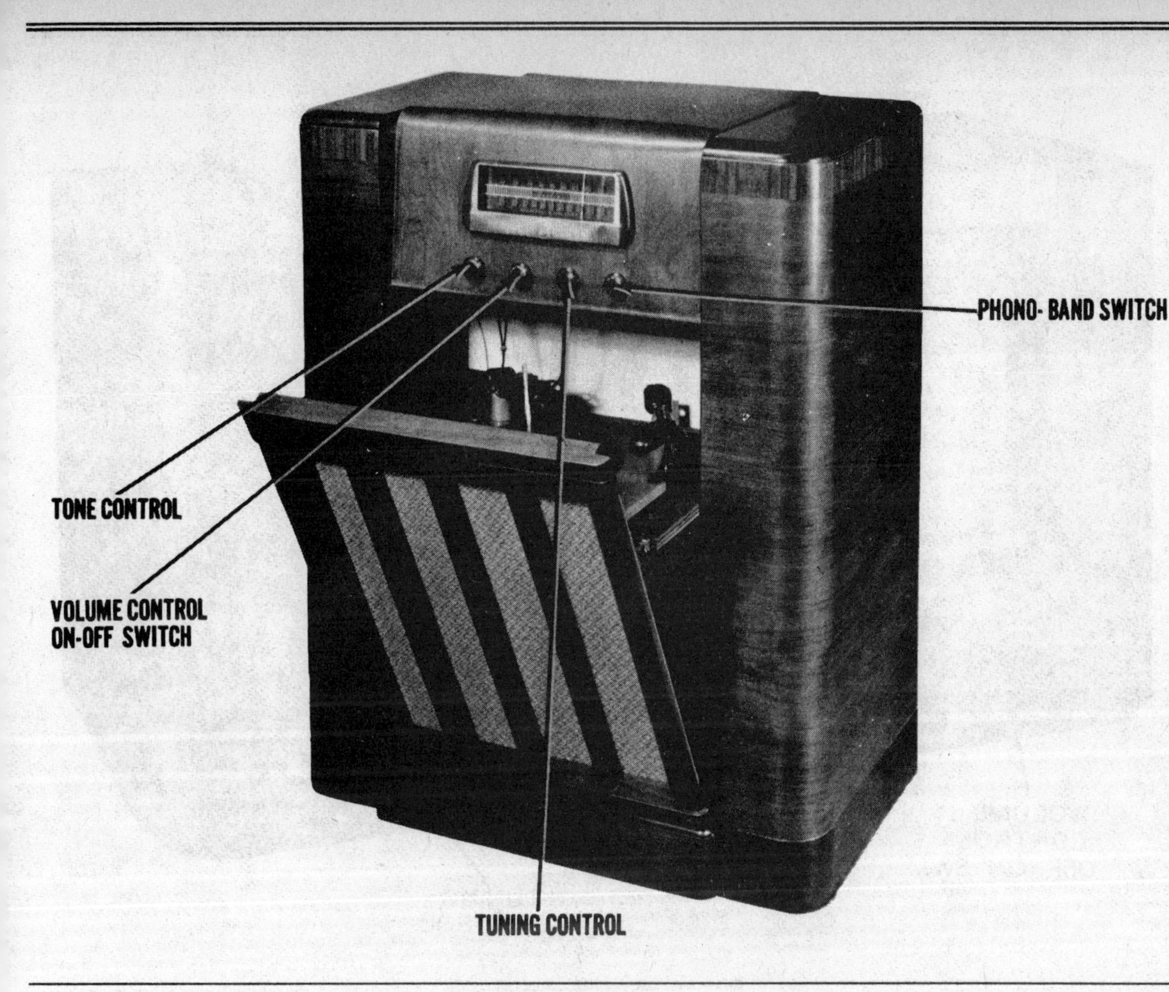

AIRLINE

MODEL PICTURED
84WG-2714A

AC operated combo phono-radio AM-FM superheterodyne receiver

TUBES
7

POWER SUPPLY
105-125 volts AC

TUNING RANGE
540-1600KC, 88-108MC

MFR/SUPPLIER
Montgomery Ward and Co.

PHOTOFACT SET
36-2

PUBLISHED
1948

AIRLINE

MODEL PICTURED
84WG-2015A

AC operated combo phono-radio AM-FM superheterodyne receiver

TUBES
7

POWER SUPPLY
105-125 volts AC

TUNING RANGE
540-1600KC, 88-108MC

MFR/SUPPLIER
Montgomery Ward and Co.

PHOTOFACT SET
38-1

PUBLISHED
1948

AIRLINE

MODEL PICTURED
74KR-1210A

Three-power operated superheterodyne receiver

TUBES
4

POWER SUPPLY
105-125 volts AC/DC or 9 volts A & 90 volts B in pack form

TUNING RANGE
540-1620KC

MFR/SUPPLIER
Montgomery Ward and Co.

PHOTOFACT SET
41-1

PUBLISHED
1948

AIRLINE

MODEL PICTURED
84WG-1060A

Three-power operated portable superheterodyne receiver with loop antenna

TUBES
4

POWER SUPPLY
105-125 volts AC/DC or 6 volts A supply & 67.5 volts B supply

TUNING RANGE
540-1600KC

MFR/SUPPLIER
Montgomery Ward and Co.

PHOTOFACT SET
42-1

PUBLISHED
1948

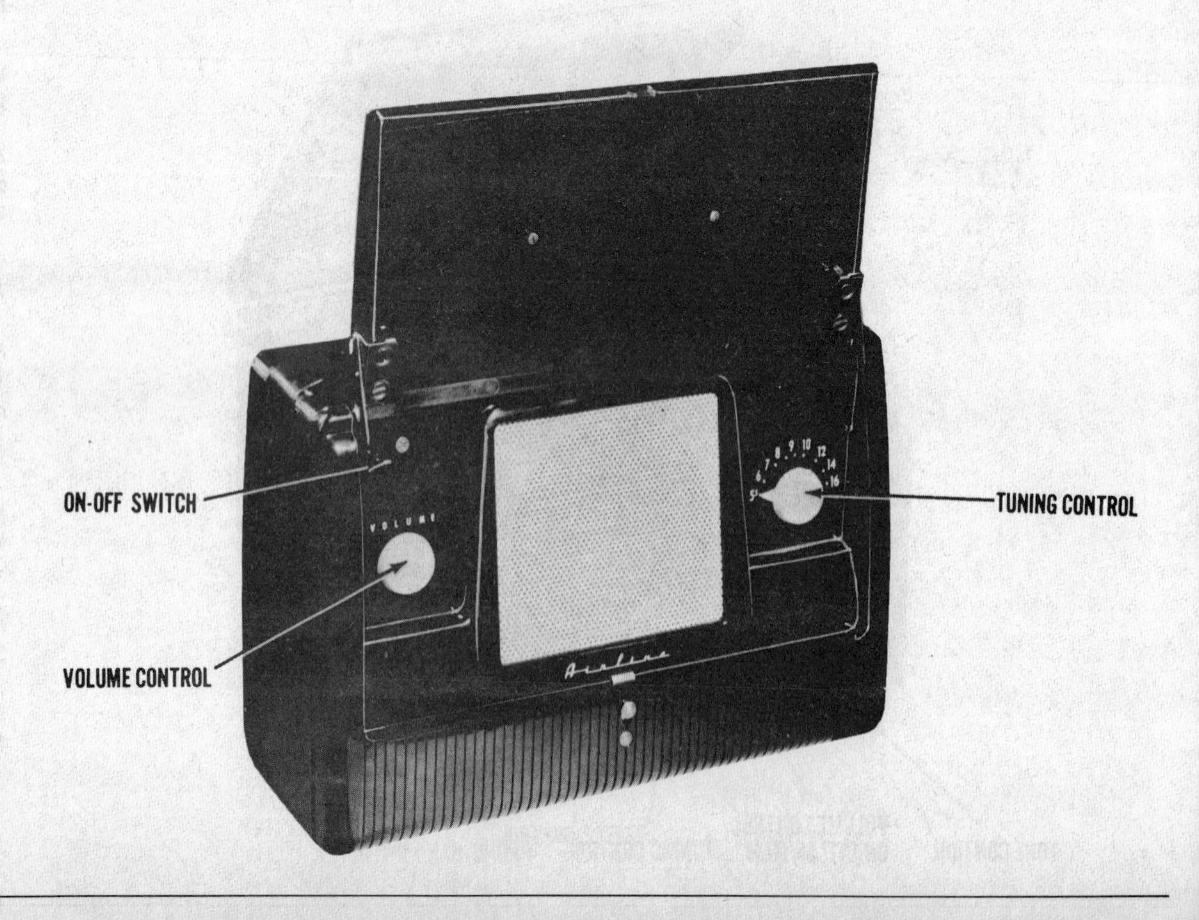

AIRLINE

MODEL PICTURED
74KR-2713A

AC operated superheterodyne receiver with loop antenna

TUBES
6

POWER SUPPLY
105-125 volts AC

TUNING RANGE
540-1620KC

MFR/SUPPLIER
Montgomery Ward and Co.

PHOTOFACT SET
43-2

PUBLISHED
1948

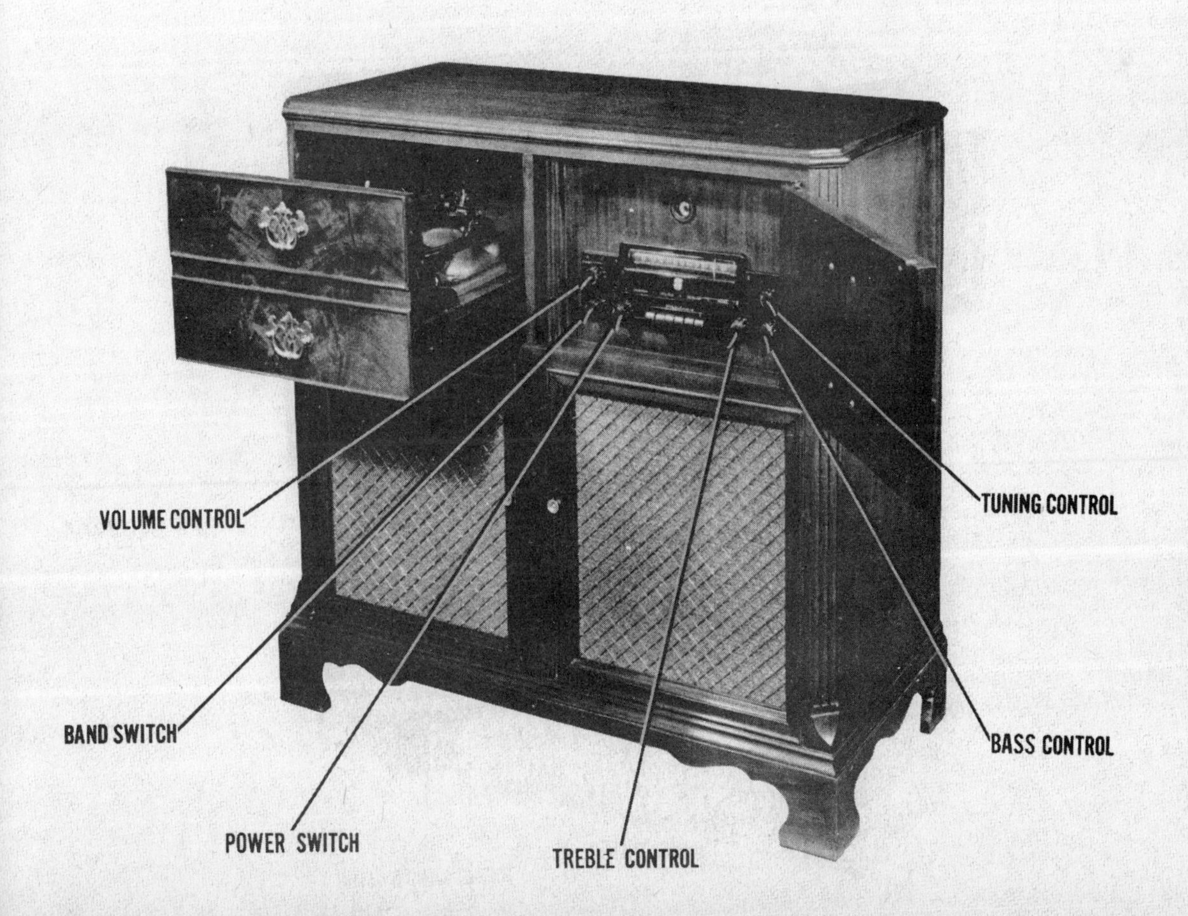

AIRLINE

MODEL PICTURED
84WG-2712A

AC operated combo radio phono two-band AM-FM superheterodyne receiver

TUBES
12

POWER SUPPLY
105-125 volts AC

TUNING RANGE
540-1600KC, 88-108MC, 5.8-18MC

MFR/SUPPLIER
Montgomery Ward and Co.

PHOTOFACT SET
43-3

PUBLISHED
1948

AIRLINE

MODEL PICTURED
84WG-2720A

AC operated phono-radio two-band AM-FM superheterodyne receiver with loop antenna

TUBES
10

POWER SUPPLY
105-125 volts AC

TUNING RANGE
540-1600KC, 88-108MC, 5.8-18MC

MFR/SUPPLIER
Montgomery Ward and Co.

PHOTOFACT SET
45-5

PUBLISHED
1948

AIRLINE

MODEL PICTURED
84WG-2721A

AC operated combo phono-radio AM/FM superheterodyne receiver with loop antenna

TUBES
7

POWER SUPPLY
105-125 volts AC

TUNING RANGE
540-1600KC

MFR/SUPPLIER
Montgomery Ward and Co.

PHOTOFACT SET
46-3

PUBLISHED
1948

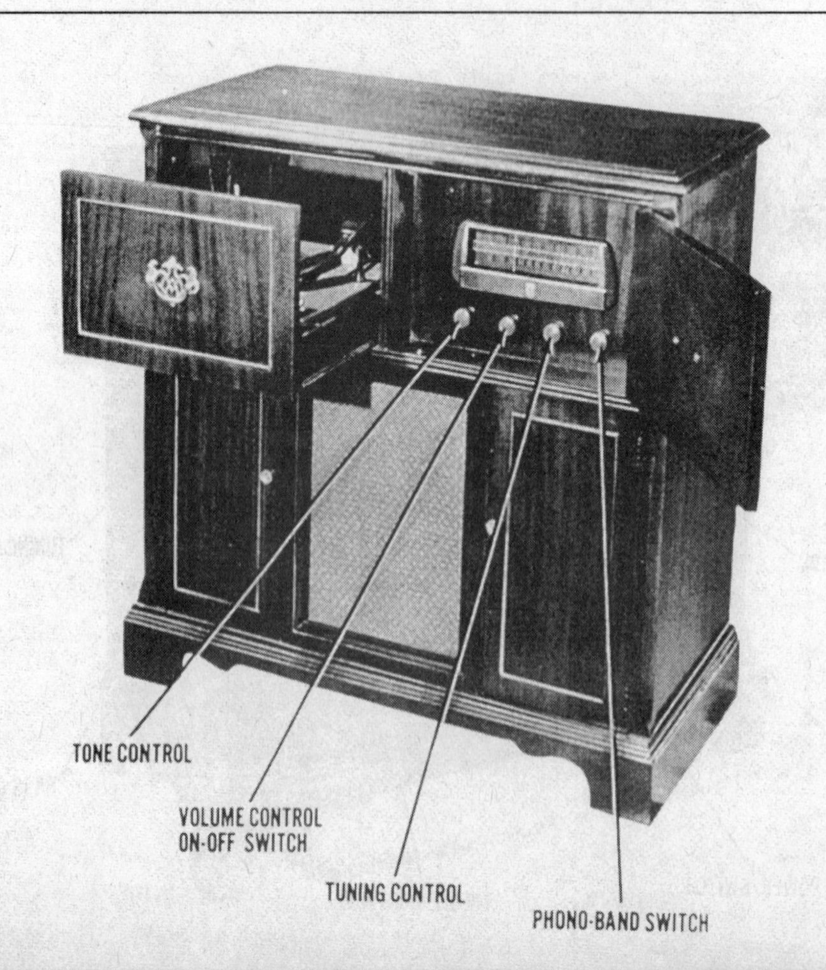

AIRLINE

MODEL PICTURED
74WG-2004A

AC operated radio-phono combo superheterodyne receiver with loop antenna

TUBES
5

POWER SUPPLY
105-125 volts AC

TUNING RANGE
540-1600KC

MFR/SUPPLIER
Montgomery Ward and Co.

PHOTOFACT SET
47-2

PUBLISHED
1947

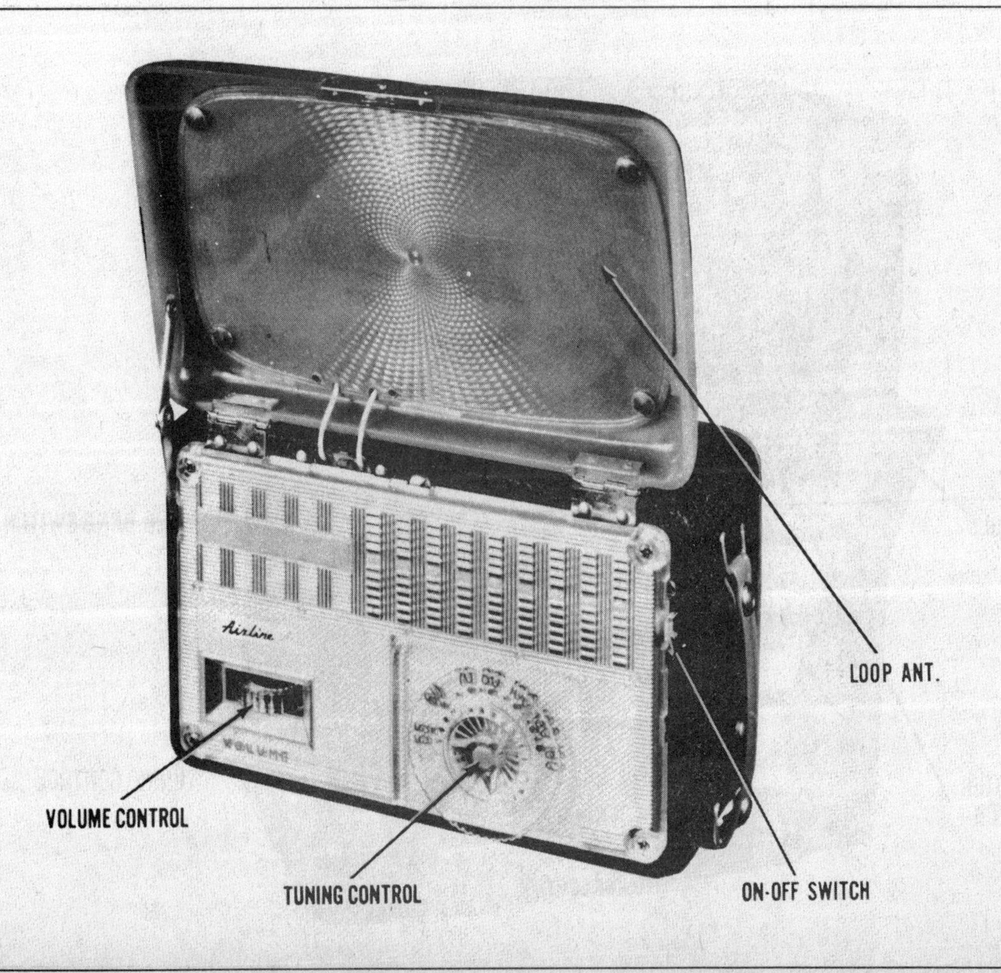

AIRLINE

MODEL PICTURED
84GCB-1062A

Battery operated portable superheterodyne receiver with loop antenna

TUBES
4

POWER SUPPLY
1.5 volts A & 67.5 volts B supply

TUNING RANGE
540-1650KC

MFR/SUPPLIER
Montgomery Ward and Co.

PHOTOFACT SET
52-26

PUBLISHED
1948

AIRLINE

MODEL PICTURED
84KR-1520A

AC/DC operated TRF receiver

TUBES
4

POWER SUPPLY
105-125 volts AC/DC

TUNING RANGE
540-1620KC

MFR/SUPPLIER
Montgomery Ward and Co.

PHOTOFACT SET
56-4

PUBLISHED
1949

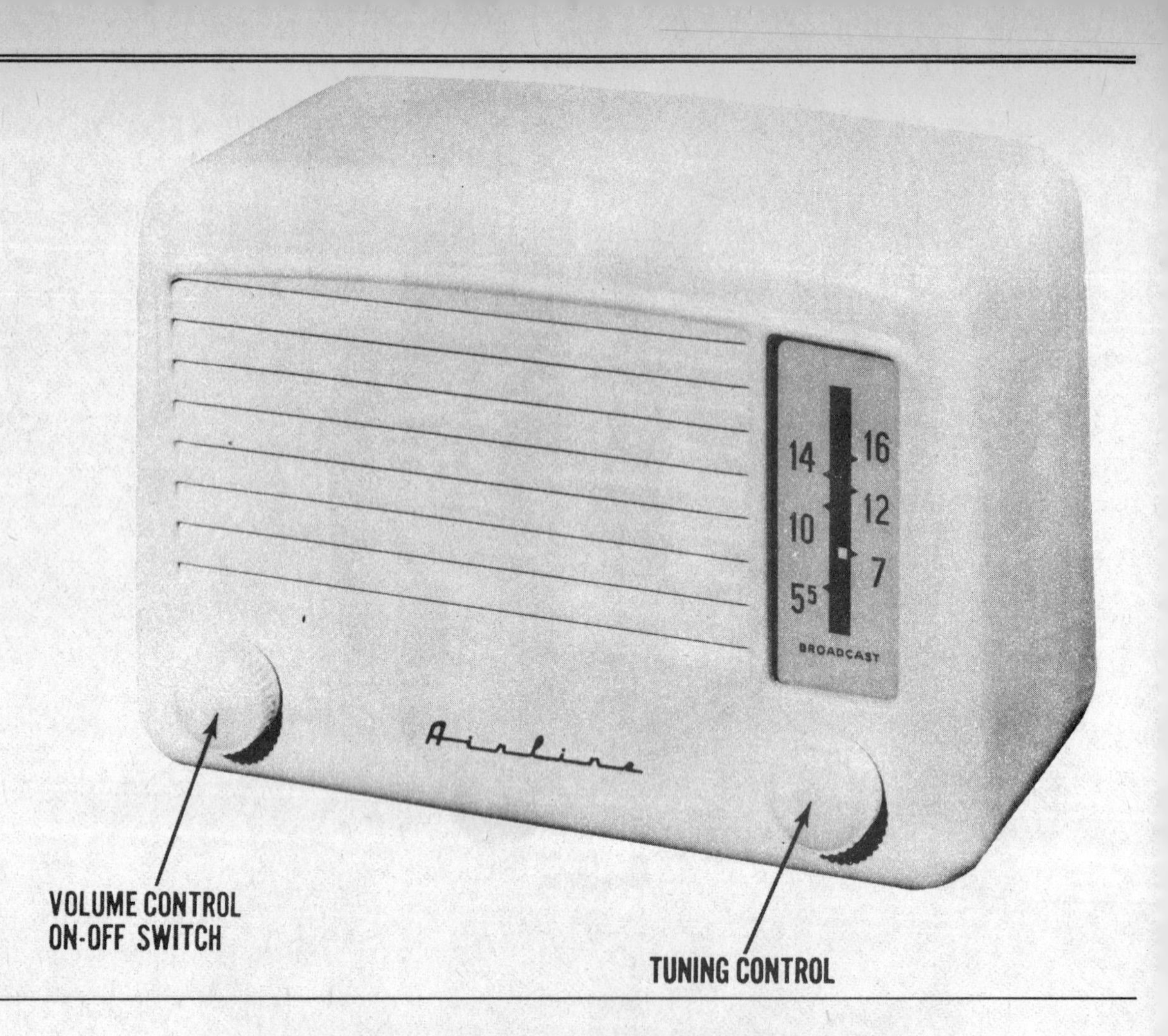

AIRLINE

MODEL PICTURED
84WG-2714F

AC operated phono-radio AM-FM superheterodyne receiver with loop antenna

TUBES
7

POWER SUPPLY
105-125 volts AC

TUNING RANGE
540-1600KC, 88-108MC

MFR/SUPPLIER
Montgomery Ward and Co.

PHOTOFACT SET
56-5

PUBLISHED
1949

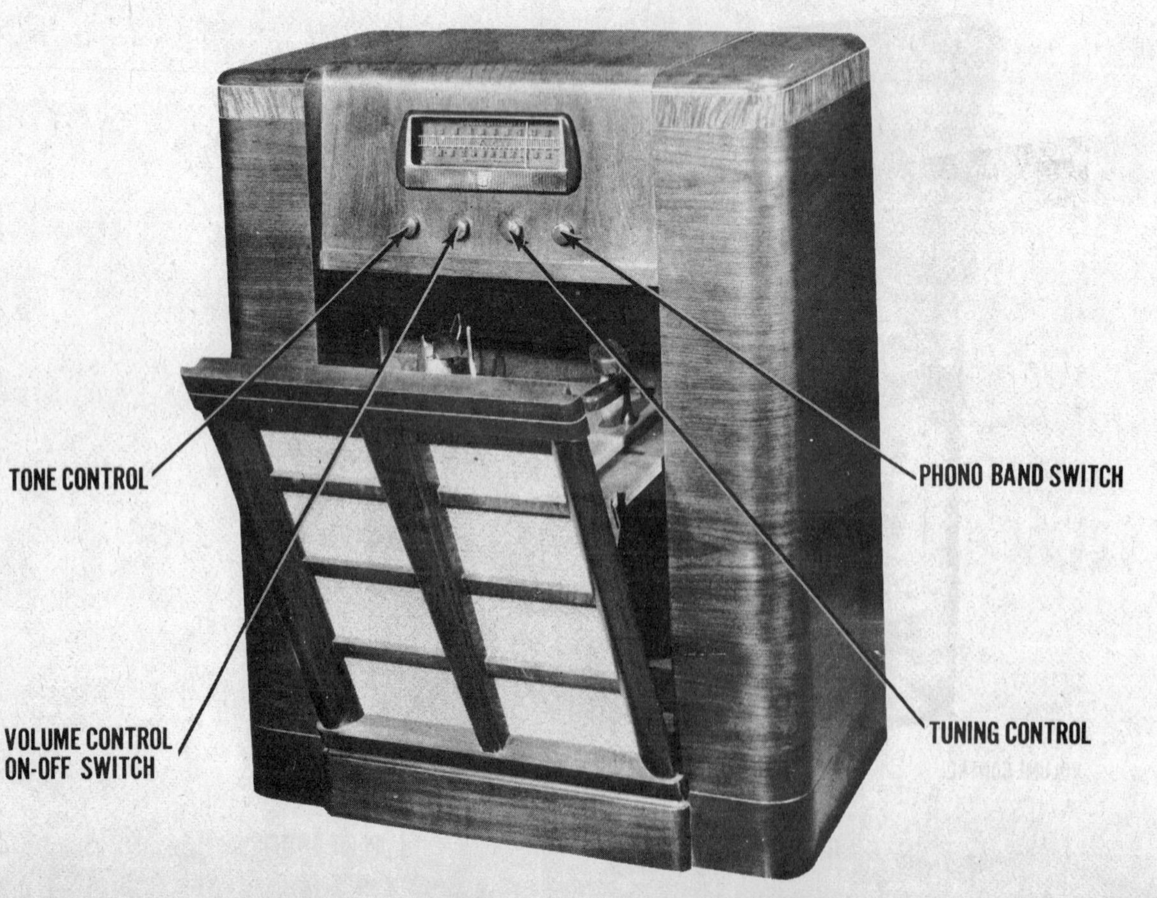

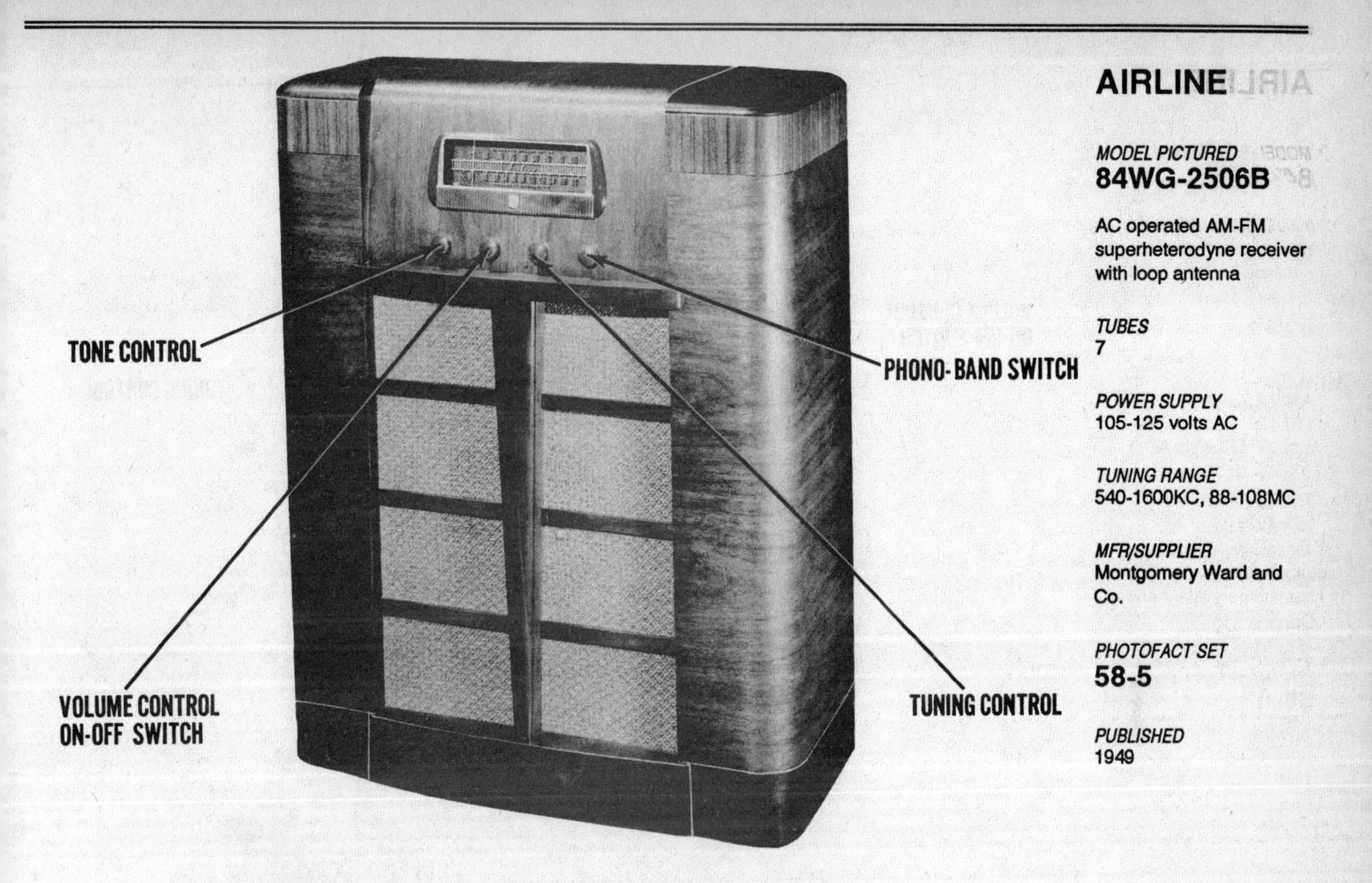

AIRLINE

MODEL PICTURED
84WG-2506B

AC operated AM-FM superheterodyne receiver with loop antenna

TUBES
7

POWER SUPPLY
105-125 volts AC

TUNING RANGE
540-1600KC, 88-108MC

MFR/SUPPLIER
Montgomery Ward and Co.

PHOTOFACT SET
58-5

PUBLISHED
1949

AIRLINE

MODEL PICTURED
94HA1528C

AC/DC operated superheterodyne receiver with loop antenna

TUBES
5

POWER SUPPLY
105-125 volts AC/DC

TUNING RANGE
540-1620KC

MFR/SUPPLIER
Montgomery Ward and Co.

PHOTOFACT SET
67-3

PUBLISHED
1949

AIRLINE

MODEL PICTURED
84KR2511A

AC/DC operated superheterodyne receiver with loop antenna

TUBES
5

POWER SUPPLY
110-120 volts AC, .24 amp @ 117 volts AC

TUNING RANGE
540-1620KC

MFR/SUPPLIER
Montgomery Ward and Co.

PHOTOFACT SET
68-4

PUBLISHED
1949

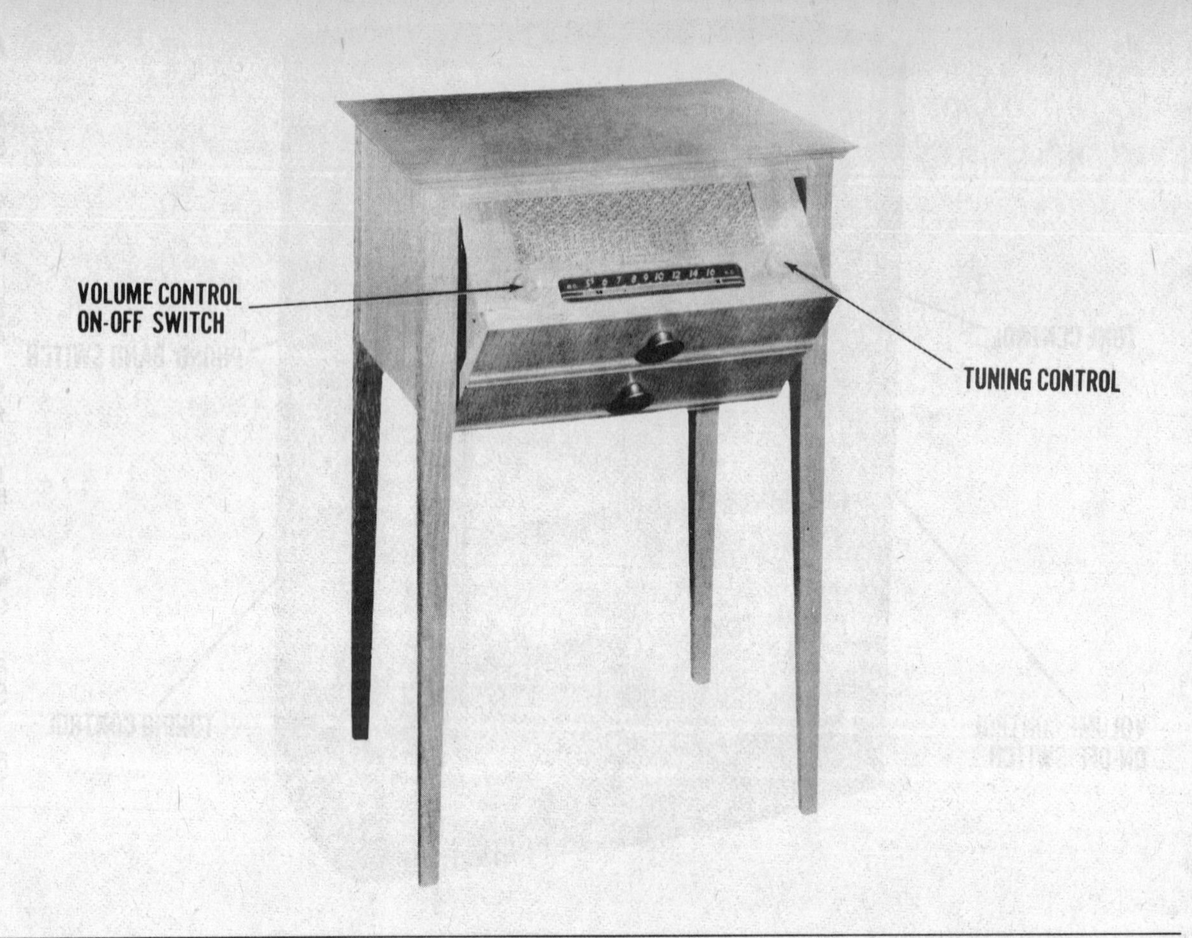

AIRLINE

MODEL PICTURED
84HA-1810C

AC/DC operated AM-FM superheterodyne receiver with loop antenna

TUBES
7

POWER SUPPLY
105-125 volts AC/DC, .26 amp @ 117 volts AC

TUNING RANGE
540-1600KC, 88-108MC

MFR/SUPPLIER
Montgomery Ward and Co.

PHOTOFACT SET
69-2

PUBLISHED
1949

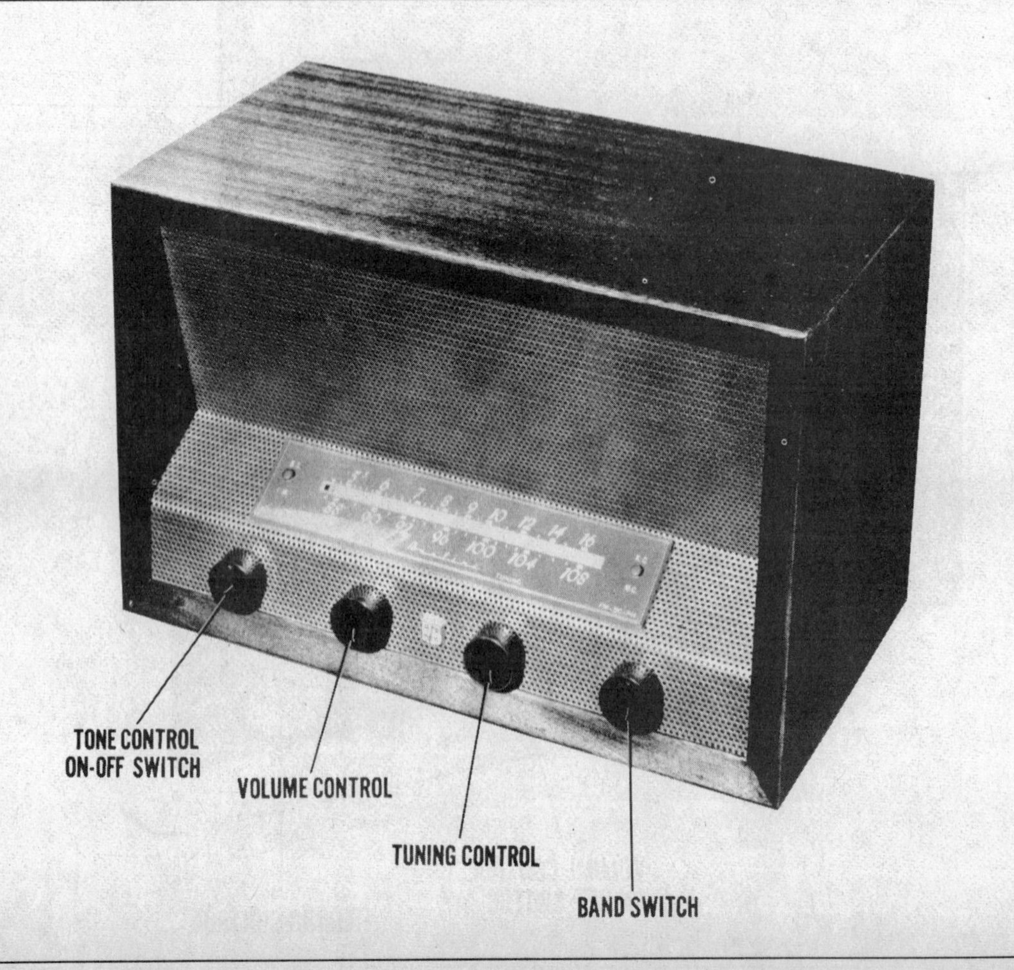

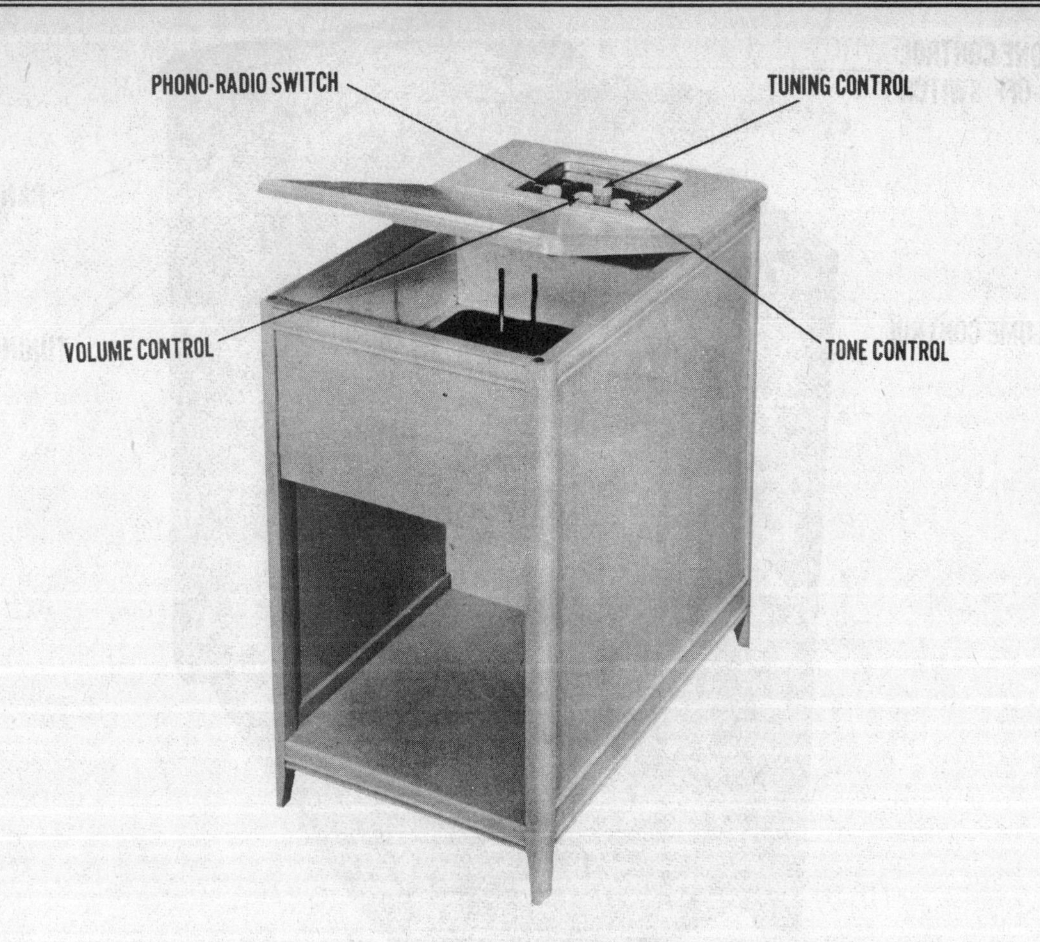

AIRLINE

MODEL PICTURED
84GSE2731A

AC operated combo phono-radio superheterodyne receiver with loop antenna

TUBES
5

POWER SUPPLY
105-125 volts AC

TUNING RANGE
530-1650KC

MFR/SUPPLIER
Montgomery Ward and Co.

PHOTOFACT SET
70-1

PUBLISHED
1949

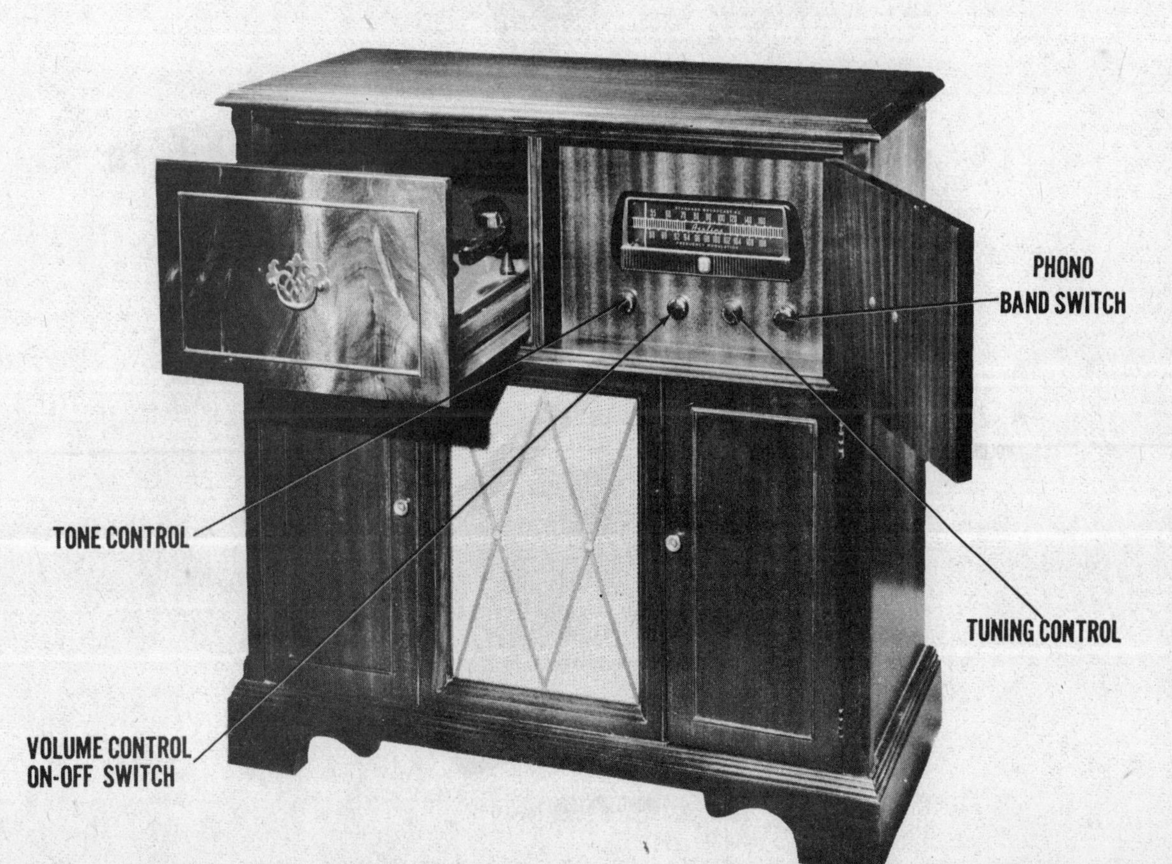

AIRLINE

MODEL PICTURED
94WG-2742A

AC operated combo phono-radio AM/FM superheterodyne receiver with loop antenna

TUBES
7

POWER SUPPLY
105-125 volts AC

TUNING RANGE
540-1600KC, 88-108MC

MFR/SUPPLIER
Montgomery Ward and Co.

PHOTOFACT SET
71-5

PUBLISHED
1949

AIRLINE

MODEL PICTURED
94GSE-2735A

AC operated combo phono-radio AM/FM superheterodyne receiver with loop antenna

TUBES
8

POWER SUPPLY
105-125 volts AC

TUNING RANGE
530-1650KC, 87-109MC

MFR/SUPPLIER
Montgomery Ward and Co.

PHOTOFACT SET
72-3

PUBLISHED
1949

AIRLINE

MODEL PICTURED
94WG-1059A

Three-power operated portable superheterodyne receiver with loop antenna

TUBES
5

POWER SUPPLY
105-125 volts AC/DC or 9 volts A & 90 volts B supply in pack form

TUNING RANGE
540-1600KC

MFR/SUPPLIER
Montgomery Ward and Co.

PHOTOFACT SET
75-3

PUBLISHED
1949

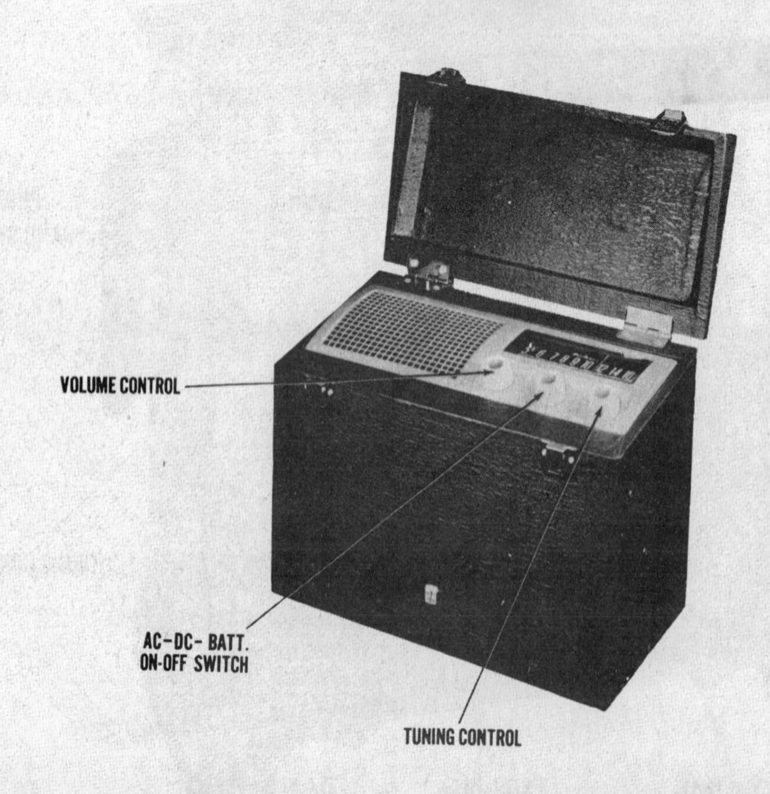

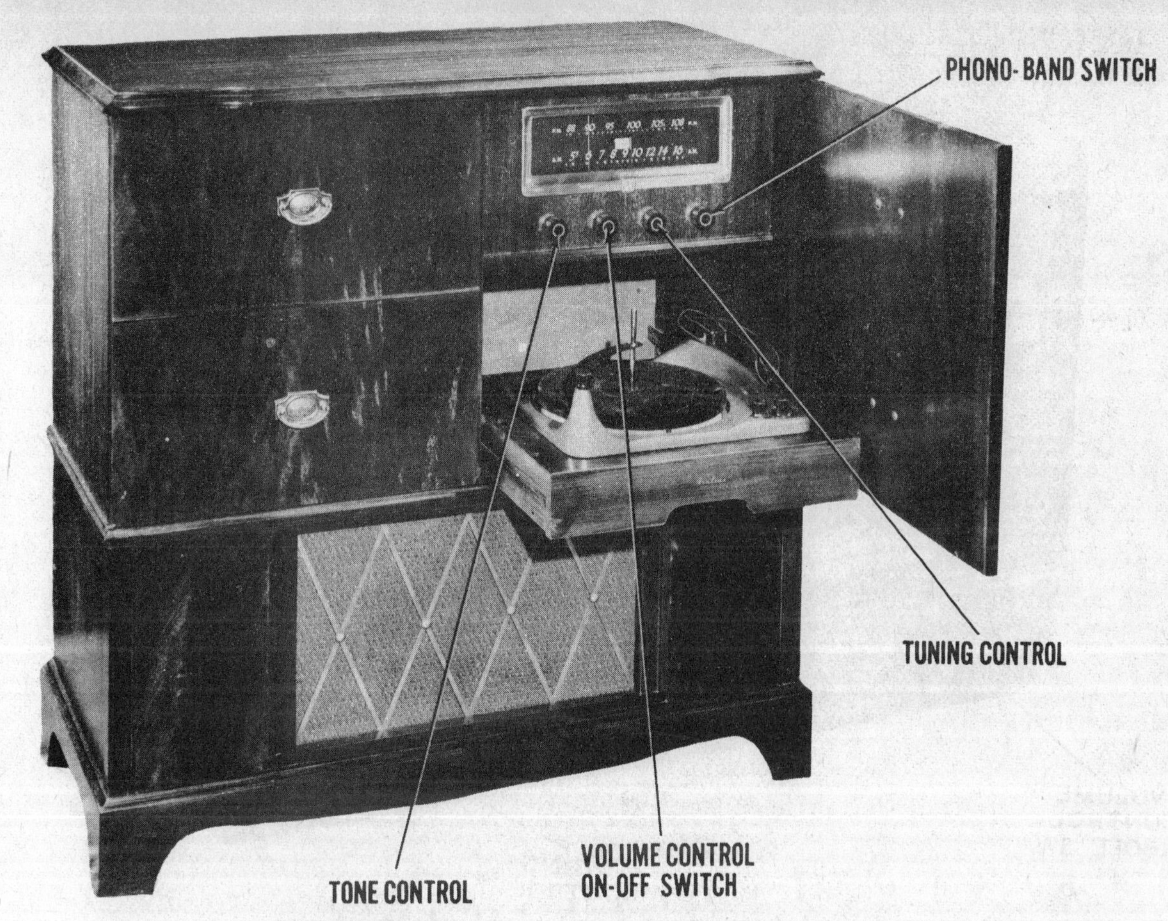

AIRLINE

MODEL PICTURED
94WG-2745A

AC operated combo phono-radio AM-FM superheterodyne receiver with loop antenna

TUBES
10

POWER SUPPLY
105-125 volts AC

TUNING RANGE
540-1600KC, 88-108MC

MFR/SUPPLIER
Montgomery Ward and Co.

PHOTOFACT SET
76-4

PUBLISHED
1949

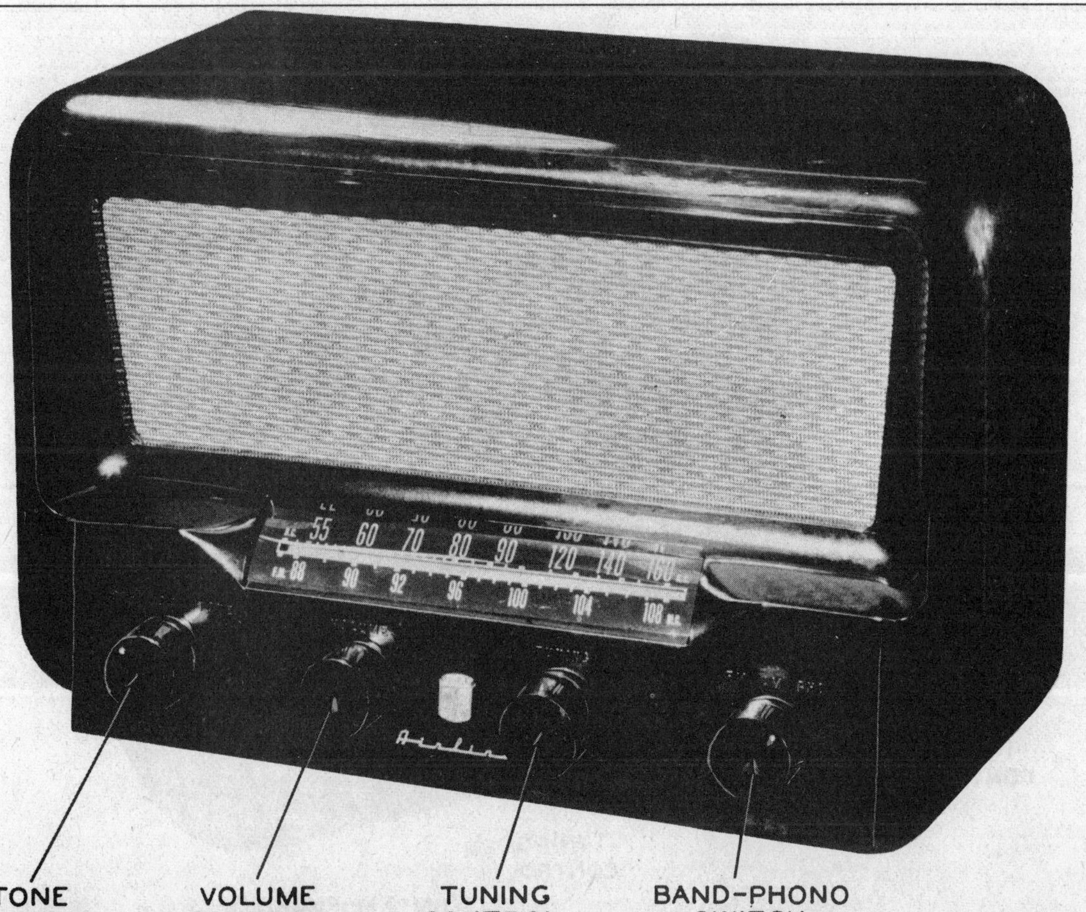

AIRLINE

MODEL PICTURED
94HA1529A

AC/DC operated AM-FM superheterodyne receiver with loop antenna

TUBES
7

POWER SUPPLY
105-125 volts AC/DC

TUNING RANGE
540-1600KC

MFR/SUPPLIER
Montgomery Ward and Co.

PHOTOFACT SET
85-2

PUBLISHED
1950

AIRLINE

MODEL PICTURED
94WG-1804D

AC/DC operated superheterodyne receiver with loop antenna

TUBES
6

POWER SUPPLY
105-125 volts AC/DC

TUNING RANGE
540-1600KC

MFR/SUPPLIER
Montgomery Ward and Co.

PHOTOFACT SET
86-2

PUBLISHED
1950

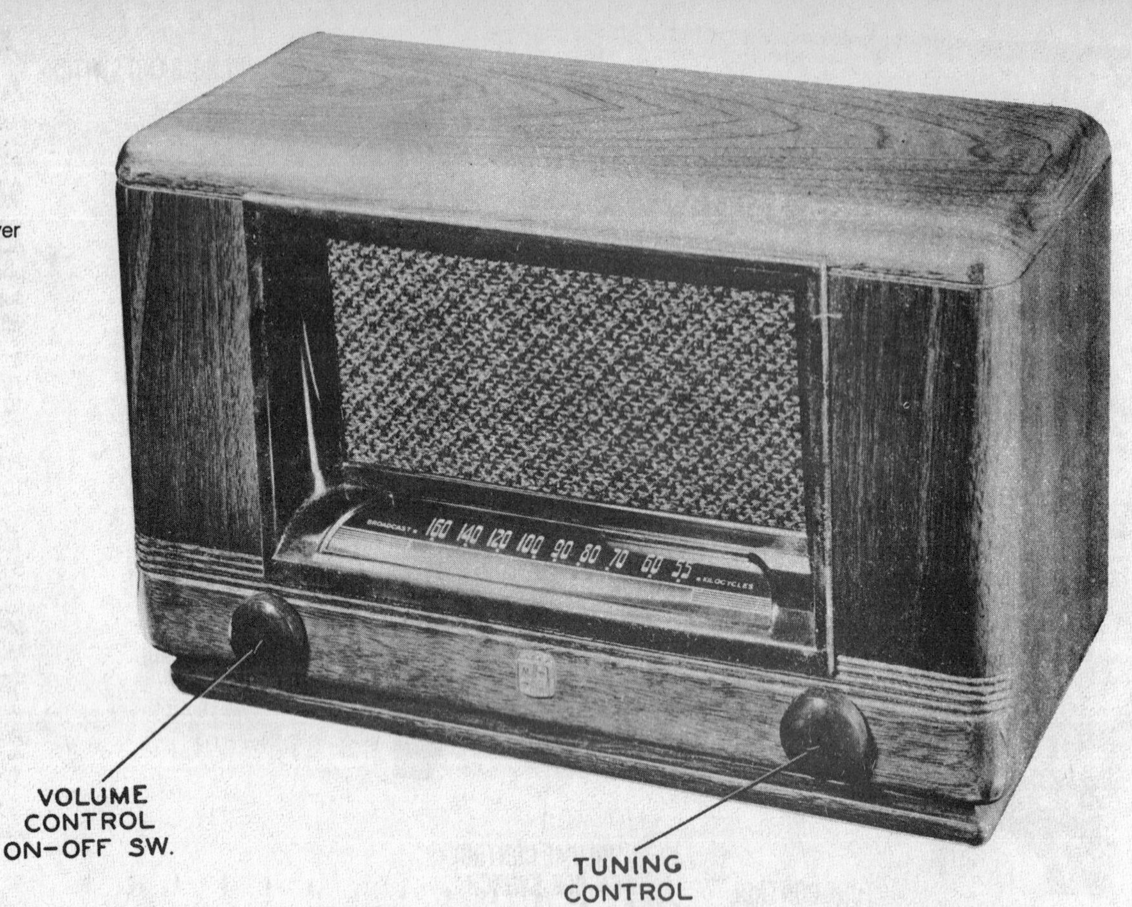

AIRLINE

MODEL PICTURED
94BR-1533A

AC/DC operated AM-FM superheterodyne receiver with loop antenna

TUBES
7

POWER SUPPLY
105-125 volts AC/DC

TUNING RANGE
545-1620KC

MFR/SUPPLIER
Montgomery Ward and Co.

PHOTOFACT SET
88-1

PUBLISHED
1950

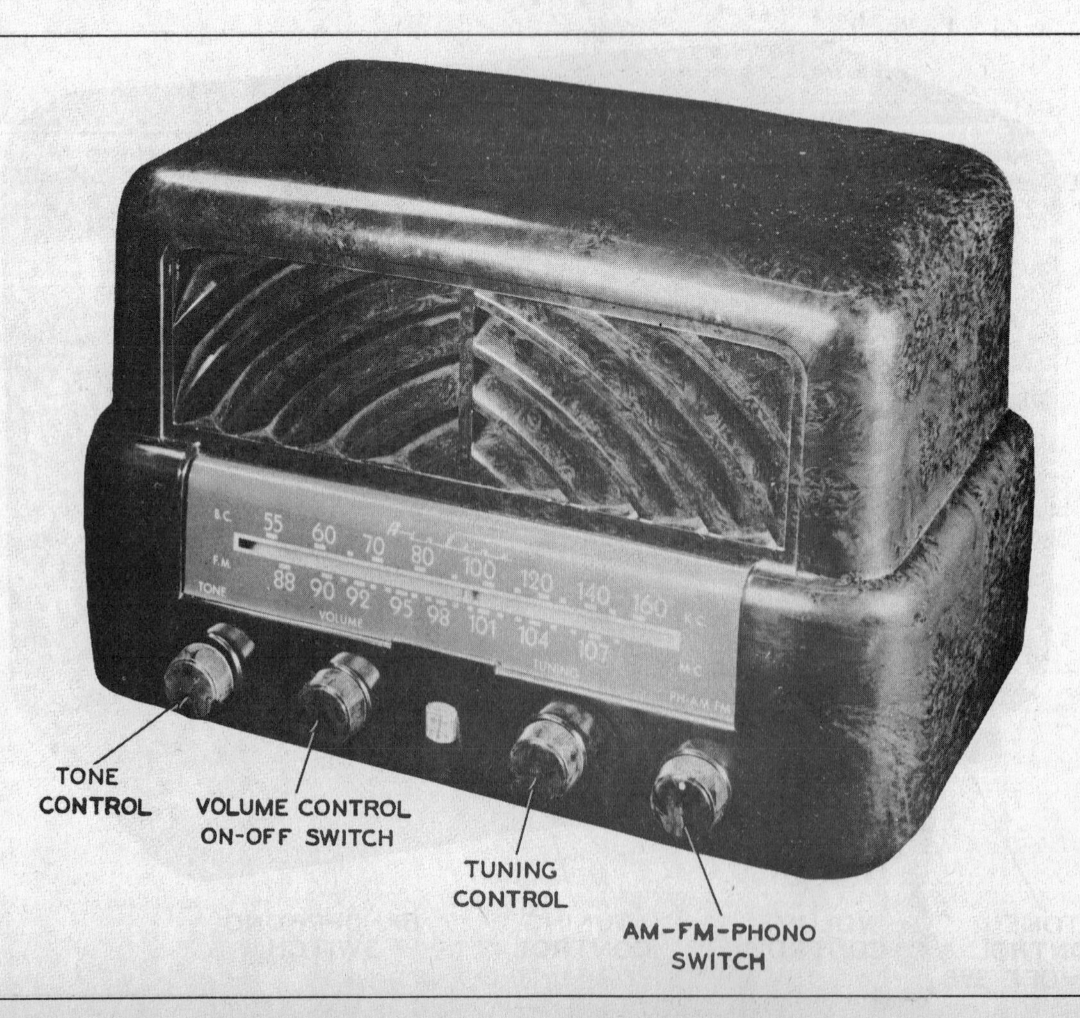

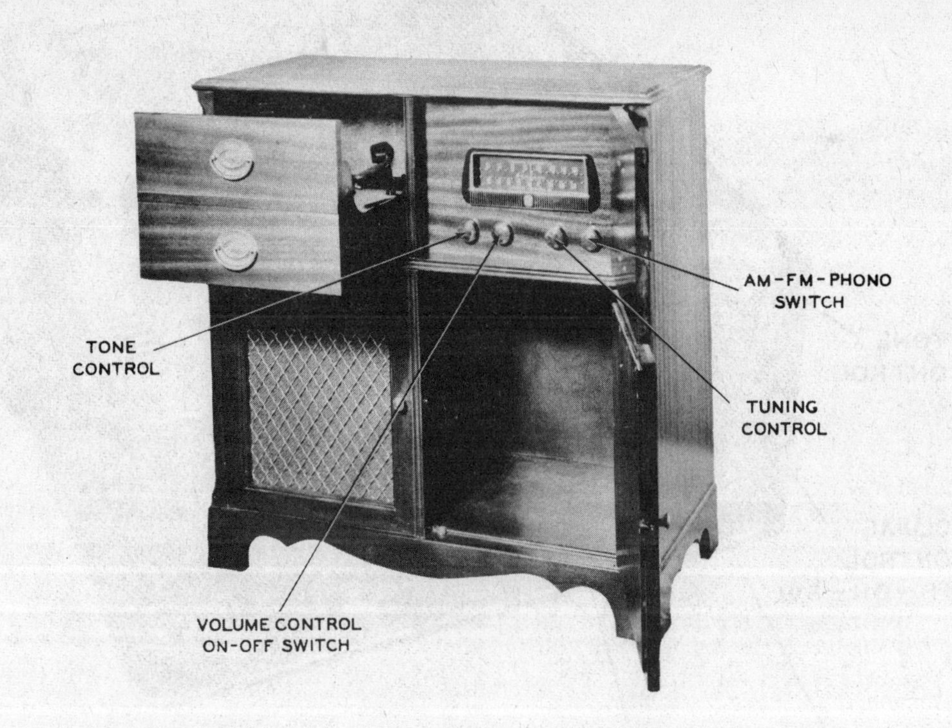

AIRLINE

MODEL PICTURED
94BR-2740A

AC operated phono-radio AM-FM superheterodyne receiver with loop antenna

TUBES
7

POWER SUPPLY
105-125 volts AC

TUNING RANGE
535-1620KC, 88-108MC

MFR/SUPPLIER
Montgomery Ward and Co.

PHOTOFACT SET
89-1

PUBLISHED
1950

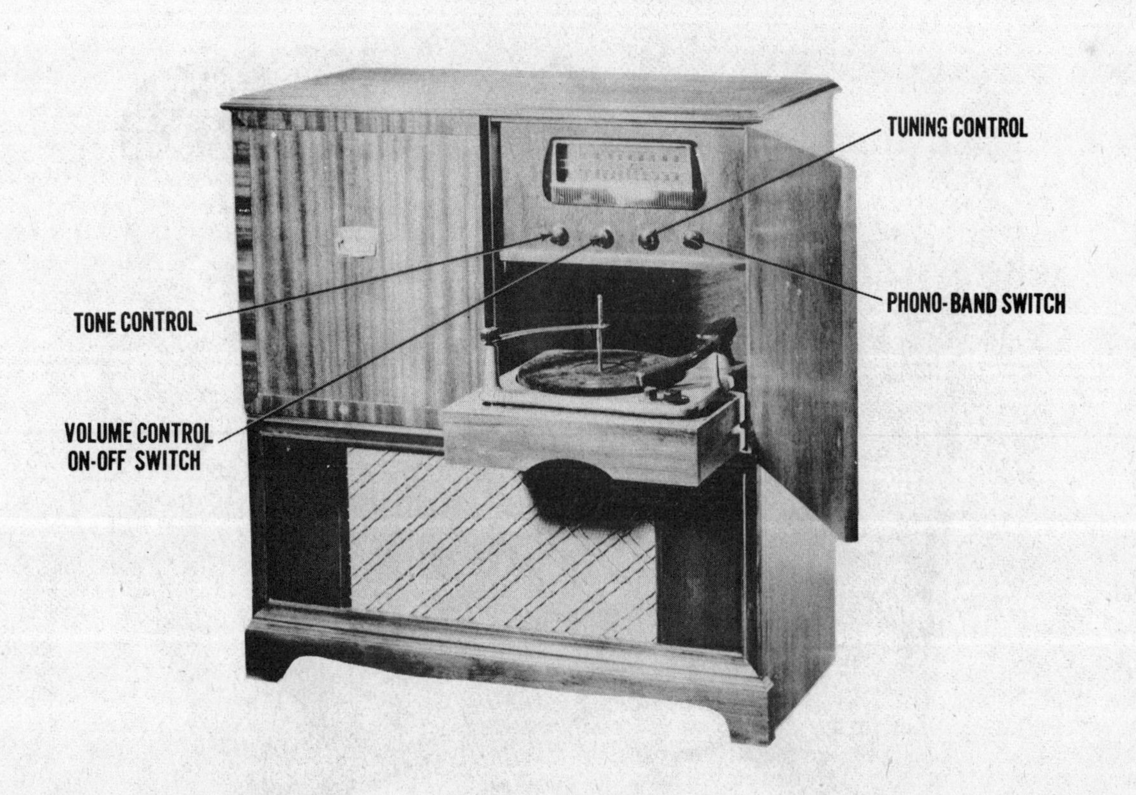

AIRLINE

MODEL PICTURED
94WG-2748A

AC operated phono-radio AM-FM superheterodyne receiver with loop antenna

TUBES
8

POWER SUPPLY
105-125 volts AC

TUNING RANGE
540-1600KC, 88-108MC

MFR/SUPPLIER
Montgomery Ward and Co.

PHOTOFACT SET
90-1

PUBLISHED
1950

AIRLINE

MODEL PICTURED
84GAA3967A

AC operated combo recorder-phono-radio superheterodyne receiver with loop antenna.

TUBES
6

POWER SUPPLY
105-125 volts AC

TUNING RANGE
535-1620KC

MFR/SUPPLIER
Montgomery Ward and Co.

PHOTOFACT SET
91-3

PUBLISHED
1950

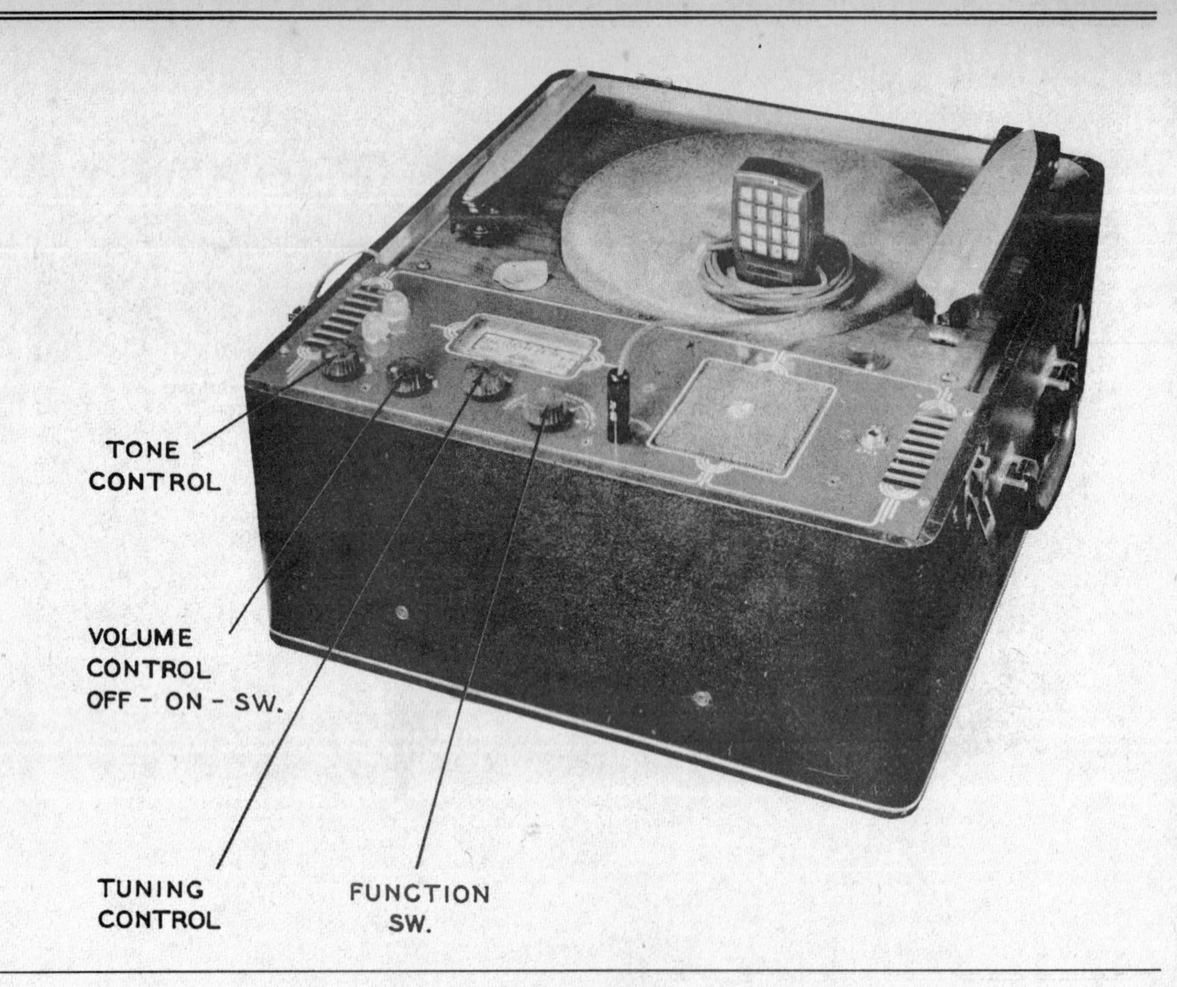

AIRLINE

MODEL PICTURED
94GCB-1064A

Battery operated portable superheterodyne receiver with loop antenna

TUBES
4

POWER SUPPLY
1.5 volts A supply & 67.5 volts B supply

TUNING RANGE
540-1650KC

MFR/SUPPLIER
Montgomery Ward and Co.

PHOTOFACT SET
96-2

PUBLISHED
1950

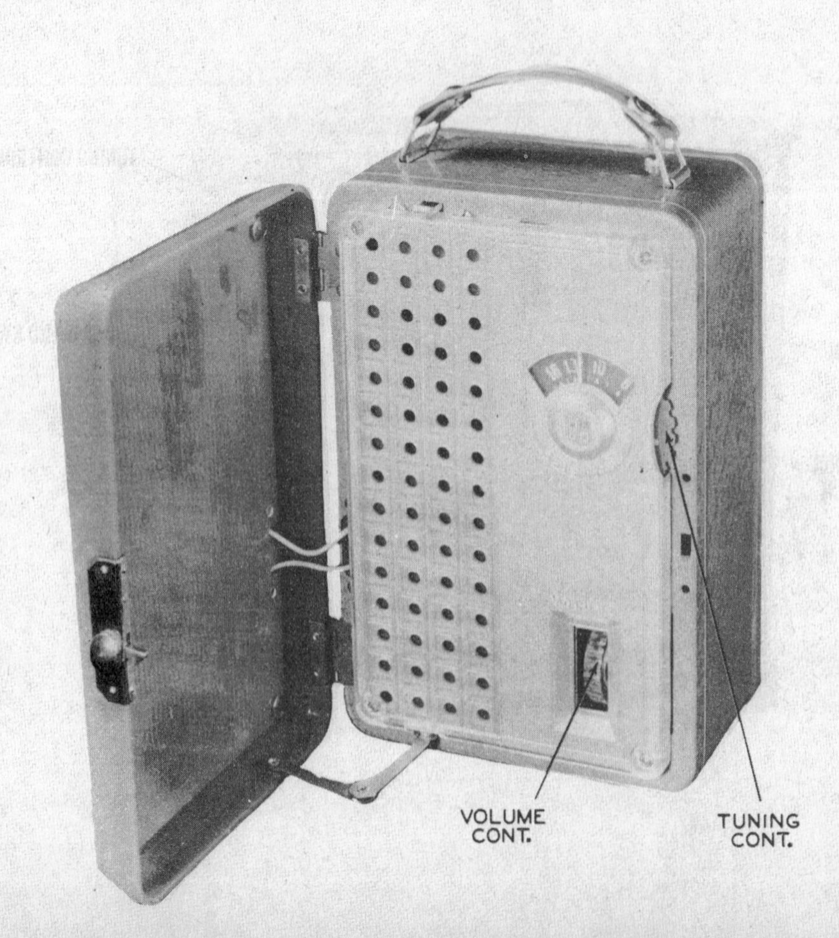

AIRLINE

MODEL PICTURED
94WG-1811A

AC operated AM-FM superheterodyne receiver with loop antenna

TUBES
8

POWER SUPPLY
105-125 volts AC, .35 amp @ 117 volts AC

TUNING RANGE
540-1600KC, 88-108MC

MFR/SUPPLIER
Montgomery Ward and Co.

PHOTOFACT SET
99-4

PUBLISHED
1950

AIRLINE

MODEL PICTURED
05WG-2752

AC operated AM-FM phono combo superheterodyne receiver with loop antenna

TUBES
7

POWER SUPPLY
110-120 volts AC

TUNING RANGE
540-1600KC, 88-108MC

MFR/SUPPLIER
Montgomery Ward and Co.

PHOTOFACT SET
100-3

PUBLISHED
1950

AIRLINE

MODEL PICTURED
05GAA-992A

Radio-phono superheterodyne receiver with loop antenna

TUBES
5

POWER SUPPLY
110-120 volts AC

TUNING RANGE
535-1620KC

MFR/SUPPLIER
Montgomery Ward and Co.

PHOTOFACT SET
125-2

PUBLISHED
1951

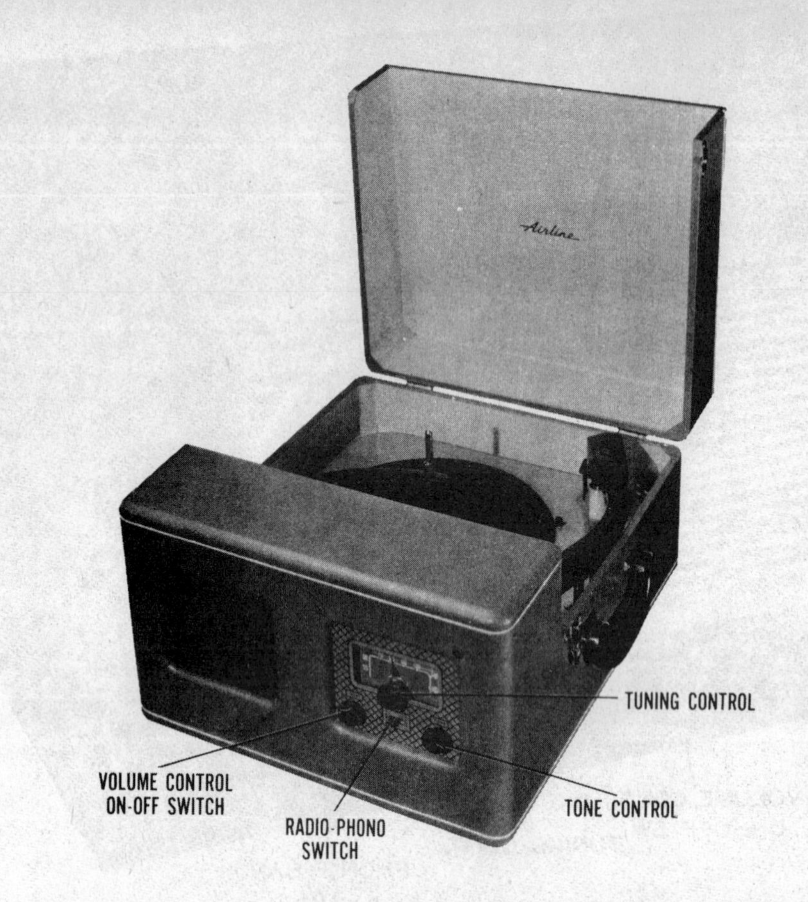

AIRLINE

MODEL PICTURED
05WG-1813A

AC operated AM-FM superheterodyne receiver with loop antenna

TUBES
9

POWER SUPPLY
105-125 volts AC

TUNING RANGE
540-1600, 88-108MC

MFR/SUPPLIER
Montgomery Ward and Co.

PHOTOFACT SET
127-4

PUBLISHED
1951

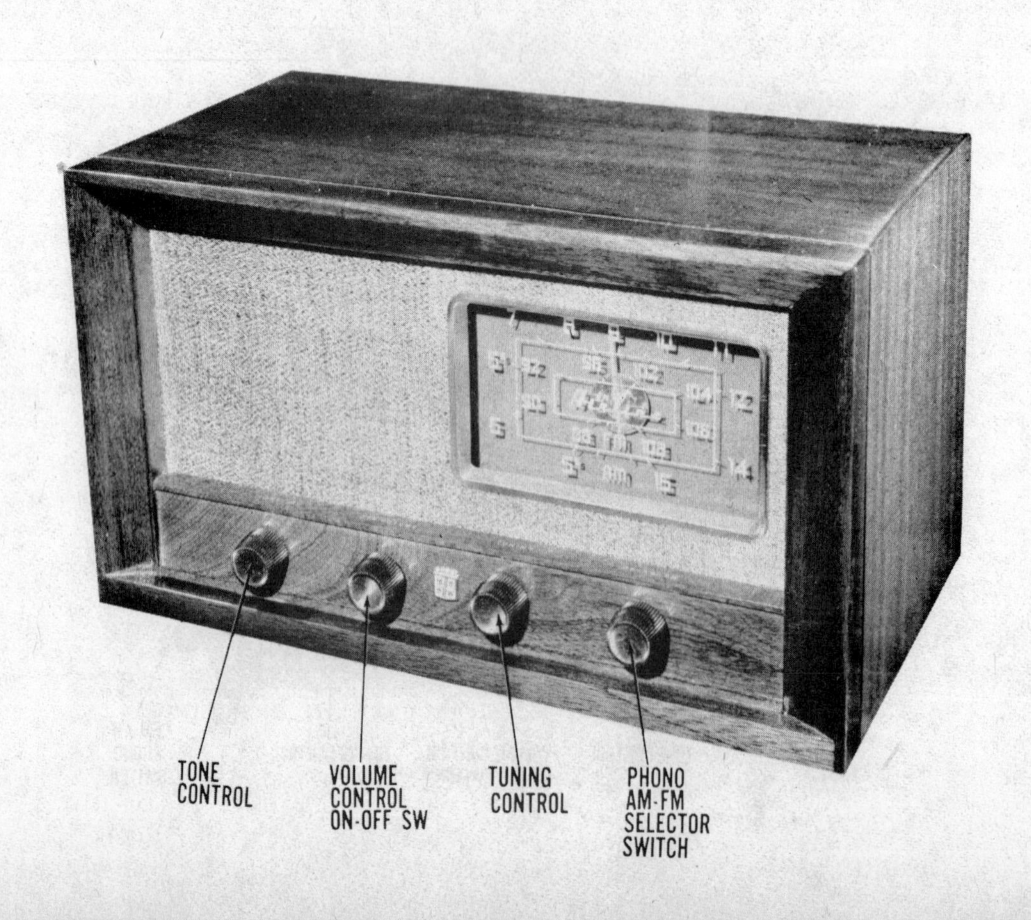

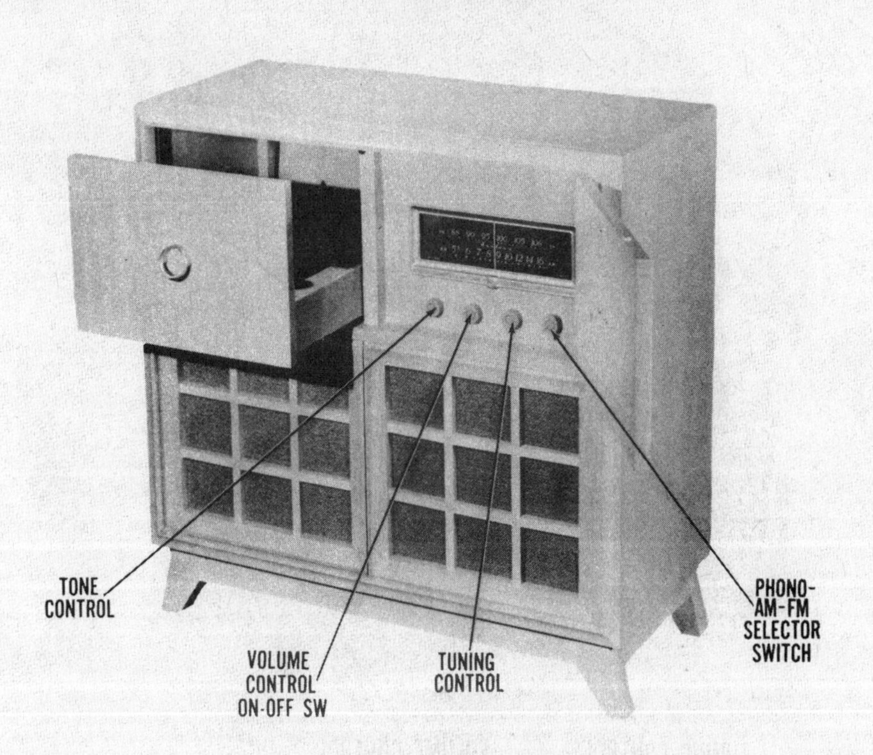

AIRLINE

MODEL PICTURED
05WG-2749D

AC operated phono-radio AM-FM superheterodyne receiver

TUBES
9

POWER SUPPLY
110-120 volts AC

TUNING RANGE
540-1600KC, 88-108MC

MFR/SUPPLIER
Montgomery Ward and Co.

PHOTOFACT SET
129-3

PUBLISHED
1951

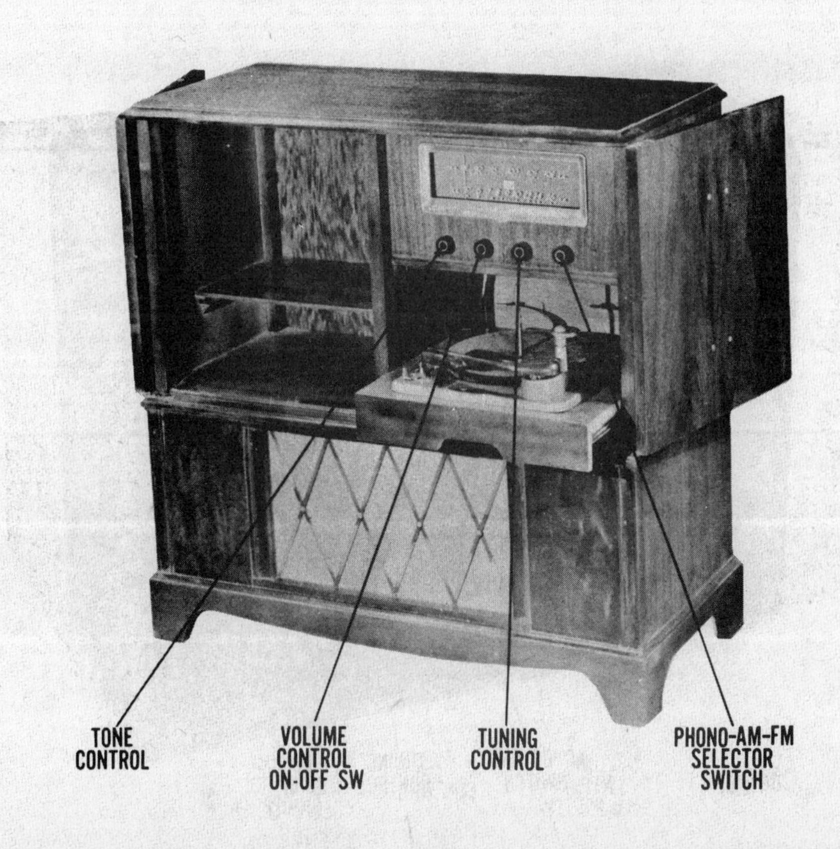

AIRLINE

MODEL PICTURED
15WG-2745C

AC operated combo phono-radio AM-FM superheterodyne receiver with loop antenna

TUBES
10

POWER SUPPLY
105-125 volts AC

TUNING RANGE
540-1600KC, 88-108MC

MFR/SUPPLIER
Montgomery Ward and Co.

PHOTOFACT SET
130-2

PUBLISHED
1951

AIRLINE

MODEL PICTURED
05GCB-1541A

AC/DC operated superheterodyne receiver with loop antenna

TUBES
4

POWER SUPPLY
105-125 volts AC

TUNING RANGE
540-1650KC

MFR/SUPPLIER
Montgomery Ward & Co.

PHOTOFACT SET
131-2

PUBLISHED
1951

AIRLINE

MODEL PICTURED
05GHM-1061A

Three-power operated portable superheterodyne receiver with loop antenna

TUBES
4

POWER SUPPLY
105-125 volts AC/DC or 7.5 volts A supply & 90 volts B supply in pack form

TUNING RANGE
540-1620KC

MFR/SUPPLIER
Montgomery Ward and Co.

PHOTOFACT SET
133-3

PUBLISHED
1951

AIRLINE

MODEL PICTURED
05WG-2748F

AC operated phono-radio AM-FM superheterodyne receiver with loop antenna

TUBES
9

POWER SUPPLY
110-120 volts AC

TUNING RANGE
540-1600KC, 88-108MC

MFR/SUPPLIER
Montgomery Ward and Co.

PHOTOFACT SET
139-4

PUBLISHED
1951

AIRLINE

MODEL PICTURED
15BR-1547A

AC/DC operated superheterodyne receiver with loop antenna

TUBES
4

POWER SUPPLY
110-120 volts AC/DC, .25 amp @ 117 volts AC

TUNING RANGE
540-1600KC

MFR/SUPPLIER
Montgomery Ward and Co.

PHOTOFACT SET
143-3

PUBLISHED
1951

AIRLINE

MODEL PICTURED
15WG-2758A

AC operated combo phono-radio AM-FM superheterodyne receiver

TUBES
8

POWER SUPPLY
105-125 volts AC

TUNING RANGE
540-1600KC, 88-108MC

MFR/SUPPLIER
Montgomery Ward and Co.

PHOTOFACT SET
144-2

PUBLISHED
1951

AIRLINE

MODEL PICTURED
15BR-1544A

AC/DC operated superheterodyne receiver with loop antenna

TUBES
5

POWER SUPPLY
110-120 volts AC/DC, .25 amp @ 117 volts AC

TUNING RANGE
540-1600KC

MFR/SUPPLIER
Montgomery Ward and Co.

PHOTOFACT SET
145-2

PUBLISHED
1951

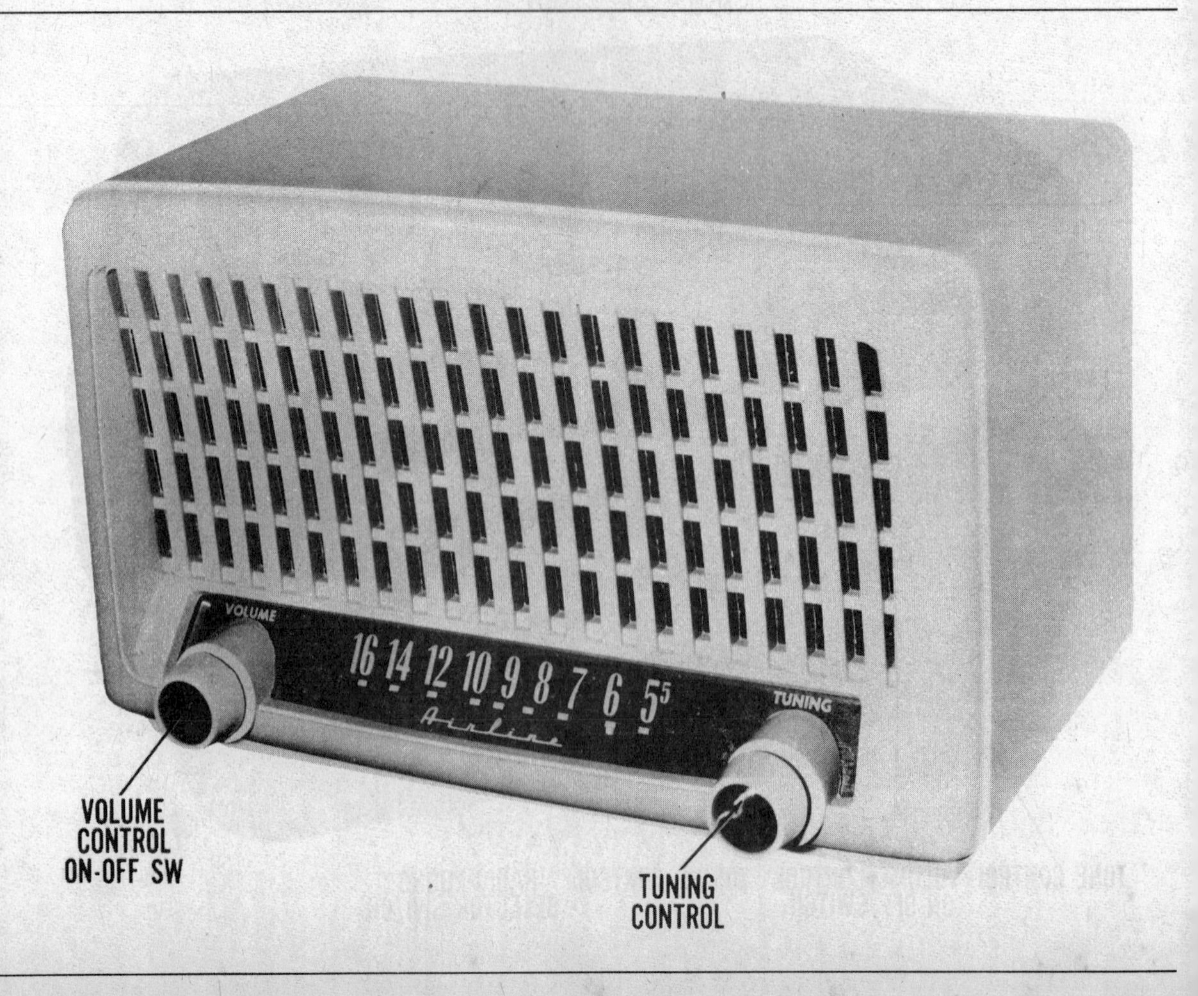

Dear Reader

Please complete this short survey...and become eligible for a free TV!

As part of our continuing effort to provide information you can use, we'd like your views on the books we publish. Your answers to these questions will help us serve you better in the future. And when you mail this filled-out card to Sams (postage paid), we'll put your name in our hat for a drawing. We'll give away one TV as soon as we receive the first 2000 responses.

First, some questions about *Radios of the Baby Boom Era.*

1 Where did you obtain this book?

____

2 Which volume(s) did you purchase? 1 2 3 4 5 6

3 Was the information useful? ___Yes ___No

4 Was the information easy to find? ___Yes ___No

5 What did you like most about the book?

____

6 What did you like least?

____

7 Is there any other information you'd like included?

____

Now some general questions about yourself:

8 Where do you most often buy books?

____

9 What are your hobbies and interests?

____

10 Do you repair things? What kinds?

____

11 Which subjects would you like us to publish more books about?

____

To include you in the drawing for a free TV, we need your name and address:

Name______________________________

Address______________________________

____

Phone______________________________

Thank you for helping us make our books better for all our readers. Now please drop this postage-paid card in the nearest mailbox.

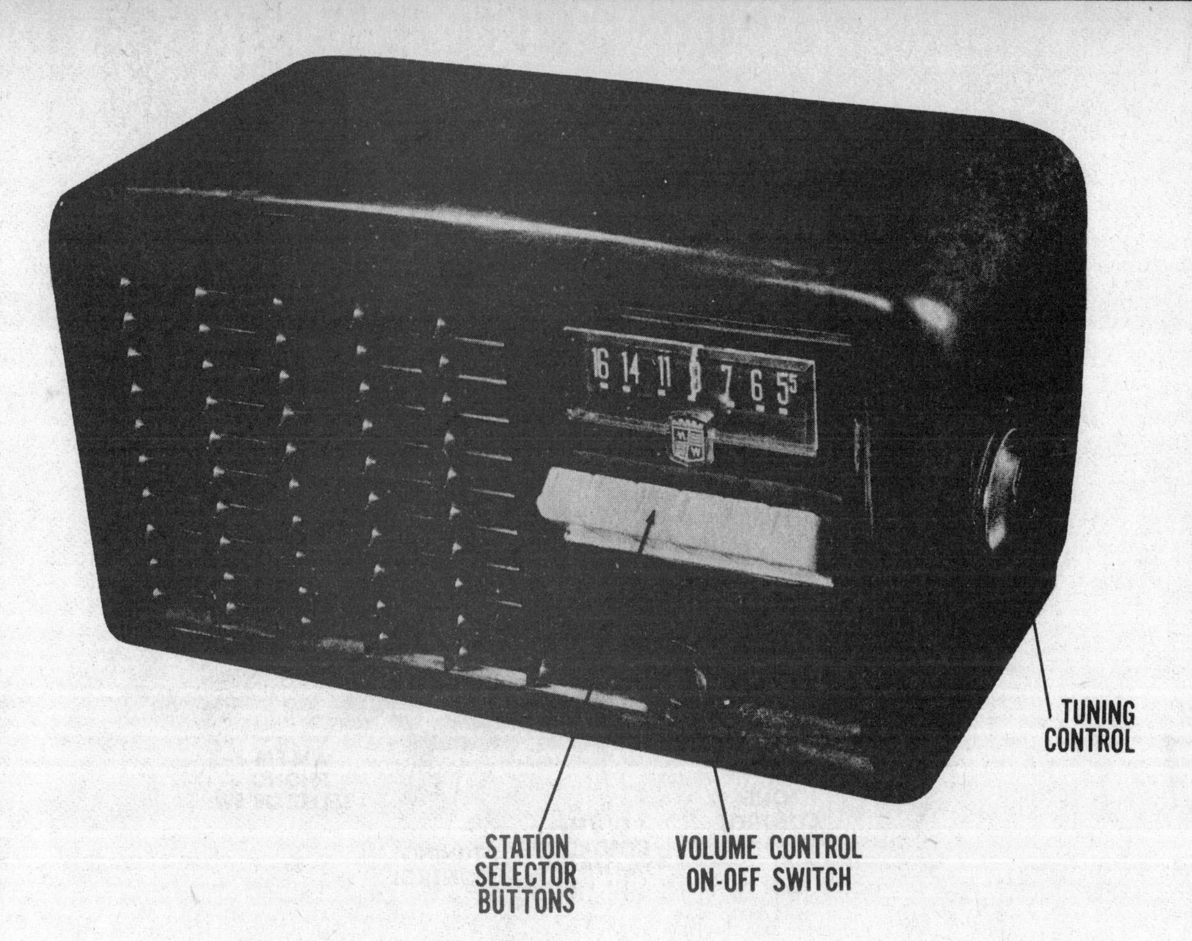

AIRLINE

MODEL PICTURED
15BR-1536B

AC/DC operated superheterodyne receiver with loop antenna

TUBES
5

POWER SUPPLY
110-120 volts AC/DC, .25 amp @ 117 volts AC

TUNING RANGE
540-1600KC

MFR/SUPPLIER
Montgomery Ward and Co.

PHOTOFACT SET
146-2

PUBLISHED
1951

AIRLINE

MODEL PICTURED
15BR-2756B

AC operated combo phono-radio AM-FM superheterodyne receiver with loop antenna

TUBES
8

POWER SUPPLY
110-120 volts AC

TUNING RANGE
540-1600KC, 88-108MC

MFR/SUPPLIER
Montgomery Ward and Co.

PHOTOFACT SET
148-3

PUBLISHED
1951

AIRLINE

MODEL PICTURED
15WG-2749F

AC operated combo phono-radio AM-FM superheterodyne receiver with loop antenna

TUBES
12

POWER SUPPLY
105-125 volts AC

TUNING RANGE
540-1600KC, 88-108MC

MFR/SUPPLIER
Montgomery Ward and Co.

PHOTOFACT SET
151-4

PUBLISHED
1951

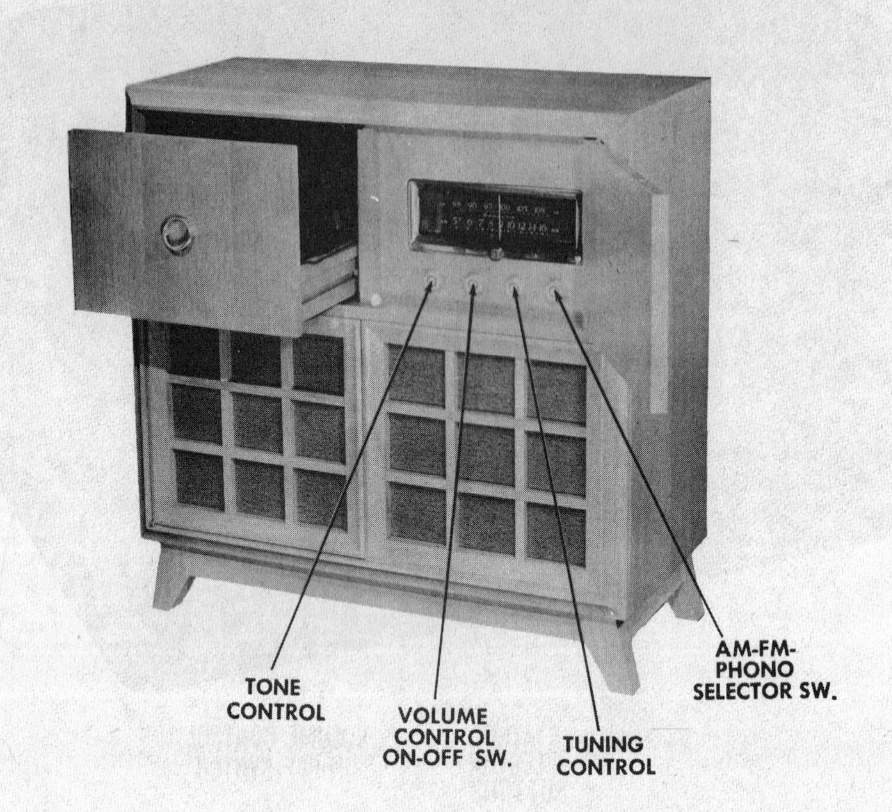

AIRLINE

MODEL PICTURED
15GSE-2764A

AC operated radio-phono combo superheterodyne receiver

TUBES
6

POWER SUPPLY
110-120 volts AC

TUNING RANGE
530-1650KC

MFR/SUPPLIER
Montgomery Ward and Co.

PHOTOFACT SET
165-4

PUBLISHED
1952

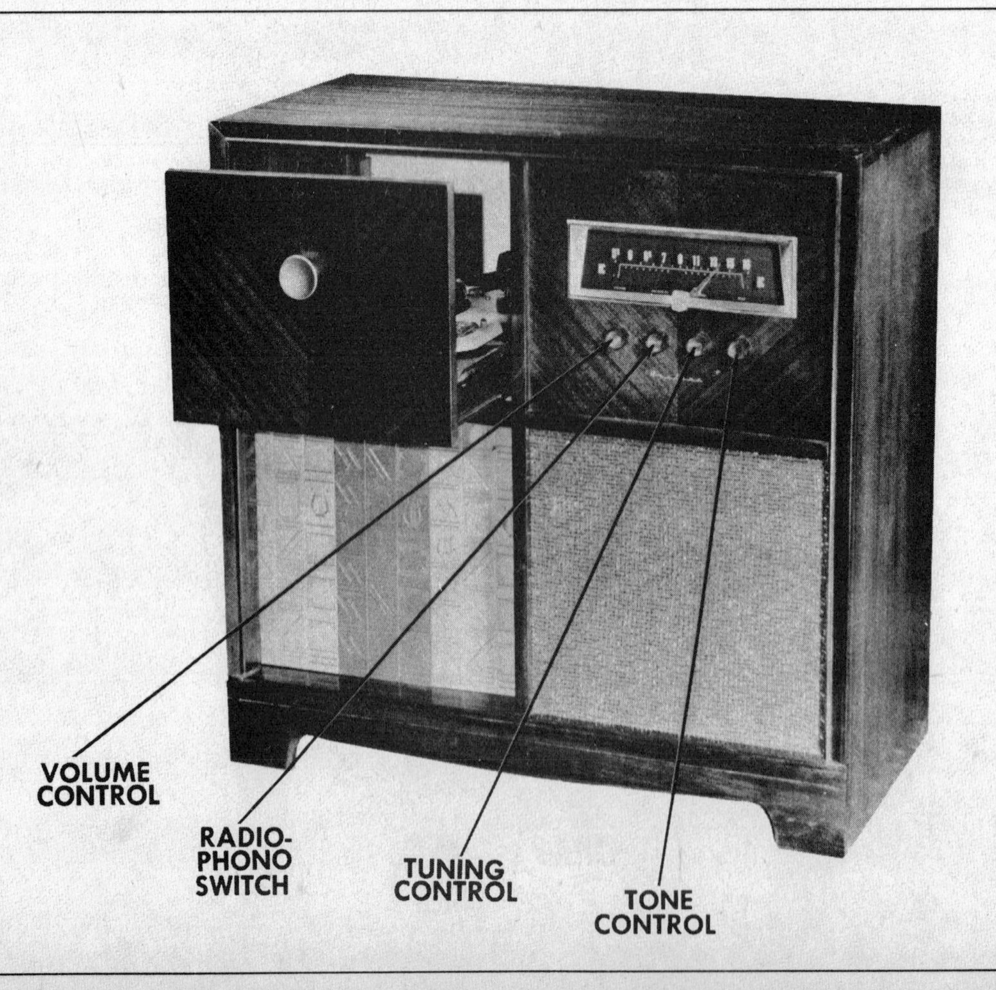

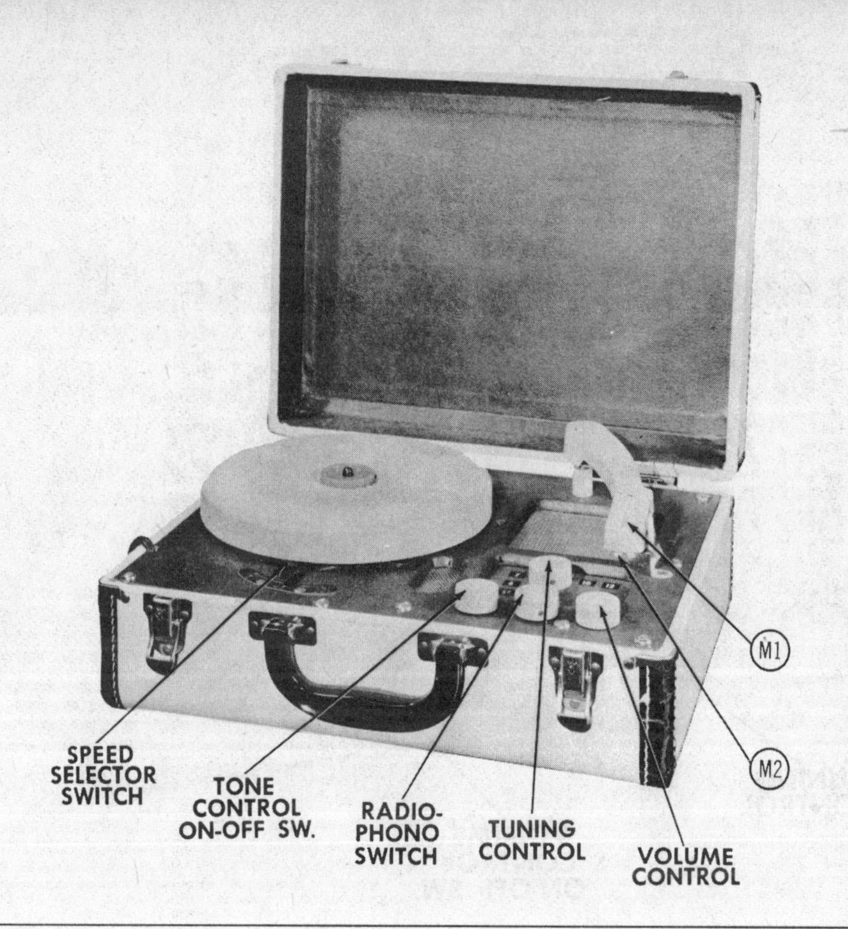

AIRLINE

MODEL PICTURED
15GHM-934A

AC operated radio-phono combo superheterodyne receiver with three-speed manual record player

TUBES
5

POWER SUPPLY
110-120 volts AC, 60 cycle

TUNING RANGE
540-1640KC

MFR/SUPPLIER
Montgomery Ward and Co.

PHOTOFACT SET
167-3

PUBLISHED
1952

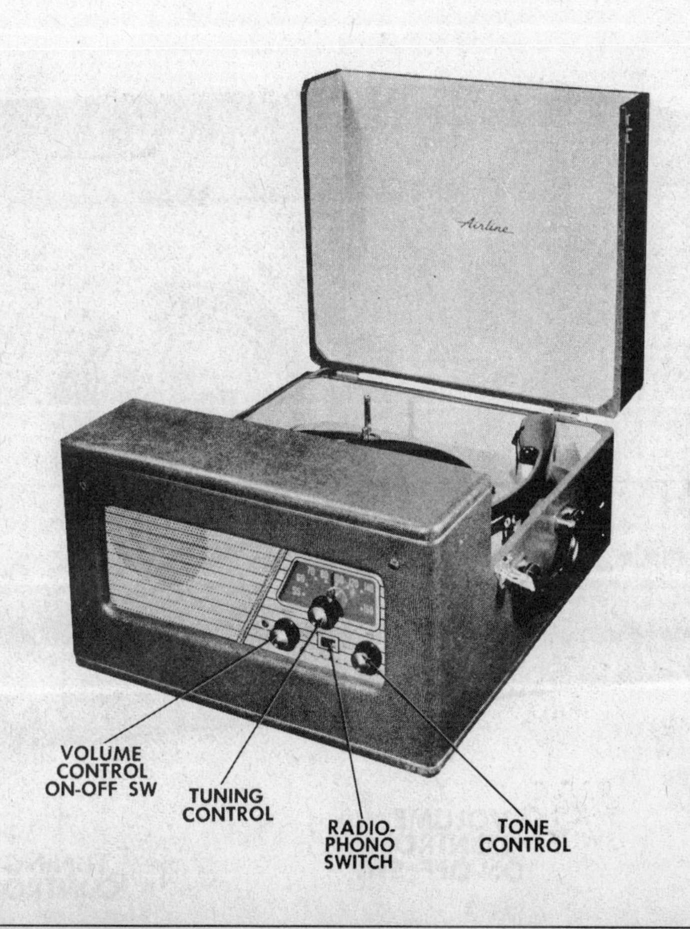

AIRLINE

MODEL PICTURED
15GAA-995A

AC operated radio-phono superheterodyne receiver

TUBES
5

POWER SUPPLY
110-120 volts AC, 60 cycle

TUNING RANGE
535-1620KC

MFR/SUPPLIER
Montgomery Ward and Co.

PHOTOFACT SET
168-3

PUBLISHED
1952

AIRLINE

MODEL PICTURED
15GSL-1564A

AC/DC operated AM superheterodyne receiver

TUBES
4

POWER SUPPLY
110-120 volts AC/DC, .22 amp @ 117 volts AC

TUNING RANGE
540-1620KC

MFR/SUPPLIER
Montgomery Ward and Co.

PHOTOFACT SET
169-3

PUBLISHED
1952

AIRLINE

MODEL PICTURED
25GSE-1555A

AC/DC operated AM superheterodyne receiver

TUBES
5

POWER SUPPLY
110-120 volts AC/DC

TUNING RANGE
540-1660KC

MFR/SUPPLIER
Montgomery Ward and Co.

PHOTOFACT SET
174-3

PUBLISHED
1952

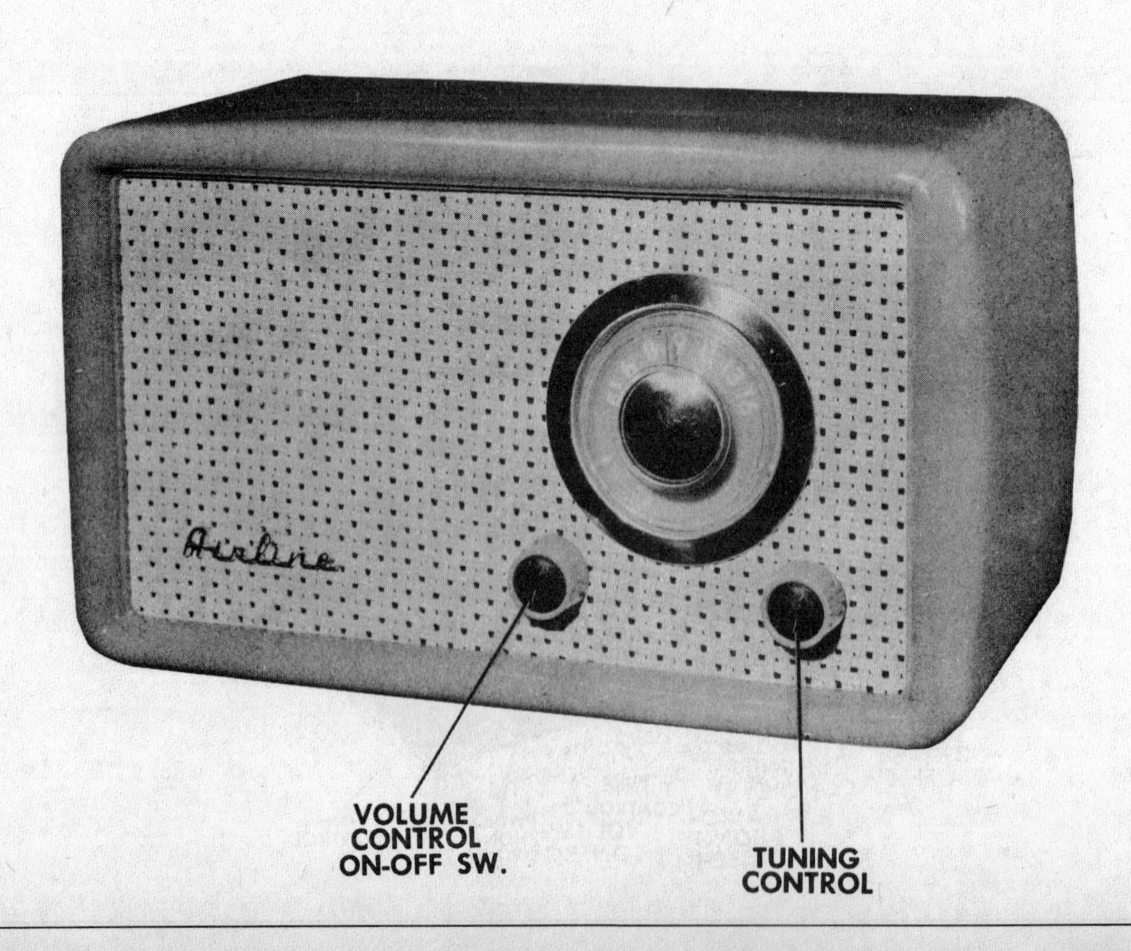

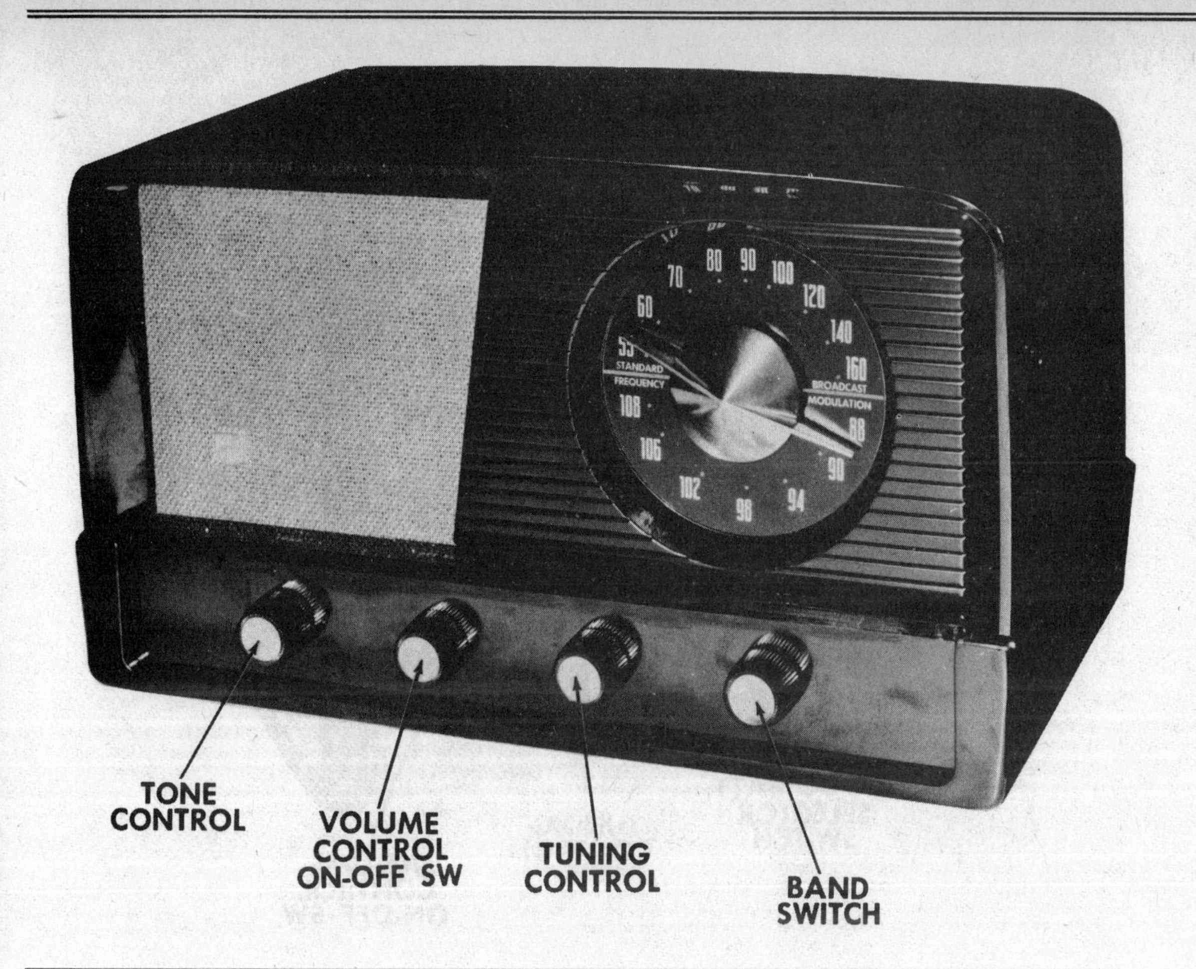

AIRLINE

MODEL PICTURED
25WG-1570A

AC operated AM-FM superheterodyne receiver

TUBES
8

POWER SUPPLY
105-125 volts AC, 50/60 cycles

TUNING RANGE
540-1600KC, 88-108MC

MFR/SUPPLIER
Montgomery Ward and Co.

PHOTOFACT SET
177-4

PUBLISHED
1952

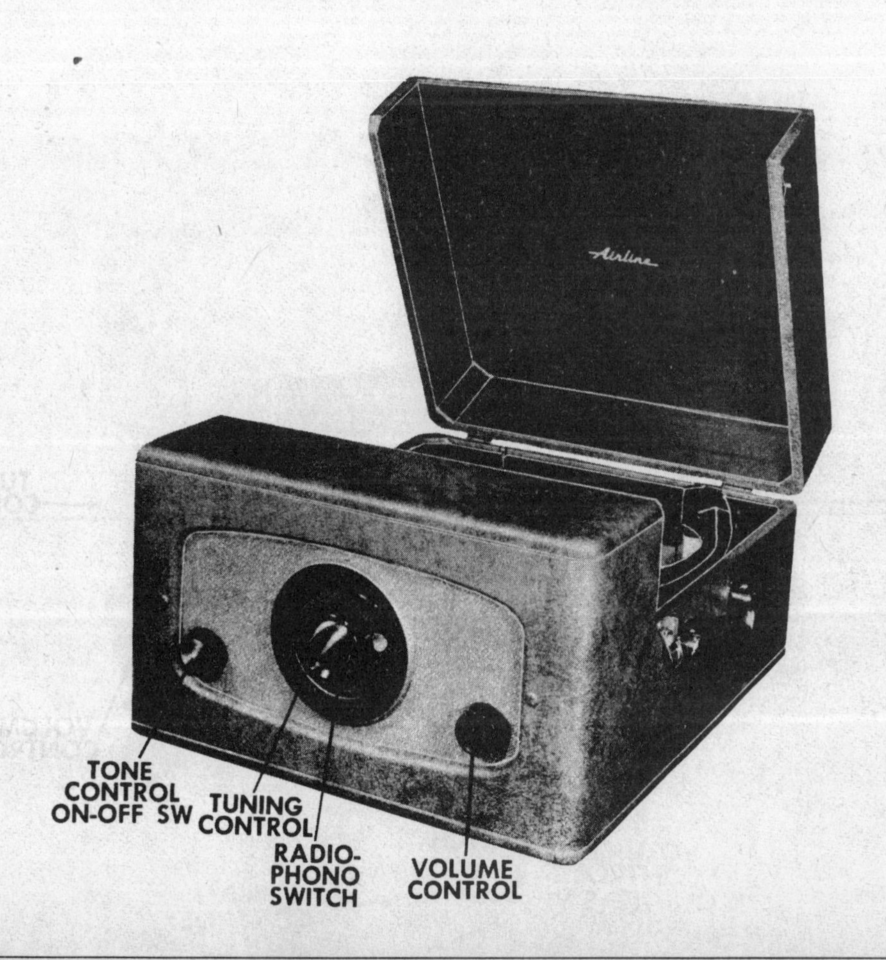

AIRLINE

MODEL PICTURED
25GAA-996A

AC operated AM-phono combo superheterodyne receiver with three-speed automatic record changer

TUBES
5

POWER SUPPLY
105-125 volts AC, 60 cycle

TUNING RANGE
535-1620KC

MFR/SUPPLIER
Montgomery Ward and Co.

PHOTOFACT SET
182-2

PUBLISHED
1952

AIRLINE

MODEL PICTURED
15GHM-1070A

Three-power portable superheterodyne receiver

TUBES
4

POWER SUPPLY
110-120 volts AC/DC or 7.5 volts A & 90 volts B battery in pack form

TUNING RANGE
540-1640KC

MFR/SUPPLIER
Montgomery Ward and Co.

PHOTOFACT SET
184-3

PUBLISHED
1952

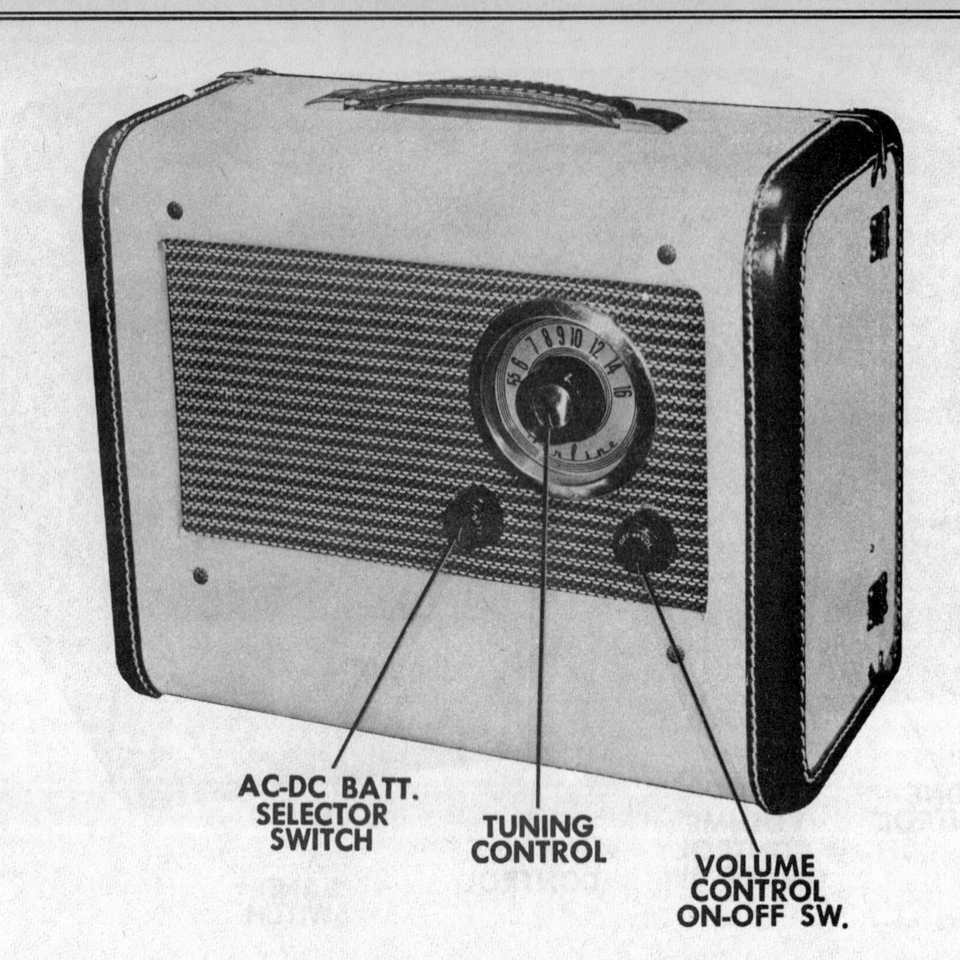

AIRLINE

MODEL PICTURED
25GSL-1560A

AC operated AM superheterodyne receiver with electric clock

TUBES
4

POWER SUPPLY
110-120 volts AC, 60 cycle

TUNING RANGE
540-1600KC

MFR/SUPPLIER
Montgomery Ward and Co.

PHOTOFACT SET
189-2

PUBLISHED
1952

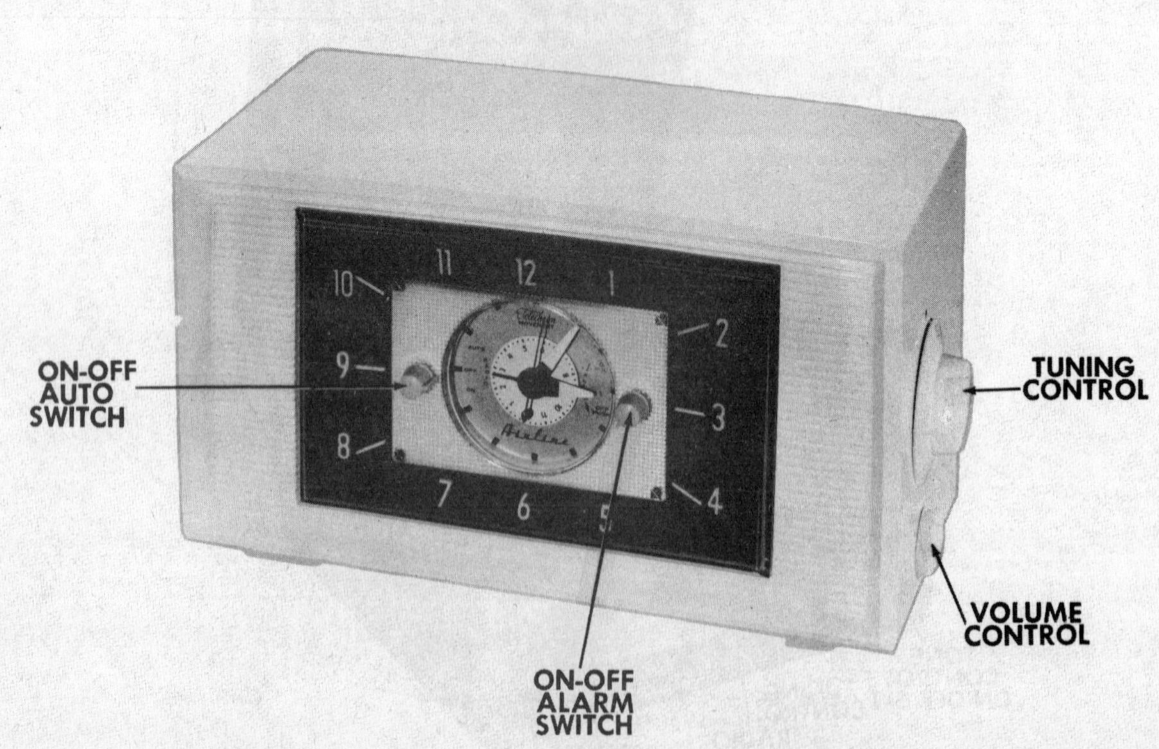

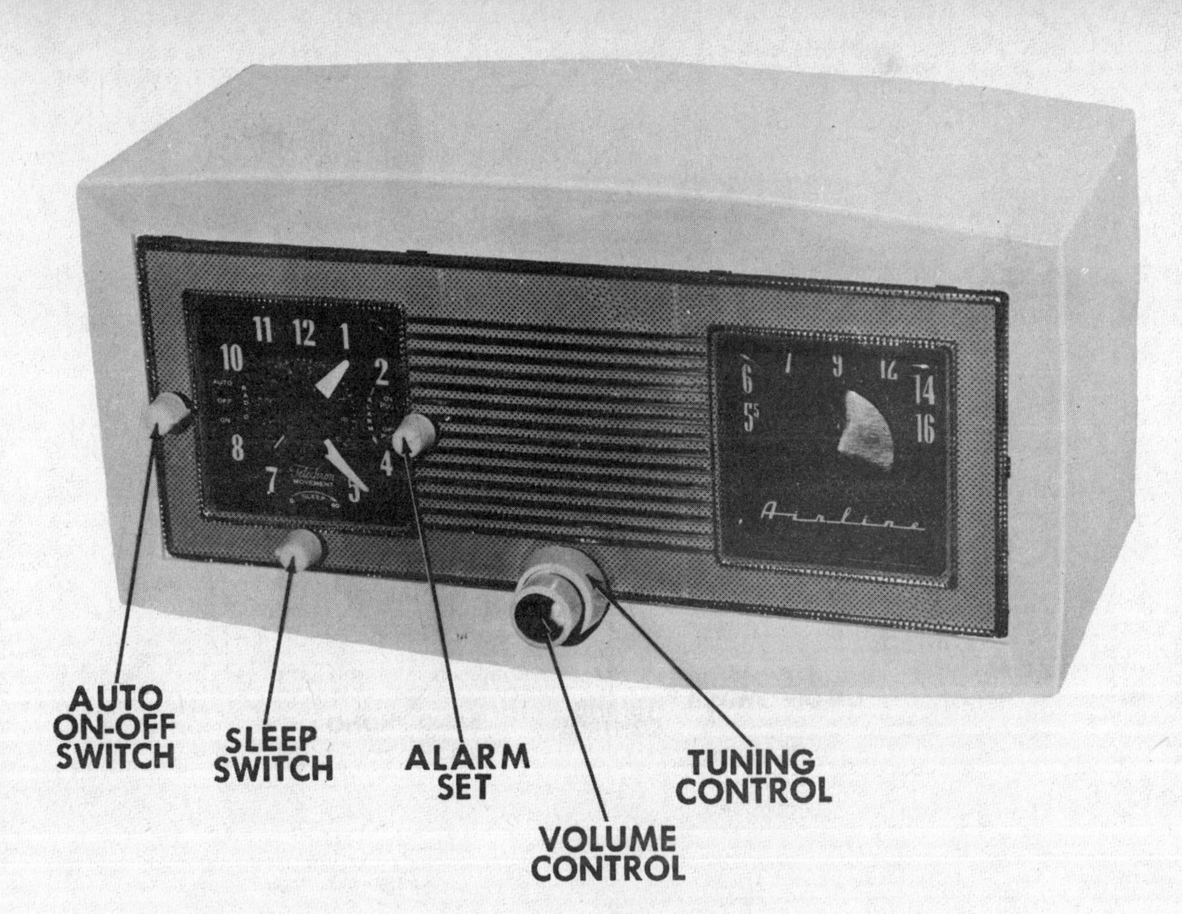

AIRLINE

MODEL PICTURED
25BR-1549B

AC operated AM superheterodyne receiver with electric clock

TUBES
4

POWER SUPPLY
105-115 volts AC

TUNING RANGE
540-1600KC

MFR/SUPPLIER
Montgomery Ward and Co.

PHOTOFACT SET
191-3

PUBLISHED
1953

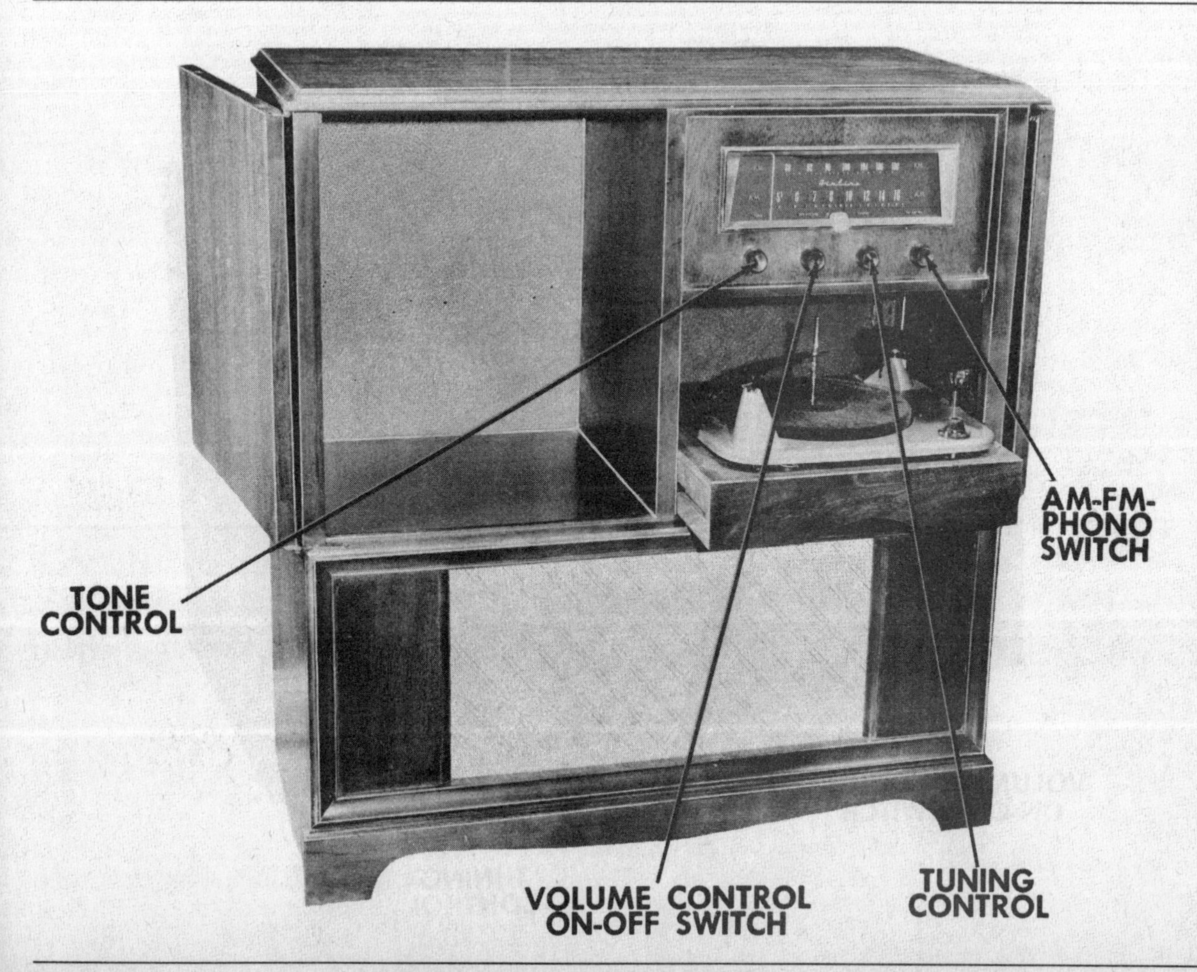

AIRLINE

MODEL PICTURED
25WG-2758C

AC operated AM-FM phono combo superheterodyne receiver

TUBES
8

POWER SUPPLY
105-125 volts AC, 60 cycle

TUNING RANGE
540-1600KC, 88-108MC

MFR/SUPPLIER
Montgomery Ward and Co.

PHOTOFACT SET
195-3

PUBLISHED
1953

AIRLINE

MODEL PICTURED
25WG-1573A

AC operated two-band AM superheterodyne receiver

TUBES
6

POWER SUPPLY
110-120 volts AC, 60 cycles

TUNING RANGE
540-1600KC, 9-15.6MC

MFR/SUPPLIER
Montgomery Ward and Co.

PHOTOFACT SET
196-2

PUBLISHED
1953

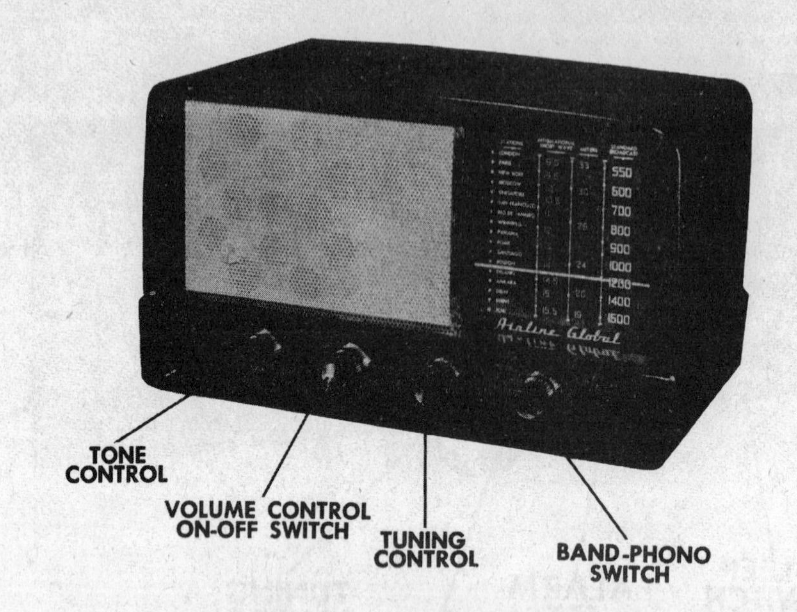

AIRLINE

MODEL PICTURED
25GSL-1814A

AC/DC operated AM superheterodyne receiver

TUBES
6

POWER SUPPLY
110-120 volts AC/DC, .21 amp @ 117 volts AC

TUNING RANGE
540-1600KC

MFR/SUPPLIER
Montgomery Ward and Co.

PHOTOFACT SET
198-1

PUBLISHED
1953

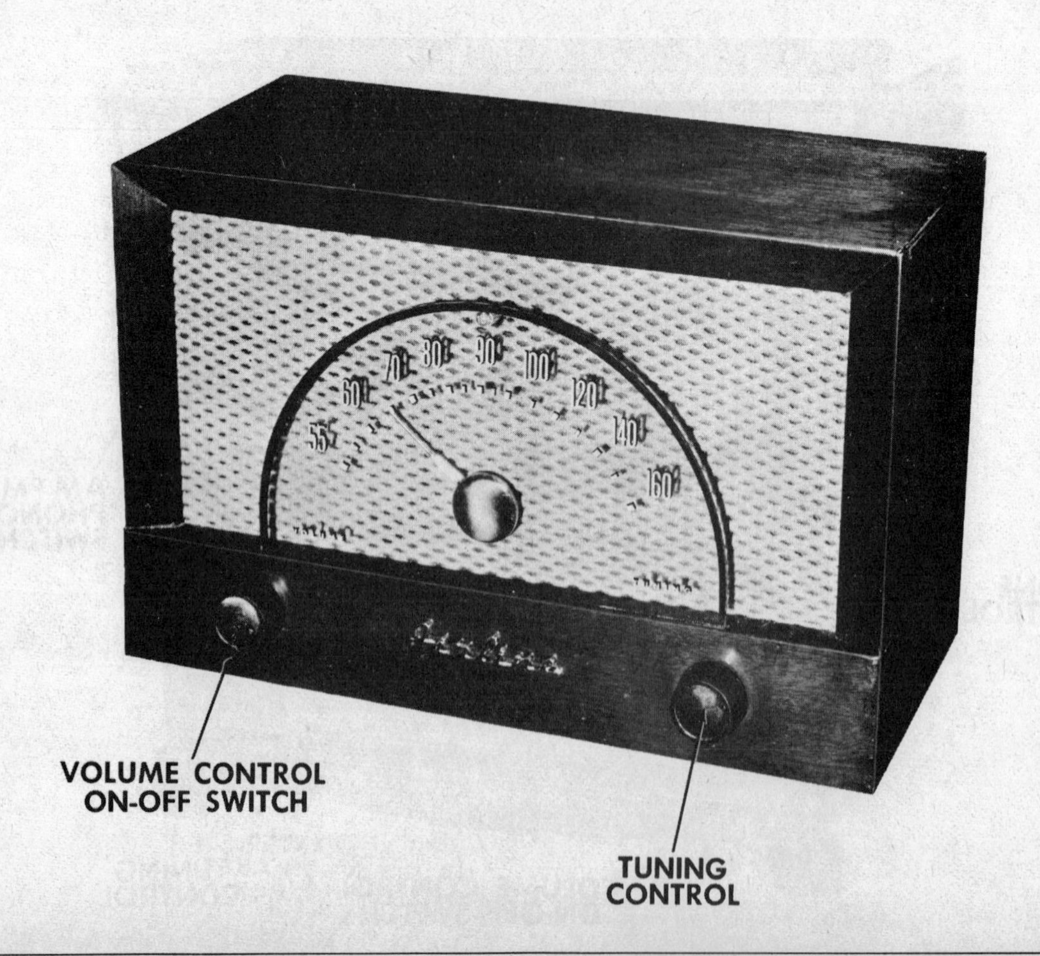

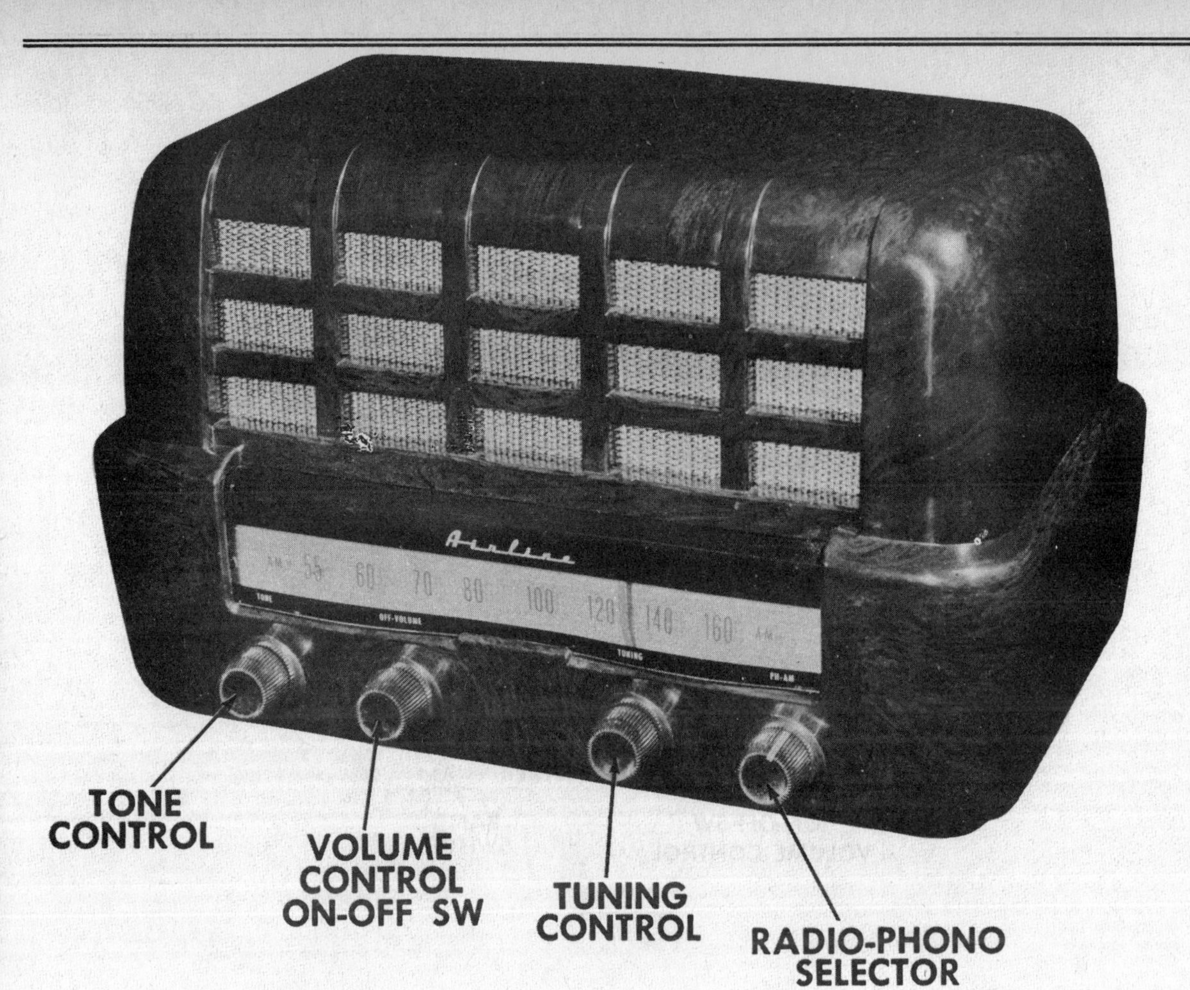

AIRLINE

MODEL PICTURED
25BR-1542A

AC/DC operated AM superheterodyne receiver

TUBES
4

POWER SUPPLY
110-120 volts AC/DC, .33 amp @ 117 volts AC

TUNING RANGE
540-1600KC

MFR/SUPPLIER
Montgomery Ward and Co.

PHOTOFACT SET
203-3

PUBLISHED
1953

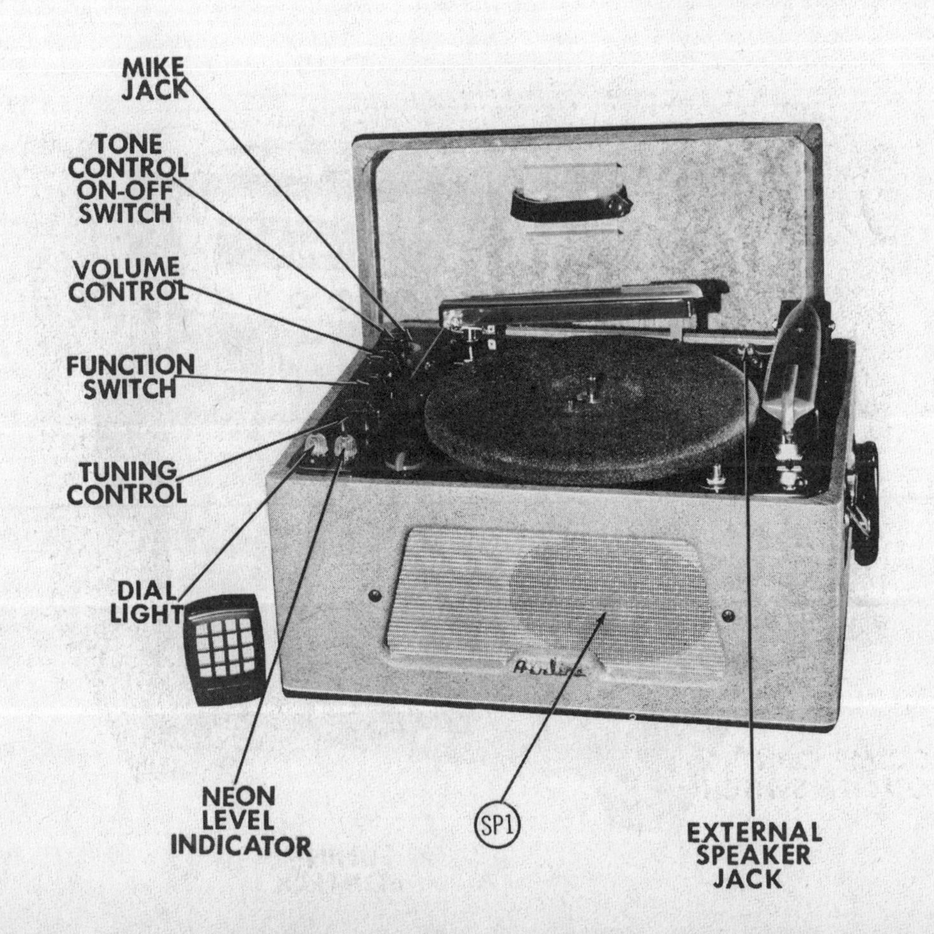

AIRLINE

MODEL PICTURED
35GAA-3969A

AC operated AM superheterodyne receiver with disc recorder and playback unit

TUBES
6

POWER SUPPLY
105-125 volts AC, 60 cycle

TUNING RANGE
535-1620KC

MFR/SUPPLIER
Montgomery Ward and Co.

PHOTOFACT SET
227-1

PUBLISHED
1954

AIRLINE

MODEL PICTURED
35WG-1573B

AC operated two-band AM superheterodyne receiver

TUBES
6

POWER SUPPLY
105-125 volts AC, 50/60 cycle, .28 amp @ 117volts AC

TUNING RANGE
540-1600KC, 9-15.6MC

MFR/SUPPLIER
Montgomery Ward and Co.

PHOTOFACT SET
228-1

PUBLISHED
1954

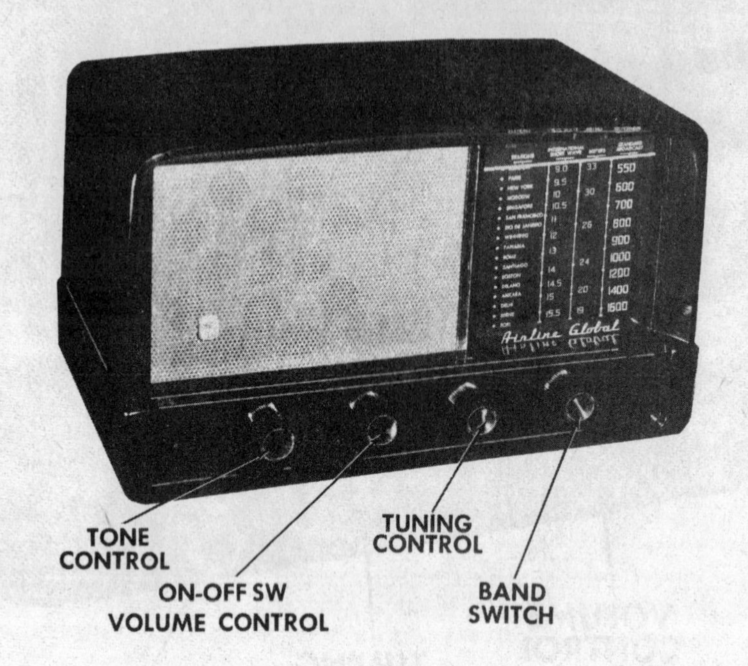

AIRLINE

MODEL PICTURED
35WG-2767A

AC operated AM-FM phono combo receiver with three-speed auto record changer

TUBES
12

POWER SUPPLY
105-125 volts AC, 60 cycle

TUNING RANGE
535-1620KC, 88-108MC

MFR/SUPPLIER
Montgomery Ward and Co.

PHOTOFACT SET
241-2

PUBLISHED
1954

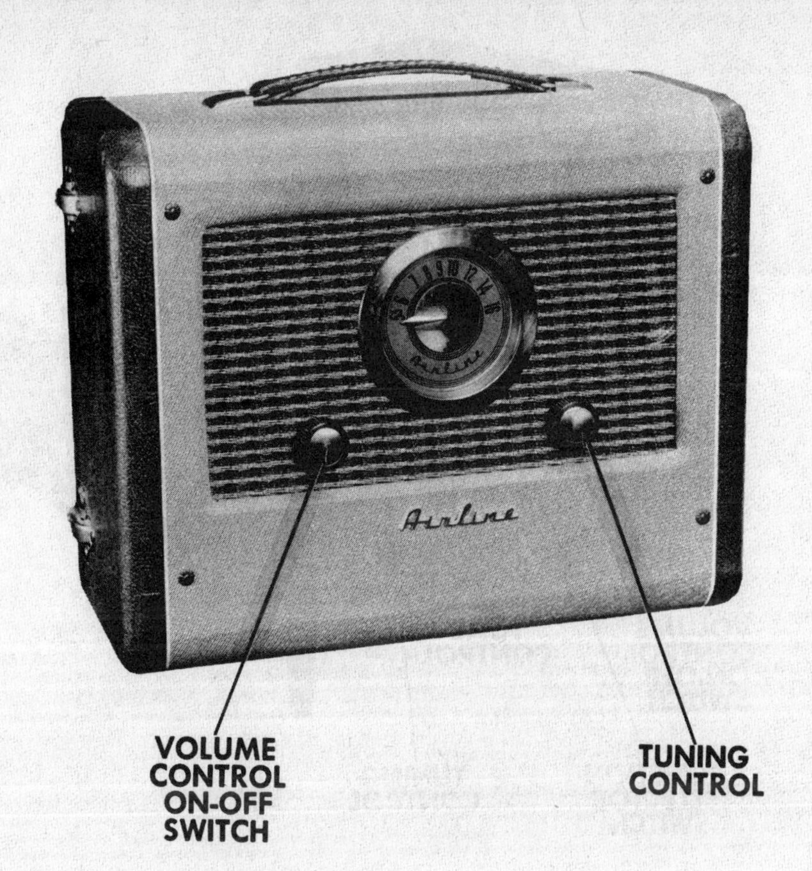

AIRLINE

MODEL PICTURED
25GHM-1073A

Three-power portable superheterodyne receiver

TUBES
5

POWER SUPPLY
105-125 volts AC/DC or 9 volts A supply & 90 volts B supply in pack form

TUNING RANGE
540-1620KC

MFR/SUPPLIER
Montgomery Ward and Co.

PHOTOFACT SET
242-2

PUBLISHED
1954

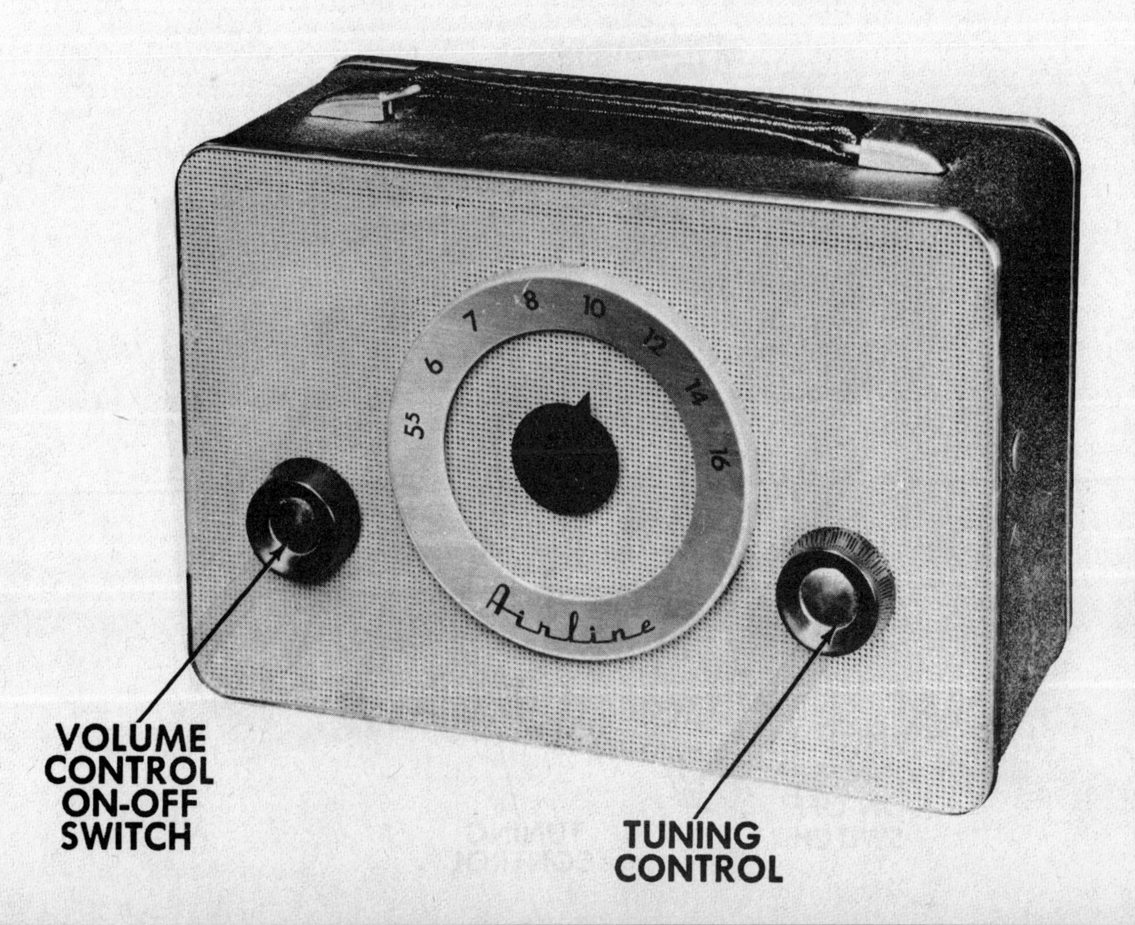

AIRLINE

MODEL PICTURED
35GHM-1074A

Three-power portable superheterodyne receiver

TUBES
4

POWER SUPPLY
105-125 volts AC/DC or 4.5 volts A supply & 75 volts B supply

TUNING RANGE
540-1620KC

MFR/SUPPLIER
Montgomery Ward and Co.

PHOTOFACT SET
243-2

PUBLISHED
1954

AIRLINE

MODEL PICTURED
35GSL-2770A

AC operated radio-phono combo superheterodyne receiver

TUBES
6

POWER SUPPLY
110-120 volts AC, 60 cycle

TUNING RANGE
540-1600KC

MFR/SUPPLIER
Montgomery Ward and Co.

PHOTOFACT SET
249-3

PUBLISHED
1954

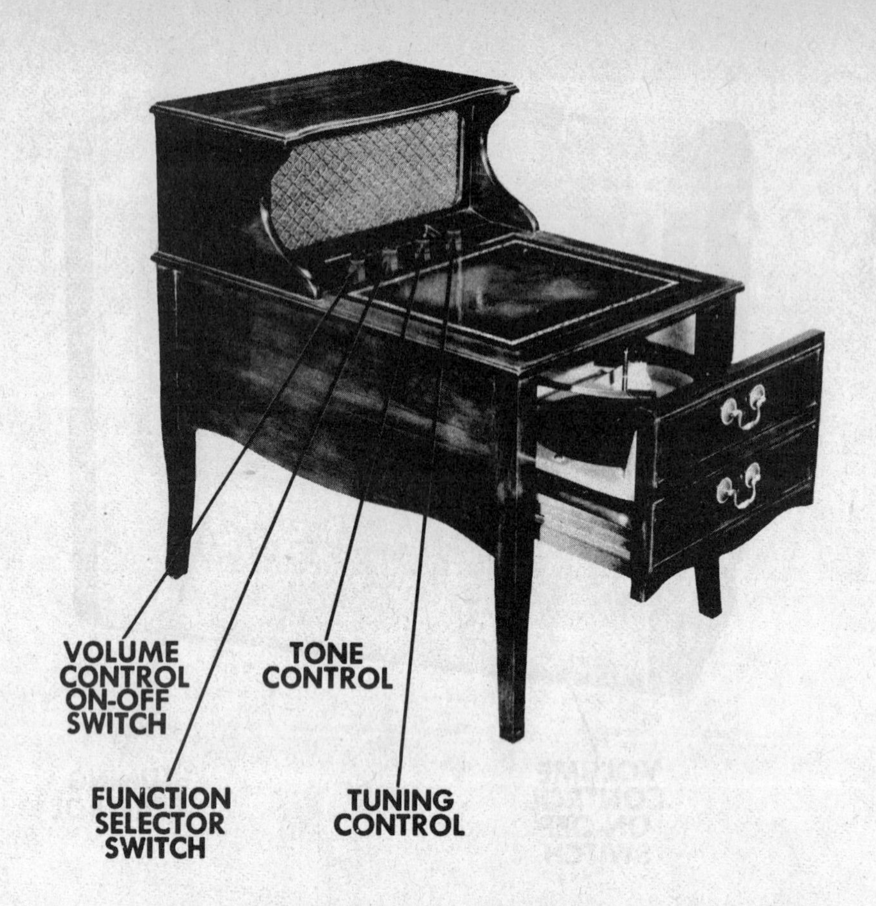

AIRLINE

MODEL PICTURED
GSE-1077A

Three-power portable AM superheterodyne receiver

TUBES
4

POWER SUPPLY
110-120 volts AC/DC or 3 volts A supply & 67.5 volts B supply

TUNING RANGE
530-1650KC

MFR/SUPPLIER
Montgomery Ward and Co.

PHOTOFACT SET
250-3

PUBLISHED
1954

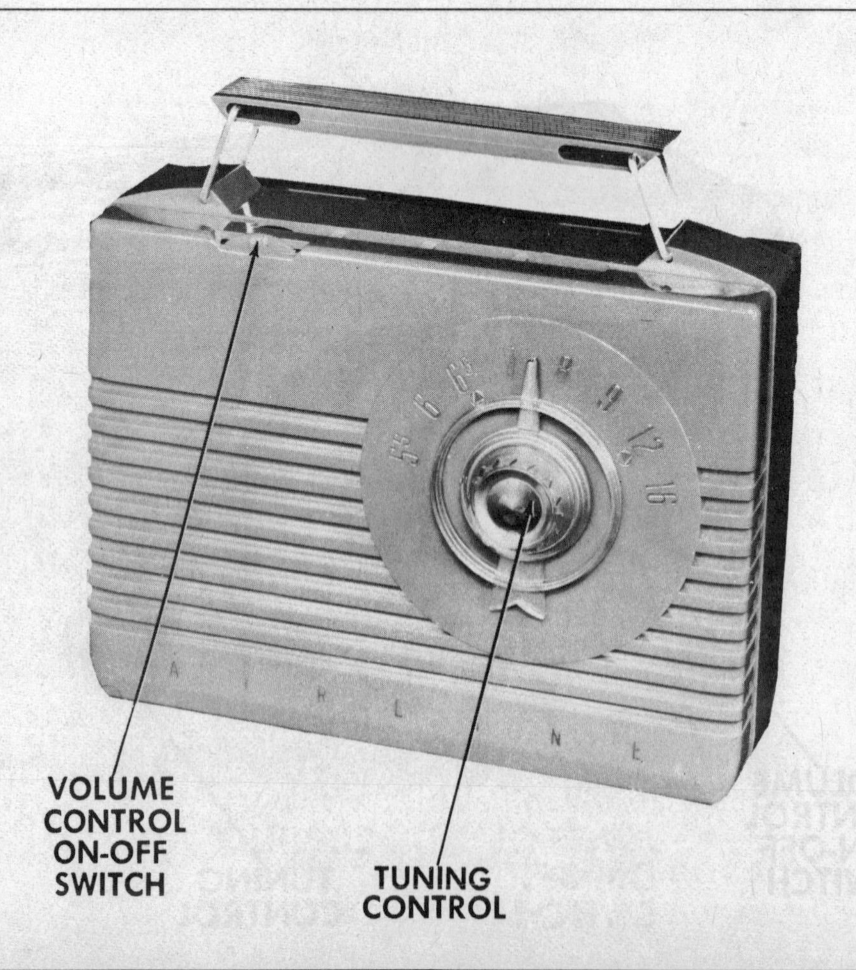

AIRLINE

MODEL PICTURED
WG-1572C

AC operated AM-FM superheterodyne receiver

TUBES
8

POWER SUPPLY
105-125 volts AC, 50/60 cycles, .32 amp @ 117 volts AC

TUNING RANGE
540-1600KC, 88-108MC

MFR/SUPPLIER
Montgomery Ward and Co.

PHOTOFACT SET
251-1

PUBLISHED
1954

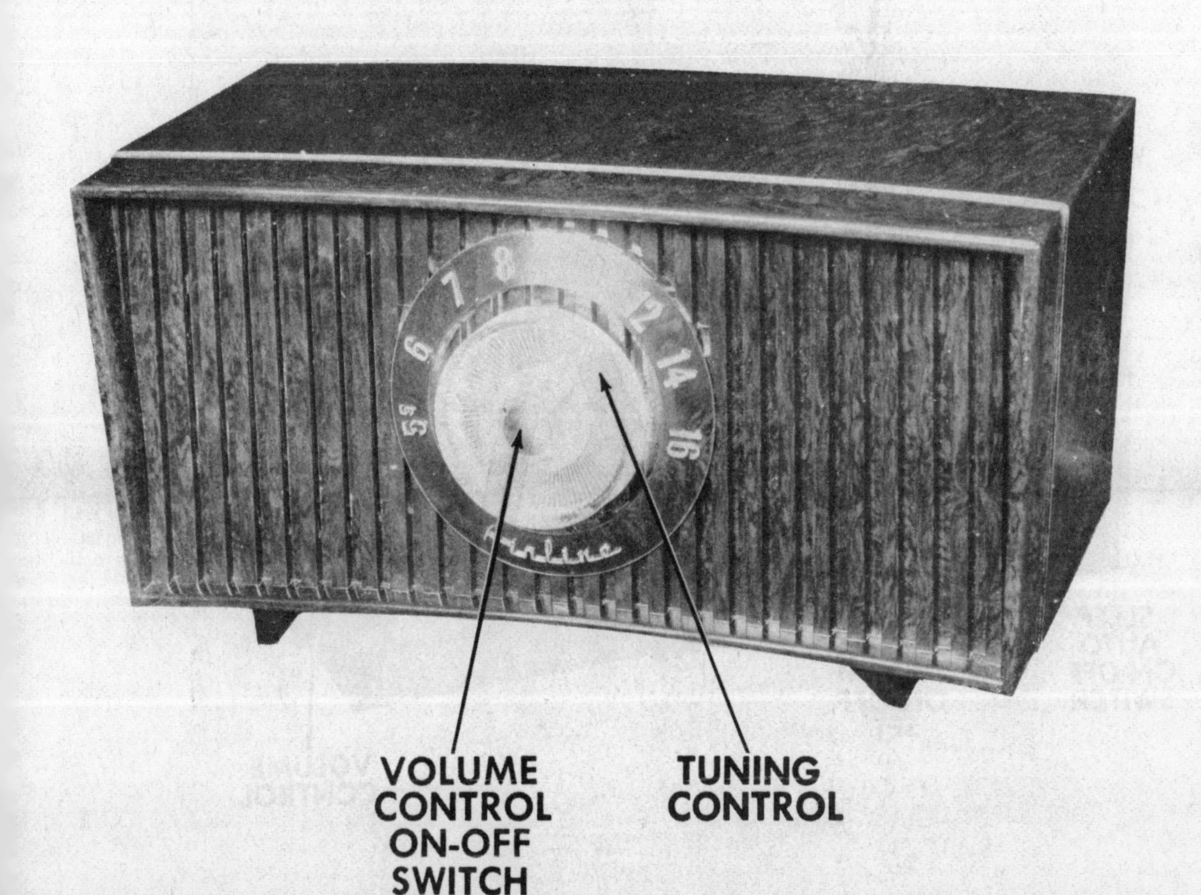

AIRLINE

MODEL PICTURED
35BR-1557A

AC/DC operated AM superheterodyne receiver

TUBES
5

POWER SUPPLY
115 volts AC/DC, .25 amp @ 115 volts AC

TUNING RANGE
535-1620KC

MFR/SUPPLIER
Montgomery Ward and Co.

PHOTOFACT SET
251-2

PUBLISHED
1954

AIRLINE

MODEL PICTURED
25GHM-2012A

AC operated radio-phono superheterodyne receiver with three-speed automatic record changer

TUBES
5

POWER SUPPLY
110-120 volts AC, 60 cycle

TUNING RANGE
540-1620KC

MFR/SUPPLIER
Montgomery Ward and Co.

PHOTOFACT SET
256-4

PUBLISHED
1954

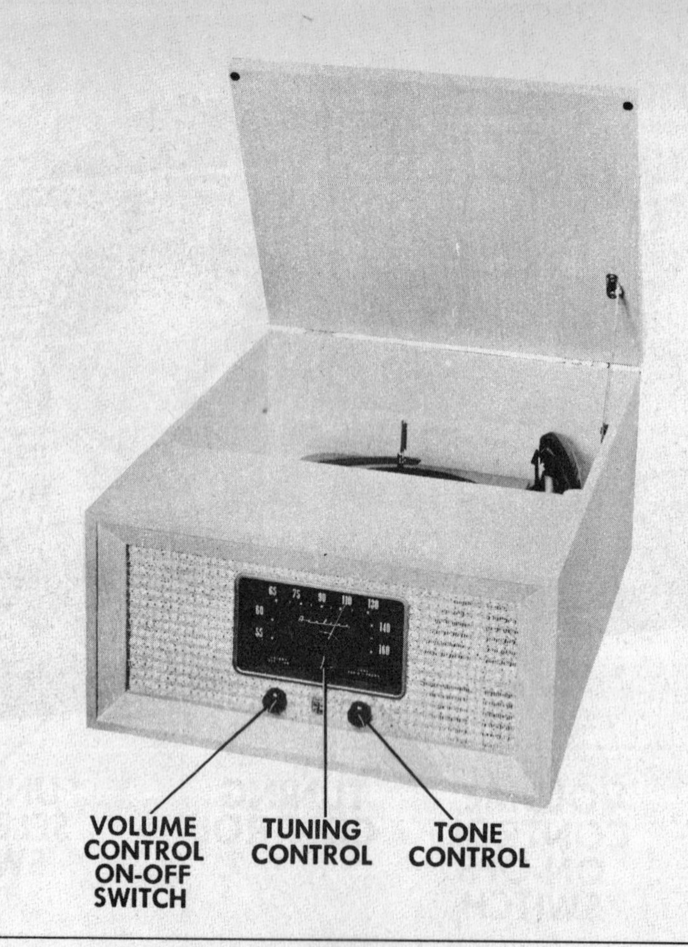

AIRLINE

MODEL PICTURED
GSL-1581A

AC operated AM superheterodyne receiver with electric clock

TUBES
5

POWER SUPPLY
110-120 volts AC, 60 cycle

TUNING RANGE
540-1600KC

MFR/SUPPLIER
Montgomery Ward and Co.

PHOTOFACT SET
280-3

PUBLISHED
1955

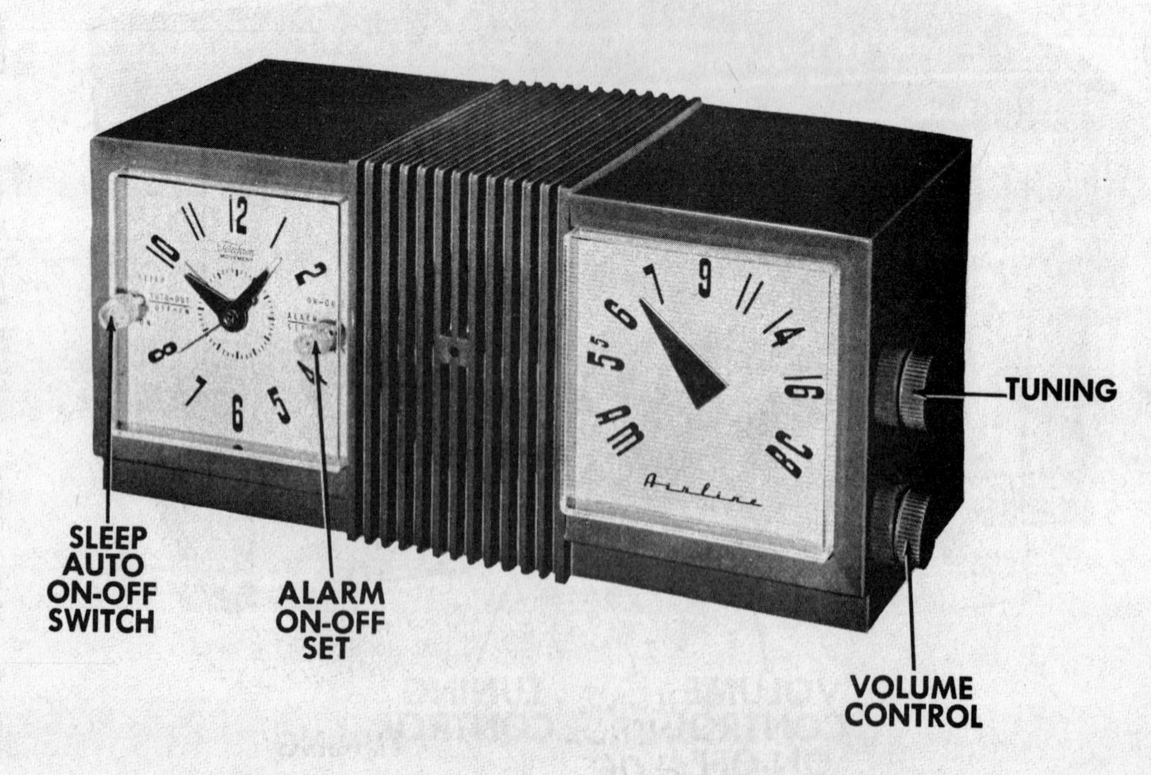

AIRLINE

MODEL PICTURED
GSL-1614A

AC/DC operated AM superheterodyne receiver

TUBES
4

POWER SUPPLY
110-120 volts AC/DC

TUNING RANGE
540-1600KC

MFR/SUPPLIER
Montgomery Ward and Co.

PHOTOFACT SET
289-2

PUBLISHED
1955

AIRLINE

MODEL PICTURED
GSE-1606A

AC/DC operated AM superheterodyne receiver

TUBES
5

POWER SUPPLY
110-120 volts AC/DC, .22 amp @ 117 volts AC

TUNING RANGE
540KC-1660KC

MFR/SUPPLIER
Montgomery Ward and Co.

PHOTOFACT SET
292-2

PUBLISHED
1955

AIRLINE

MODEL PICTURED
GSL-1079-A

Three-power portable two-band AM superheterodyne receiver

TUBES
5

POWER SUPPLY
110-120 volts AC/DC or 9 volts A & 90 volts B supply

TUNING RANGE
540-1620KC, 5.8-18.3MC

MFR/SUPPLIER
Montgomery Ward and Co.

PHOTOFACT SET
294-2

PUBLISHED
1955

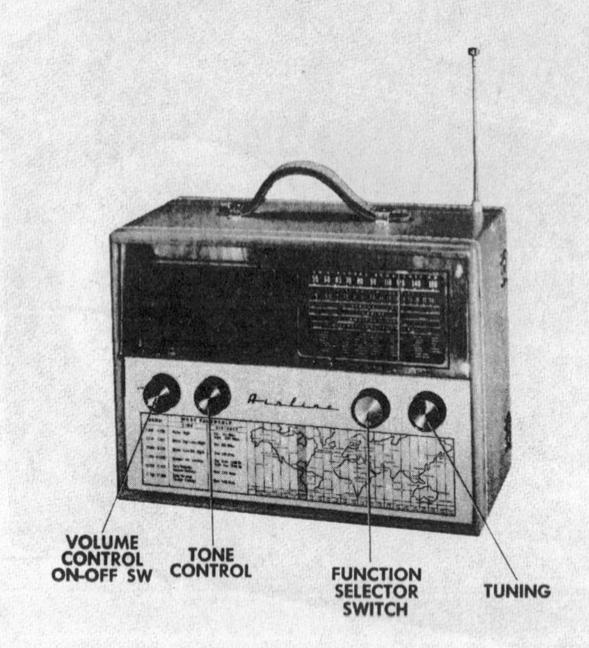

AIRLINE

MODEL PICTURED
WG-1635A

AC/DC operated AM superheterodyne receiver

TUBES
6

POWER SUPPLY
105-125 volts AC/DC, .25 amp @ 117 volts AC

TUNING RANGE
540KC-1600KC

MFR/SUPPLIER
Montgomery Ward and Co.

PHOTOFACT SET
306-2

PUBLISHED
1956

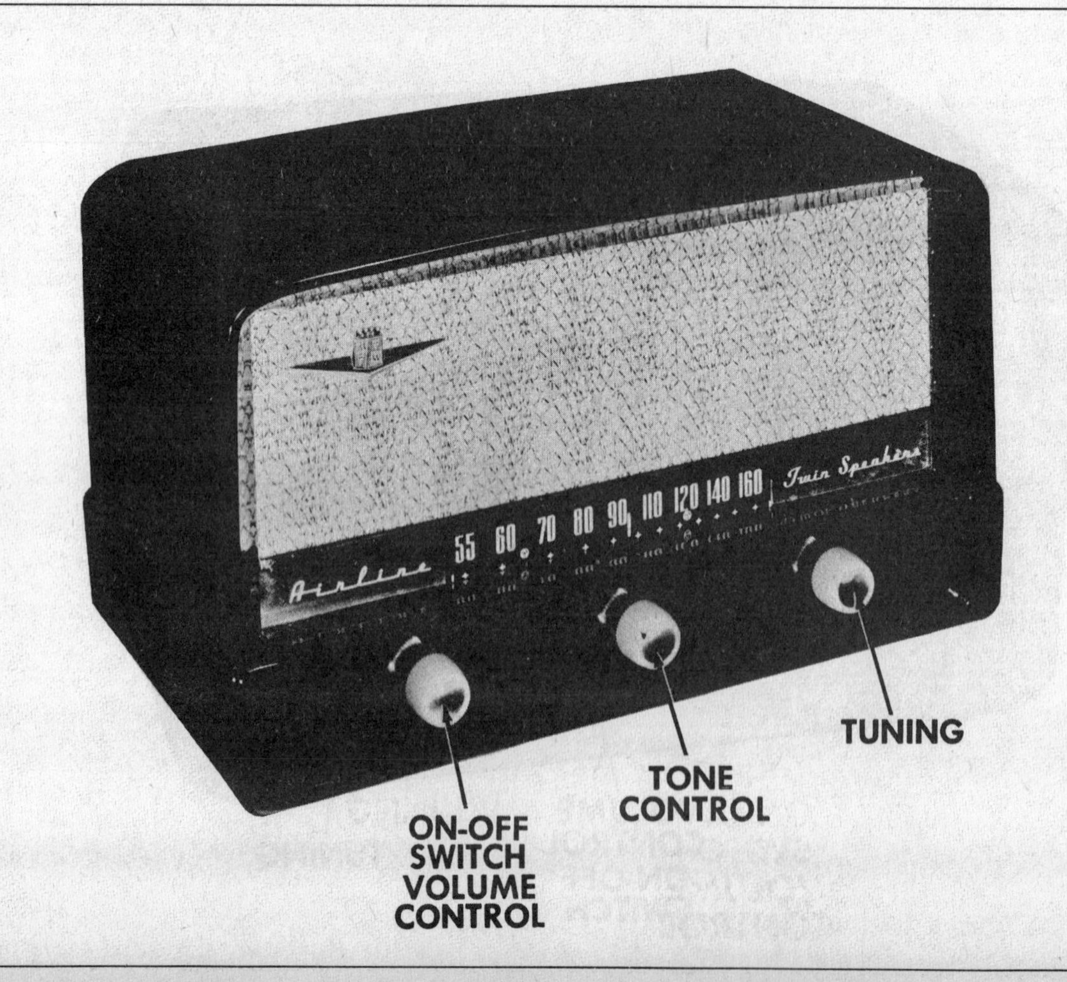

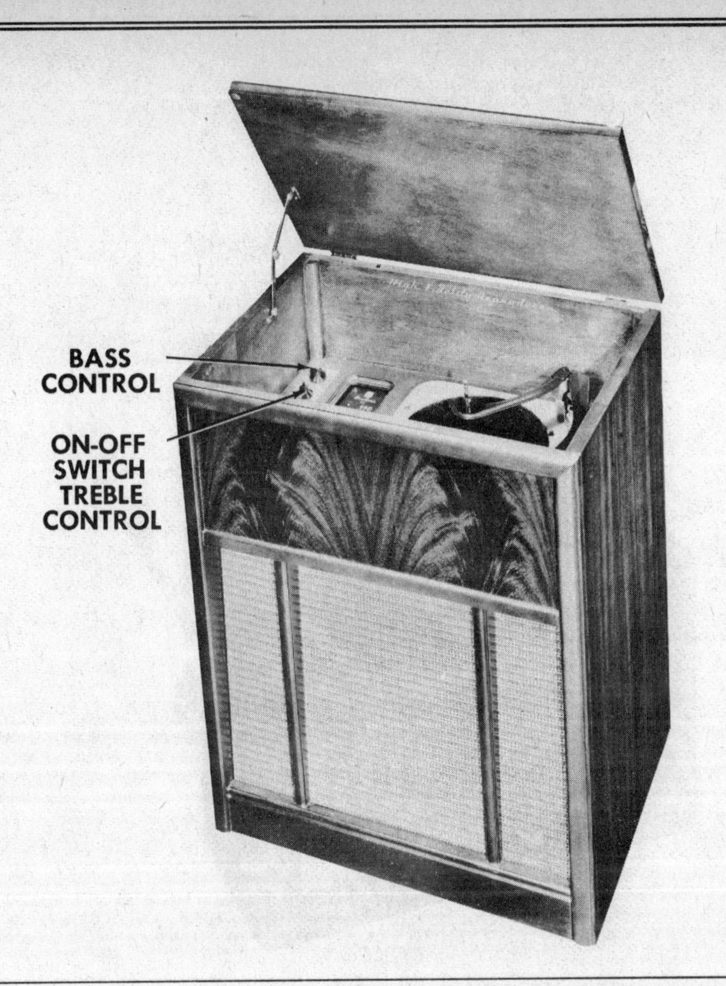

AIRLINE

MODEL PICTURED
WG-2602A

AC operated AM-phono superheterodyne receiver

TUBES
8

POWER SUPPLY
105-120 volts AC, 60 cycle

TUNING RANGE
540-1600KC

MFR/SUPPLIER
Montgomery Ward and Co.

PHOTOFACT SET
307-3

PUBLISHED
1956

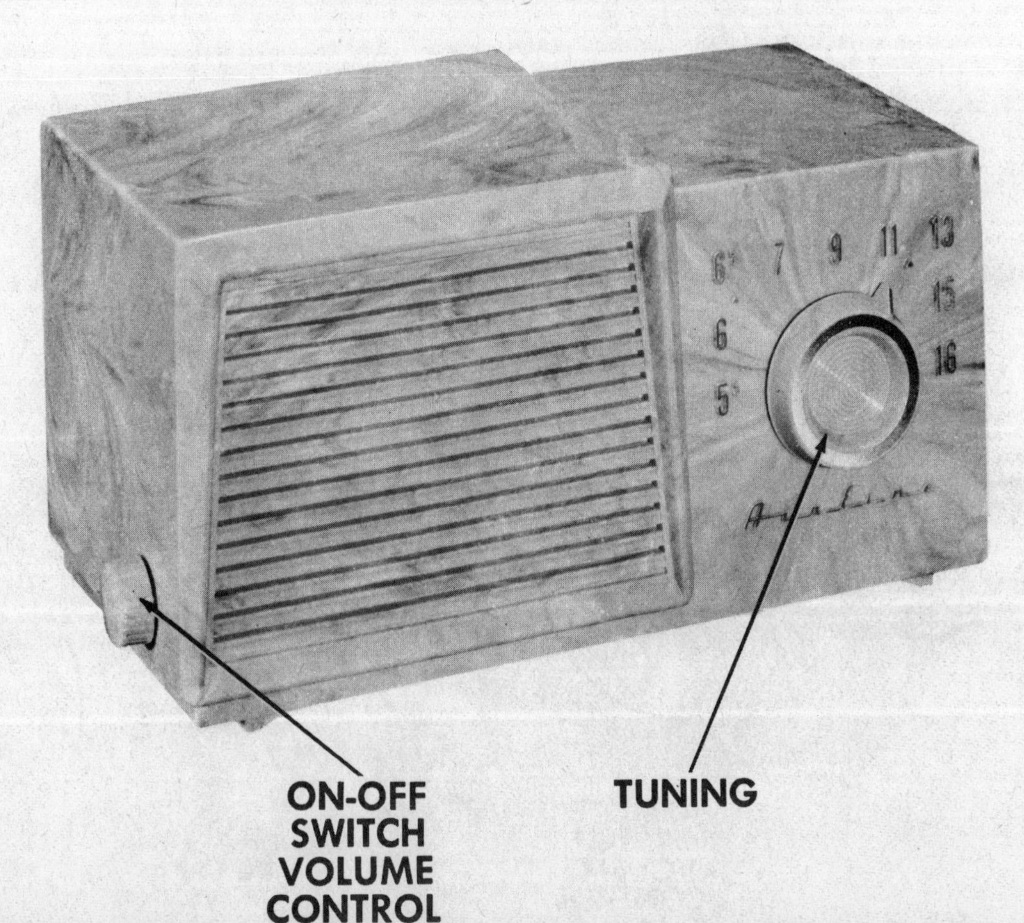

AIRLINE

MODEL PICTURED
GSE-1620A

AC/DC operated AM superheterodyne receiver

TUBES
5

POWER SUPPLY
110-120 volts AC/DC

TUNING RANGE
540-1620KC

MFR/SUPPLIER
Montgomery Ward and Co.

PHOTOFACT SET
317-2

PUBLISHED
1956

AIRLINE

MODEL PICTURED
GAA-990A

AC operated AM superheterodyne receiver with three-speed auto record changer

TUBES
5

POWER SUPPLY
105-125 volts AC, 60 cycle

TUNING RANGE
535-1620KC

MFR/SUPPLIER
Montgomery Ward and Co.

PHOTOFACT SET
320-3

PUBLISHED
1956

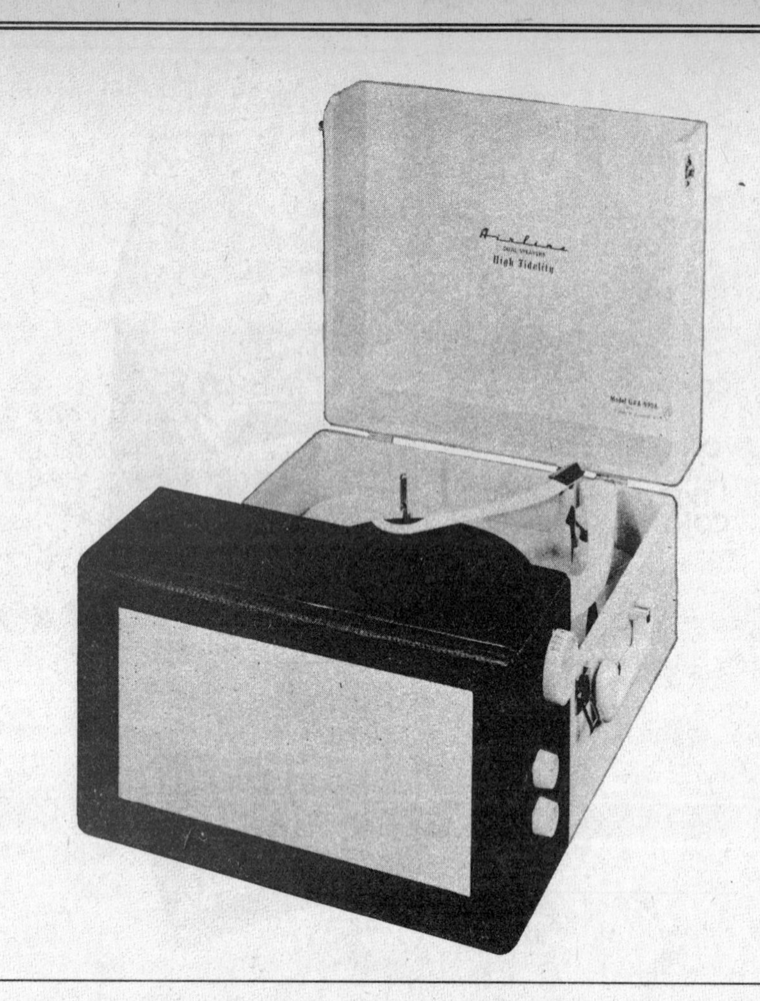

AIRLINE

MODEL PICTURED
GSL-1575A

AC operated AM superheterodyne receiver with electric clock

TUBES
4

POWER SUPPLY
110-120 volts AC, 60 cycle

TUNING RANGE
540-1600KC

MFR/SUPPLIER
Montgomery Ward and Co.

PHOTOFACT SET
323-3

PUBLISHED
1956

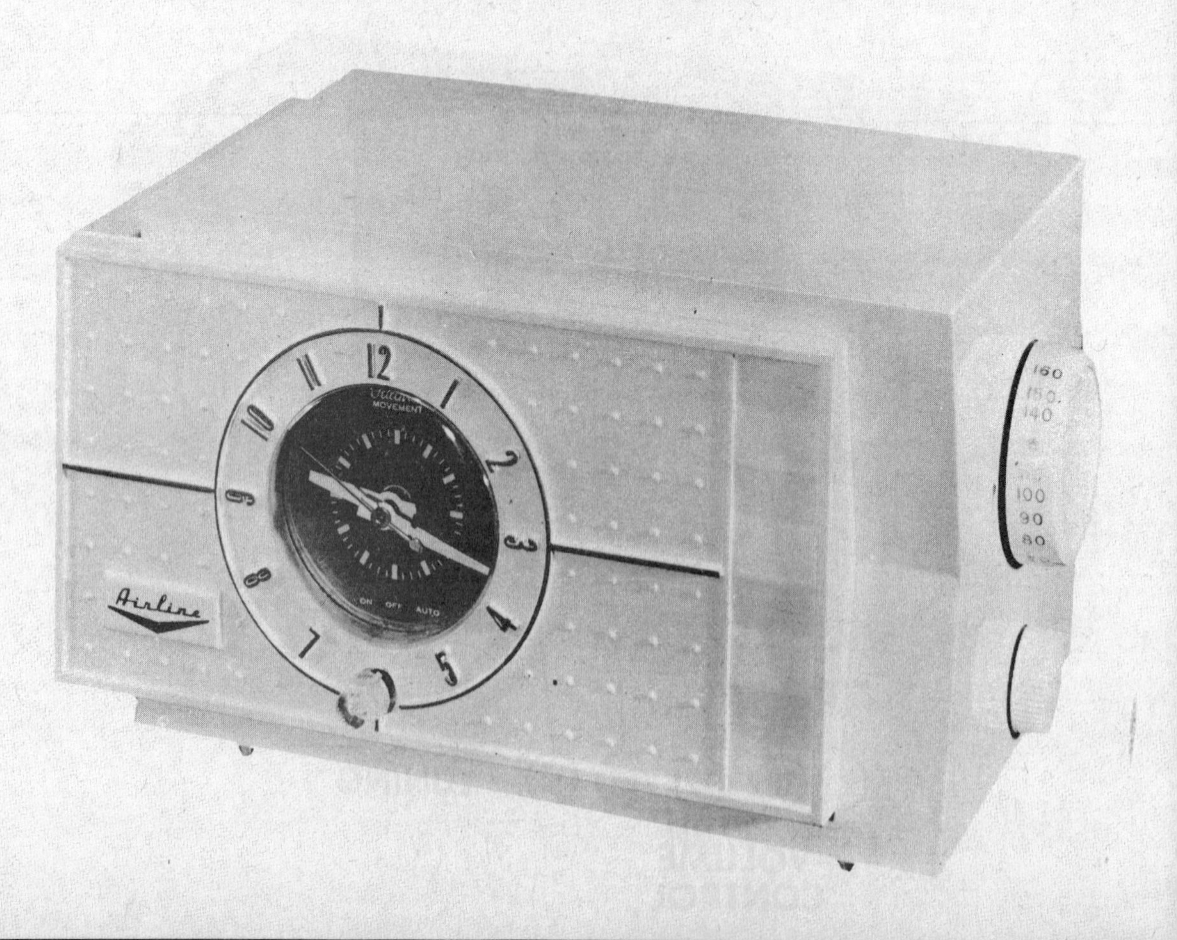

AIRLINE

MODEL PICTURED
GSE-1625A

AC/DC operated AM superheterodyne receiver

TUBES
5

POWER SUPPLY
110-120 volts AC/DC, .23 amp @ 117 volts AC

TUNING RANGE
540KC-1620KC

MFR/SUPPLIER
Montgomery Ward and Co.

PHOTOFACT SET
325-3

PUBLISHED
1956

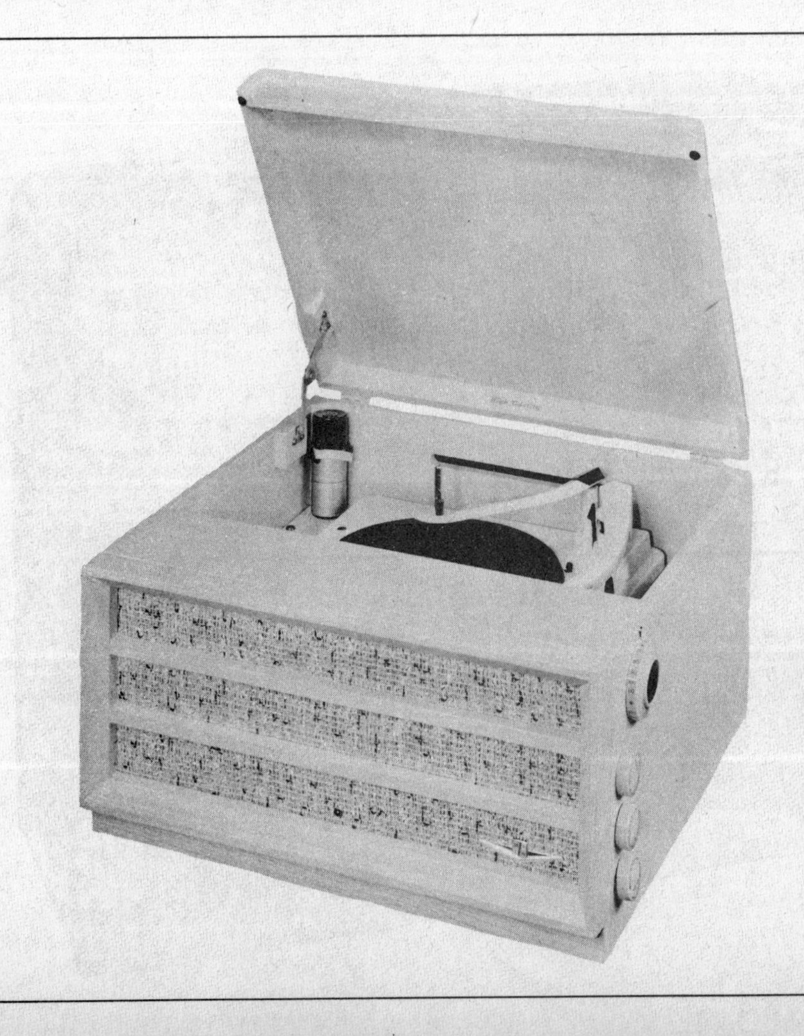

AIRLINE

MODEL PICTURED
2509

AC operated AM superheterodyne receiver with four-speed auto record changer

TUBES
6

POWER SUPPLY
110-120 volts AC, 60 cycle

TUNING RANGE
540KC-1625KC

MFR/SUPPLIER
Montgomery Ward and Co.

PHOTOFACT SET
328-2

PUBLISHED
1956

AIRLINE

MODEL PICTURED
GEN-1103A

Three-power portable AM receiver

TUBES
4

POWER SUPPLY
110-120 volts AC/DC or 7.5 volts A & 67.5 volts B supply

TUNING RANGE
535-1620KC

MFR/SUPPLIER
Montgomery Ward and Co.

PHOTOFACT SET
349-2

PUBLISHED
1957

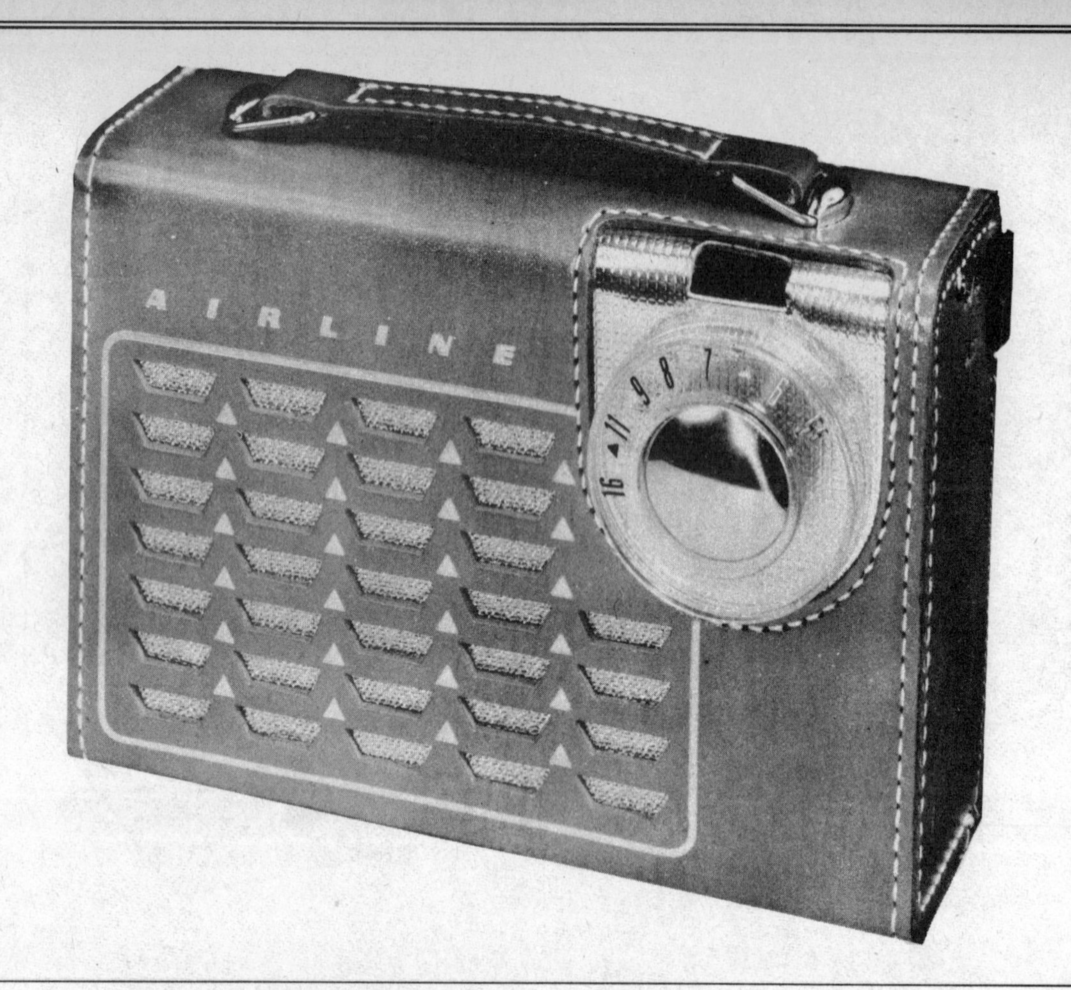

AIRLINE

MODEL PICTURED
GEN-1090A

Battery operated portable AM receiver

TUBES
4

POWER SUPPLY
1.5 volts A supply & 67.5 volts B supply, .245A @ 1.5 volts DC and 7mA @ 67.5 volts DC

TUNING RANGE
535KC-1640KC

MFR/SUPPLIER
Montgomery Ward and Co.

PHOTOFACT SET
350-2

PUBLISHED
1957

AIRLINE

MODEL PICTURED
GTC-1085A

Three-power operated portable AM receiver

TUBES
4

POWER SUPPLY
105-125 volts AC/DC or 3 volts A & 90 volts B supply

TUNING RANGE
535-1630KC

MFR/SUPPLIER
Montgomery Ward and Co.

PHOTOFACT SET
356-2

PUBLISHED
1957

AIRLINE

MODEL PICTURED
GEN-1655A

AC/DC operated AM receiver

TUBES
5

POWER SUPPLY
110-120 AC/DC

TUNING RANGE
540-1600KC

MFR/SUPPLIER
Montgomery Ward and Co.

PHOTOFACT SET
358-1

PUBLISHED
1957

AIRLINE

MODEL PICTURED
GSL-1650A

AC operated AM receiver with electric clock

TUBES
5

POWER SUPPLY
110-120 volts AC, 60 cycle

TUNING RANGE
535-1620KC

MFR/SUPPLIER
Montgomery Ward and Co.

PHOTOFACT SET
359-2

PUBLISHED
1957

AIRLINE

MODEL PICTURED
WG-1637A

AC/DC operated AM receiver with phono provision

TUBES
6

POWER SUPPLY
105-125 volts AC/DC

TUNING RANGE
540-1600KC

MFR/SUPPLIER
Montgomery Ward and Co.

PHOTOFACT SET
363-3

PUBLISHED
1957

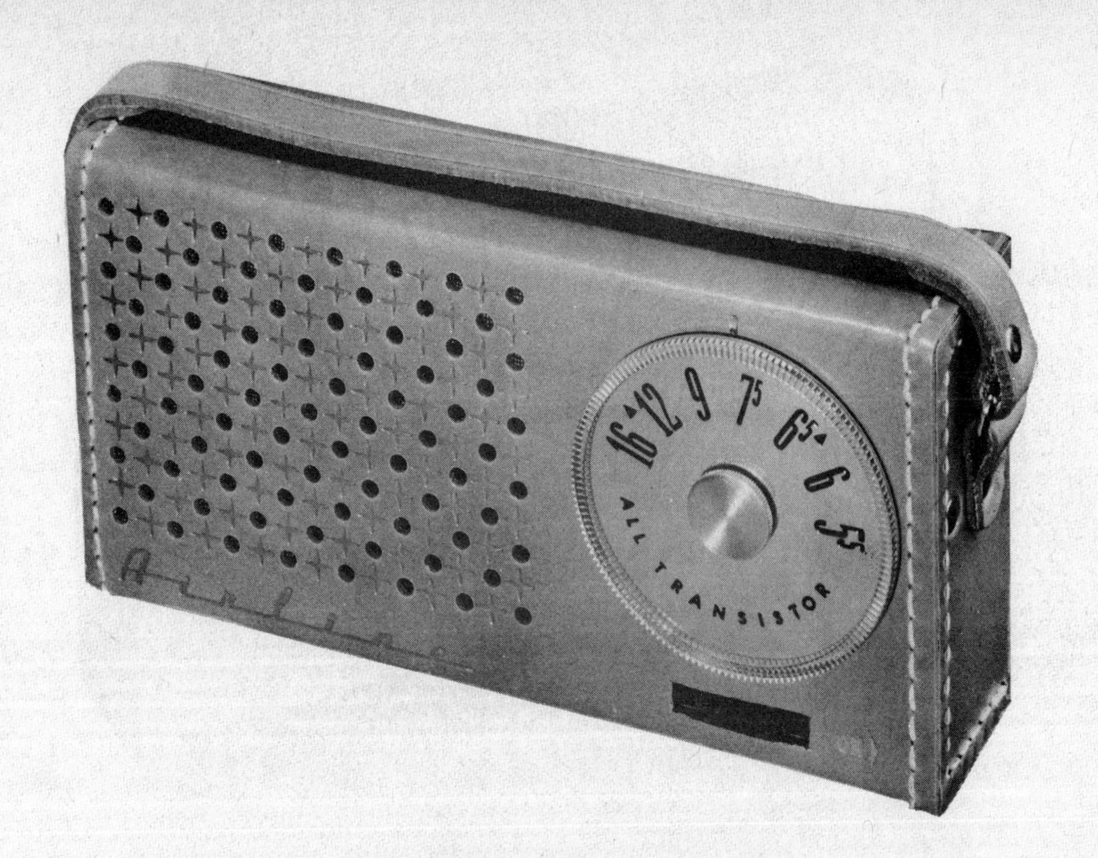

AIRLINE

MODEL PICTURED
GTM-1108A

Battery operated portable AM transistorized receiver

TUBES
0

POWER SUPPLY
9 volts DC

TUNING RANGE
540-1600KC

MFR/SUPPLIER
Montgomery Ward and Co.

PHOTOFACT SET
379-5

PUBLISHED
1957

AIRLINE

MODEL PICTURED
GAA-2620A

AC operated AM-FM receiver with four-speed auto record changer

TUBES
14

POWER SUPPLY
105-125 volts AC, 60 cycle

TUNING RANGE
540-1600KC, 88-108MC

MFR/SUPPLIER
Montgomery Ward and Co.

PHOTOFACT SET
384-6

PUBLISHED
1958

AIRLINE

MODEL PICTURED
GTM-1109A

Battery operated portable transistorized receiver

TUBES
0

POWER SUPPLY
9 volts DC, 6mA @ 9 volts DC

TUNING RANGE
540-1600KC

MFR/SUPPLIER
Montgomery Ward and Co.

PHOTOFACT SET
392-6

PUBLISHED
1958

AIRLINE

MODEL PICTURED
GAA-1003A

AC operated AM receiver with four-speed auto record changer

TUBES
5

POWER SUPPLY
105-125 volts AC, 60 cycle

TUNING RANGE
535-1620KC

MFR/SUPPLIER
Montgomery Ward and Co.

PHOTOFACT SET
404-4

PUBLISHED
1958

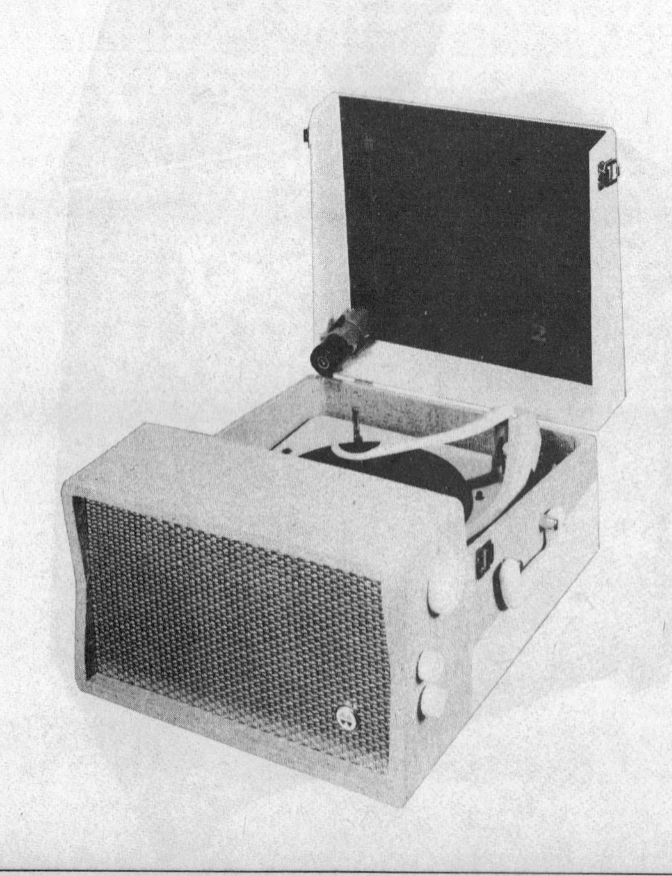

AIRLINE

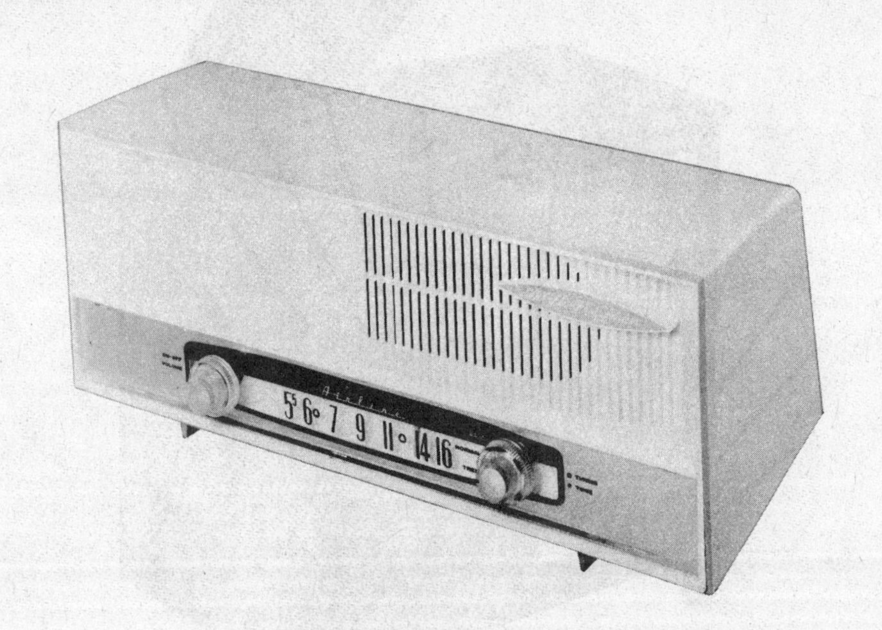

MODEL PICTURED
GTM-1639B

AC/DC operated AM receiver

TUBES
6

POWER SUPPLY
105-120 volts AC/DC

TUNING RANGE
540-1600KC

MFR/SUPPLIER
Montgomery Ward and Co.

PHOTOFACT SET
411-5

PUBLISHED
1958

AIRLINE

MODEL PICTURED
GEN-1670A

AC operated AM receiver with electric clock

TUBES
4

POWER SUPPLY
105-120 volts AC, 60 cycles, 26 WATTS, .275 amp @ 117 volts AC (less clock)

TUNING RANGE
540-1640KC

MFR/SUPPLIER
Montgomery Ward and Co.

PHOTOFACT SET
428-3

PUBLISHED
1959

AIRLINE

MODEL PICTURED
GEN-2645A

AC operated FM/AM receiver with four-speed auto record changer

TUBES
16

POWER SUPPLY
105-125 volts AC, 60 cycles

TUNING RANGE
540-1640KC, 88-108MC

MFR/SUPPLIER
Montgomery Ward and Co.

PHOTOFACT SET
452-4

PUBLISHED
1959

AIRLINE

MODEL PICTURED
WG-2673A

AC operated FM/AM tuner/amplifier with four-speed auto record changer

TUBES
17

POWER SUPPLY
105-125 volts AC, 60 cycles

TUNING RANGE
535-1620KC, 88-108MC

MFR/SUPPLIER
Montgomery Ward and Co.

PHOTOFACT SET
453-3

PUBLISHED
1959

AIRLINE

MODEL PICTURED
GEN-1120C

Battery operated portable transistorized AM receiver

TUBES
0

POWER SUPPLY
9 volts DC

TUNING RANGE
537-1640KC

MFR/SUPPLIER
Montgomery Ward and Co.

PHOTOFACT SET
458-4

PUBLISHED
1959

AIRLINE

MODEL PICTURED
WG-2684A

AC operated FM-AM tuner, stereo preamp, stereo amp, four-speed record changer

TUBES
9

POWER SUPPLY
110-120 volts AC, 60 cycle

TUNING RANGE
535-1620KC, 88-108MC

MFR/SUPPLIER
Montgomery Ward and Co.

PHOTOFACT SET
466-4

PUBLISHED
1959

AIRLINE

MODEL PICTURED
GTM-1666A

AC operated seven-tube FM-AM receiver

TUBES
7

POWER SUPPLY
105-125 Volts AC, 60 cycle

TUNING RANGE
540-1625KC, 88-108MC

MFR/SUPPLIER
Montgomery Ward and Co.

PHOTOFACT SET
467-3

PUBLISHED
1959

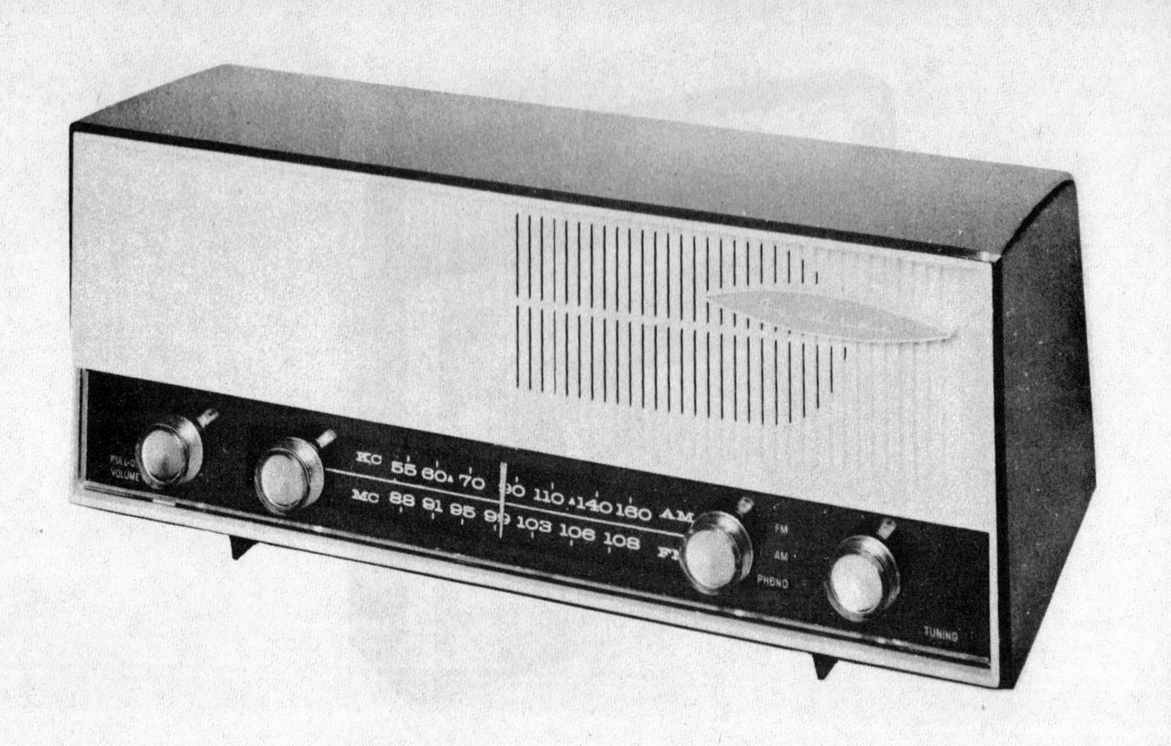

AIRLINE

MODEL PICTURED
GTM-1117A

Three-power portable AM receiver

TUBES
5

POWER SUPPLY
110-120 volts AC/DC or 9 volts A supply & 90 volts B supply

TUNING RANGE
540-1600KC

MFR/SUPPLIER
Montgomery Ward and Co.

PHOTOFACT SET
473-4

PUBLISHED
1960

AIRLINE

MODEL PICTURED
GTM-1201A

Battery operated transistorized portable AM receiver

TUBES
0

POWER SUPPLY
6 volts DC, 8.5mA @ 6 volts DC (no signal, minimum volume), 13mA @ 6 volts DC (signal, normal volume)

TUNING RANGE
535-1625KC

MFR/SUPPLIER
Montgomery Ward and Co.

PHOTOFACT SET
478-6

PUBLISHED
1960

AIRLINE

MODEL PICTURED
GAA-2640A

AC operated FM/AM receiver with four-speed auto stereo record changer

TUBES
15

POWER SUPPLY
110-120 volts AC, 60 cycle

TUNING RANGE

MFR/SUPPLIER
Montgomery Ward and Co.

PHOTOFACT SET
496-5

PUBLISHED
1960

ALGENE

MODEL PICTURED
AR5U

AC/DC operated superheterodyne receiver with loop antenna

TUBES
5

POWER SUPPLY
110-120 volts AC/DC, .240 amp @ 117 volts AC

TUNING RANGE
530-1700KC

MFR/SUPPLIER
Algene Radio Corp.

PHOTOFACT SET
22-3

PUBLISHED
1947

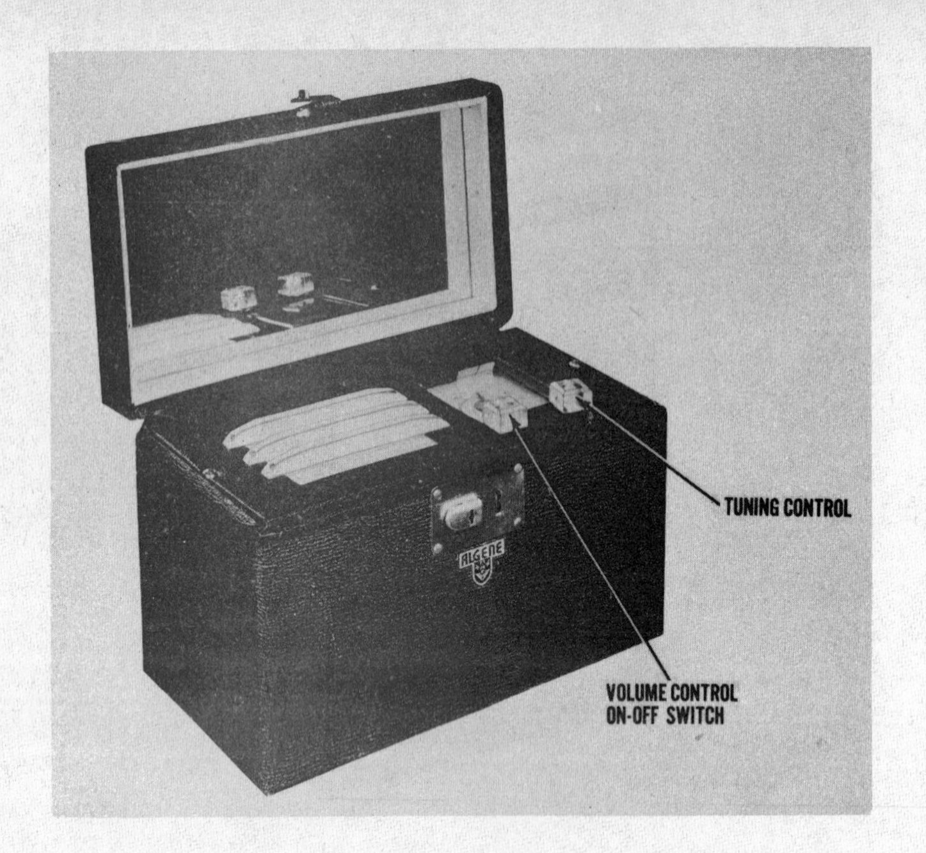

ALGENE

MODEL PICTURED
AR-6U

Three-power portable superheterodyne receiver with loop antenna

TUBES
6

POWER SUPPLY
110-120 volts AC/DC or 7.5 volts A supply & 90 volts B supply in pack form

TUNING RANGE
530-1700KC

MFR/SUPPLIER
Algene Radio Corp.

PHOTOFACT SET
22-4

PUBLISHED
1947

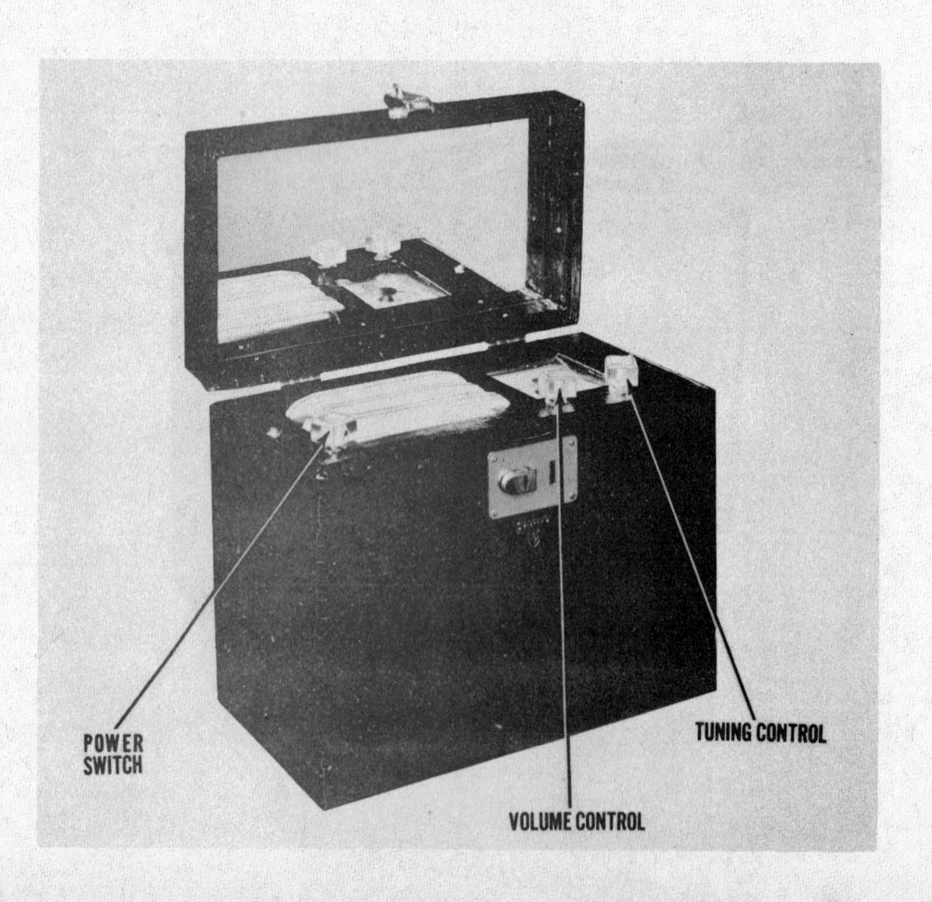

AMC

MODEL PICTURED
126

AC/DC operated superheterodyne receiver, self-contained loop antenna

TUBES
6

POWER SUPPLY
105-125 volts AC/DC, .260 amp @ 117 volts AC

TUNING RANGE
540-1650KC

MFR/SUPPLIER
Associated Merchandising Corp.

PHOTOFACT SET
16-1

PUBLISHED
1947

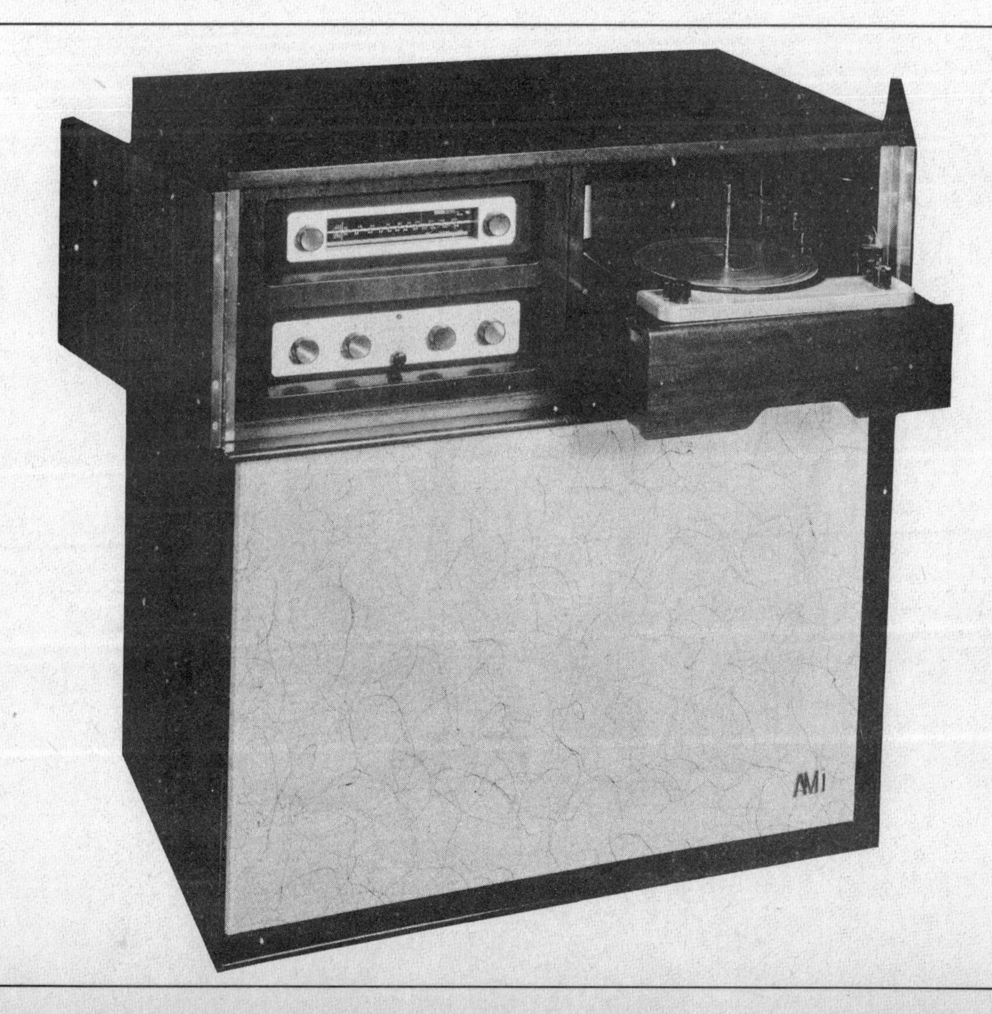

A.M.I.

MODEL PICTURED
PBA (Mark 1)

AC operated AM-FM tuner with 5-channel audio amp & four-speed auto record changer

TUBES
18

POWER SUPPLY
110-120 volts AC, 60 cycle

TUNING RANGE
530-1650KC, 88-108MC

MFR/SUPPLIER
Automatic Musical Instruments

PHOTOFACT SET
361-3

PUBLISHED
1957

ANDREA

MODEL PICTURED
P-163

Three-power portable 3-band superheterodyne receiver with loop antenna

TUBES
6

POWER SUPPLY
110-130 volts AC/DC or 9 volts A Battery & 90 volts B battery

TUNING RANGE
530-1600, 2.1MC, 6.4MC, 6.2-18.5MC

MFR/SUPPLIER
Andrea Radio Corp.

PHOTOFACT SET
18-8

PUBLISHED
1947

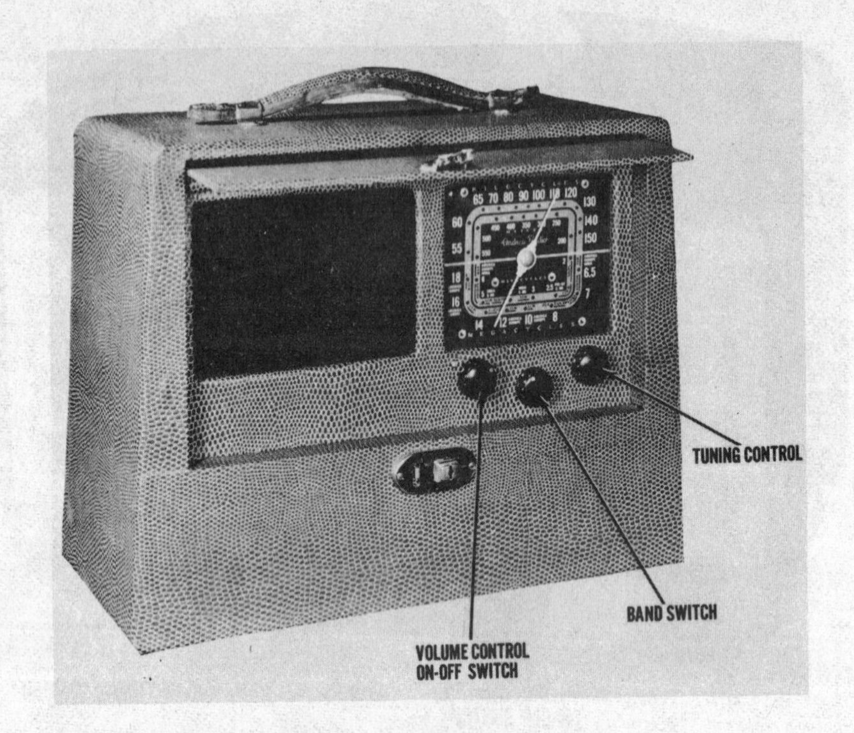

ANDREA

MODEL PICTURED
T-16

AC operated three-band superheterodyne receiver with loop antenna

TUBES
6

POWER SUPPLY
110-240 volts AC, .540 amp @ 117 volts AC

TUNING RANGE
540-1620KC, 2.2-7.3MC, 7.6-22.4MC

MFR/SUPPLIER
Andrea Radio Corp.

PHOTOFACT SET
21-2

PUBLISHED
1947

ANDREA

MODEL PICTURED
T-U16

AC/DC operated three-band superheterodyne receiver with loop antenna

TUBES
6

POWER SUPPLY
100-250 volts AC/DC, .400 amp @ 117 volts AC

TUNING RANGE
540-1620KC, 2.2-7.3MC, 7.6-22.4MC

MFR/SUPPLIER
Andrea Radio Corp.

PHOTOFACT SET
21-3

PUBLISHED
1947

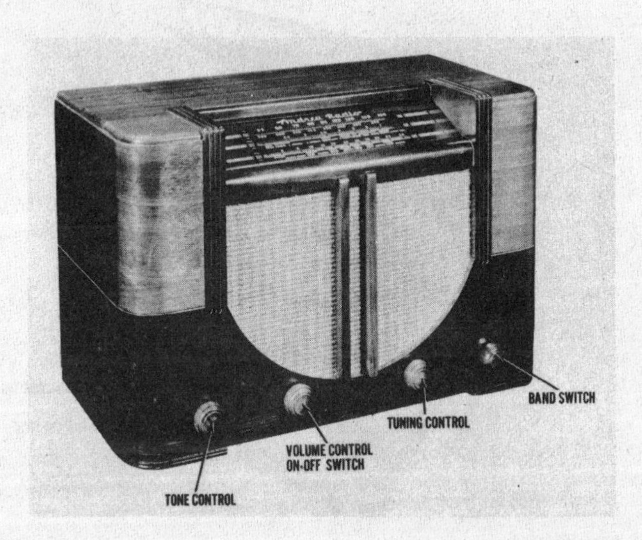

ANDREA

MODEL PICTURED
T-U15

AC/DC operated two-band superheterodyne receiver with loop antenna

TUBES
5

POWER SUPPLY
105-130 volts AC/DC

TUNING RANGE
540-1620KC, 5.8-22MC

MFR/SUPPLIER
Andrea Radio Corp.

PHOTOFACT SET
24-7

PUBLISHED
1947

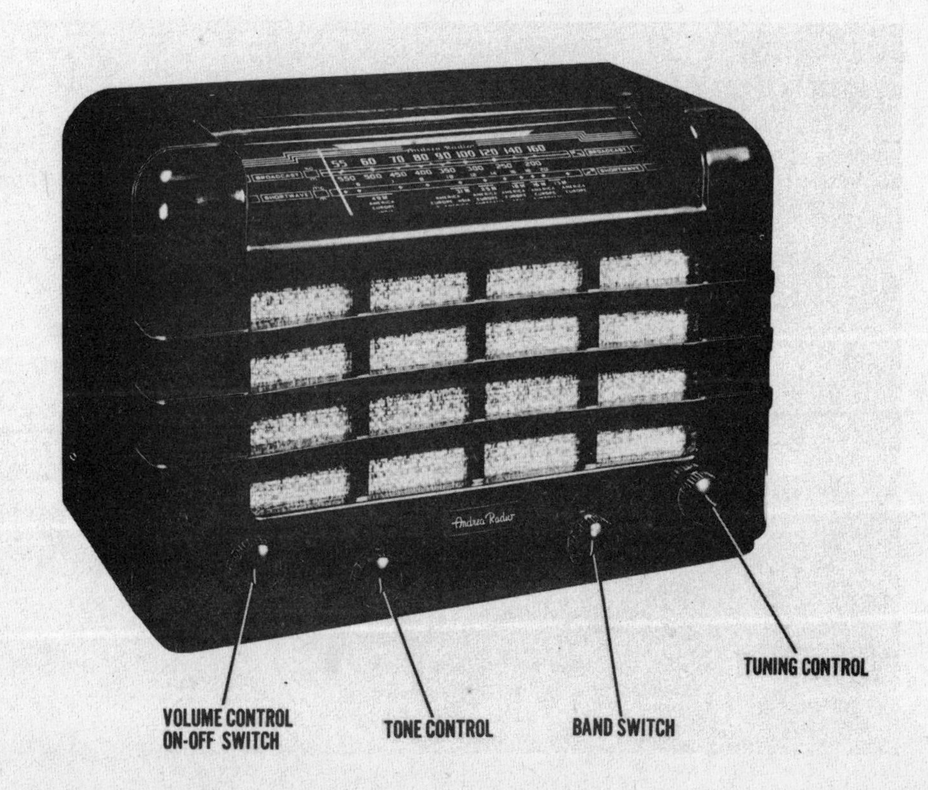

ANDREA

MODEL PICTURED
CO-U15

AC operated radio-phono combo superheterodyne receiver with loop antenna

TUBES
5

POWER SUPPLY
103-130/220-250 volts AC, .250 amp @ 117 volts AC

TUNING RANGE
540-1620KC, 5.8-22MC

MFR/SUPPLIER
Andrea Radio Corp.

PHOTOFACT SET
27-3

PUBLISHED
1947

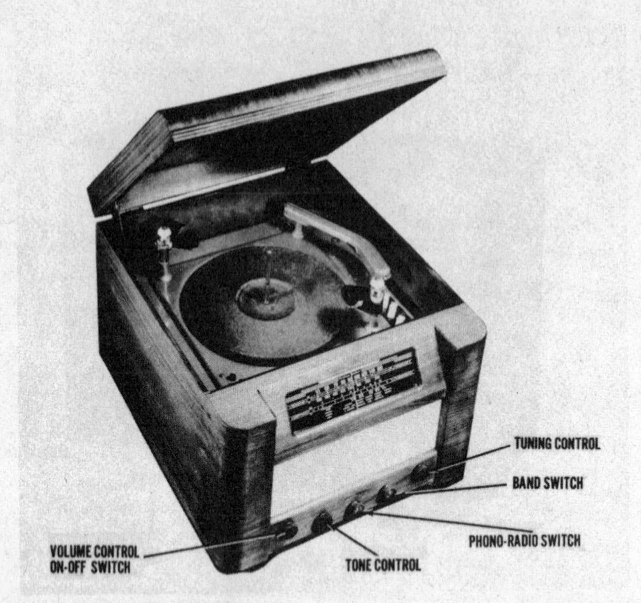

ANDREA

MODEL PICTURED
CRP-24W

AC operated AM-FM tuner with amplifier chassis & three-speed auto record changer

TUBES
16

POWER SUPPLY
110-120 volts AC, 60 cycle

TUNING RANGE
540-1700KC, 88-108MC

MFR/SUPPLIER
Andrea Radio Corp.

PHOTOFACT SET
367-4

PUBLISHED
1957

ANDREA

MODEL PICTURED
W69P Spacemaster Deluxe

Three-power portable nine-band AM receiver

TUBES
6

POWER SUPPLY
105-125 volts AC/DC or 9 volts A & 90 volts B supply in pack form

TUNING RANGE
Nine-bands

MFR/SUPPLIER
Andrea Radio Corp.

PHOTOFACT SET
371-1

PUBLISHED
1957

ANDREA

MODEL PICTURED
CRP-24WA

AC operated AM-FM tuner with amplifier and four-speed automatic record changer

TUBES
24

POWER SUPPLY
110-120 volts AC, 60 cycles

TUNING RANGE
540-1700KC, 88-108MC

MFR/SUPPLIER
Andrea Radio Corp.

PHOTOFACT SET
416-4

PUBLISHED
1958

ANSLEY

MODEL PICTURED
41 Paneltone

AC operated
superheterodyne receiver

TUBES
7

POWER SUPPLY
110-120 volts AC

TUNING RANGE
540-1730KC

MFR/SUPPLIER
Ansley Radio Corp.

PHOTOFACT SET
4-38

PUBLISHED
1946

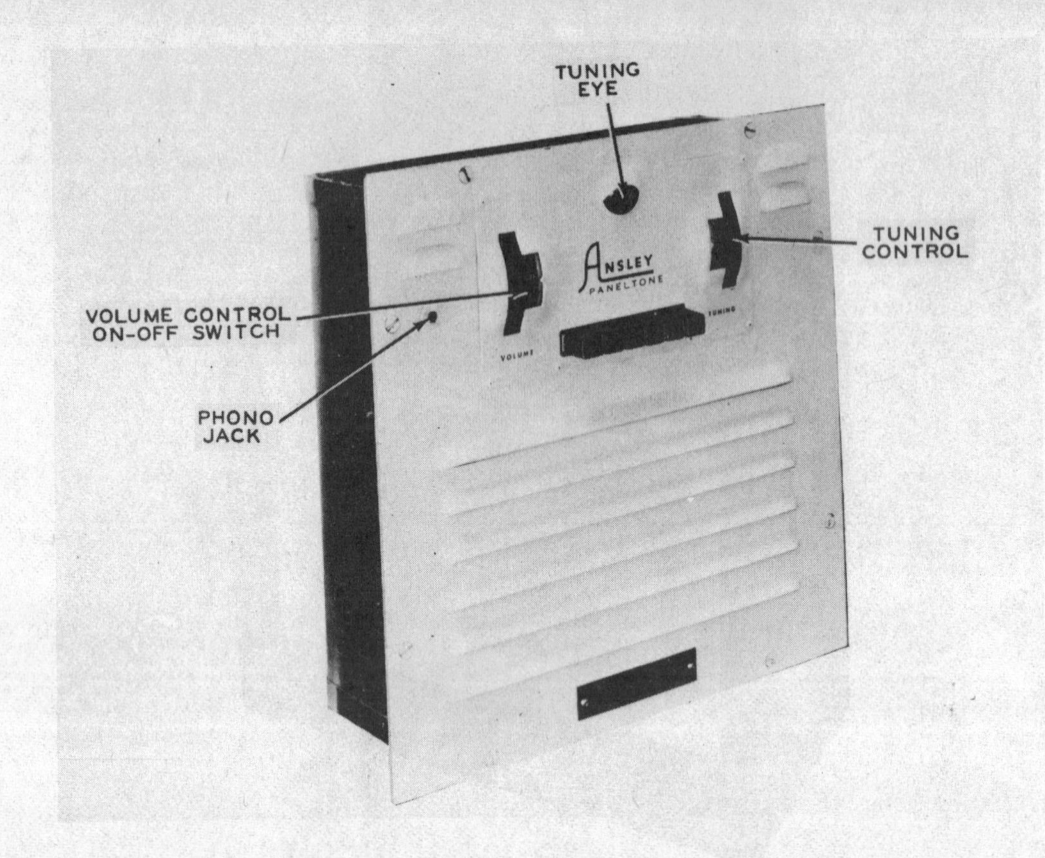

ANSLEY

MODEL PICTURED
32

Two-band
superheterodyne receiver
with self-contained loop
antenna

TUBES
9

POWER SUPPLY
115 volts AC

TUNING RANGE
540-1630KC, 5.7 -
18.3MC

MFR/SUPPLIER
Ansley Radio Corp.

PHOTOFACT SET
5-27

PUBLISHED
1946

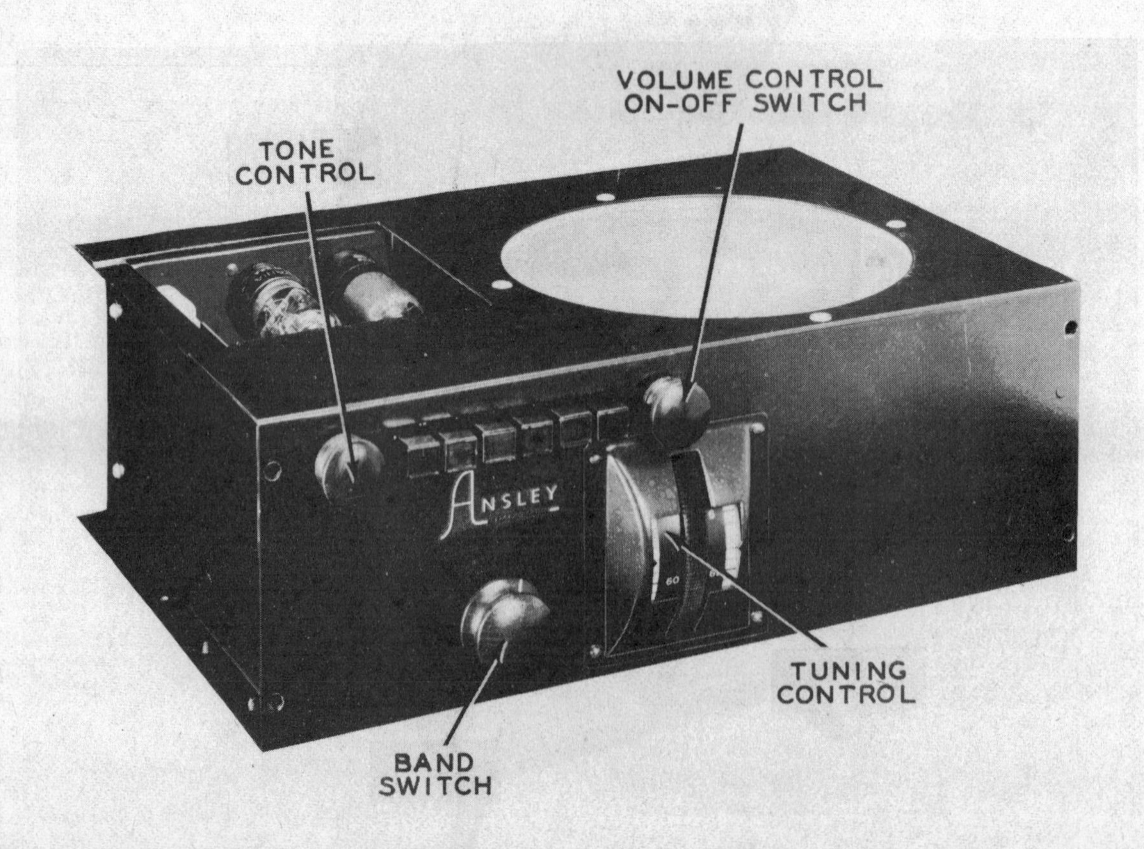

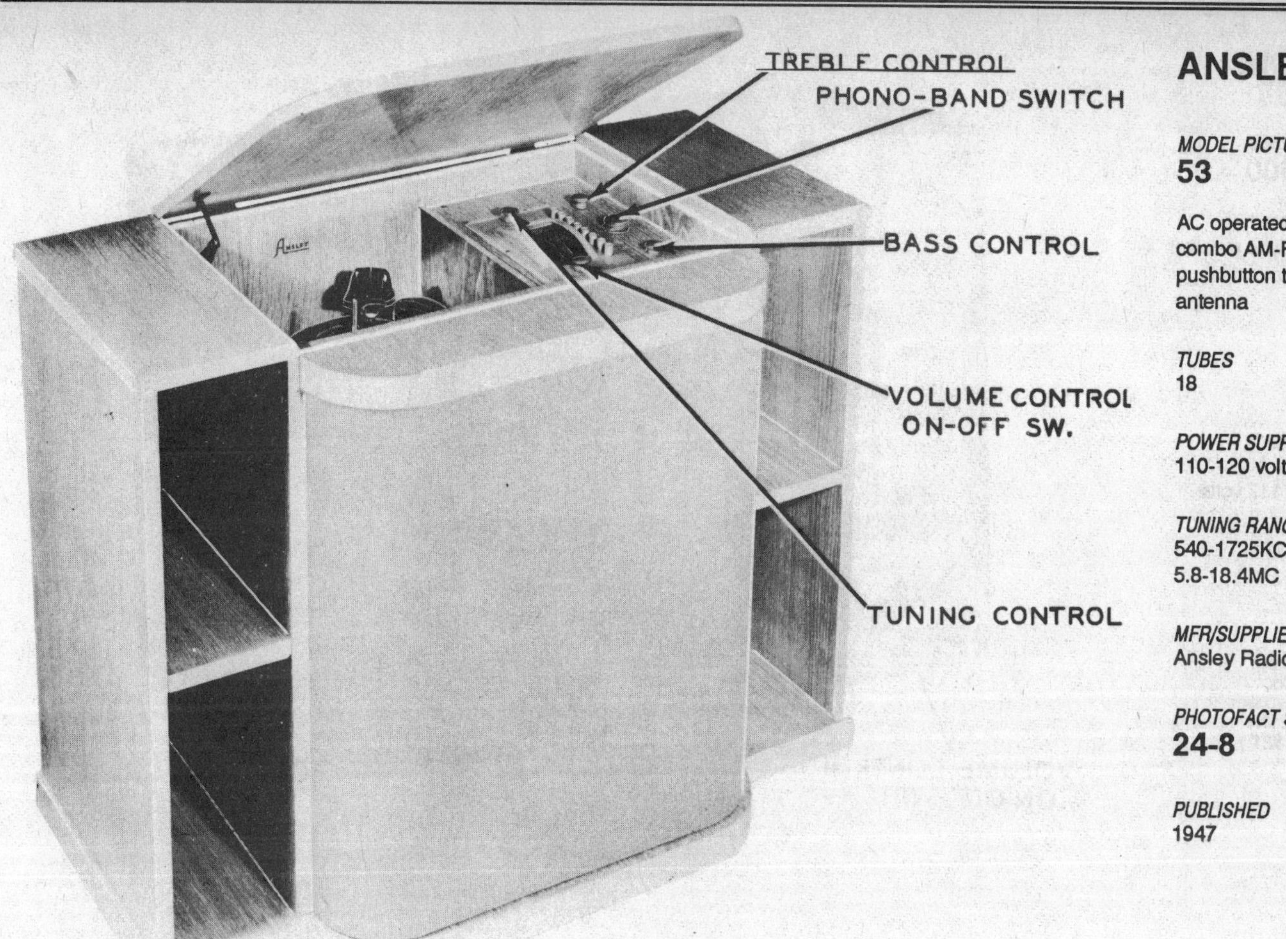

ANSLEY

MODEL PICTURED
53

AC operated phono-radio combo AM-FM super pushbutton tuning, loop antenna

TUBES
18

POWER SUPPLY
110-120 volts AC

TUNING RANGE
540-1725KC, 88-108MC, 5.8-18.4MC

MFR/SUPPLIER
Ansley Radio Corp.

PHOTOFACT SET
24-8

PUBLISHED
1947

APEX INDUSTRIES

MODEL PICTURED
4B5

AC/DC operated superheterodyne receiver with loop antenna

TUBES
5

POWER SUPPLY
110-120 volts AC/DC

TUNING RANGE
540-1720KC

MFR/SUPPLIER
Apex Industries

PHOTOFACT SET
37-2

PUBLISHED
1948

ARCADIA

MODEL PICTURED
37D14-600

AC/DC operated superheterodyne receiver with self-contained loop antenna

TUBES
7

POWER SUPPLY
105-125 volts AC/DC, .370 amp @ 117 volts

TUNING RANGE
528-1600KC, 5.75-183MC

MFR/SUPPLIER
Whitney & Co.

PHOTOFACT SET
9-3

PUBLISHED
1946

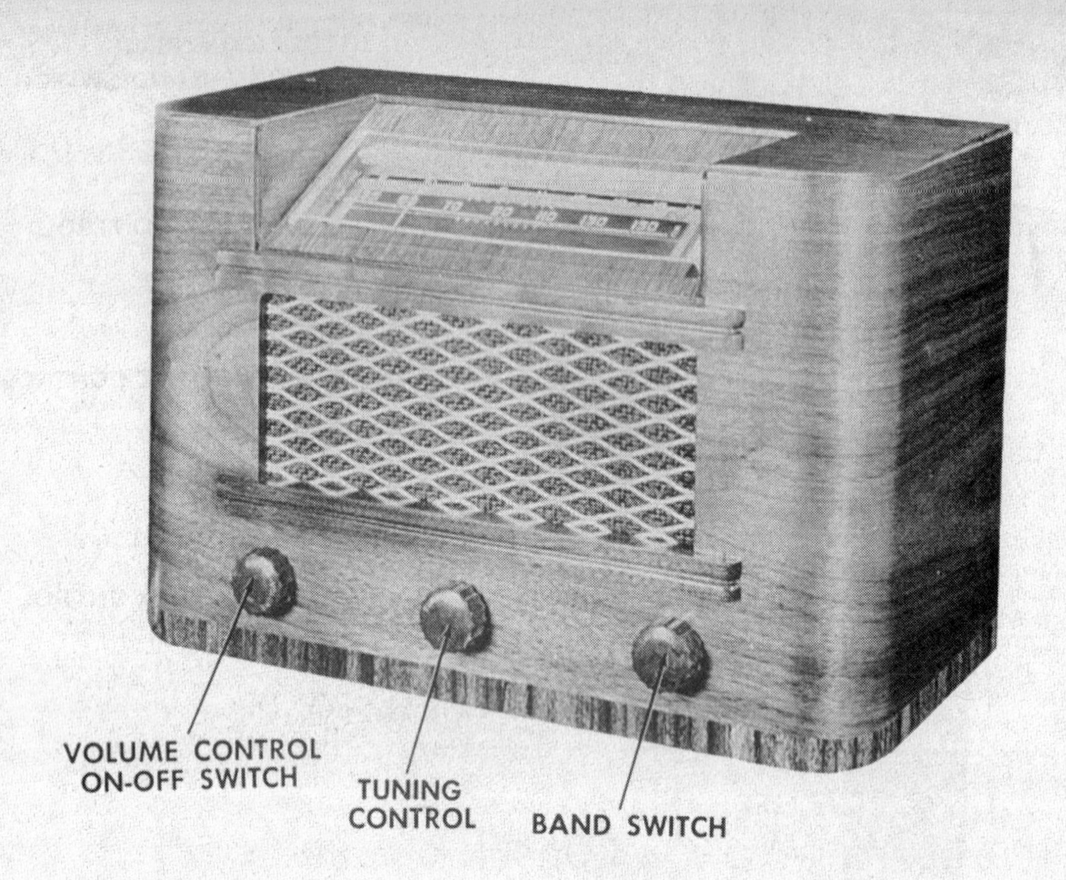

ARIA

MODEL PICTURED
554-1-61A

AC operated combo auto phono superheterodyne receiver, self-contained loop antenna

TUBES
6

POWER SUPPLY
105-125 volts AC

TUNING RANGE
540-1600KC

MFR/SUPPLIER
International Detrola Co.

PHOTOFACT SET
7-2

PUBLISHED
1946

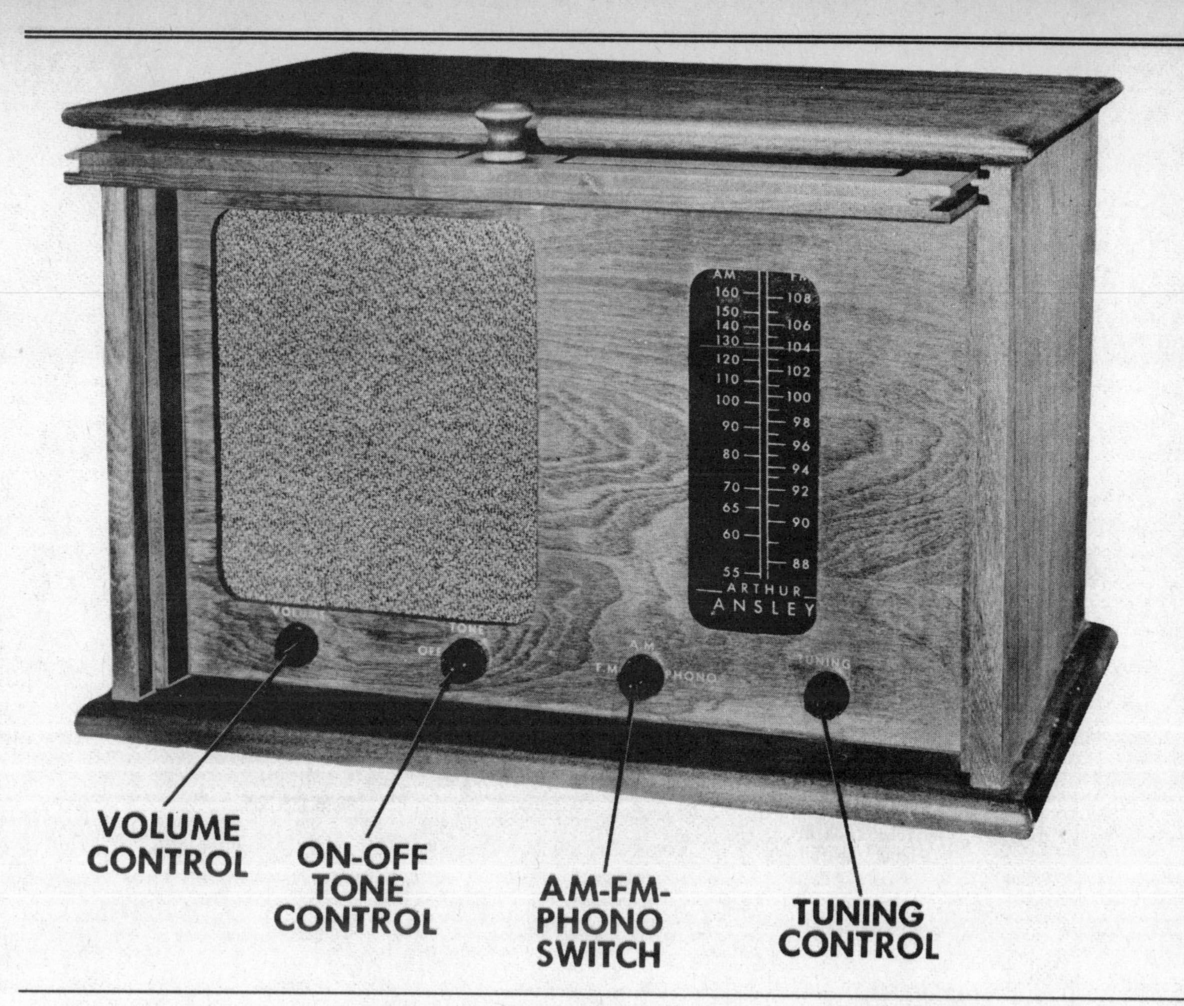

ARTHUR ANSLEY

MODEL PICTURED
R-1

AC operated AM-FM superheterodyne receiver

TUBES
10

POWER SUPPLY
110-120 volts AC

TUNING RANGE
540-1620KC, 88-108MC

MFR/SUPPLIER
Arthur Ansley Mfg. Co.

PHOTOFACT SET
200-2

PUBLISHED
1953

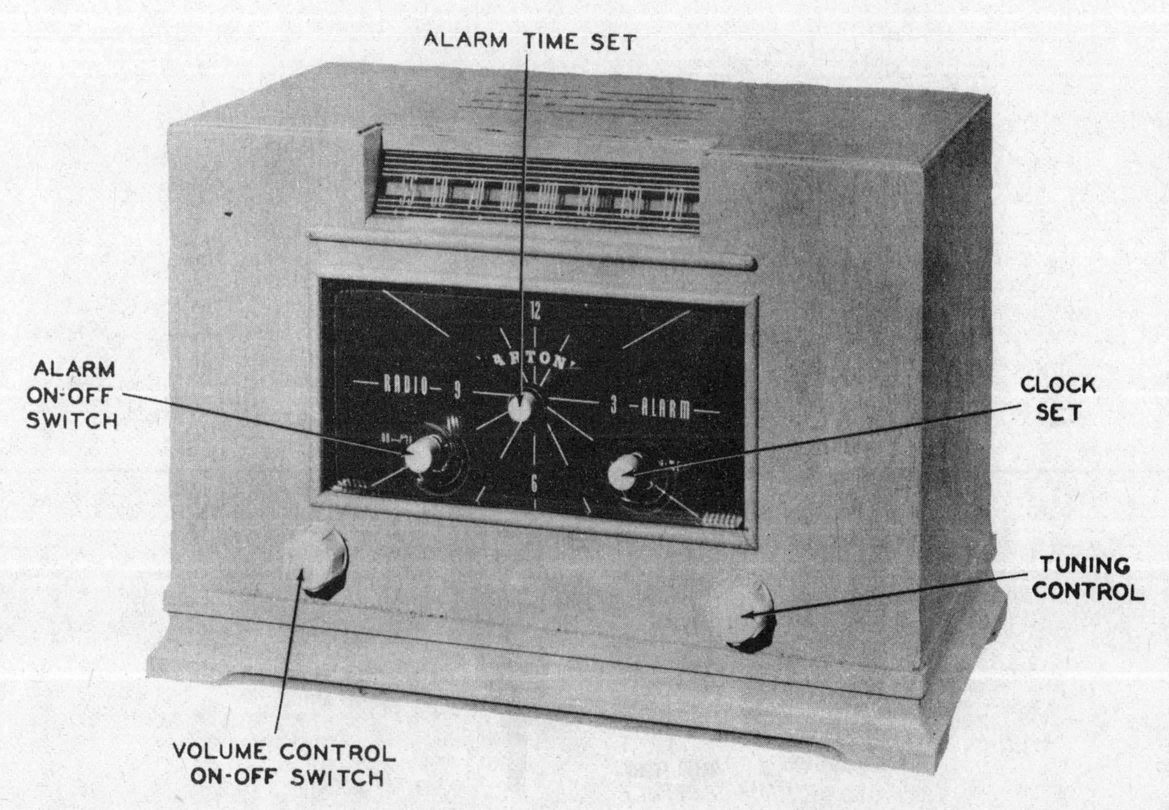

ARTONE

MODEL PICTURED
524

AC operated superheterodyne receiver with electric clock

TUBES
5

POWER SUPPLY
110-120 volts AC

TUNING RANGE
540-1720KC

MFR/SUPPLIER
Affiliated Retailers

PHOTOFACT SET
76-6

PUBLISHED
1949

ARVIN

MODEL PICTURED
555

AC/DC operated superheterodyne receiver, self-contained loop antenna

TUBES
5

POWER SUPPLY
105-125 volts AC/DC, .270 amp @ 117 volts AC

TUNING RANGE
540-1600KC

MFR/SUPPLIER
Noblitt-Sparks Industries

PHOTOFACT SET
13-9

PUBLISHED
1947

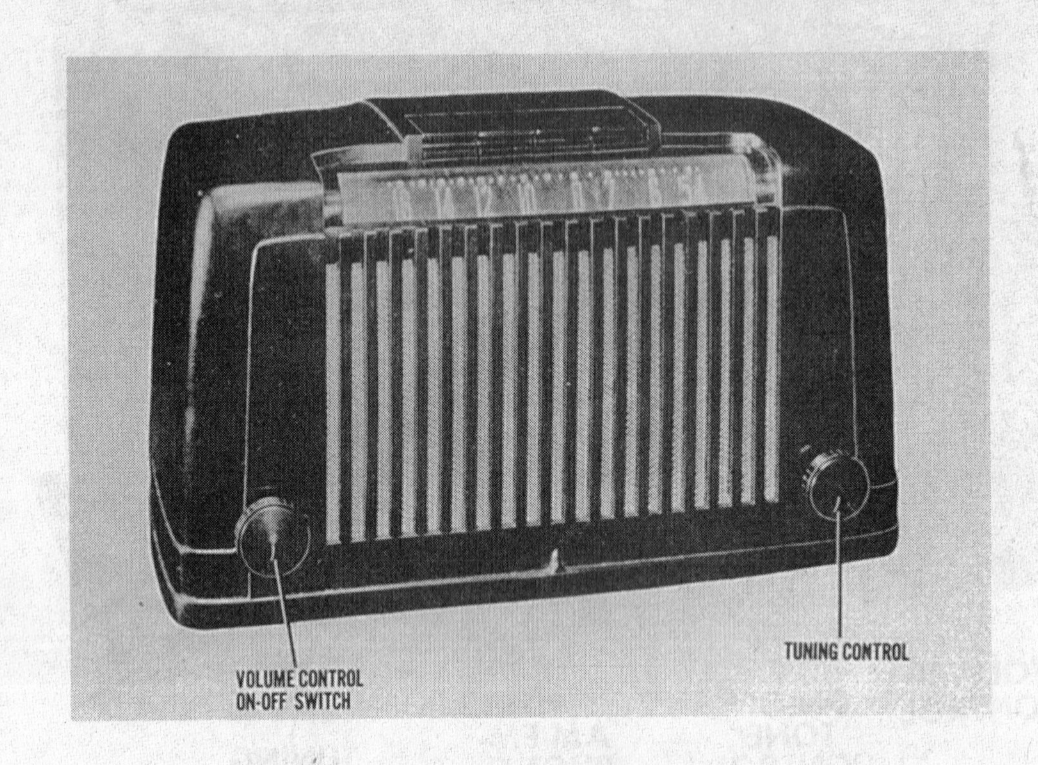

ARVIN

MODEL PICTURED
665

AC operated phono radio combo superheterodyne with loop antenna

TUBES
6

POWER SUPPLY
110-120 volts AC

TUNING RANGE
540-1600KC

MFR/SUPPLIER
Noblitt-Sparks Industries

PHOTOFACT SET
18-10

PUBLISHED
1947

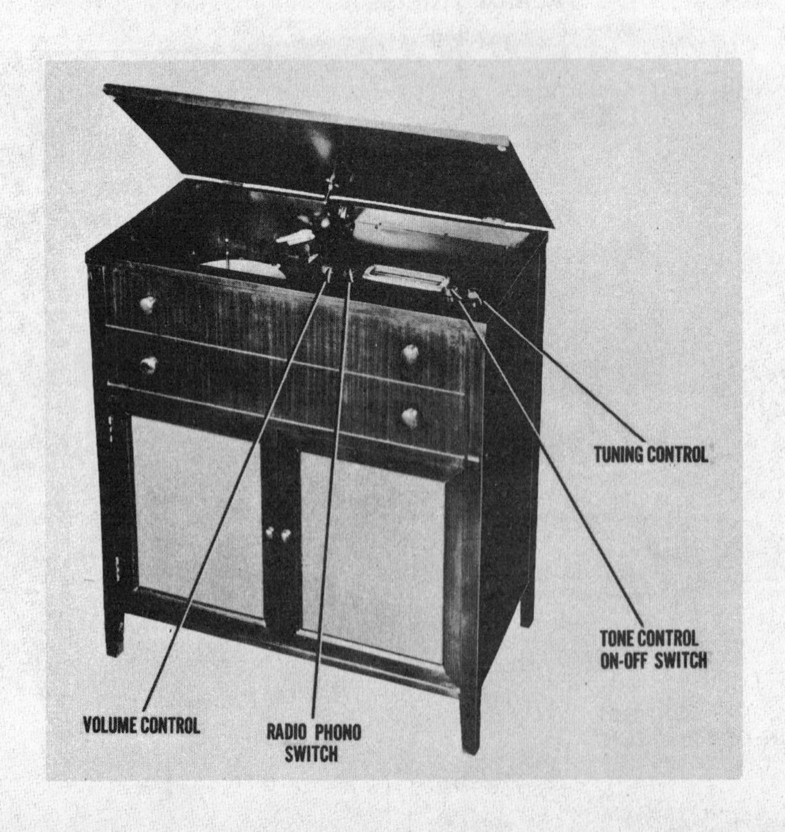

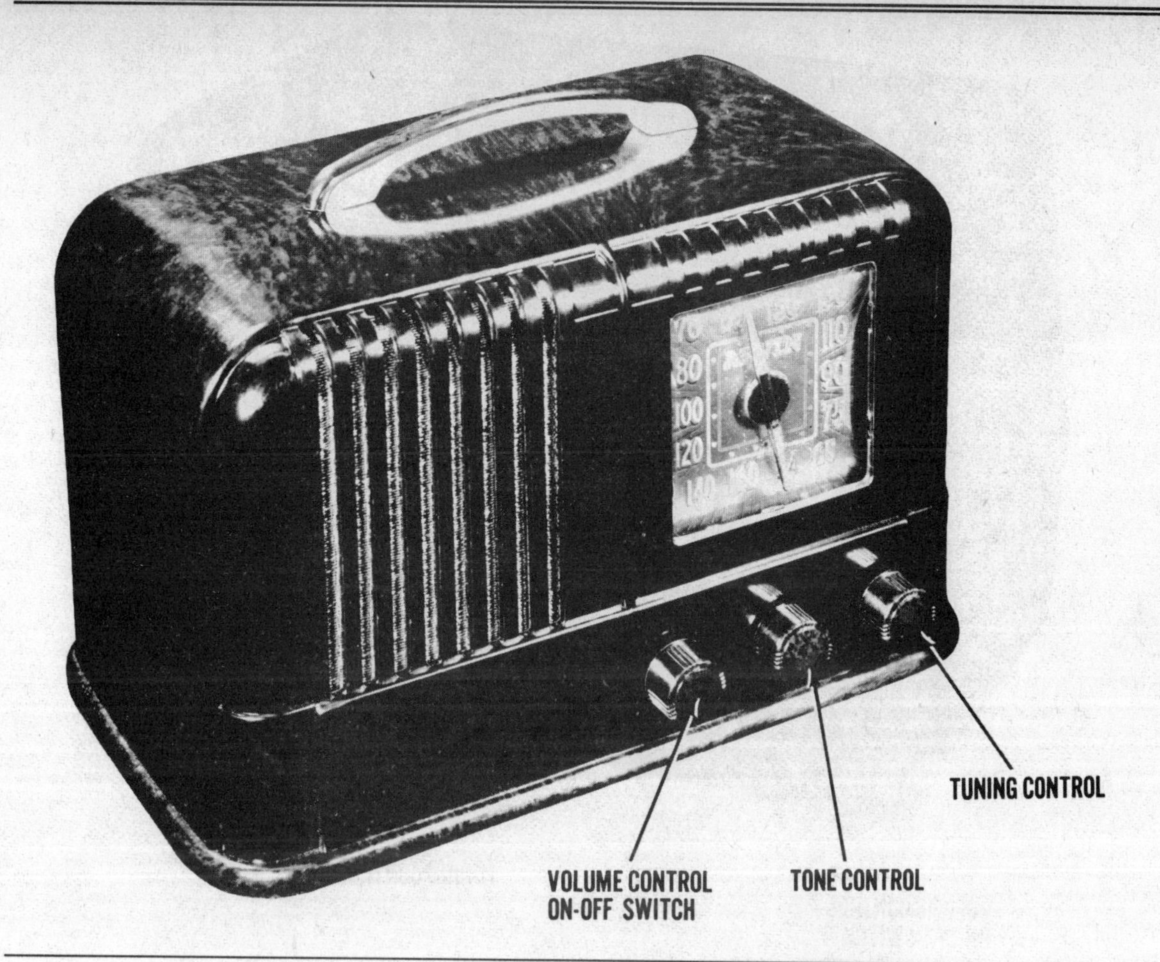

ARVIN

MODEL PICTURED
664

AC/DC operated superheterodyne receiver with loop antenna

TUBES
6

POWER SUPPLY
105-125 volts AC/DC

TUNING RANGE
540-1600KC

MFR/SUPPLIER
Noblitt-Sparks Industries

PHOTOFACT SET
29-2

PUBLISHED
1947

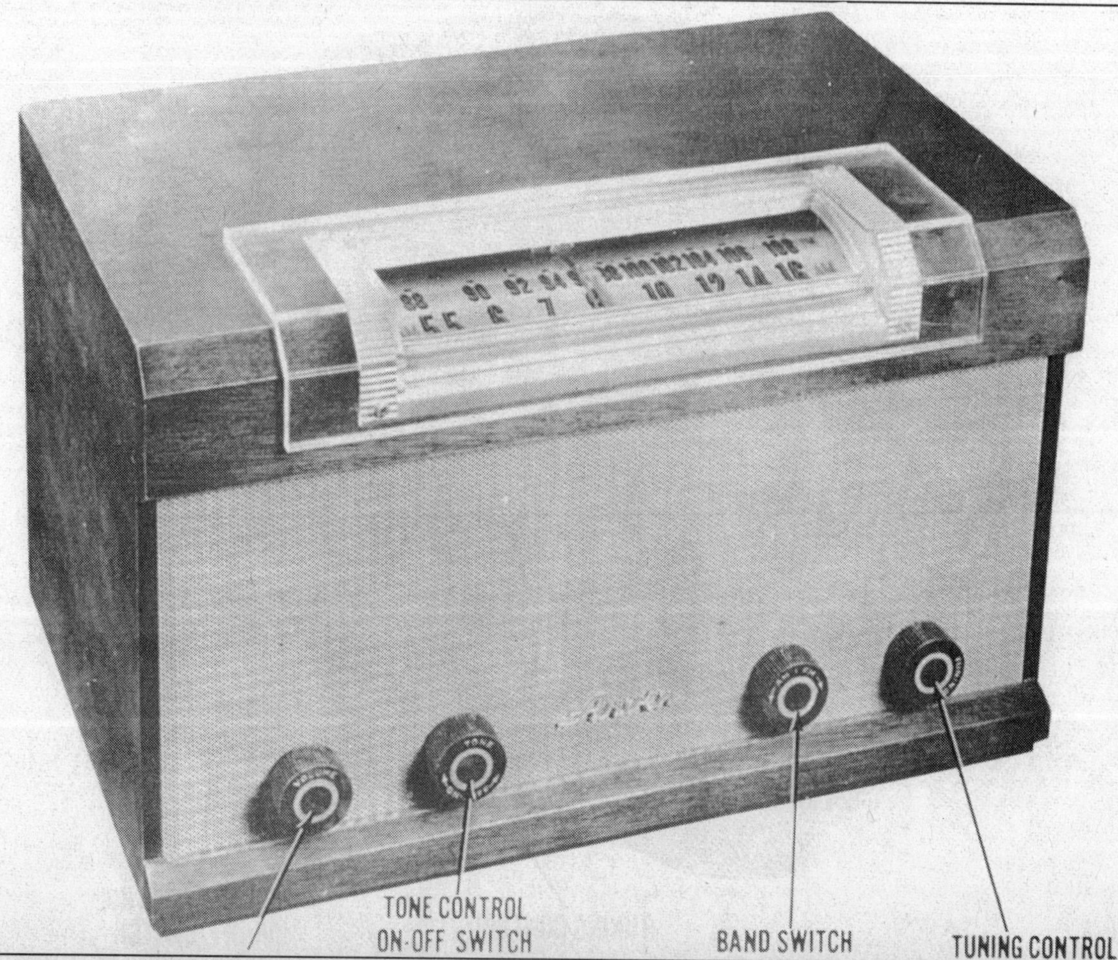

ARVIN

MODEL PICTURED
182TFM

AC/DC operated AM-FM superheterodyne receiver with loop antenna

TUBES
8

POWER SUPPLY
105-125 volts AC/DC, .38 amp @ 117 volts AC

TUNING RANGE
540-1600KC, 88-108MC

MFR/SUPPLIER
Noblitt-Sparks Industries

PHOTOFACT SET
32-3

PUBLISHED
1948

ARVIN

MODEL PICTURED
152T

AC/DC operated superheterodyne receiver with loop antenna

TUBES
5

POWER SUPPLY
110-120 volts AC/DC, .25 amp @ 117 volts AC

TUNING RANGE
540-1620KC

MFR/SUPPLIER
Noblitt-Sparks Industries

PHOTOFACT SET
33-1

PUBLISHED
1948

ARVIN

MODEL PICTURED
442

AC/DC operated superheterodyne receiver

TUBES
4

POWER SUPPLY
110-120 volts AC/DC

TUNING RANGE
540-1700KC

MFR/SUPPLIER
Noblitt-Sparks Industries

PHOTOFACT SET
34-2

PUBLISHED
1948

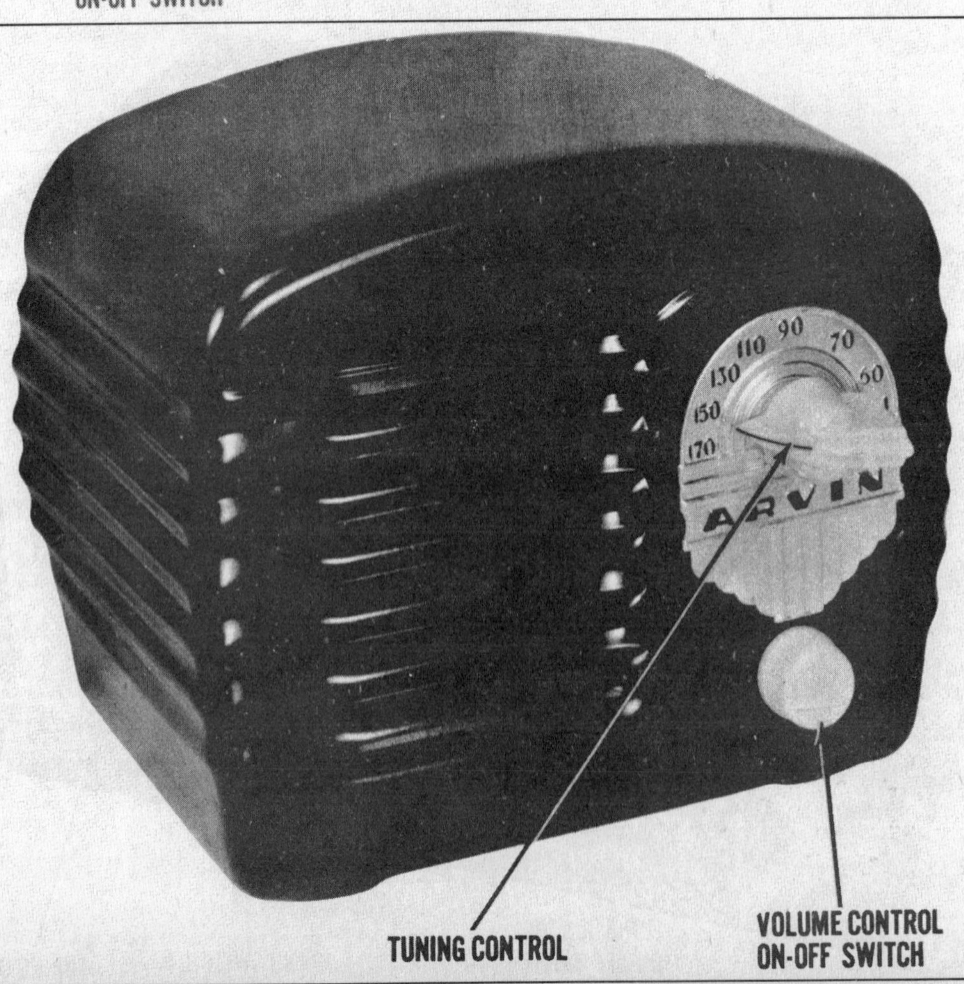

ARVIN

MODEL PICTURED
151TC

AC operated phono radio superheterodyne receiver with loop antenna

TUBES
5

POWER SUPPLY
105-125 volts AC

TUNING RANGE
540-1600KC

MFR/SUPPLIER
Noblitt-Sparks Industries

PHOTOFACT SET
39-2

PUBLISHED
1948

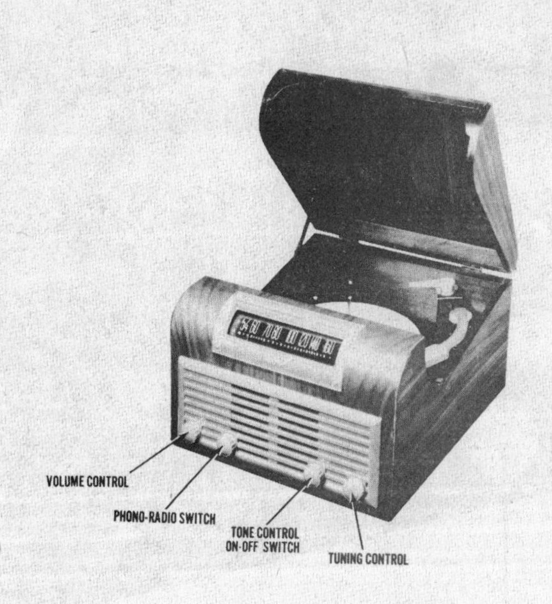

ARVIN

MODEL PICTURED
240-P

Battery operated portable superheterodyne receiver with loop antenna

TUBES
4

POWER SUPPLY
1.5 volts A supply & 67.5 volts B supply

TUNING RANGE
540-1600KC

MFR/SUPPLIER
Noblitt-Sparks Industries

PHOTOFACT SET
42-2

PUBLISHED
1948

ARVIN

MODEL PICTURED
547

AC/DC operated superheterodyne receiver with loop antenna

TUBES
5

POWER SUPPLY
110-120 volts AC/DC, .25 amp @ 117 volts AC

TUNING RANGE
540-1620KC

MFR/SUPPLIER
Noblitt-Sparks Industries

PHOTOFACT SET
42-3

PUBLISHED
1948

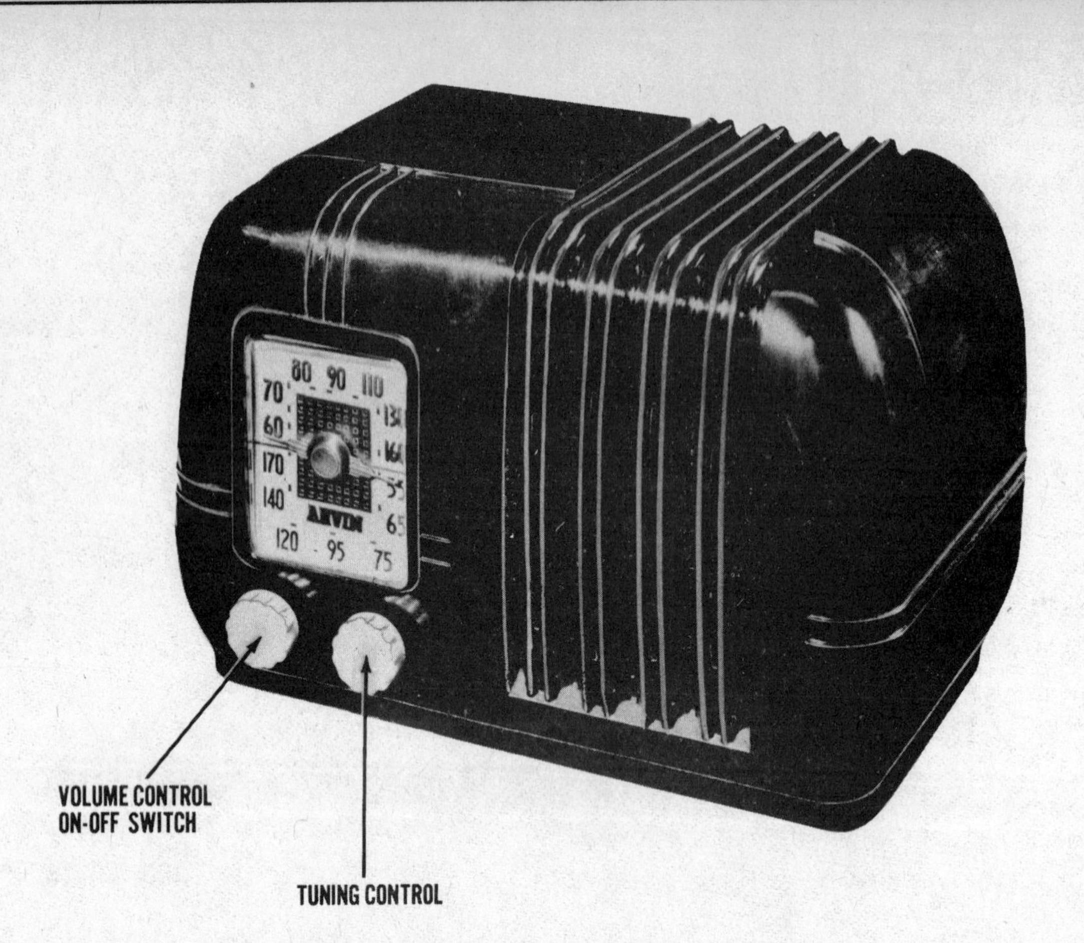

ARVIN

MODEL PICTURED
250-P

Three-power operated portable superheterodyne receiver with loop antenna

TUBES
5

POWER SUPPLY
110-120 volts AC/DC or 9 volts A and 90 volts B battery in pack form

TUNING RANGE
540-1600KC

MFR/SUPPLIER
Noblitt-Sparks Industries

PHOTOFACT SET
43-4

PUBLISHED
1948

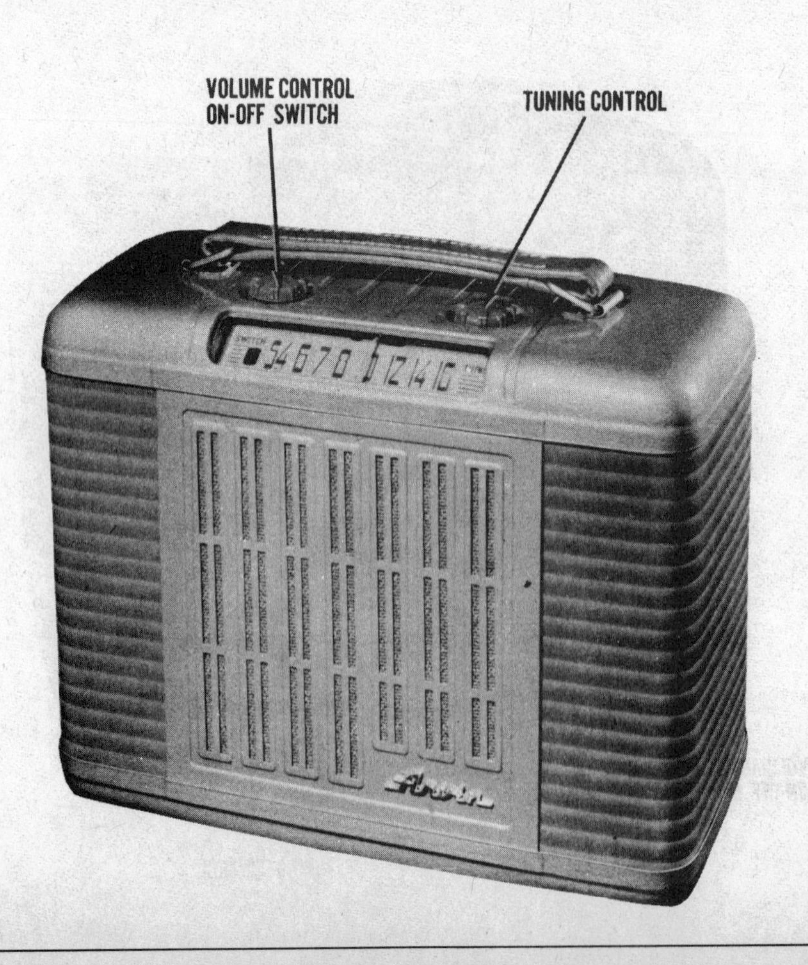

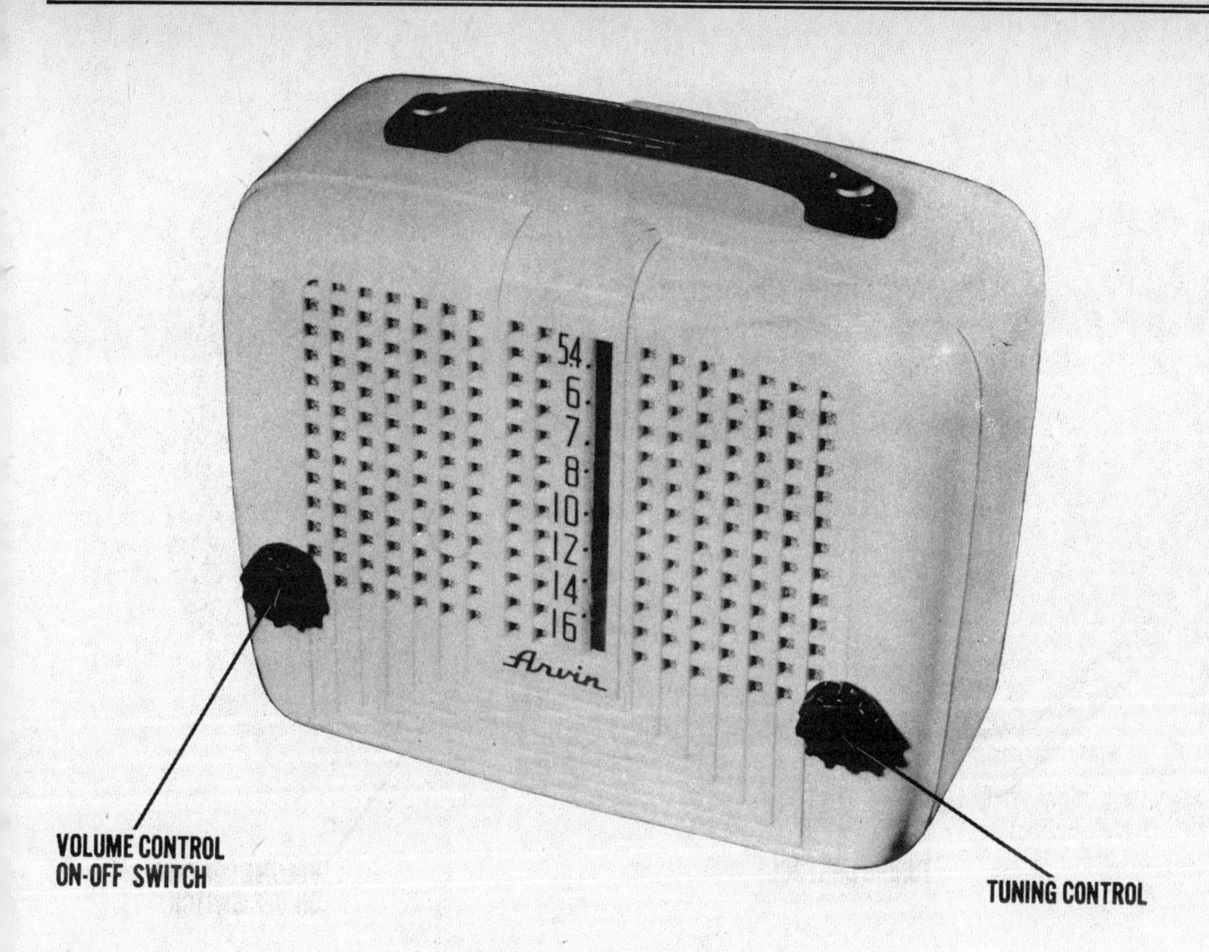

ARVIN

MODEL PICTURED
2410P

Three-power operated portable superheterodyne receiver with loop antenna

TUBES
4

POWER SUPPLY
110-120 volts AC/DC or 6 volts A supply and 67.5 volts B supply

TUNING RANGE
540-1600KC

MFR/SUPPLIER
Noblitt-Sparks Industries

PHOTOFACT SET
47-3

PUBLISHED
1948

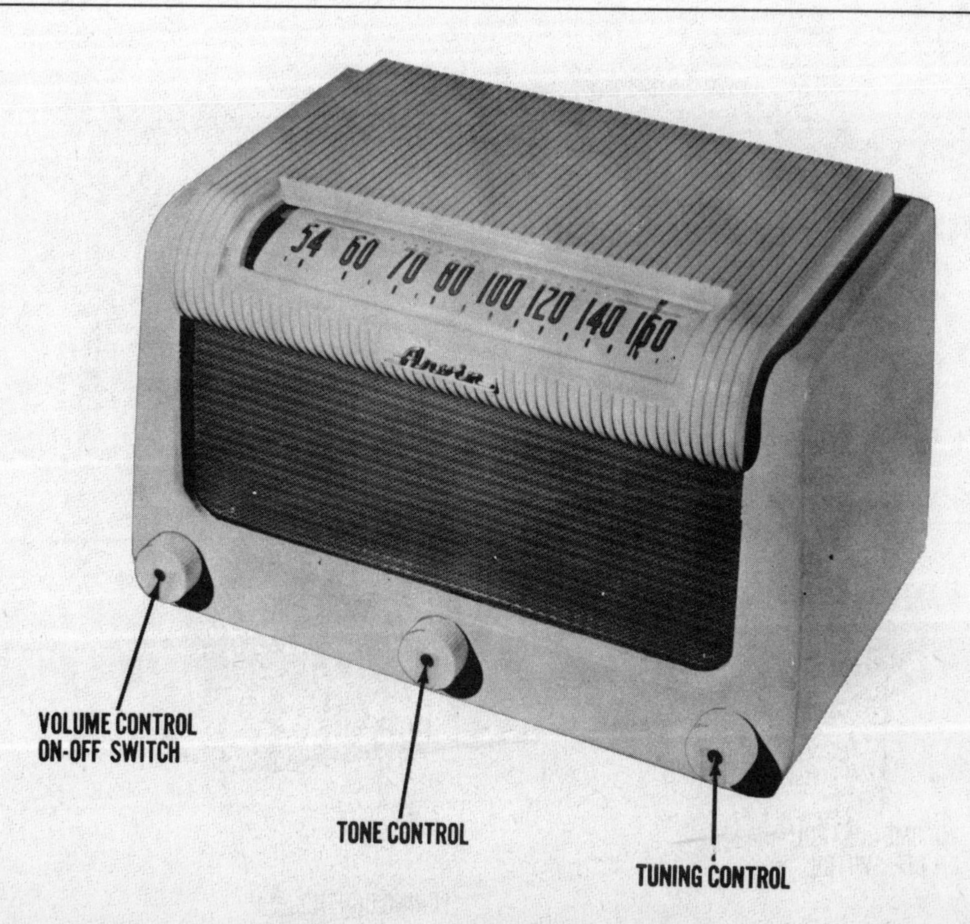

ARVIN

MODEL PICTURED
160T

AC/DC operated superheterodyne receiver with loop antenna

TUBES
6

POWER SUPPLY
110-120 volts AC/DC

TUNING RANGE
540-1620KC

MFR/SUPPLIER
Noblitt-Sparks Industries

PHOTOFACT SET
49-5

PUBLISHED
1948

ARVIN

MODEL PICTURED
242T

AC/DC operated superheterodyne receiver with loop antenna

TUBES
4

POWER SUPPLY
105-125 volts AC/DC

TUNING RANGE
540-1600KC

MFR/SUPPLIER
Noblitt-Sparks Industries

PHOTOFACT SET
52-3

PUBLISHED
1948

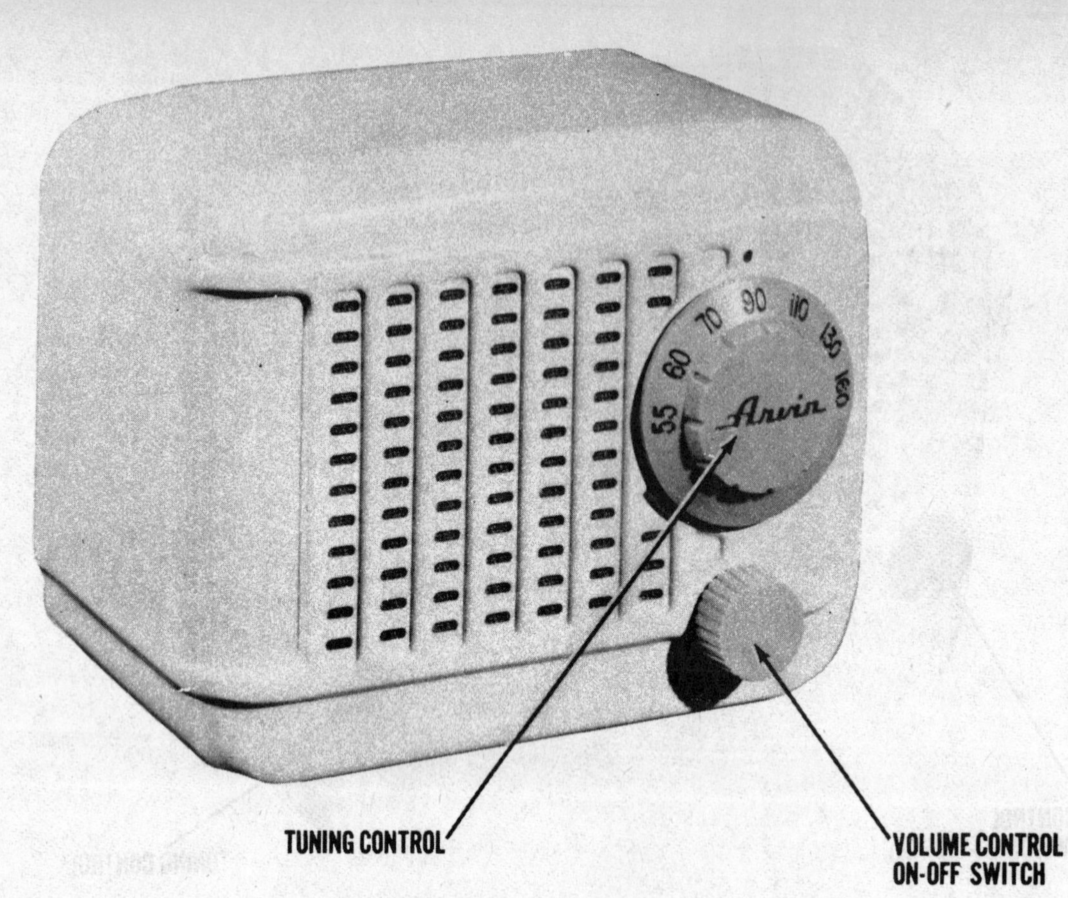

ARVIN

MODEL PICTURED
255T

AC/DC operated superheterodyne receiver with loop antenna

TUBES
5

POWER SUPPLY
105-125 volts AC/DC, .26 amp @ 117 volts AC

TUNING RANGE
540-1600KC

MFR/SUPPLIER
Noblitt-Sparks Industries

PHOTOFACT SET
53-5

PUBLISHED
1949

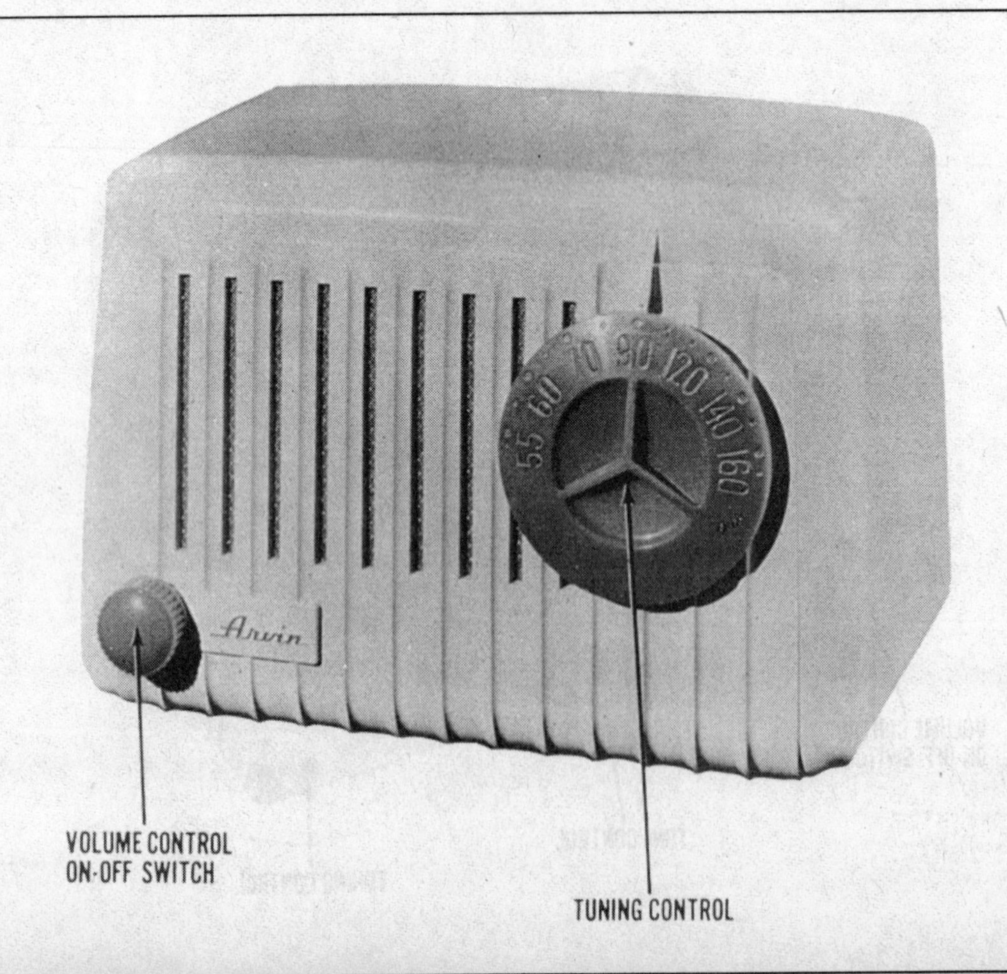

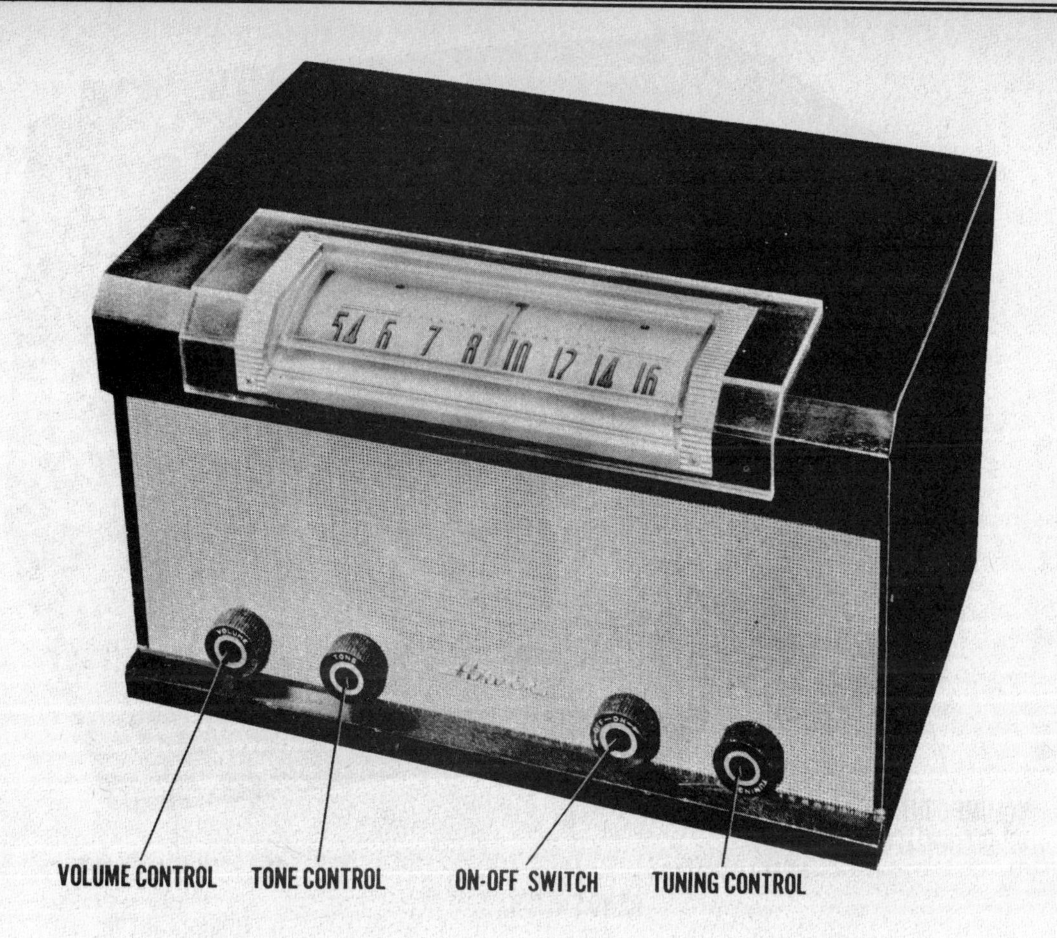

ARVIN

MODEL PICTURED
264T

AC/DC operated superheterodyne receiver with loop antenna

TUBES
6

POWER SUPPLY
110-120 volts AC/DC, .26 amp @ 117 volts AC

TUNING RANGE
540-1600KC

MFR/SUPPLIER
Noblitt-Sparks Industries

PHOTOFACT SET
64-2

PUBLISHED
1949

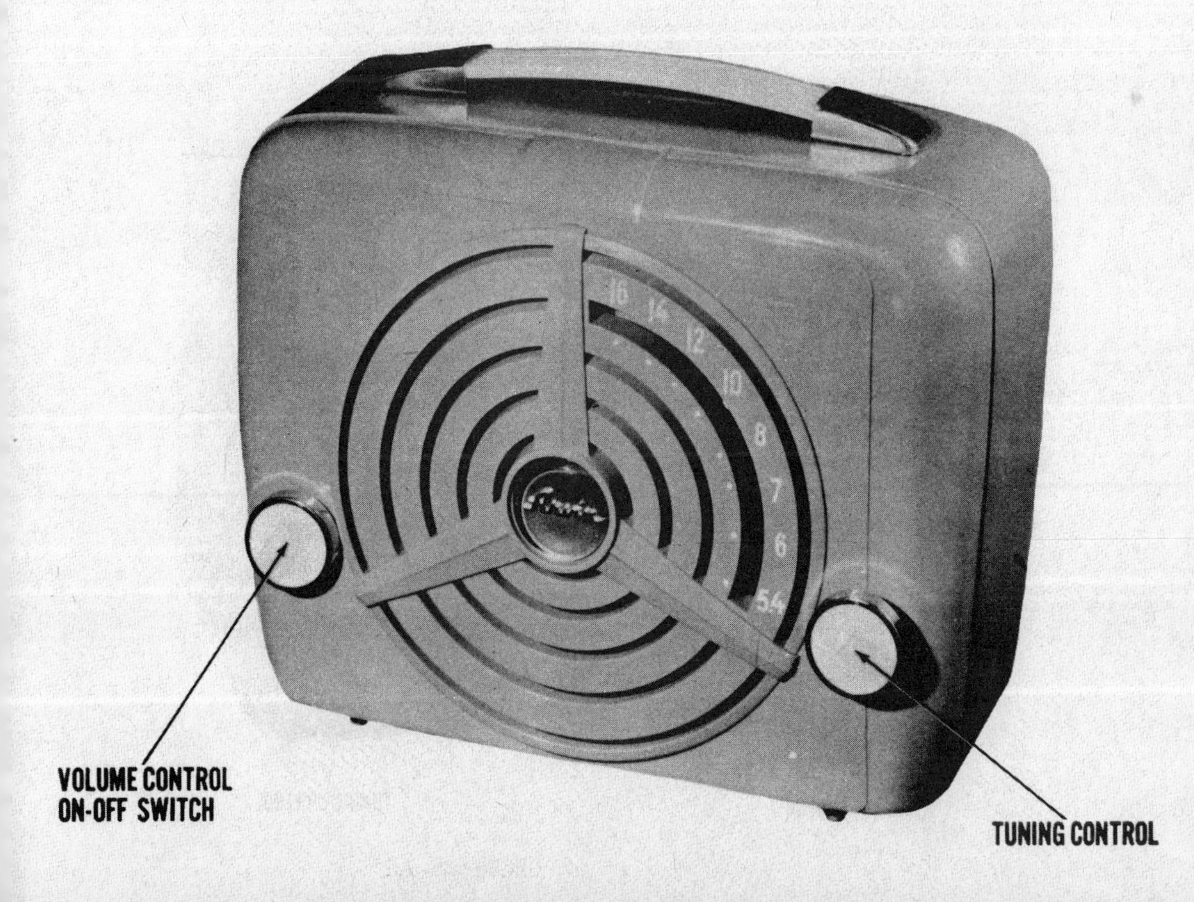

ARVIN

MODEL PICTURED
350P

Three-power operated portable superheterodyne receiver with loop antenna

TUBES
5

POWER SUPPLY
110-120 volts AC/DC or 9 volts A supply and 90 volts B supply

TUNING RANGE
540-1600KC

MFR/SUPPLIER
Noblitt-Sparks Industries

PHOTOFACT SET
69-3

PUBLISHED
1949

ARVIN

MODEL PICTURED
360TFM

AC/DC operated AM-FM superheterodyne receiver with loop antenna

TUBES
6

POWER SUPPLY
110-120 volts AC/DC, .32 amp @ 117 volts AC

TUNING RANGE
540-1600KC, 88-108MC

MFR/SUPPLIER
Noblitt-Sparks Industries

PHOTOFACT SET
70-2

PUBLISHED
1949

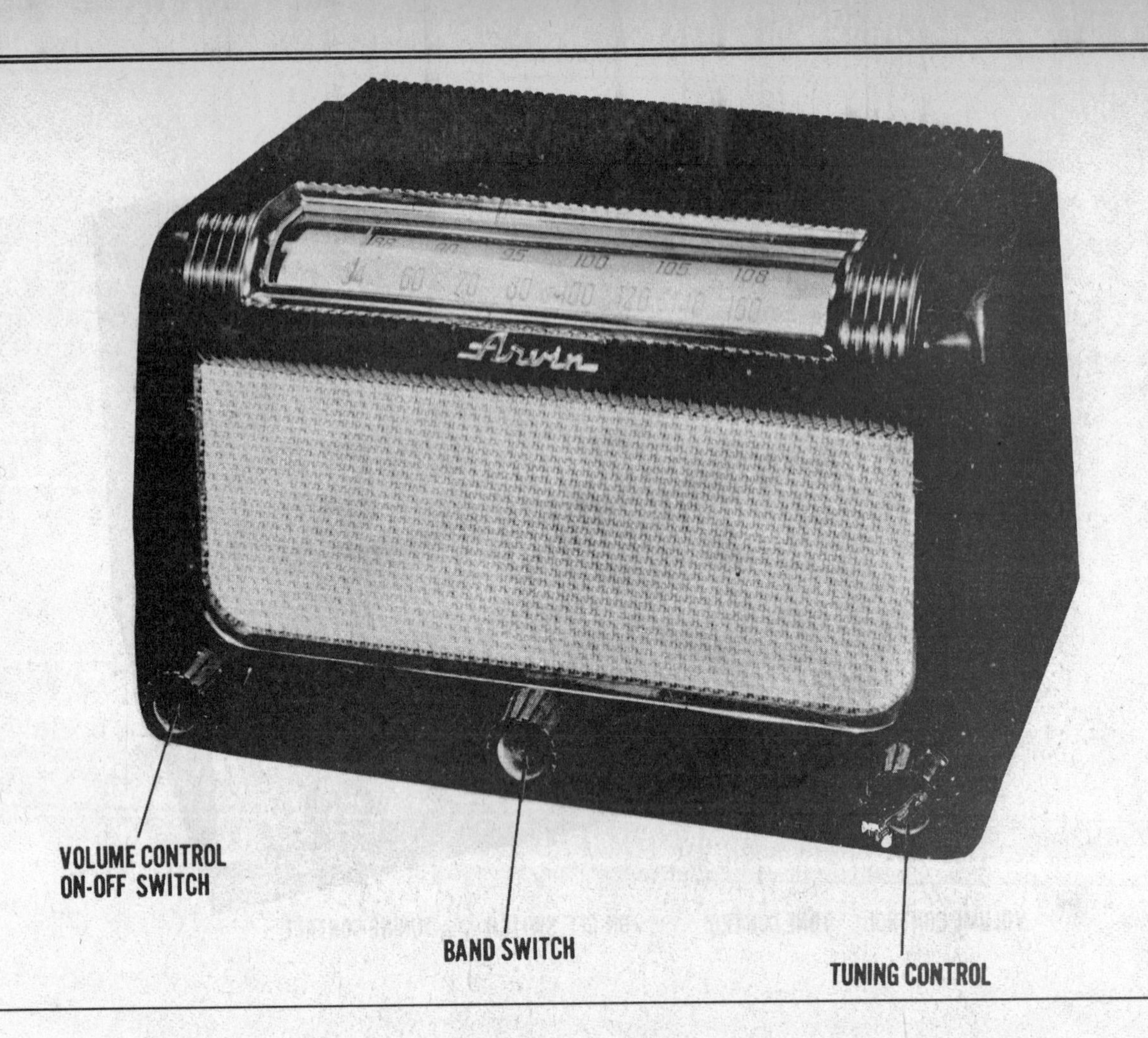

ARVIN

MODEL PICTURED
356T

AC/DC operated superheterodyne receiver with loop antenna

TUBES
5

POWER SUPPLY
110-120 volts AC/DC, .26 amp @ 117 volts AC

TUNING RANGE
540-1600KC

MFR/SUPPLIER
Noblitt-Sparks Industries

PHOTOFACT SET
78-2

PUBLISHED
1949

ARVIN

MODEL PICTURED
341T

AC/DC operated superheterodyne receiver

TUBES
4

POWER SUPPLY
105-125 volts AC/DC

TUNING RANGE
540-1600KC

MFR/SUPPLIER
Noblitt-Sparks Industries

PHOTOFACT SET
84-3

PUBLISHED
1950

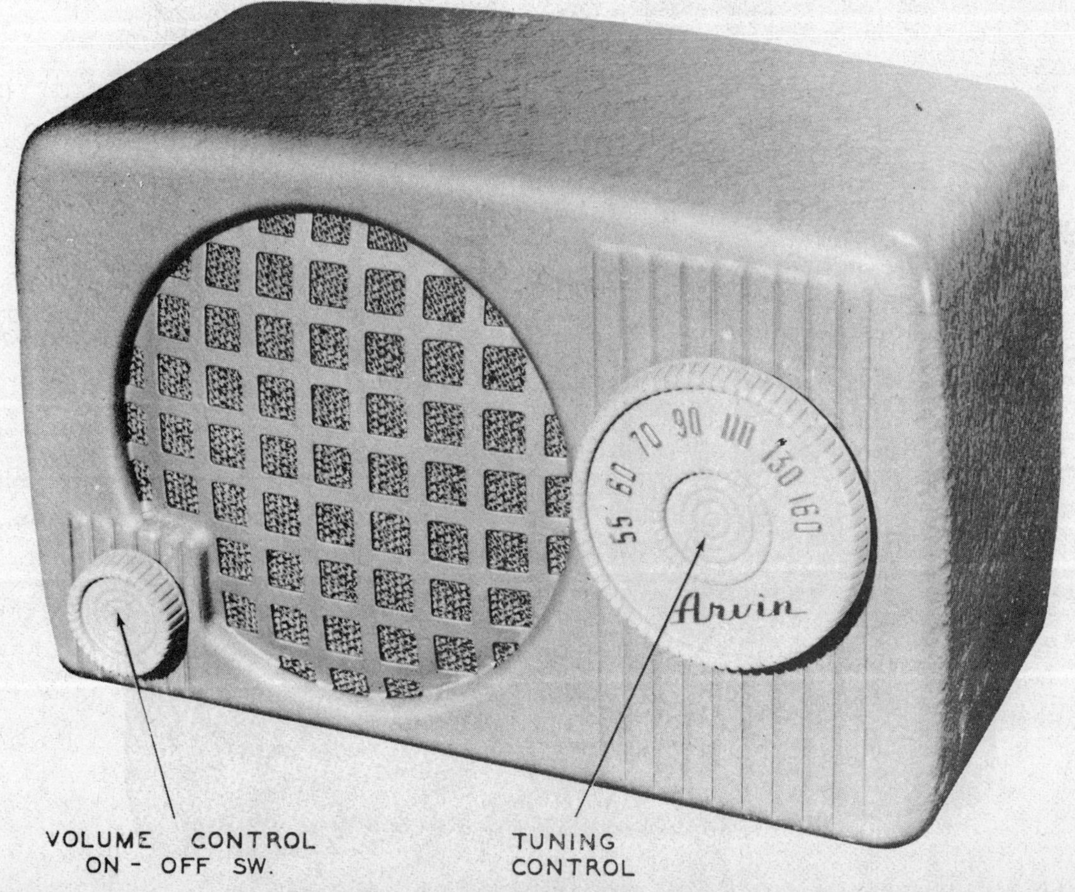

ARVIN

MODEL PICTURED
440T

AC/DC operated superheterodyne receiver

TUBES
4

POWER SUPPLY
105-125 volts AC/DC, .26 amp @ 117 volts AC

TUNING RANGE
540-1600KC

MFR/SUPPLIER
Noblitt-Sparks Industries

PHOTOFACT SET
96-3

PUBLISHED
1950

ARVIN

MODEL PICTURED
350-PL

Three-power portable superheterodyne receiver

TUBES
5

POWER SUPPLY
110-120 volts AC/DC or 9volts A supply & 90 volts B supply

TUNING RANGE
540-1600KC

MFR/SUPPLIER
Noblitt-Sparks Industries

PHOTOFACT SET
100-4

PUBLISHED
1950

ARVIN

MODEL PICTURED
446P

Battery operated portable superheterodyne receiver with loop antenna

TUBES
4

POWER SUPPLY
1.5 volts A supply and 67.5 volts B supply

TUNING RANGE
540-1600KC

MFR/SUPPLIER
Noblitt-Sparks Industries

PHOTOFACT SET
106-2

PUBLISHED
1950

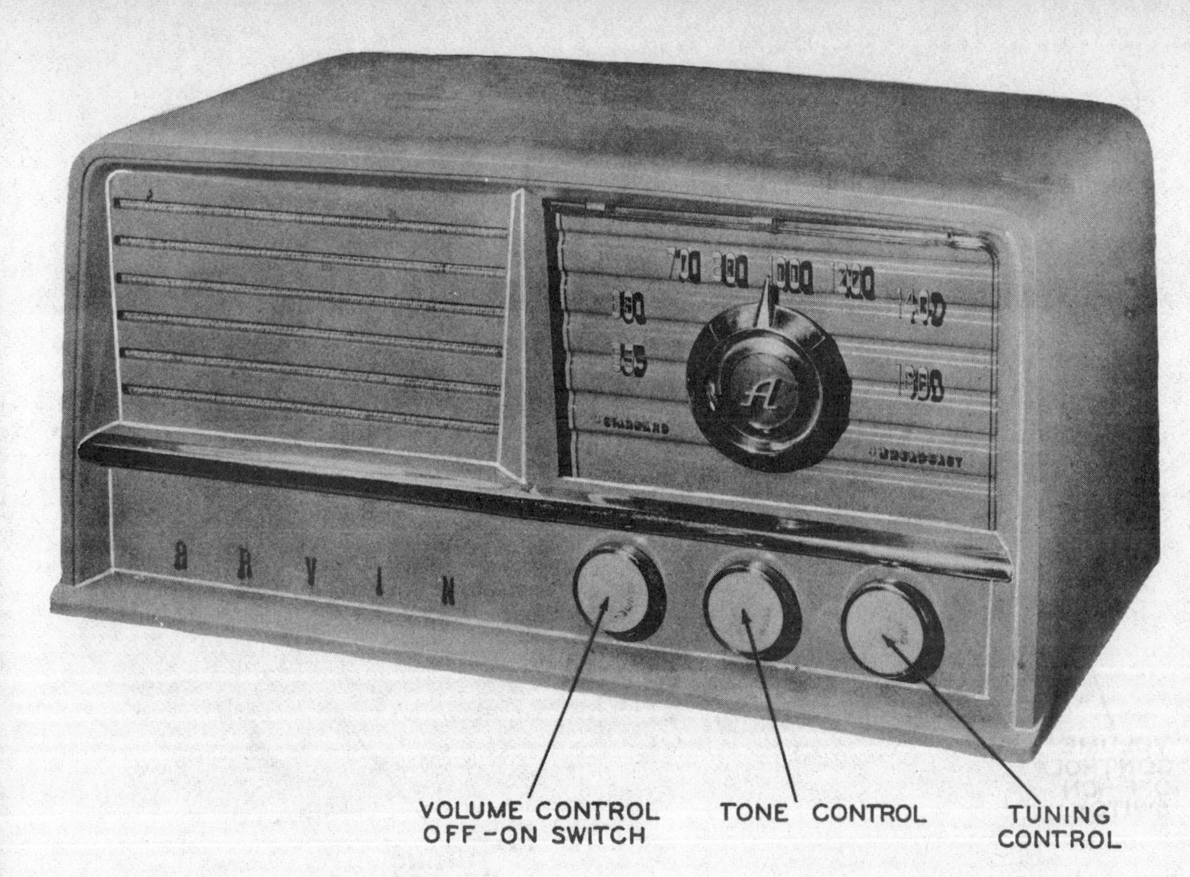

ARVIN

MODEL PICTURED
460T

AC/DC operated superheterodyne receiver with loop antenna

TUBES
6

POWER SUPPLY
110-120 volts AC/DC

TUNING RANGE
540-1600KC

MFR/SUPPLIER
Arvin Industries, Inc.

PHOTOFACT SET
107-3

PUBLISHED
1950

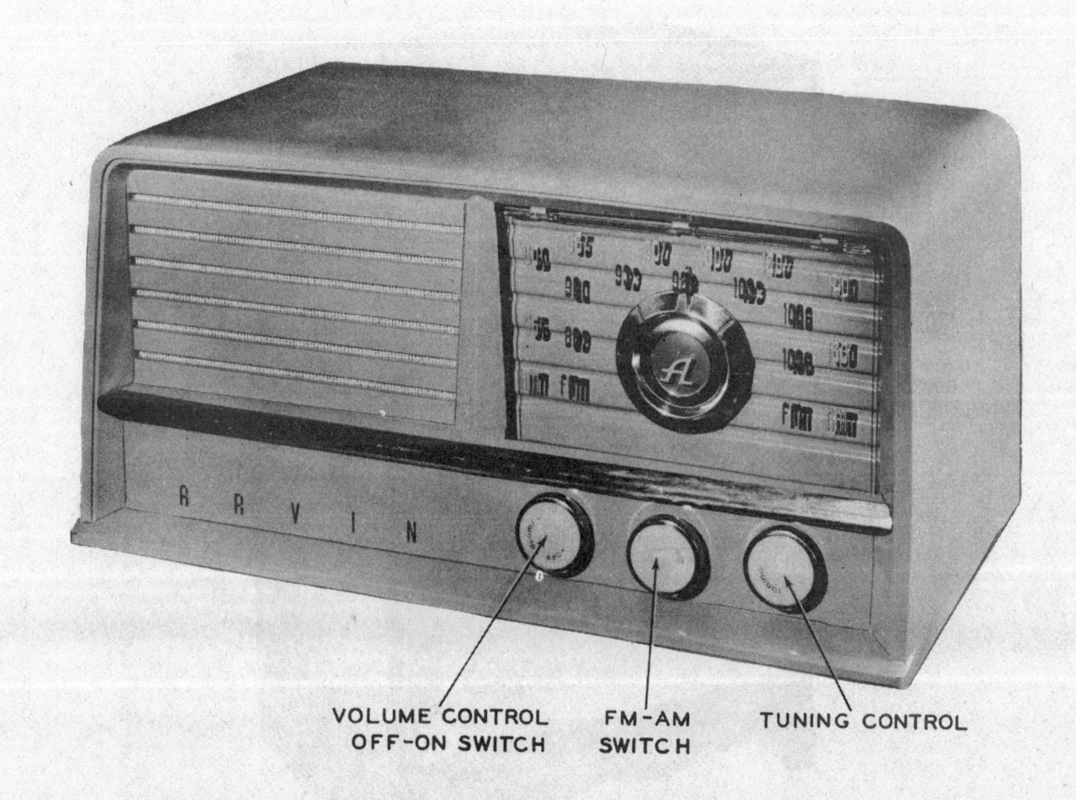

ARVIN

MODEL PICTURED
480TFM

AC/DC operated AM-FM superheterodyne receiver with loop antenna

TUBES
8

POWER SUPPLY
110-120 volts AC

TUNING RANGE
540-1600KC, 88-108MC

MFR/SUPPLIER
Arvin Industries, Inc.

PHOTOFACT SET
107-4

PUBLISHED
1950

ARVIN

MODEL PICTURED
450T

AC/DC operated superheterodyne receiver with loop antenna

TUBES
5

POWER SUPPLY
105-125 volts AC/DC, .25 amp @ 117 volts AC

TUNING RANGE
540-1600KC

MFR/SUPPLIER
Arvin Industries, Inc.

PHOTOFACT SET
110-3

PUBLISHED
1950

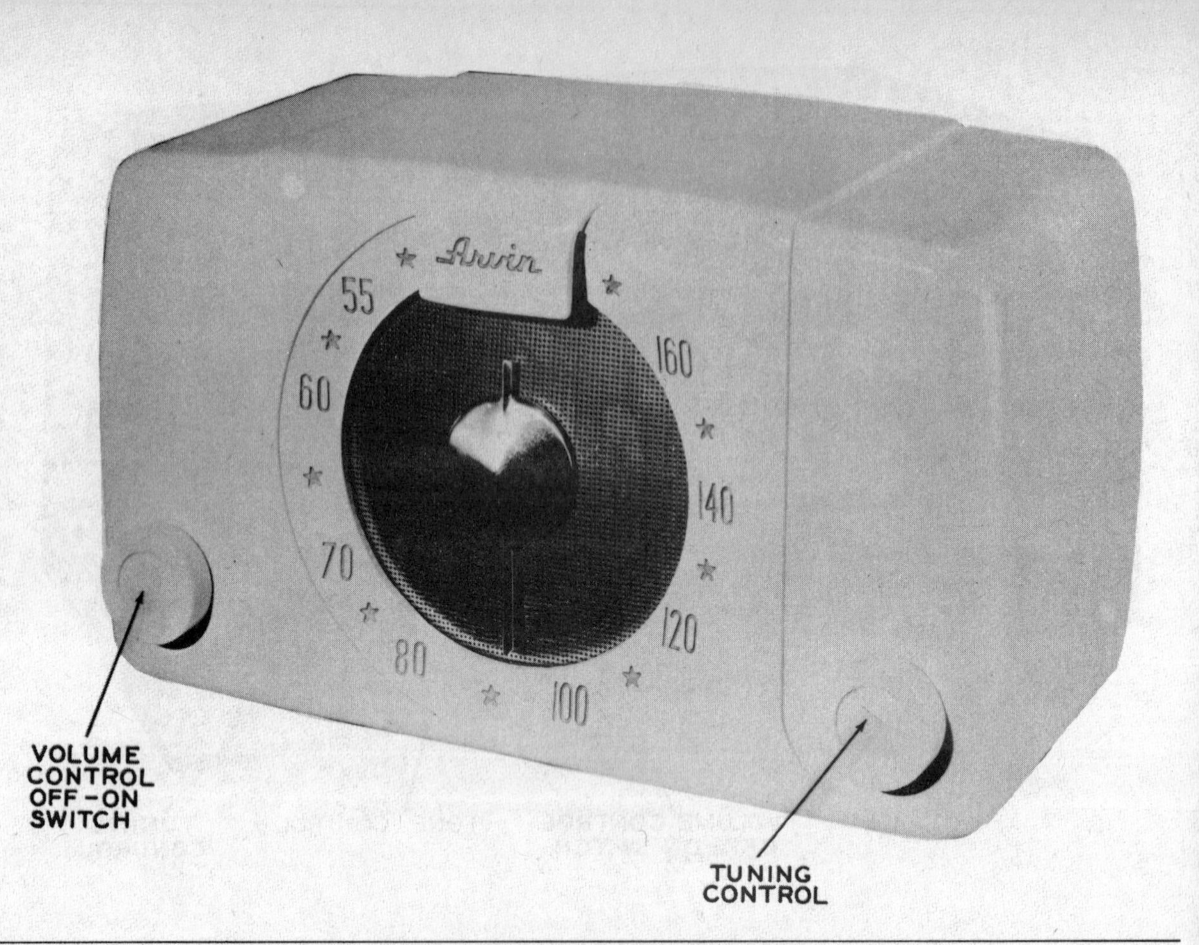

ARVIN

MODEL PICTURED
462-CM

AC operated phono radio superheterodyne receiver with loop antenna

TUBES
6

POWER SUPPLY
110-120 volts AC

TUNING RANGE
540-1600KC

MFR/SUPPLIER
Arvin Industries, Inc.

PHOTOFACT SET
116-3

PUBLISHED
1950

ARVIN

MODEL PICTURED
482CFB

AC operated phono radio AM-FM superheterodyne receiver with loop antenna

TUBES
8

POWER SUPPLY
105-120 volts AC

TUNING RANGE
540-1600KC, 88-108MC

MFR/SUPPLIER
Arvin Industries, Inc.

PHOTOFACT SET
117-4

PUBLISHED
1950

ARVIN

MODEL PICTURED
540T

AC/DC operated superheterodyne receiver with loop antenna

TUBES
4

POWER SUPPLY
105-125 volts AC, .26 amp @ 117 volts AC

TUNING RANGE
540-1600KC

MFR/SUPPLIER
Arvin Industries, Inc.

PHOTOFACT SET
143-4

PUBLISHED
1951

ARVIN

MODEL PICTURED
580TFM

AC operated AM-FM superheterodyne receiver with loop antenna

TUBES
8

POWER SUPPLY
110-120 volts AC, .35 amp @ 117 volts AC

TUNING RANGE
540-1600KC, 88-108MC

MFR/SUPPLIER
Arvin Industries, Inc.

PHOTOFACT SET
152-2

PUBLISHED
1951

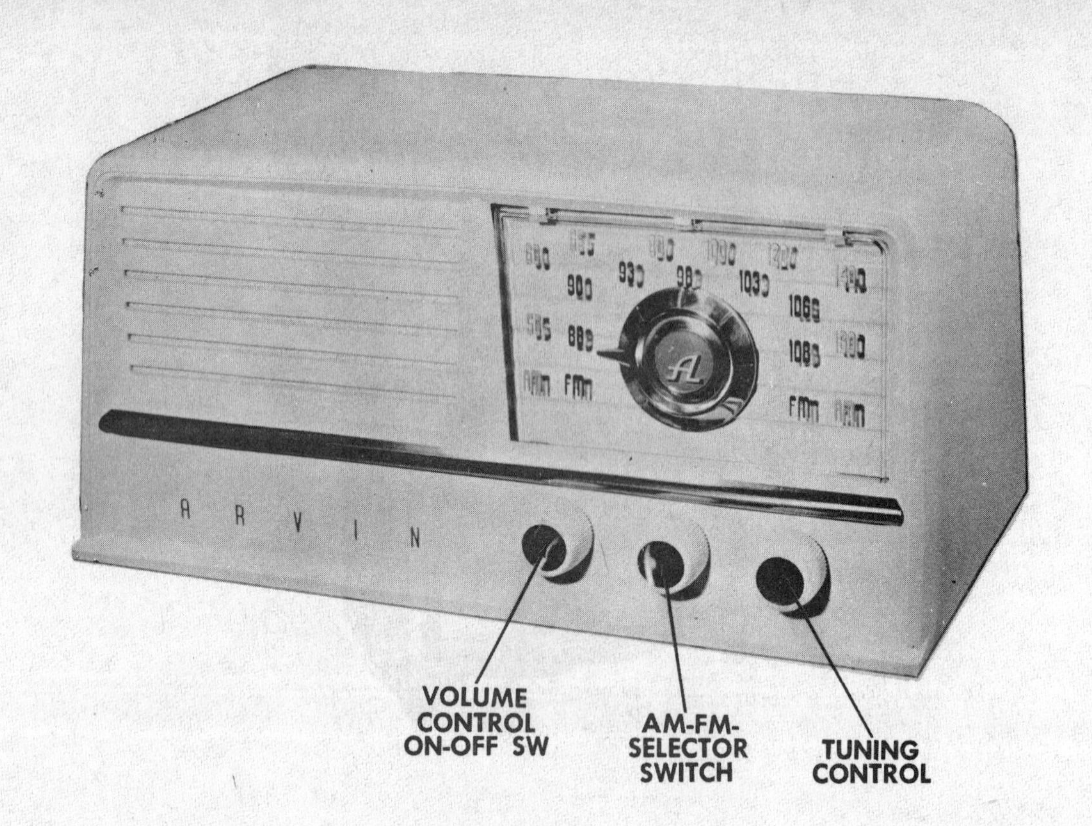

ARVIN

MODEL PICTURED
551T

AC operated superheterodyne receiver with loop antenna

TUBES
5

POWER SUPPLY
110-120 volts AC, .6 amp @ 117 volts AC

TUNING RANGE
540-1600KC

MFR/SUPPLIER
Arvin Industries, Inc.

PHOTOFACT SET
154-2

PUBLISHED
1951

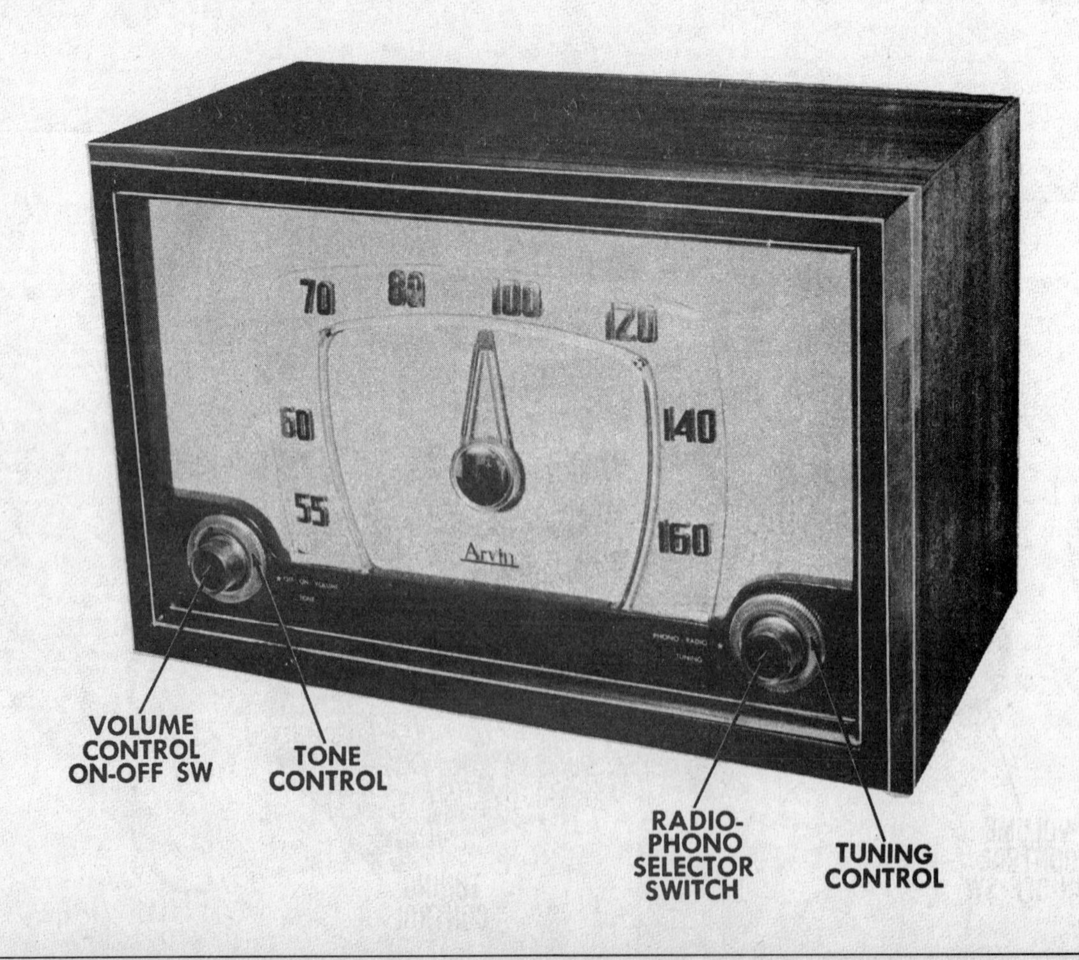

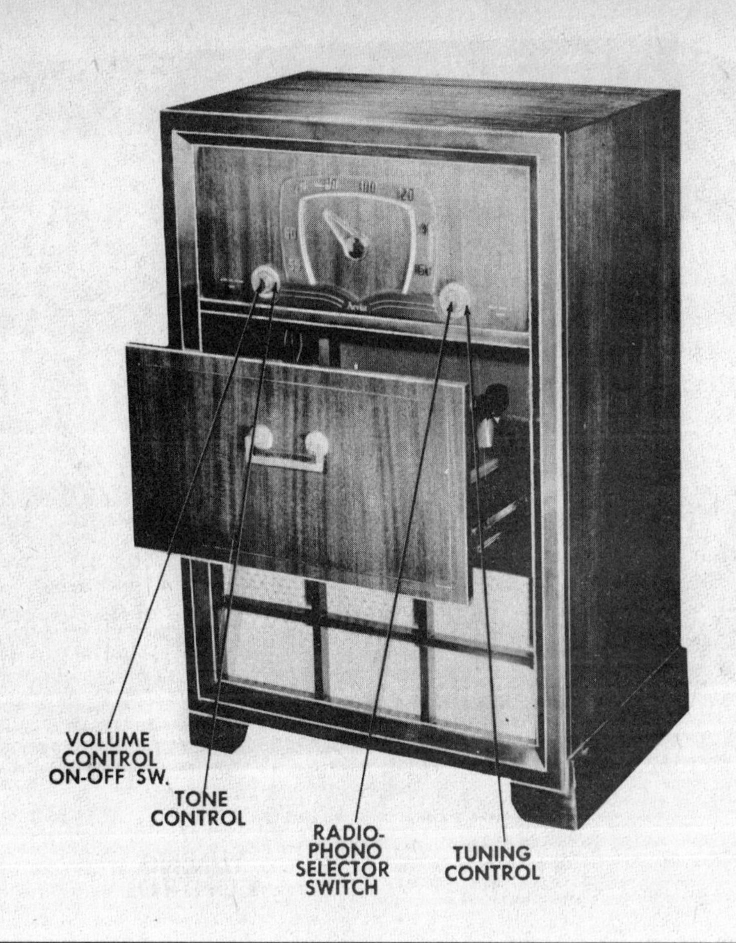

ARVIN

MODEL PICTURED
554CCM

AC operated radio phono combo superheterodyne receiver with loop antenna

TUBES
5

POWER SUPPLY
110-120 volts AC

TUNING RANGE
540-1600KC

MFR/SUPPLIER
Arvin Industries, Inc.

PHOTOFACT SET
155-3

PUBLISHED
1952

ARVIN

MODEL PICTURED
582CFB

AC operated AM-FM phono combo superheterodyne receiver

TUBES
8

POWER SUPPLY
110-120 volts AC

TUNING RANGE
540-1600KC, 88-108MC

MFR/SUPPLIER
Arvin Industries, Inc.

PHOTOFACT SET
156-4

PUBLISHED
1952

ARVIN

MODEL PICTURED
553

AC/DC operated superheterodyne receiver with loop antenna

TUBES
5

POWER SUPPLY
105-125 volts AC/DC, .25 amp @ 117 volts AC

TUNING RANGE
540-1600KC

MFR/SUPPLIER
Arvin Industries, Inc.

PHOTOFACT SET
159-4

PUBLISHED
1952

ARVIN

MODEL PICTURED
657-T

AC operated AM superheterodyne receiver with electric clock

TUBES
5

POWER SUPPLY
110-120 volts AC, 60 cycles

TUNING RANGE
530-1650KC

MFR/SUPPLIER
Arvin Industries, Inc.

PHOTOFACT SET
168-5

PUBLISHED
1952

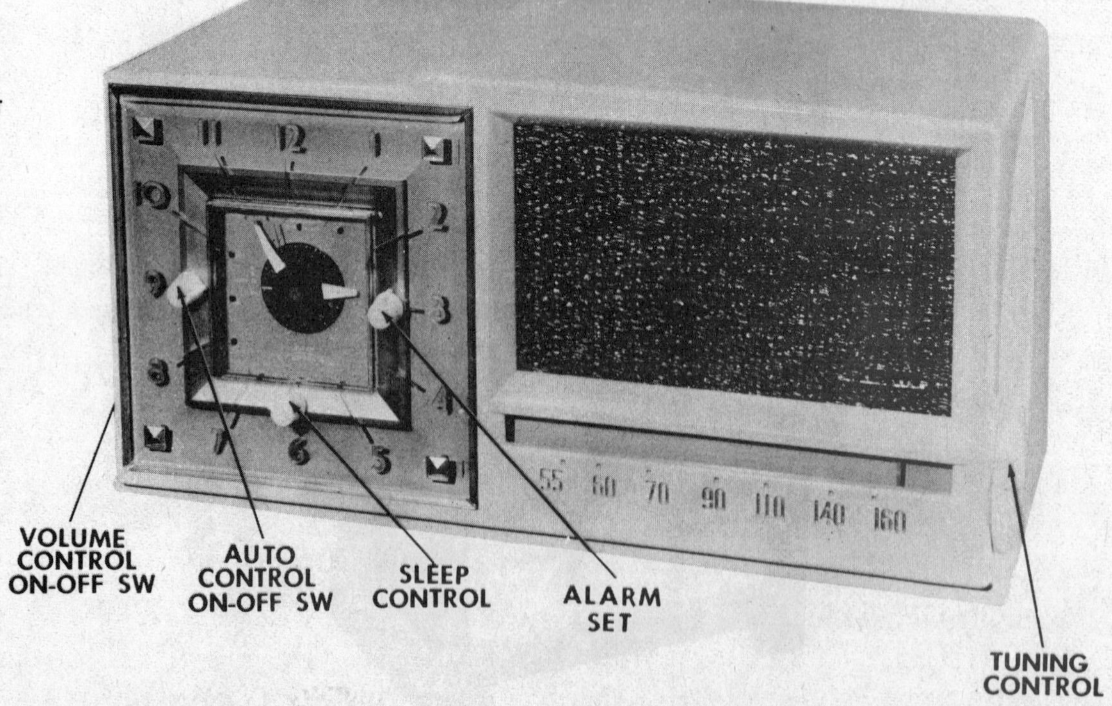

ARVIN

MODEL PICTURED
650-P

Three-power portable AM superheterodyne receiver

TUBES
5

POWER SUPPLY
110-120 volts AC/DC or 9 volts A supply & 90 volts B supply

TUNING RANGE
530-1660KC

MFR/SUPPLIER
Arvin Industries, Inc.

PHOTOFACT SET
175-6

PUBLISHED
1952

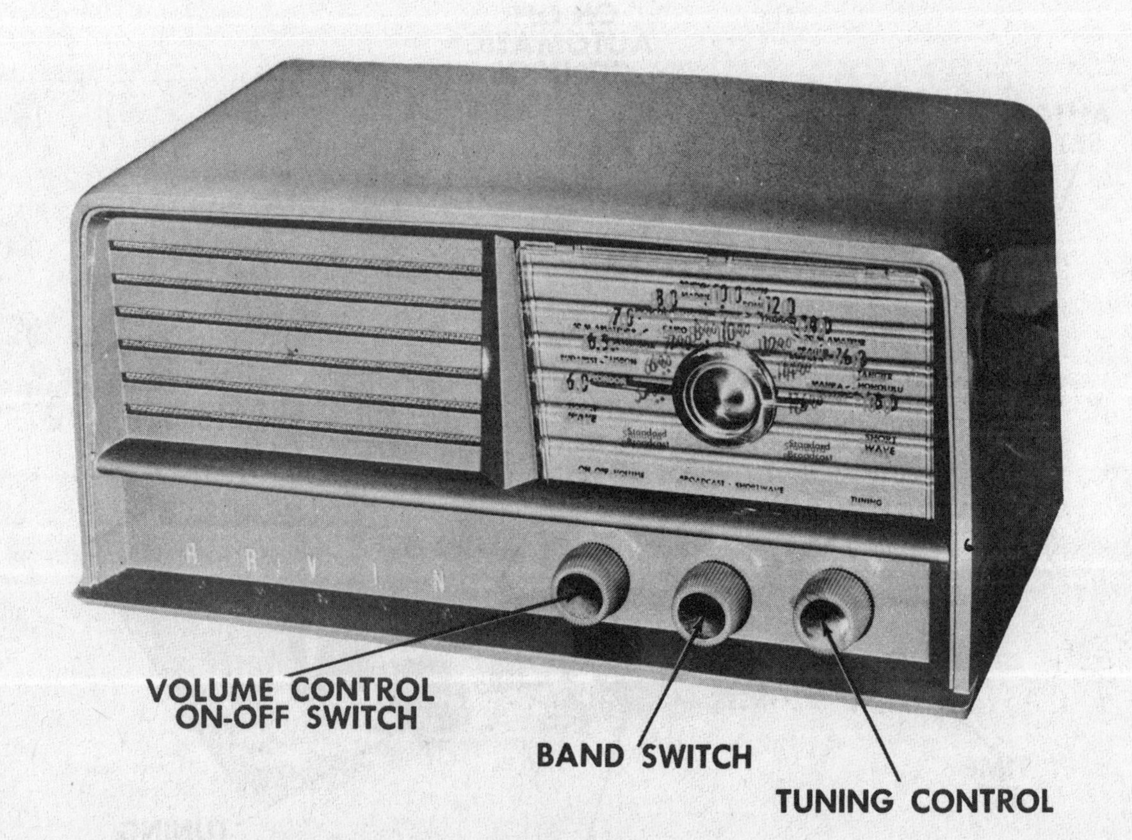

ARVIN

MODEL PICTURED
655SWT

AC/DC operated two-band AM superheterodyne receiver

TUBES
5

POWER SUPPLY
105-125 volts AC/DC, .26 amp @ 117 volts AC

TUNING RANGE
540-1620KC, 6-18MC

MFR/SUPPLIER
Arvin Industries, Inc.

PHOTOFACT SET
187-2

PUBLISHED
1952

ARVIN

MODEL PICTURED
753T

AC/DC operated AM superheterodyne receiver

TUBES
5

POWER SUPPLY
105-120 volts AC/DC, .25 amp @ 117 volts AC

TUNING RANGE
540-1650KC

MFR/SUPPLIER
Arvin Industries, Inc.

PHOTOFACT SET
220-2

PUBLISHED
1953

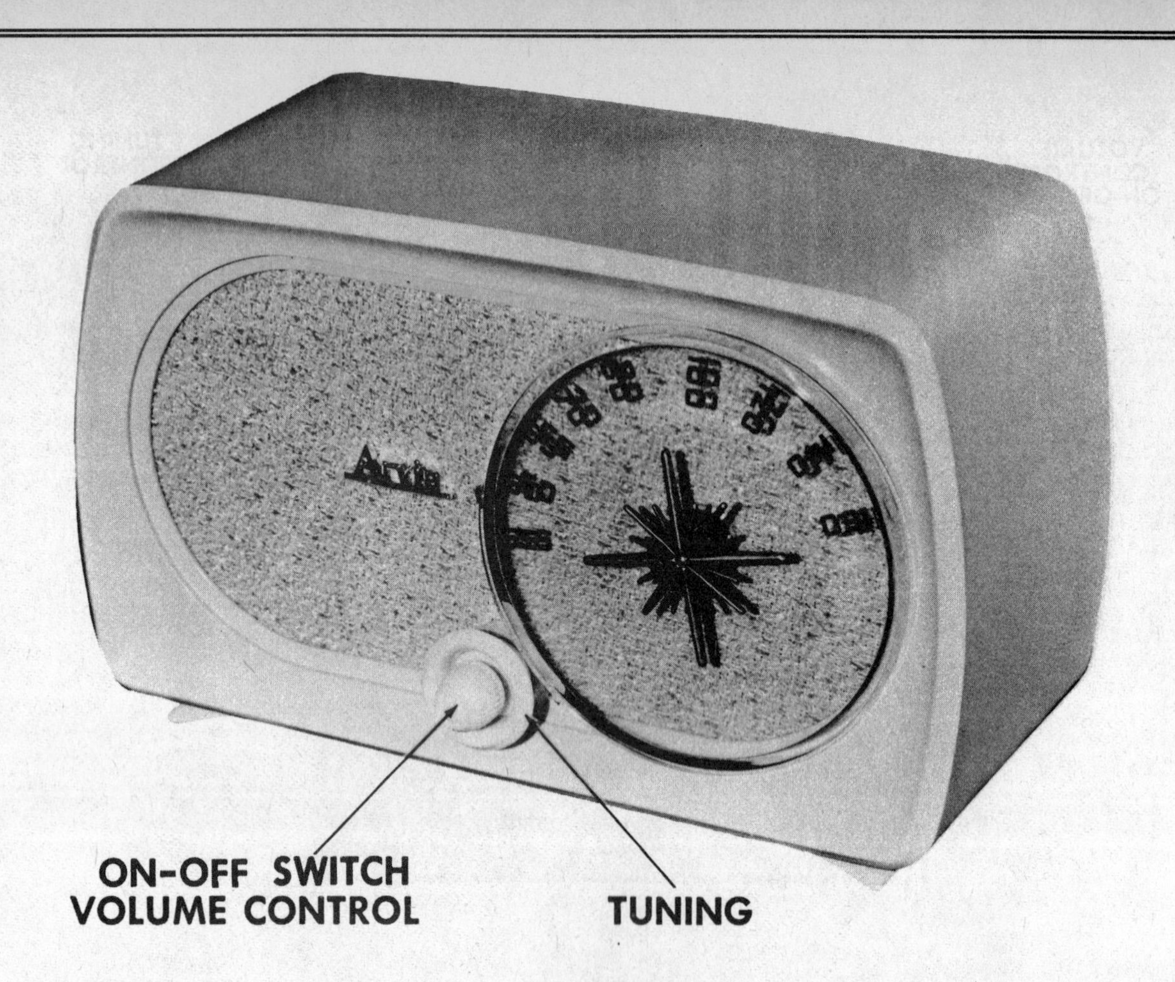

ARVIN

MODEL PICTURED
758T

AC operated AM superheterodyne receiver with electric clock

TUBES
5

POWER SUPPLY
105-120 volts AC, 60 cycles, .32 amp @ 117 volts AC

TUNING RANGE
540-1650KC

MFR/SUPPLIER
Arvin Industries, Inc.

PHOTOFACT SET
221-3

PUBLISHED
1953

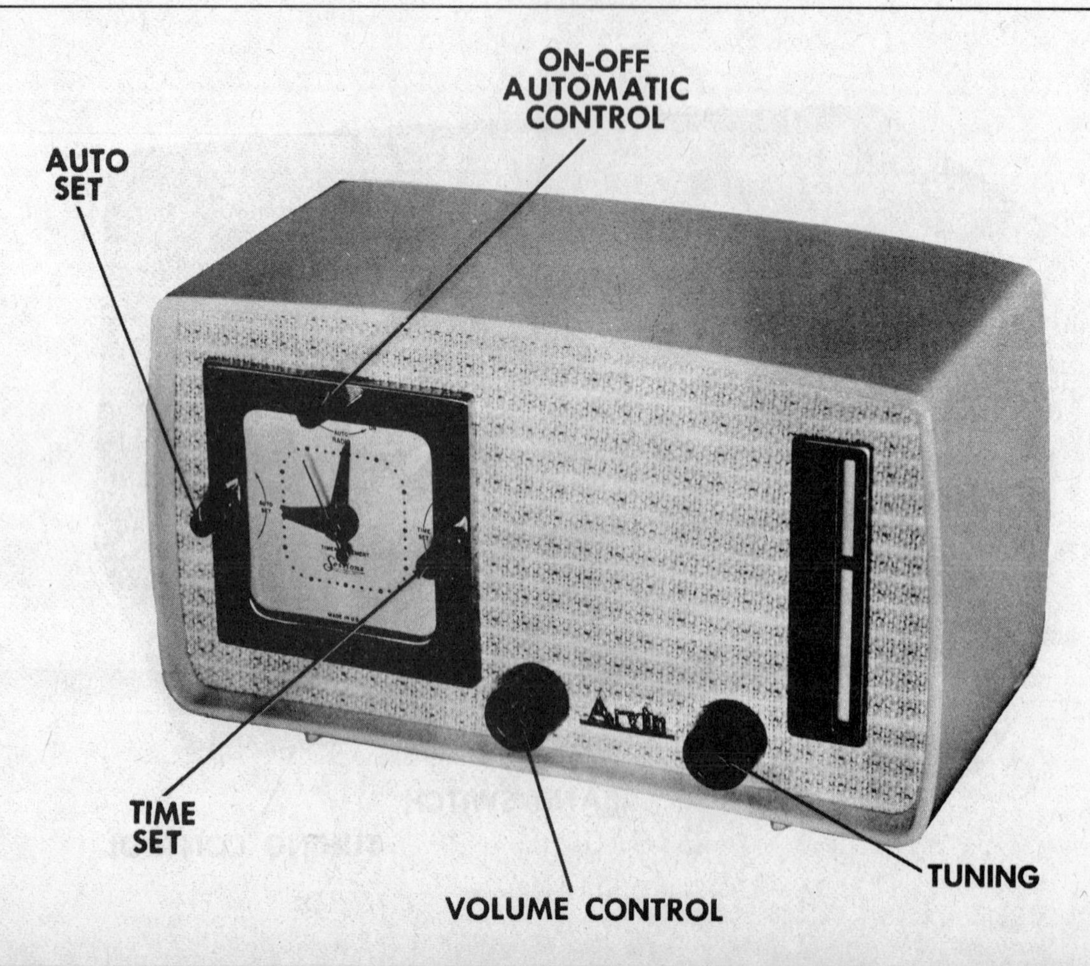

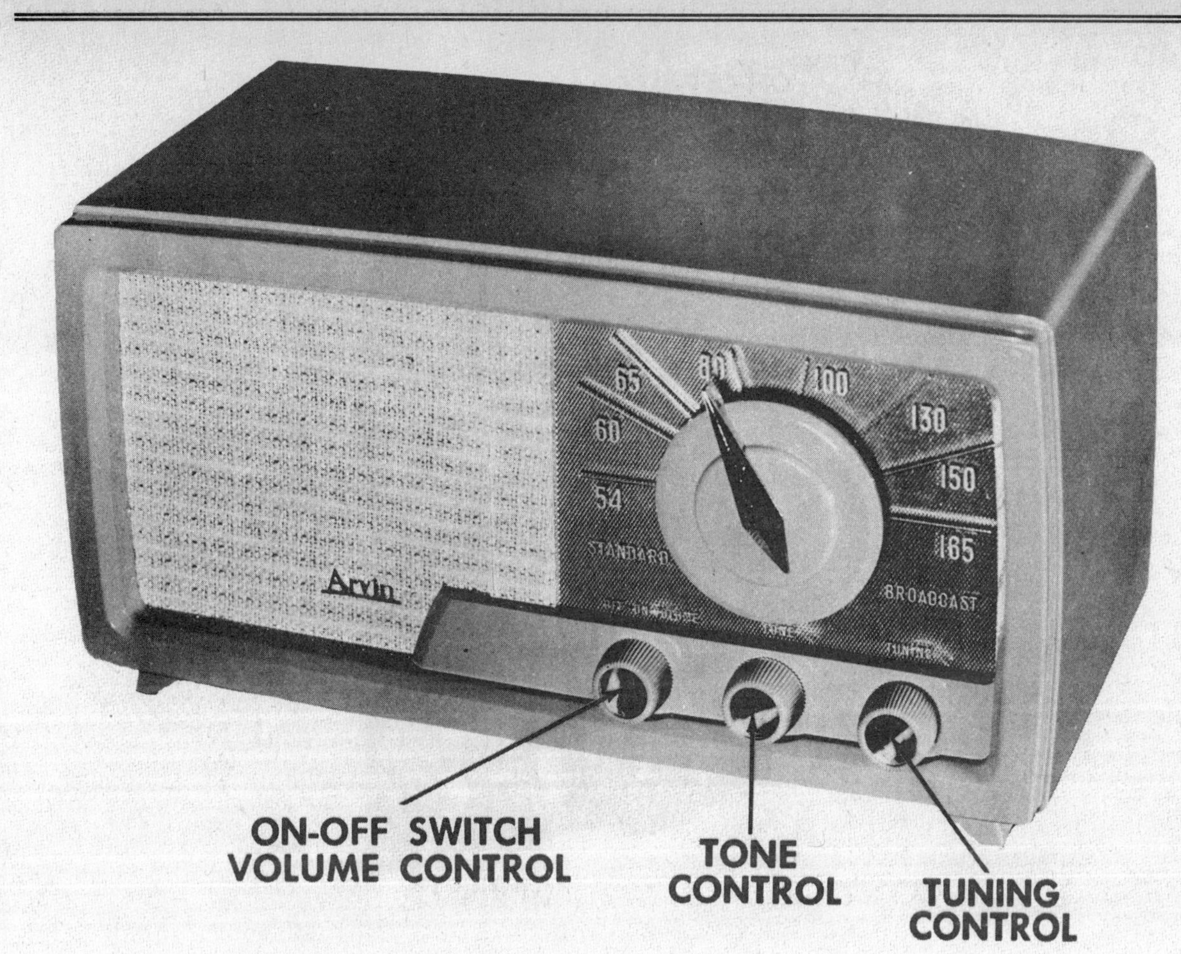

ARVIN

MODEL PICTURED
760T

AC/DC operated AM superheterodyne receiver

TUBES
6

POWER SUPPLY
105-125 volts AC/DC, .26 amp @ 117 volts AC

TUNING RANGE
540-1650KC

MFR/SUPPLIER
Arvin Industries, Inc.

PHOTOFACT SET
223-3

PUBLISHED
1953

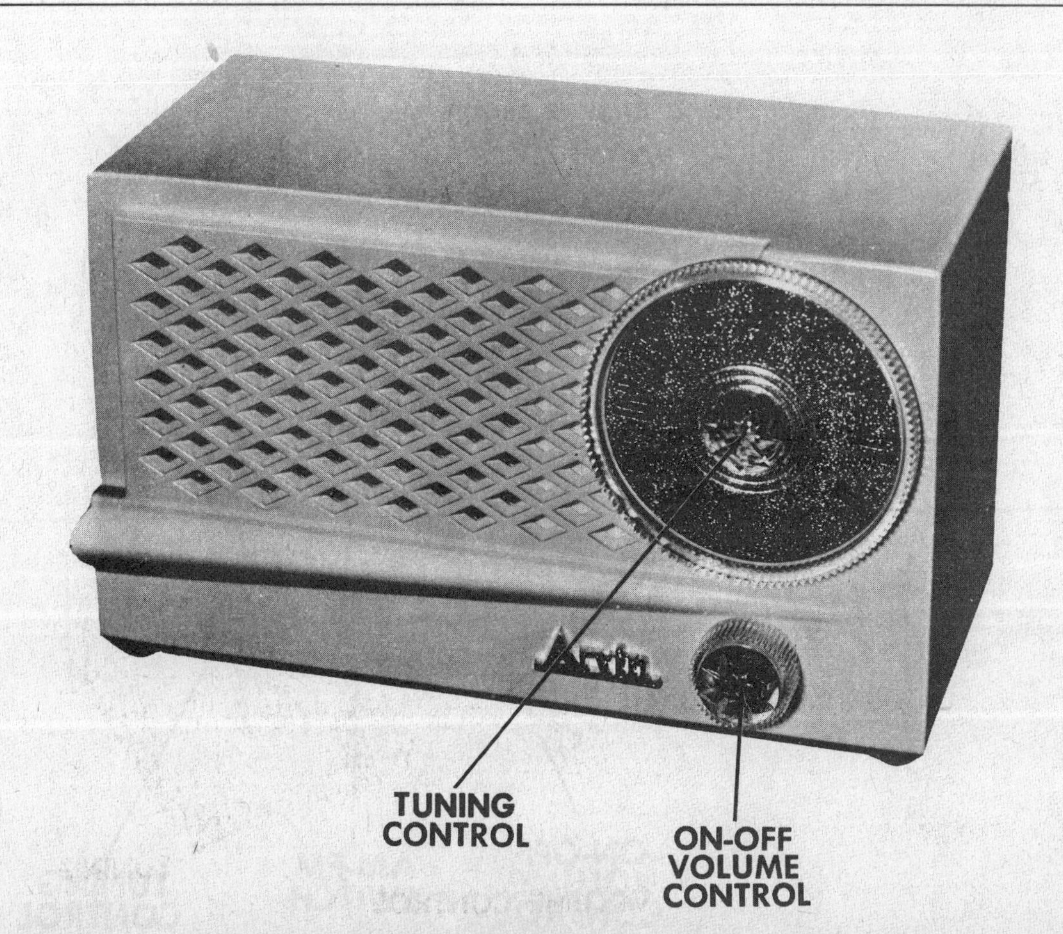

ARVIN

MODEL PICTURED
741T

AC/DC operated AM superheterodyne receiver

TUBES
4

POWER SUPPLY
110-120 volts AC/DC, .23 amp @ 117 volts AC

TUNING RANGE
540-1650KC

MFR/SUPPLIER
Arvin Industries, Inc.

PHOTOFACT SET
225-4

PUBLISHED
1953

ARVIN

MODEL PICTURED
746P

TUBES
4

POWER SUPPLY
1.5 volts A supply & 67.5 volts B supply

TUNING RANGE
540-1600KC

MFR/SUPPLIER
Arvin Industries, Inc.

PHOTOFACT SET
225-5

PUBLISHED
1953

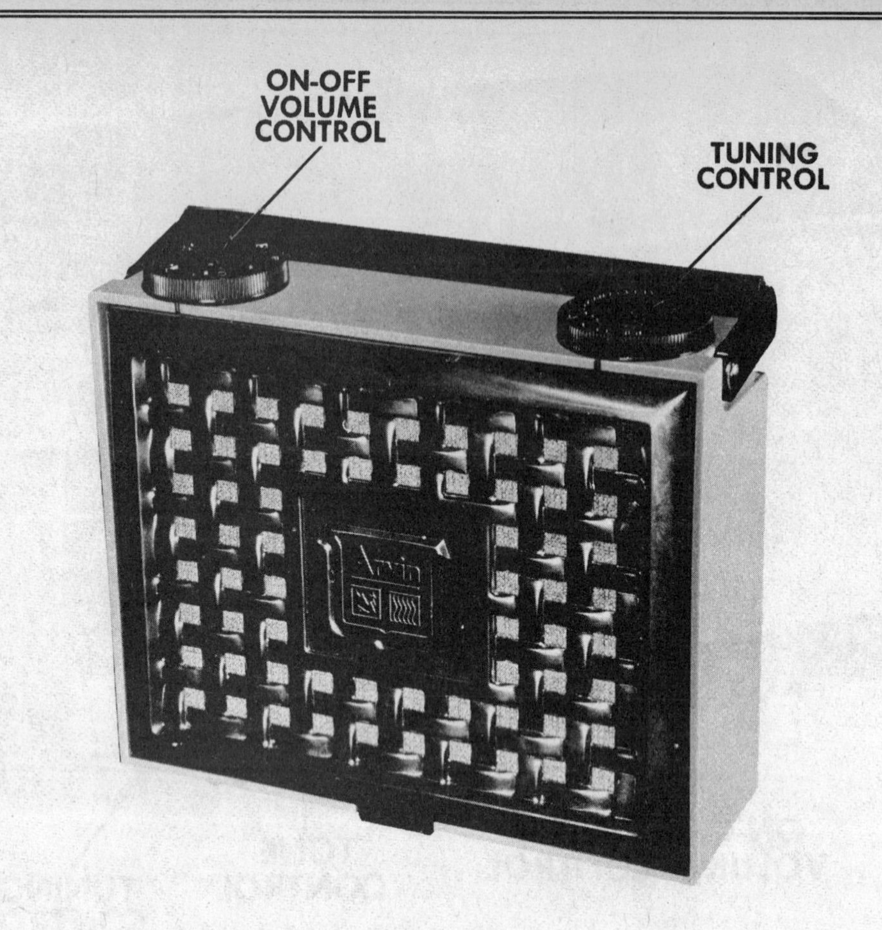

ARVIN

MODEL PICTURED
581TFM

AC operated AM-FM superheterodyne receiver

TUBES
8

POWER SUPPLY
105-120 volts AC, 60 cycle, .37 amp @ 117 volts AC

TUNING RANGE
540KC-1600KC, 88-108MC

MFR/SUPPLIER
Arvin Industries, Inc.

PHOTOFACT SET
227-2

PUBLISHED
1954

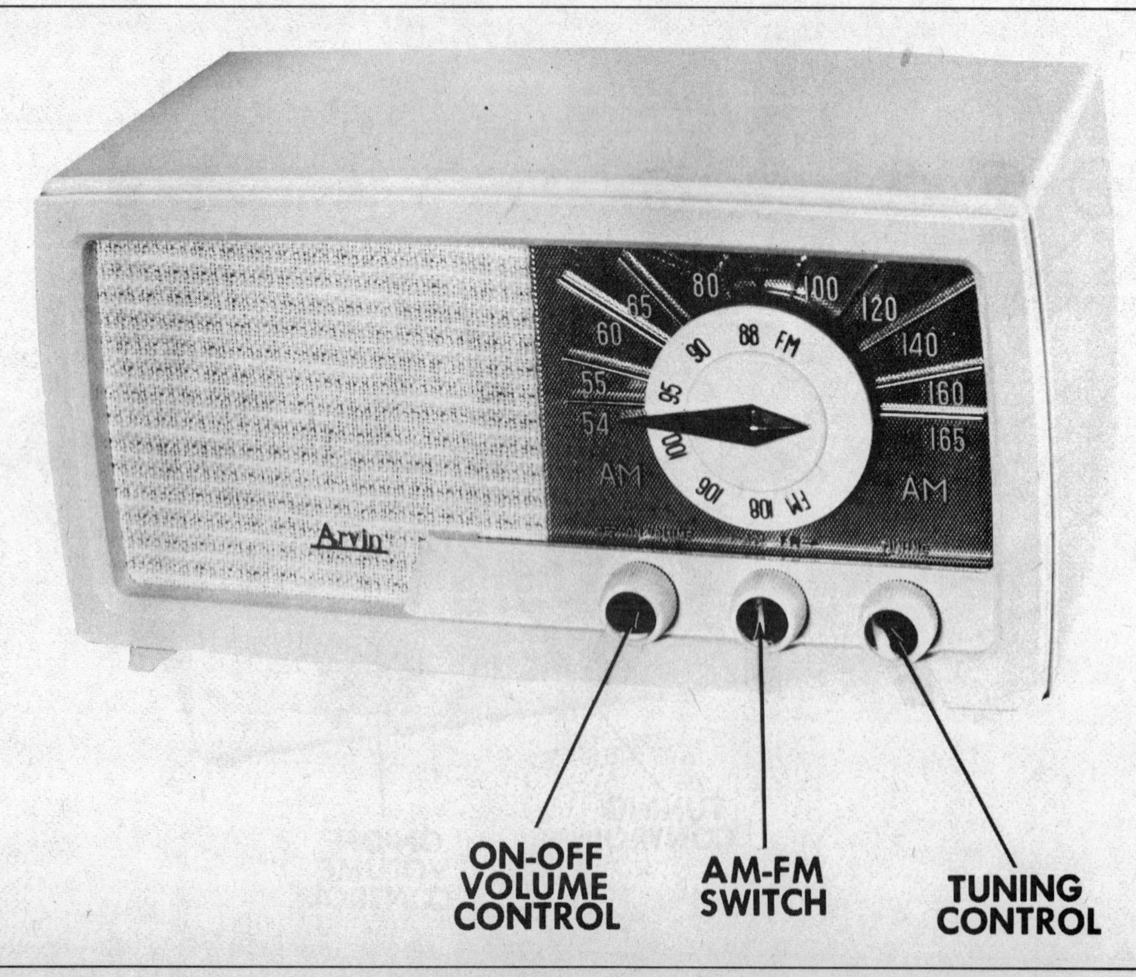

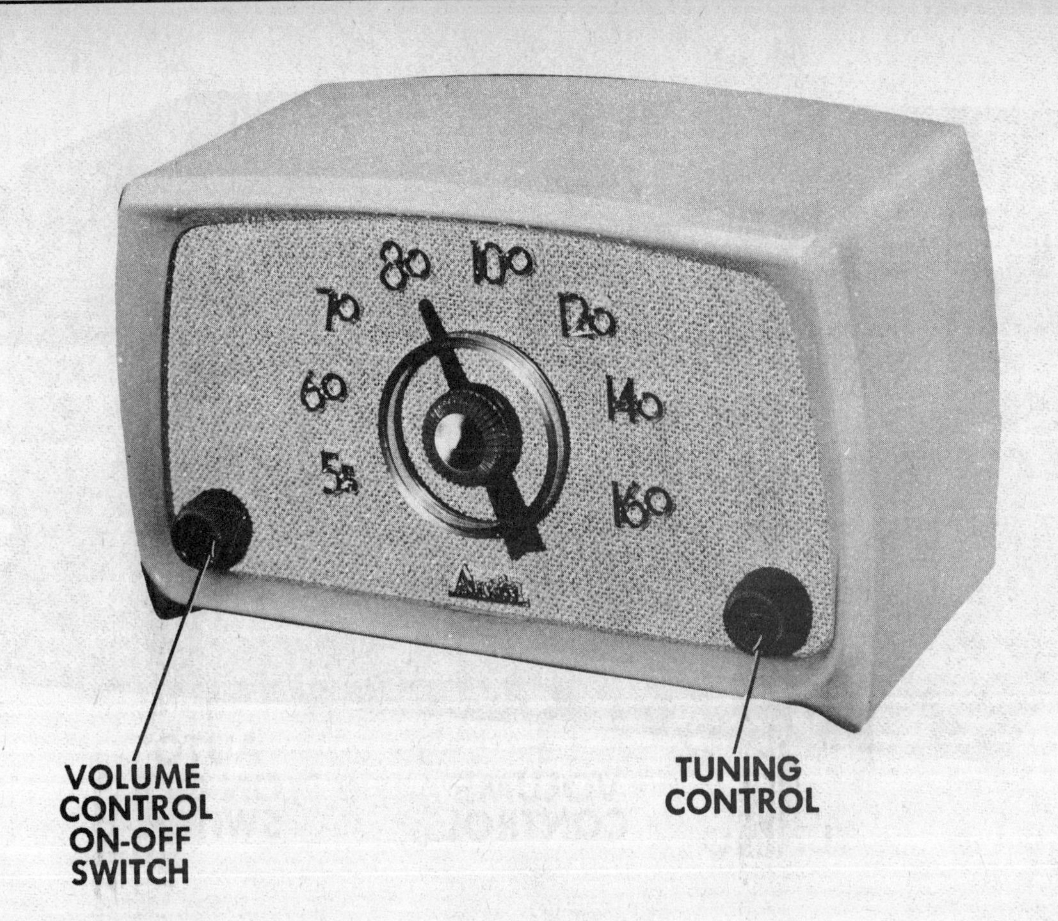

ARVIN

MODEL PICTURED
651T

AC/DC operated AM superheterodyne receiver

TUBES
5

POWER SUPPLY
105-125 volts AC/DC, .24 amp @ 117 volts AC

TUNING RANGE
540-1600KC

MFR/SUPPLIER
Arvin Industries, Inc.

PHOTOFACT SET
251-3

PUBLISHED
1954

ARVIN

MODEL PICTURED
852P

Three-power portable AM superheterodyne receiver

TUBES
4

POWER SUPPLY
110-120 volts AC/DC, or 7.5 volts A & 90 volts B supply

TUNING RANGE
540-1620KC

MFR/SUPPLIER
Arvin Industries, Inc.

PHOTOFACT SET
258-3

PUBLISHED
1954

ARVIN

MODEL PICTURED
848T

AC operated AM superheterodyne receiver with electric clock

TUBES
4

POWER SUPPLY
105-120 volts AC, 60 cycles, .26 amp @ 117 volts AC

TUNING RANGE
535-1620KC

MFR/SUPPLIER
Arvin Industries, Inc.

PHOTOFACT SET
259-2

PUBLISHED
1954

ARVIN

MODEL PICTURED
858T

AC operated AM superheterodyne receiver with electric clock

TUBES
5

POWER SUPPLY
110-120 volts AC, 60 cycle

TUNING RANGE
540-1650KC

MFR/SUPPLIER
Arvin Industries, Inc.

PHOTOFACT SET
261-2

PUBLISHED
1954

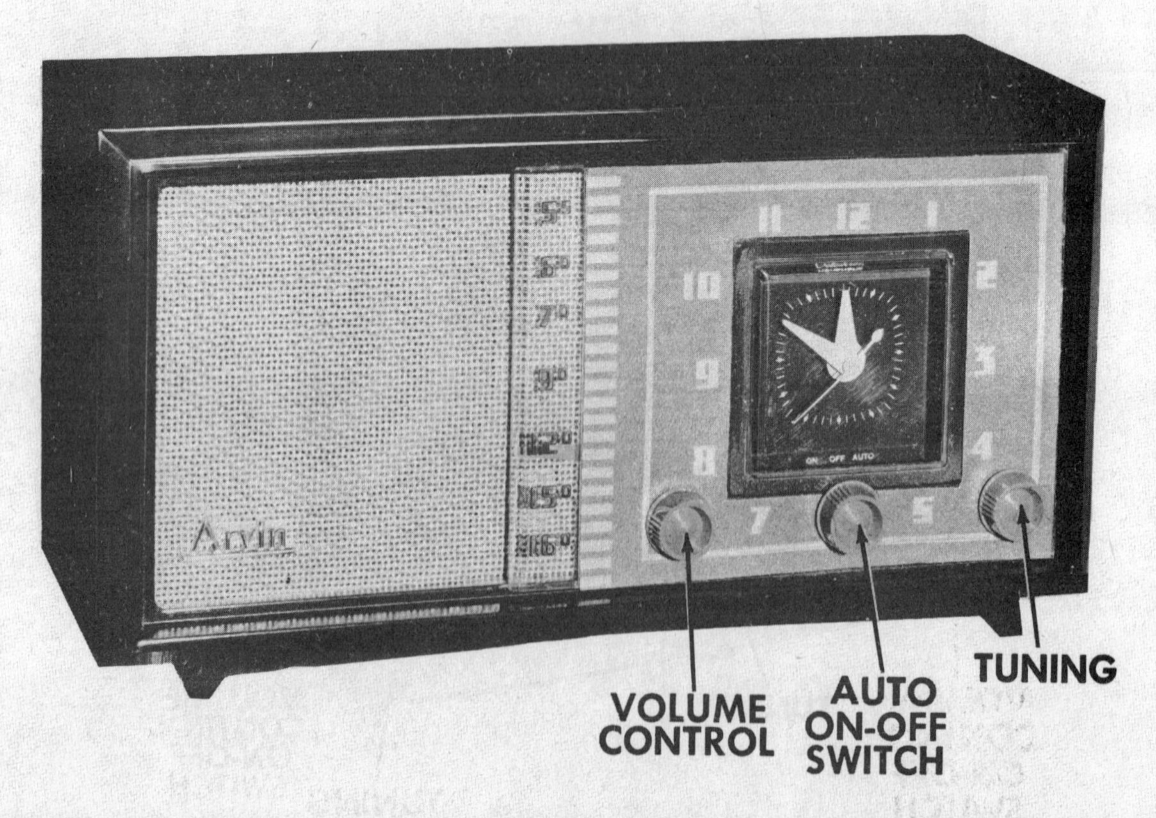

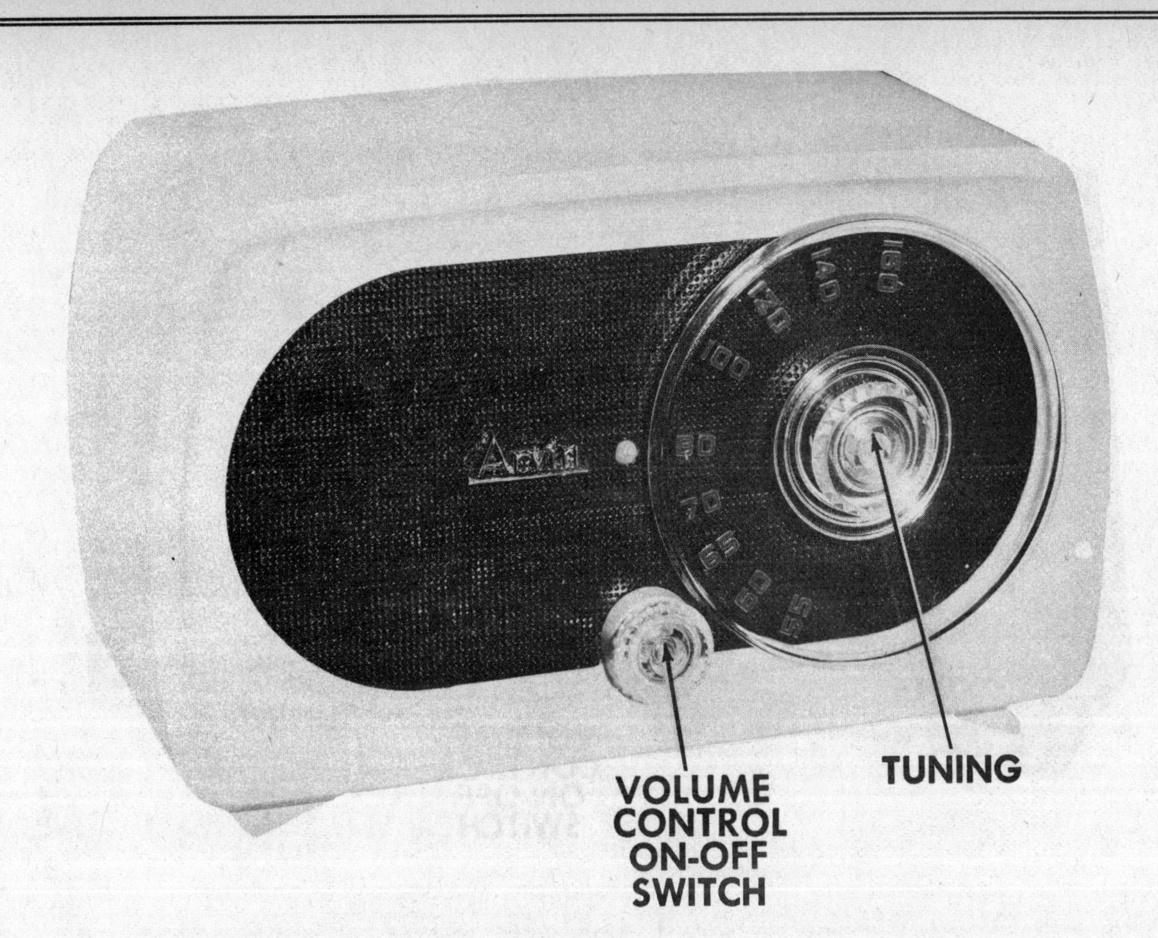

ARVIN

MODEL PICTURED
850T

AC/DC operated AM superheterodyne receiver

TUBES
5

POWER SUPPLY
110-120 volts AC/DC

TUNING RANGE
540-1650KC

MFR/SUPPLIER
Arvin Industries, Inc.

PHOTOFACT SET
262-3

PUBLISHED
1955

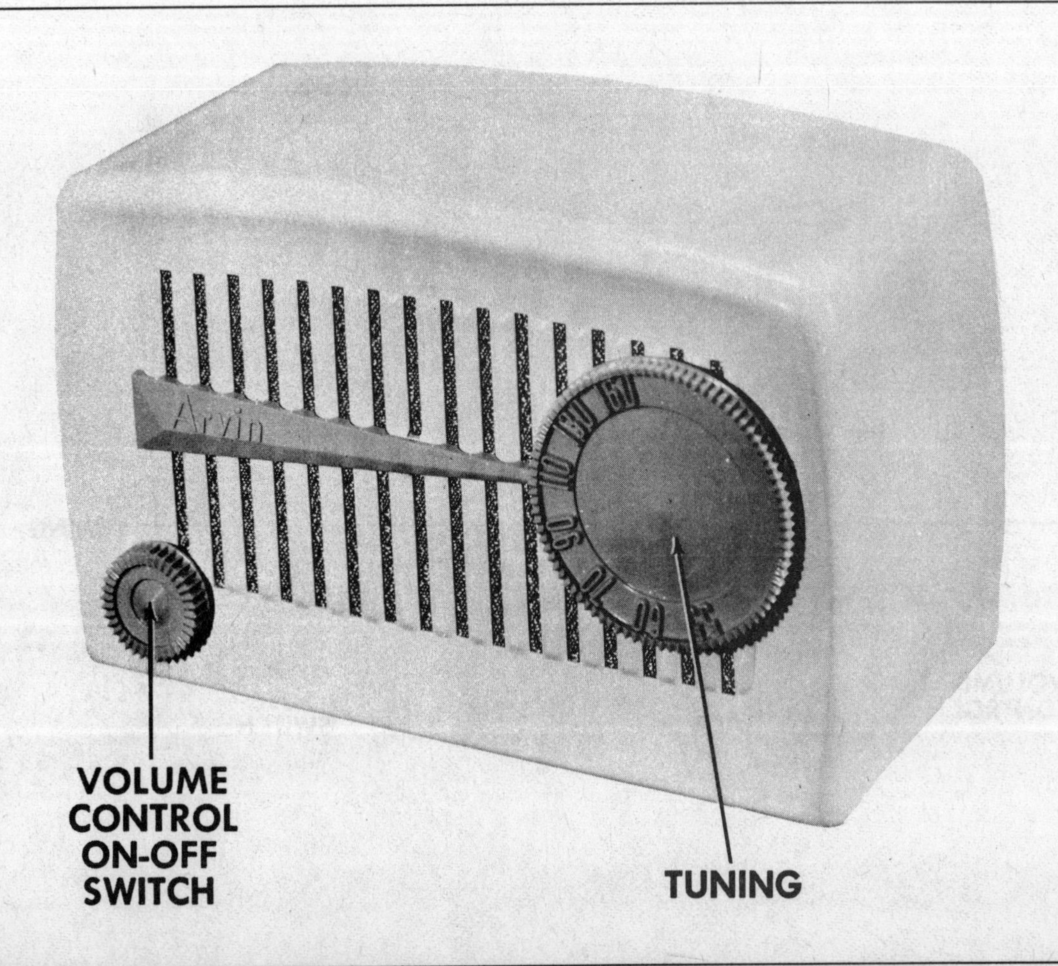

ARVIN

MODEL PICTURED
840T

AC/DC operated AM superheterodyne receiver

TUBES
4

POWER SUPPLY
110-120 volts AC/DC

TUNING RANGE
540-1600KC

MFR/SUPPLIER
Arvin Industries, Inc.

PHOTOFACT SET
263-3

PUBLISHED
1955

ARVIN

MODEL PICTURED
851T

AC/DC operated AM superheterodyne receiver

TUBES
5

POWER SUPPLY
105-125 volts AC/DC

TUNING RANGE
540-1600KC

MFR/SUPPLIER
Arvin Industries, Inc.

PHOTOFACT SET
266-3

PUBLISHED
1955

ARVIN

MODEL PICTURED
857T

AC operated AM superheterodyne receiver with electric clock

TUBES
5

POWER SUPPLY
105-125 volts AC, 60 cycles, .24 amp @ 117 volts AC

TUNING RANGE
540KC-1600KC

MFR/SUPPLIER
Arvin Industries, Inc.

PHOTOFACT SET
275-4

PUBLISHED
1955

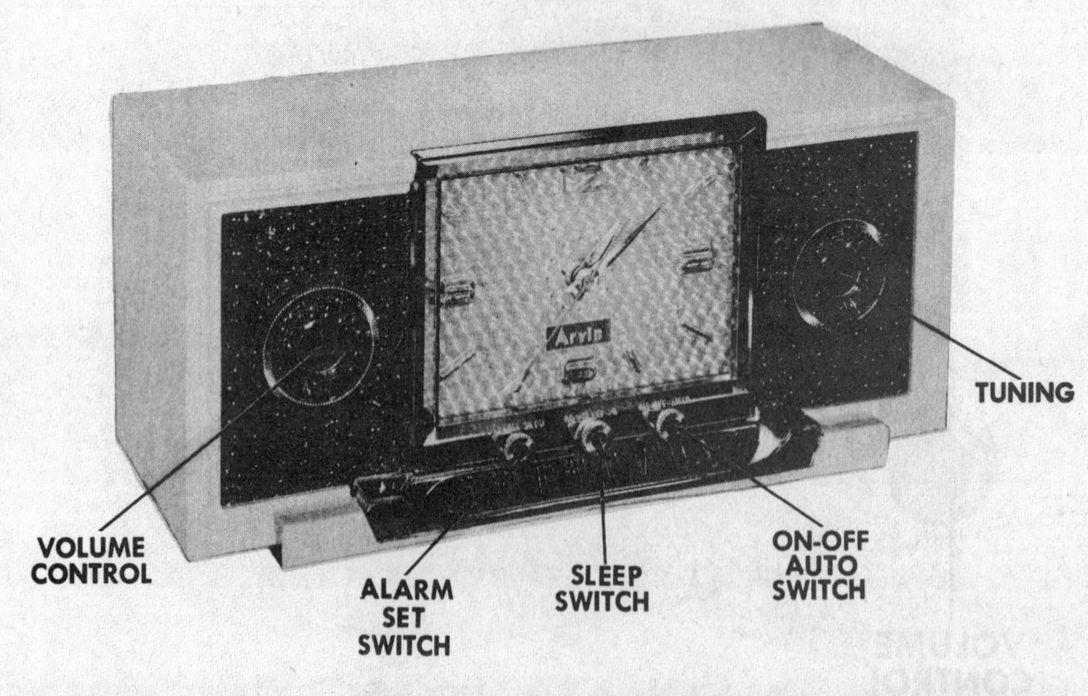

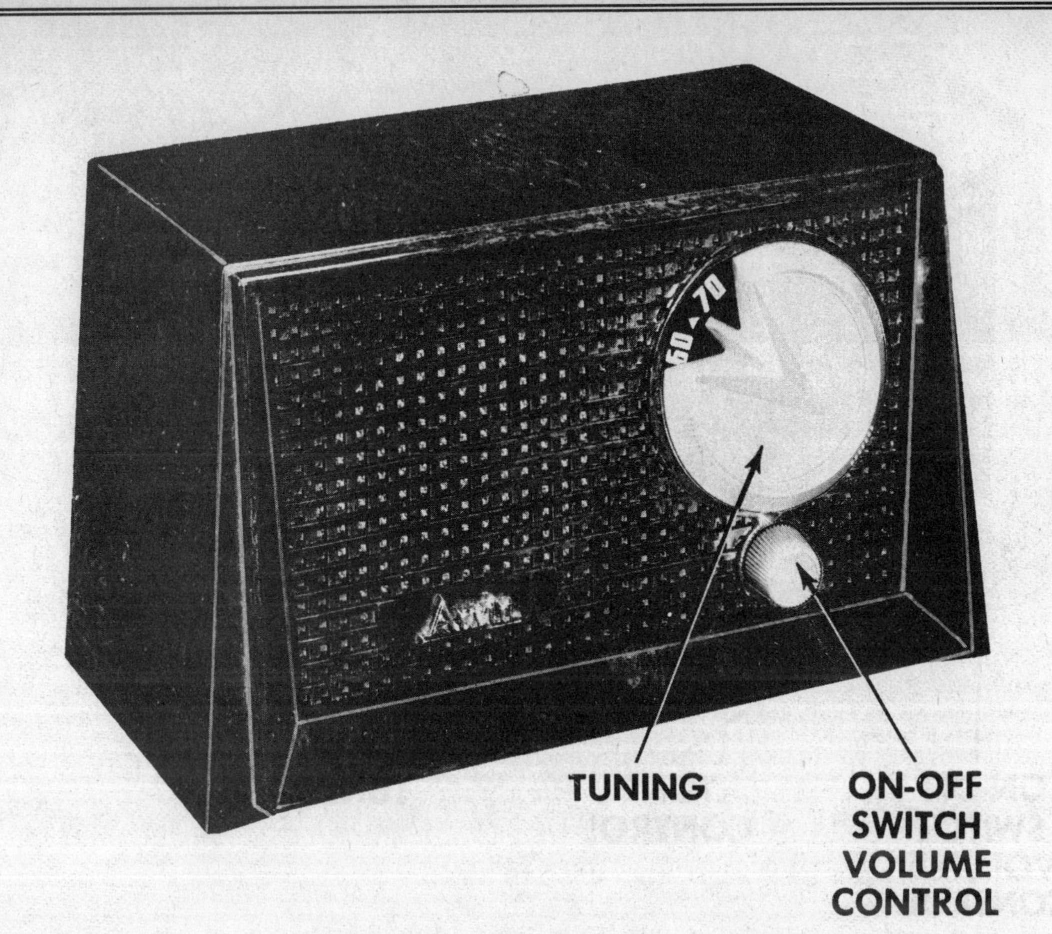

ARVIN

MODEL PICTURED
950T

AC/DC operated AM superheterodyne receiver

TUBES
5

POWER SUPPLY
105-125 volts AC/DC, .21 amp @ 117 volts AC

TUNING RANGE
540KC-1600KC

MFR/SUPPLIER
Arvin Industries, Inc.

PHOTOFACT SET
295-3

PUBLISHED
1955

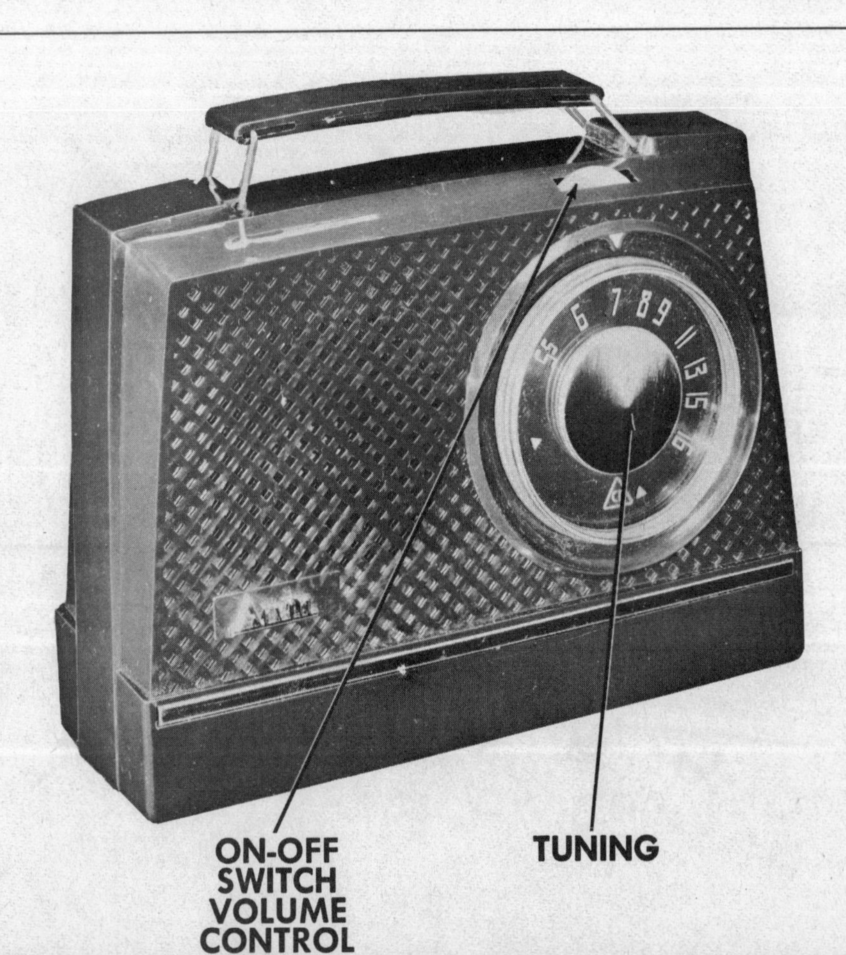

ARVIN

MODEL PICTURED
952P

Three-power portable AM superheterodyne receiver

TUBES
4

POWER SUPPLY
110-120 volts AC/DC or 7.5 volts A & 90 volts B supply

TUNING RANGE
540-1600KC

MFR/SUPPLIER
Arvin Industries, Inc.

PHOTOFACT SET
300-2

PUBLISHED
1955

ARVIN

MODEL PICTURED
955T

AC/DC operated AM superheterodyne receiver

TUBES
5

POWER SUPPLY
105-125 volts AC/DC

TUNING RANGE
540-1650KC

MFR/SUPPLIER
Arvin Industries, Inc.

PHOTOFACT SET
304-1

PUBLISHED
1956

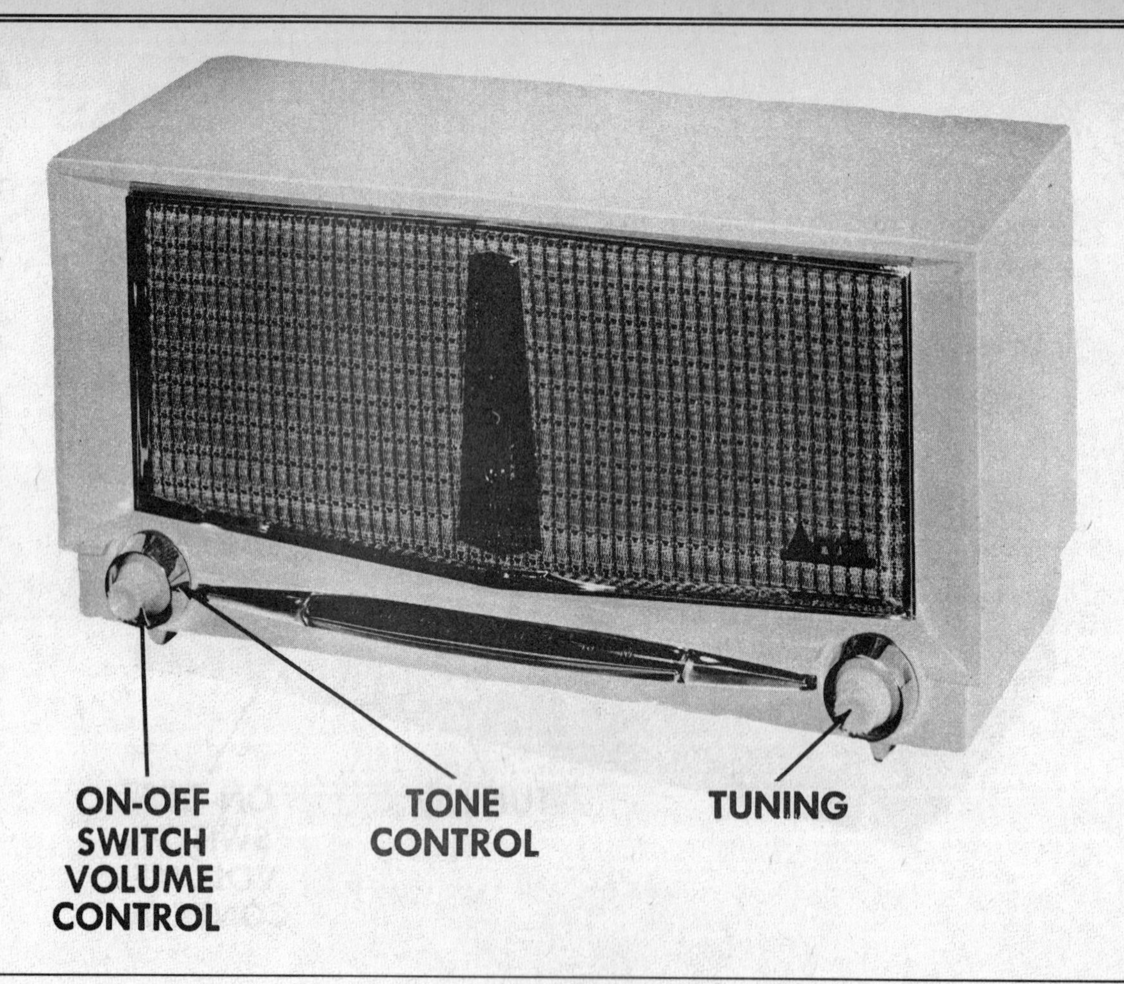

ARVIN

MODEL PICTURED
957T

AC operated AM superheterodyne receiver with electric clock

TUBES
5

POWER SUPPLY
110-120 volts AC, 60 cycle

TUNING RANGE
540-1600KC

MFR/SUPPLIER
Arvin Industries, Inc.

PHOTOFACT SET
321-2

PUBLISHED
1956

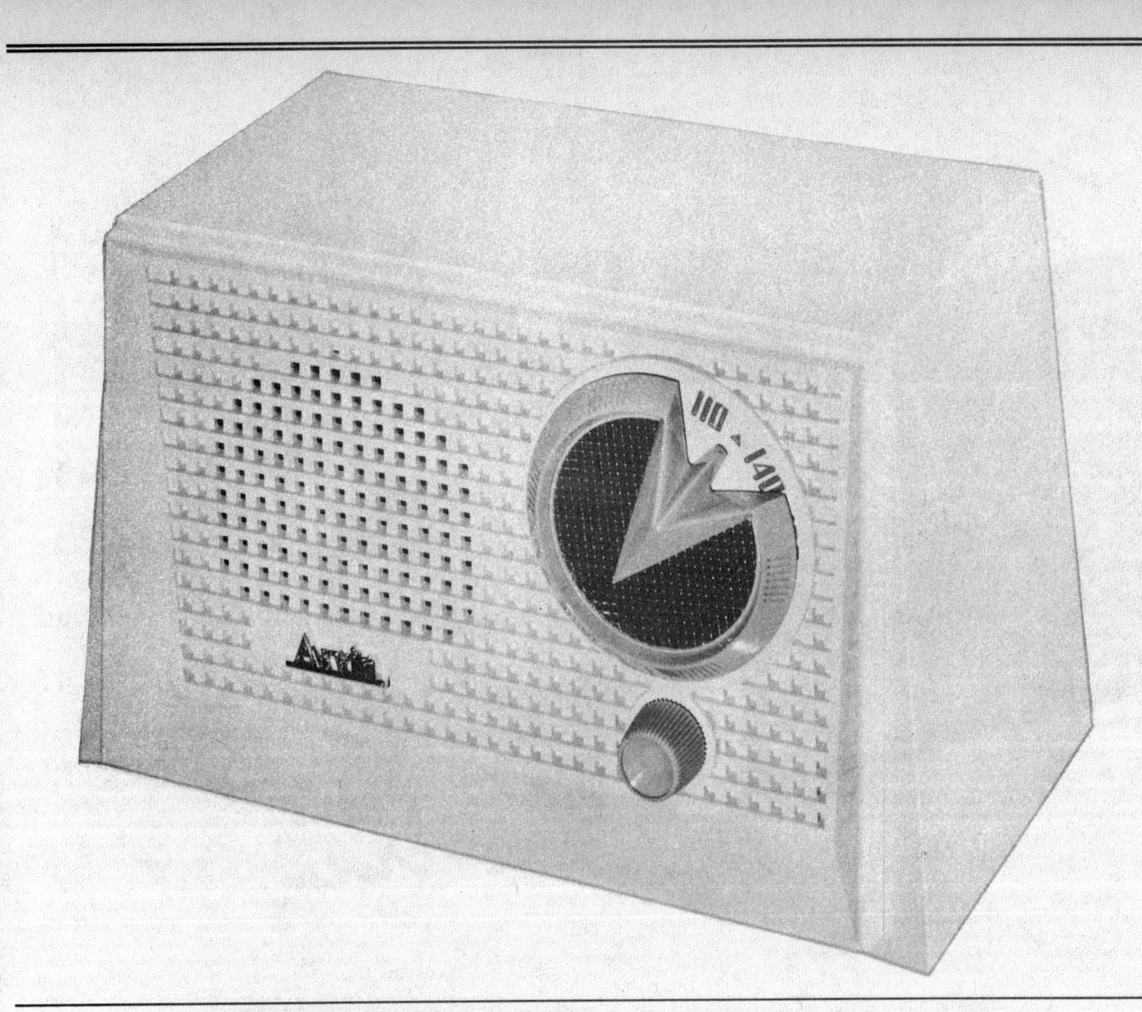

ARVIN

MODEL PICTURED
950T1

AC/DC operated AM superheterodyne receiver

TUBES
5

POWER SUPPLY
105-125 volts AC/DC

TUNING RANGE
540-1600KC

MFR/SUPPLIER
Arvin Industries, Inc.

PHOTOFACT SET
340-2

PUBLISHED
1956

ARVIN

MODEL PICTURED
9562

Battery operated portable AM transistorized receiver

TUBES
0

POWER SUPPLY
7.5 volts DC positive supply and 1.5 volts DC negative supply

TUNING RANGE
540-1670KC

MFR/SUPPLIER
Arvin Industries, Inc.

PHOTOFACT SET
348-1

PUBLISHED
1957

ARVIN

MODEL PICTURED
5561

AC operated AM receiver with electric clock

TUBES
5

POWER SUPPLY
105-120 volts AC, 60 cycle

TUNING RANGE
540-1670KC

MFR/SUPPLIER
Arvin Industries, Inc.

PHOTOFACT SET
351-3

PUBLISHED
1957

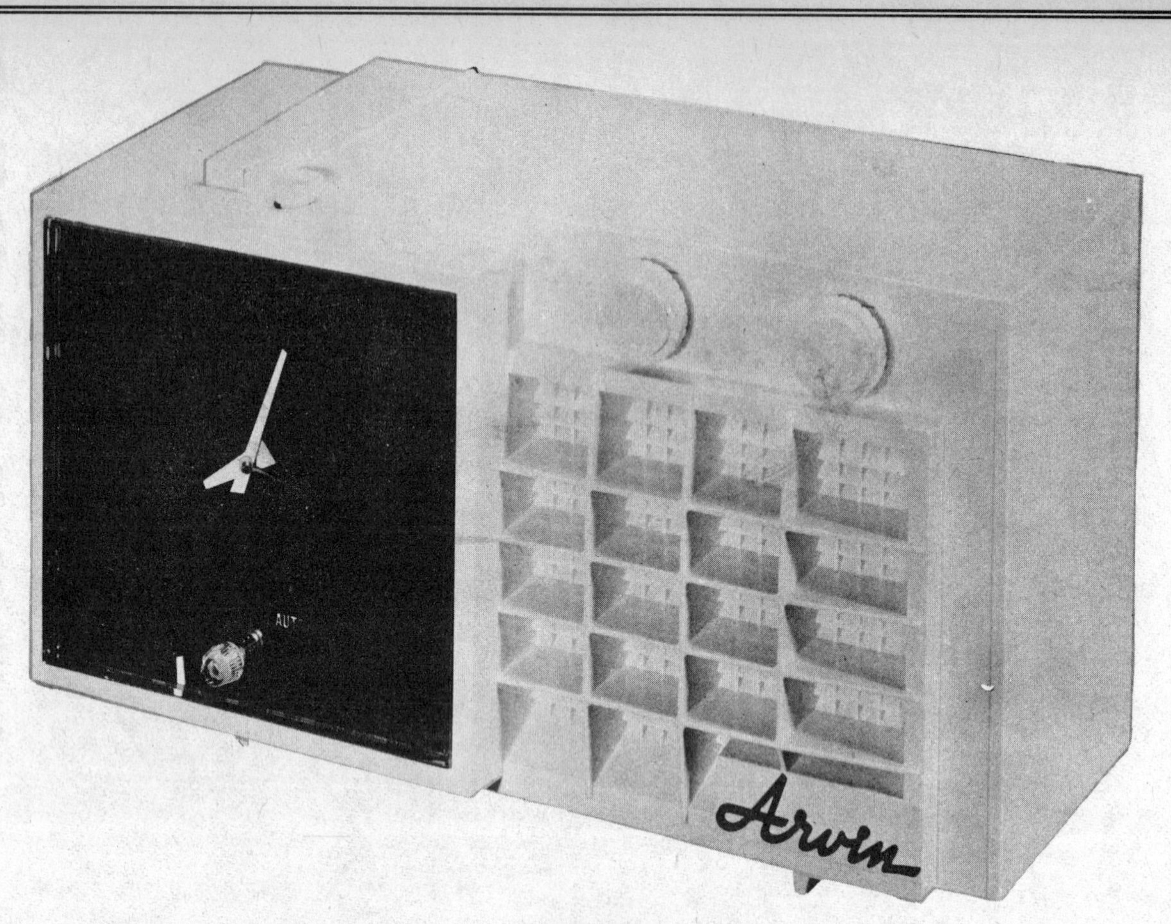

ARVIN

MODEL PICTURED
2563

AC/DC operated AM receiver

TUBES
5

POWER SUPPLY
110-120 volts AC/DC

TUNING RANGE
535-1670KC

MFR/SUPPLIER
Arvin Industries, Inc.

PHOTOFACT SET
356-3

PUBLISHED
1957

ARVIN

MODEL PICTURED
3561

AC/DC operated AM receiver

TUBES
6

POWER SUPPLY
110-120 volts AC/DC

TUNING RANGE
540-1620KC

MFR/SUPPLIER
Arvin Industries, Inc.

PHOTOFACT SET
358-2

PUBLISHED
1957

ARVIN

MODEL PICTURED
9574P

Battery operated portable AM transistorized receiver

TUBES
0

POWER SUPPLY
9 volts DC

TUNING RANGE
540-1670KC

MFR/SUPPLIER
Arvin Industries, Inc.

PHOTOFACT SET
377-6

PUBLISHED
1957

ARVIN

MODEL PICTURED
950T2

AC/DC operated AM receiver

TUBES
5

POWER SUPPLY
105-125 volts AC/DC

TUNING RANGE
540-1670KC

MFR/SUPPLIER
Arvin Industries, Inc.

PHOTOFACT SET
384-7

PUBLISHED
1958

ARVIN

MODEL PICTURED
2572

AC/DC operated AM receiver

TUBES
5

POWER SUPPLY
110-120 volts AC, 60 cycle

TUNING RANGE
540-1670KC

MFR/SUPPLIER
Arvin Industries, Inc.

PHOTOFACT SET
386-6

PUBLISHED
1958

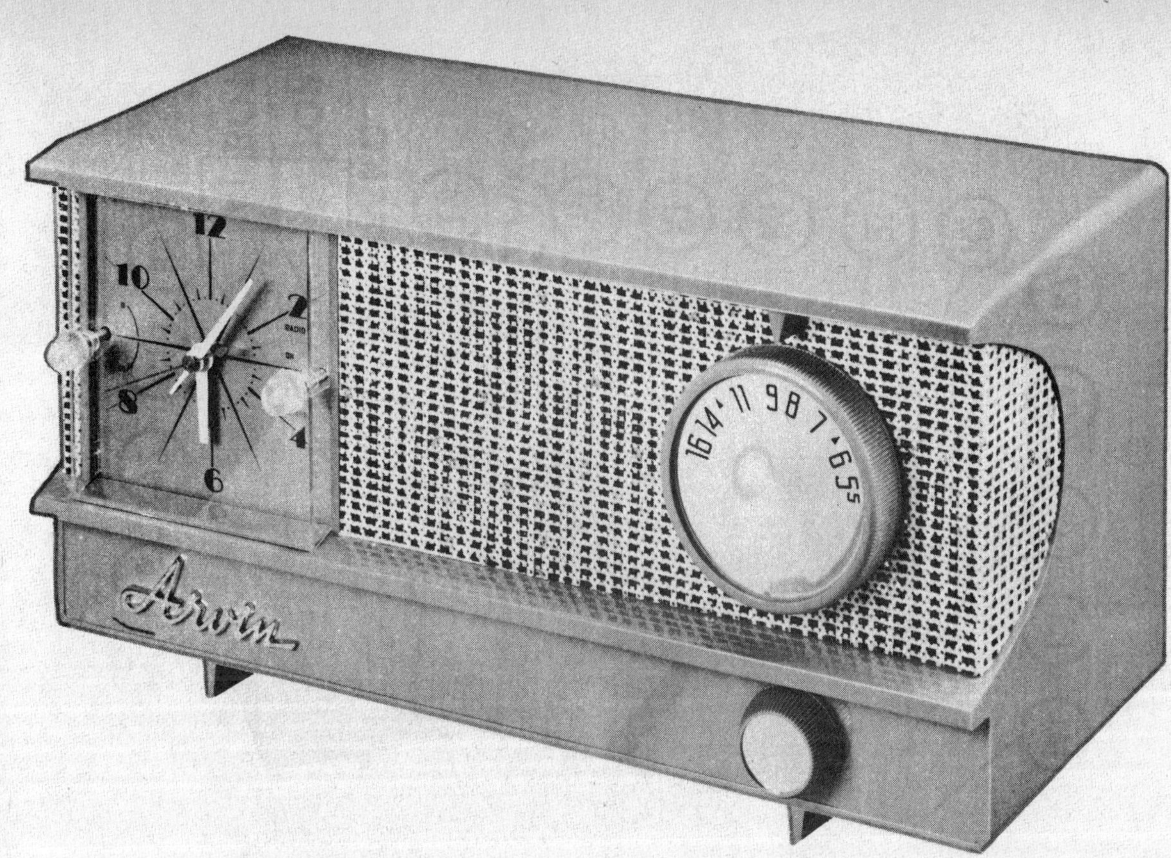

ARVIN

MODEL PICTURED
5571

AC operated AM receiver with electric clock

TUBES
5

POWER SUPPLY
110-120 volts AC, 60 cycle

TUNING RANGE
540-1670KC

MFR/SUPPLIER
Arvin Industries, Inc.

PHOTOFACT SET
387-7

PUBLISHED
1958

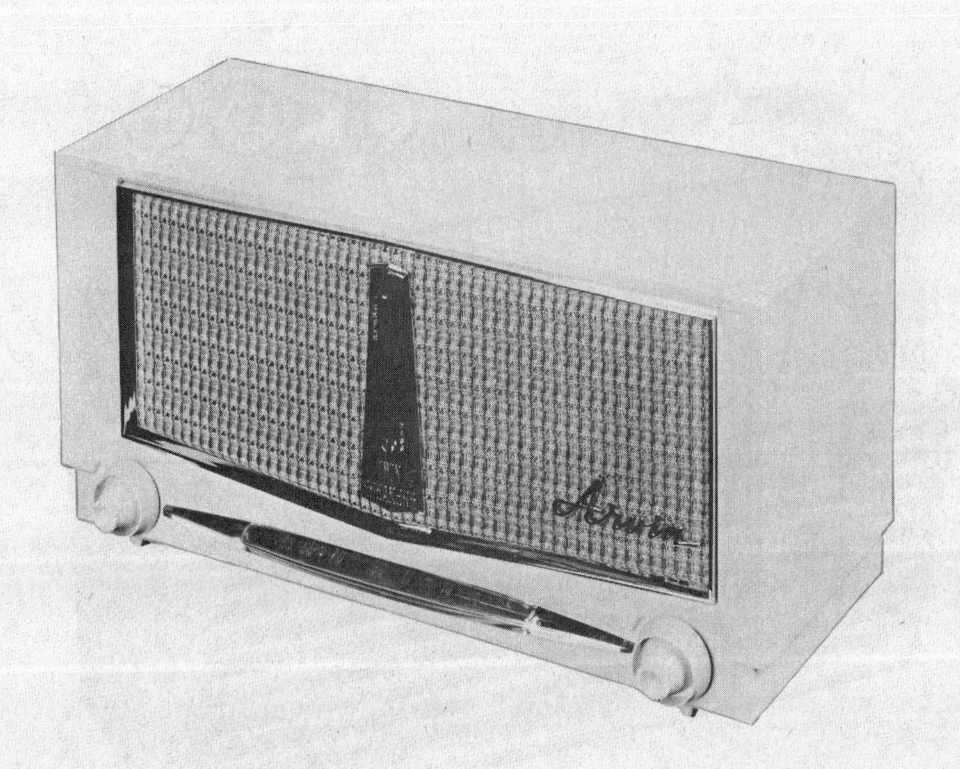

ARVIN

MODEL PICTURED
956T1

AC/DC operated AM receiver

TUBES
5

POWER SUPPLY
105-125 volts AC/DC

TUNING RANGE
540-1670KC

MFR/SUPPLIER
Arvin Industries, Inc.

PHOTOFACT SET
391-5

PUBLISHED
1958

ARVIN

MODEL PICTURED
5578

AC operated AM receiver with electric clock

TUBES
5

POWER SUPPLY
105-125 volts AC, 60 cycle

TUNING RANGE
540-1670KC

MFR/SUPPLIER
Arvin Industries, Inc.

PHOTOFACT SET
393-6

PUBLISHED
1958

ARVIN

MODEL PICTURED
8571

Three-power portable AM receiver

TUBES
4

POWER SUPPLY
110-120 volts AC/DC or 7.5 volts A & 90 volts B supply

TUNING RANGE
540-1670KC

MFR/SUPPLIER
Arvin Industries, Inc.

PHOTOFACT SET
394-7

PUBLISHED
1958

ARVIN

MODEL PICTURED
8576

Battery operated portable AM transistorized receiver

TUBES
0

POWER SUPPLY
9 volts DC battery

TUNING RANGE
540-1670KC

MFR/SUPPLIER
Arvin Industries, Inc.

PHOTOFACT SET
395-5

PUBLISHED
1958

ARVIN

MODEL PICTURED
1581

AC/DC operated AM receiver

TUBES
4

POWER SUPPLY
110-120 volts AC/DC

TUNING RANGE
540-1650KC

MFR/SUPPLIER
Arvin Industries, Inc.

PHOTOFACT SET
411-6

PUBLISHED
1958

ARVIN

MODEL PICTURED
2581

AC/DC operated AM receiver

TUBES
5

POWER SUPPLY
105-125 volts AC/DC, .24 amp @ 117 volts AC (27 watts)

TUNING RANGE
540-1670KC

MFR/SUPPLIER
Arvin Industries, Inc.

PHOTOFACT SET
427-4

PUBLISHED
1959

ARVIN

MODEL PICTURED
2585

AC/DC operated AM receiver

TUBES
5

POWER SUPPLY
110-120 volts AC/DC, 27 watts, .285 amp @ 117 volts AC

TUNING RANGE
540-1650KC

MFR/SUPPLIER
Arvin Industries, Inc.

PHOTOFACT SET
428-4

PUBLISHED
1959

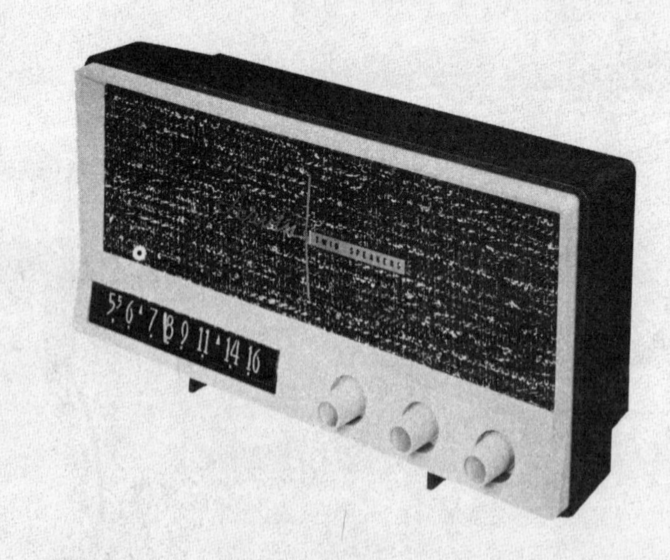

ARVIN

MODEL PICTURED
8584

Battery operated portable transistorized AM receiver

TUBES
0

POWER SUPPLY
6 volts DC, 6mA @ 6 volts DC (no signal, minimum volume), 8mA @ 6 volts DC (signal, normal volume)

TUNING RANGE
540-1620KC

MFR/SUPPLIER
Arvin Industries, Inc.

PHOTOFACT SET
429-3

PUBLISHED
1959

ARVIN

MODEL PICTURED
5583

AC operated AM receiver

TUBES
5

POWER SUPPLY
110-120 volts AC, 60 cycles, 28 watts, .29 amp @ 117 volts AC

TUNING RANGE
540-1650KC

MFR/SUPPLIER
Arvin Industries, Inc.

PHOTOFACT SET
430-4

PUBLISHED
1959

ARVIN

MODEL PICTURED
8581

Three-power portable AM receiver

TUBES
4

POWER SUPPLY
110-120 volts AC/DC or 7.5 volts DC A supply & 90 volts DC B supply

TUNING RANGE
535-1650KC

MFR/SUPPLIER
Arvin Industries, Inc.

PHOTOFACT SET
432-4

PUBLISHED
1959

ARVIN

MODEL PICTURED
3588

Battery operated transistorized AM receiver

TUBES
0

POWER SUPPLY
9 volts DC, 6.4mA @ 9 volts DC (no signal, minimum volume), 12mA @ 9 volts DC (signal, normal volume)

TUNING RANGE
540-1650KC

MFR/SUPPLIER
Arvin Industries, Inc.

PHOTOFACT SET
433-5

PUBLISHED
1959

ARVIN

MODEL PICTURED
3582

AC/DC operated AM receiver

TUBES
7

POWER SUPPLY
110-120 volts AC/DC, 44 watts, .41 amp @ 117 volts AC

TUNING RANGE
540-1670KC

MFR/SUPPLIER
Arvin Industries, Inc.

PHOTOFACT SET
444-5

PUBLISHED
1959

ARVIN

MODEL PICTURED
3586

AC operated FM/AM receiver

TUBES
9

POWER SUPPLY
105-120 volts AC, 60 cycles, 58 watts, .56 amp @ 117 volts AC

TUNING RANGE
540-1670KC, 88-108MC

MFR/SUPPLIER
Arvin Industries, Inc.

PHOTOFACT SET
454-5

PUBLISHED
1959

ARVIN

MODEL PICTURED
9595

Battery operated transistorized portable AM receiver

TUBES
0

POWER SUPPLY
6 volts DC, 5.2mA @ 6 volts DC (no signal, minimum volume), 9mA @ 6 volts DC (signal, normal volume)

TUNING RANGE
540-1670KC

MFR/SUPPLIER
Arvin Industries, Inc.

PHOTOFACT SET
470-4

PUBLISHED
1960

ARVIN

MODEL PICTURED
5591

AC operated AM receiver with electric clock

TUBES
5

POWER SUPPLY
110-120 volts AC, 60 cycles, 20 watts, .23 amp @ 117 volts AC

TUNING RANGE
540-1670KC

MFR/SUPPLIER
Arvin Industries, Inc.

PHOTOFACT SET
473-5

PUBLISHED
1960

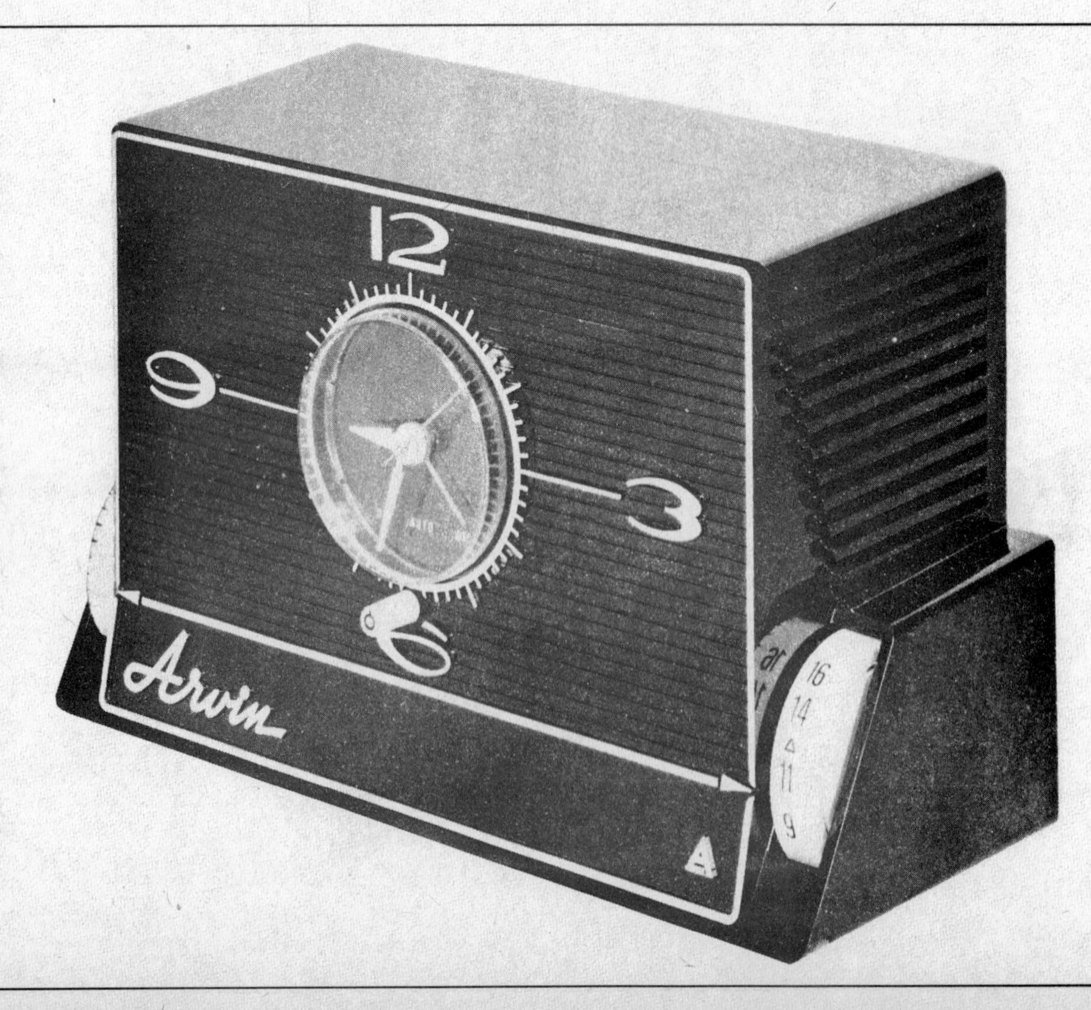

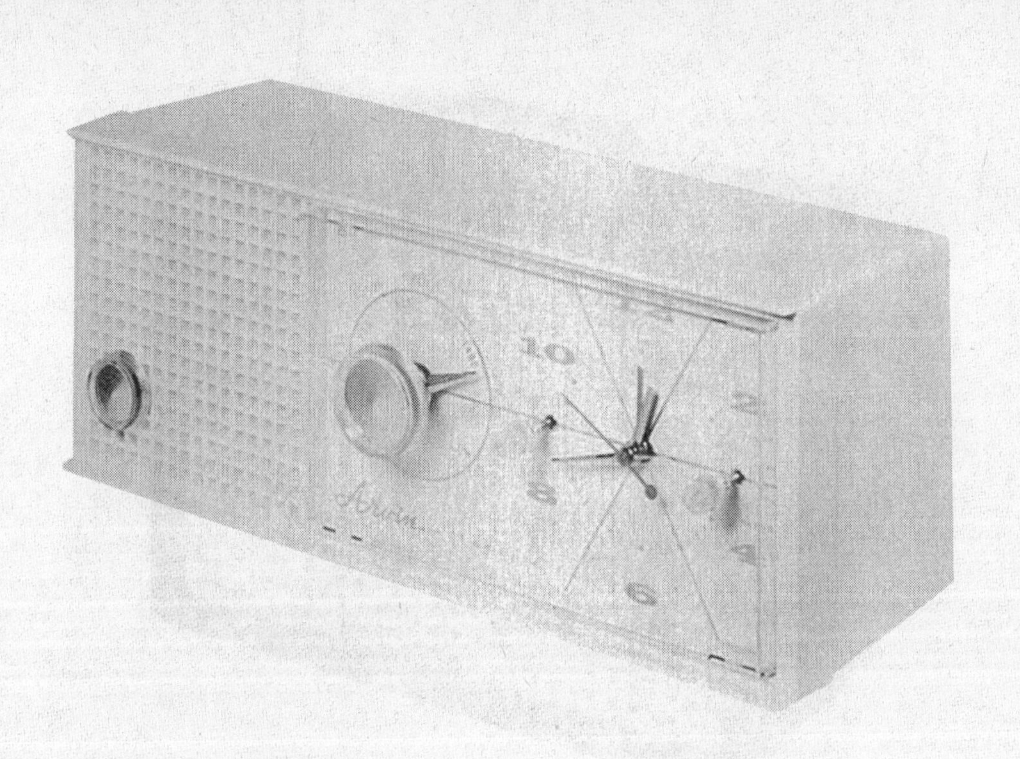

ARVIN

MODEL PICTURED
5594

AC operated AM receiver with electric clock

TUBES
5

POWER SUPPLY
110-120 volts AC, 60 cycles, 19 watts, .22 amp @ 117 volts AC (less clock)

TUNING RANGE
535-1670KC

MFR/SUPPLIER
Arvin Industries, Inc.

PHOTOFACT SET
474-5

PUBLISHED
1960

ARVIN

MODEL PICTURED
2598

Battery operated transistorized portable AM receiver

TUBES
0

POWER SUPPLY
9 volts DC, 7A @ 9 volts DC (no signal, minimum volume), 14mA @ 9 volts DC (signal, normal volume)

TUNING RANGE
540-1670KC

MFR/SUPPLIER
Arvin Industries, Inc.

PHOTOFACT SET
485-3

PUBLISHED
1960

ARVIN

MODEL PICTURED
9598

Battery operated transistorized portable BC-SW-LW receiver

TUBES
0

POWER SUPPLY
12 volts DC, 9mA @ 12 volts DC (no signal, minimum volume), 15mA @ 12 volts DC (signal, normal volume)

TUNING RANGE
Broadcast: 540-1630KC
Shortwave: 2.1-6.3MC
Longwave: 180-425KC

MFR/SUPPLIER
Arvin Industries, Inc.

PHOTOFACT SET
485-4

PUBLISHED
1960

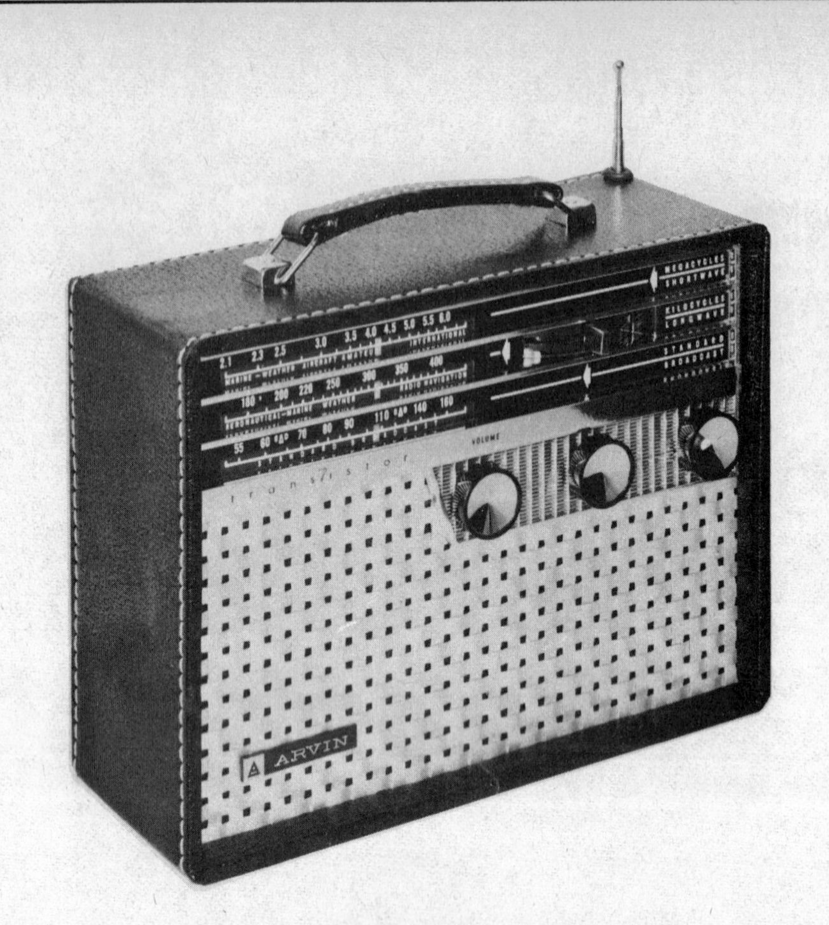

ARVIN

MODEL PICTURED
9594

Battery operated transistorized portable AM receiver

TUBES
0

POWER SUPPLY
6 volts DC, 5 mA @ 6 volts DC (no signal, minimum volume), 11mA @ 6 volts DC (signal, normal volume)

TUNING RANGE
540-1670KC

MFR/SUPPLIER
Arvin Industries, Inc.

PHOTOFACT SET
486-5

PUBLISHED
1960

ARVIN

MODEL PICTURED
60R49

Battery operated transistorized portable AM receiver

TUBES
0

POWER SUPPLY
6 volts DC

TUNING RANGE
540-1670KC

MFR/SUPPLIER
Arvin Industries, Inc.

PHOTOFACT SET
506-6

PUBLISHED
1960

ARVIN

MODEL PICTURED
60R33

Battery operated transistorized portable AM receiver

TUBES
0

POWER SUPPLY
6 volts DC

TUNING RANGE
540-1670KC

MFR/SUPPLIER
Arvin Industries, Inc.

PHOTOFACT SET
507-4

PUBLISHED
1960

ARVIN

MODEL PICTURED
60R23

Battery operated transistorized portable AM receiver

TUBES
0

POWER SUPPLY
6 volts DC

TUNING RANGE
540-1670KC

MFR/SUPPLIER
Arvin Industries, Inc.

PHOTOFACT SET
508-5

PUBLISHED
1960

ASTRA-SONIC

MODEL PICTURED
748

AC operated combo phono-wire recorder, superheterodyne receiver with loop antenna.

TUBES
7

POWER SUPPLY
110-120 volts AC, .45 amp @ 117 volts AC

TUNING RANGE
540-1620KC

MFR/SUPPLIER
Pentron Corp.

PHOTOFACT SET
53-6

PUBLISHED
1949

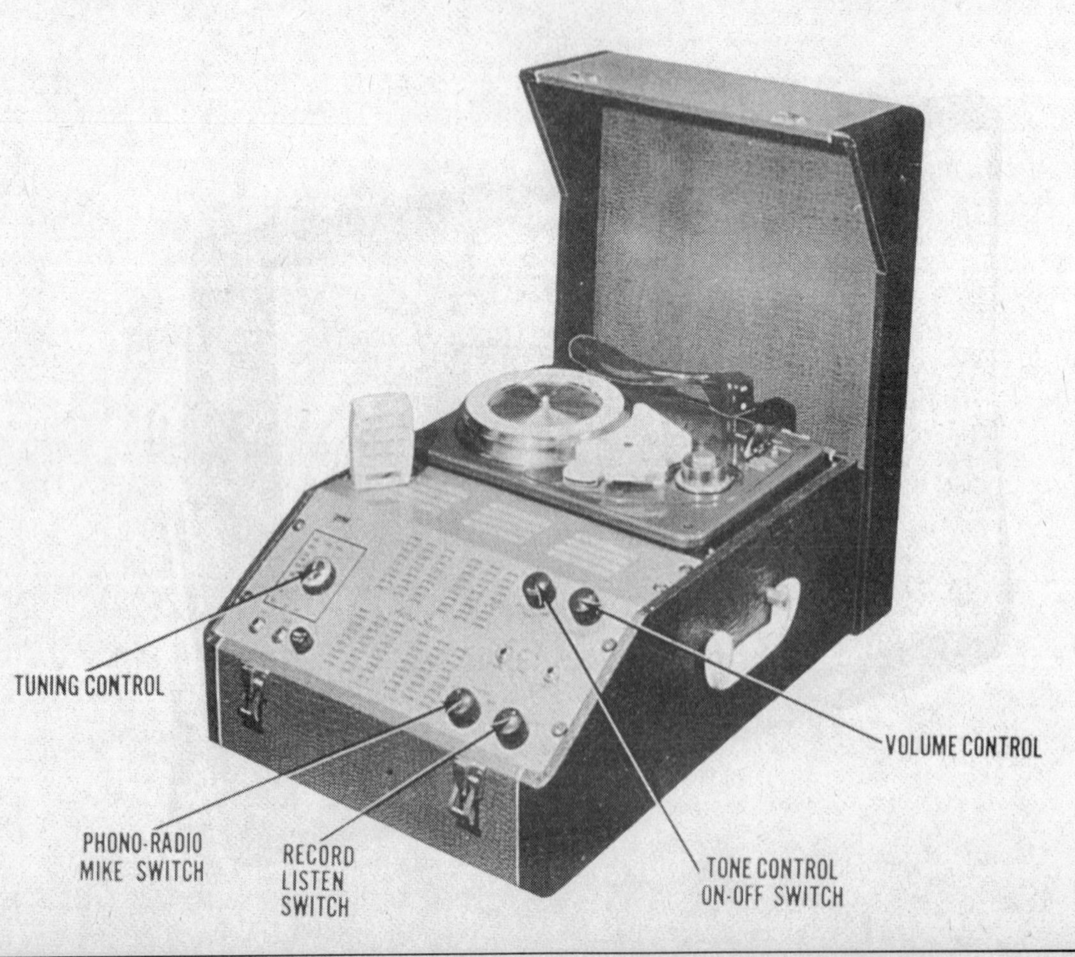

ATLAS

MODEL PICTURED
AB-45

AC/DC operated two-band superheterodyne receiver, self-contained loop antenna

TUBES
6

POWER SUPPLY
105-125 volts AC/DC

TUNING RANGE
540-1750KC
5.8-17.3MC

MFR/SUPPLIER
Atomic Heater & Radio

PHOTOFACT SET
14-5

PUBLISHED
1947

AUDAR

MODEL PICTURED
PR-6

AC operated radio-phono combo superheterodyne receiver

TUBES
4

POWER SUPPLY
110-120 volts AC

TUNING RANGE
540-1680KC

MFR/SUPPLIER
Audar, Inc.

PHOTOFACT SET
13-10

PUBLISHED
1947

AUDAR

MODEL PICTURED
(Telvar) RER-9

AC operated combo phono-recorder-radio, superheterodyne receiver

TUBES
8

POWER SUPPLY
110-120 volts AC

TUNING RANGE
540-1720KC

MFR/SUPPLIER
Audar, Inc.

PHOTOFACT SET
65-2

PUBLISHED
1949

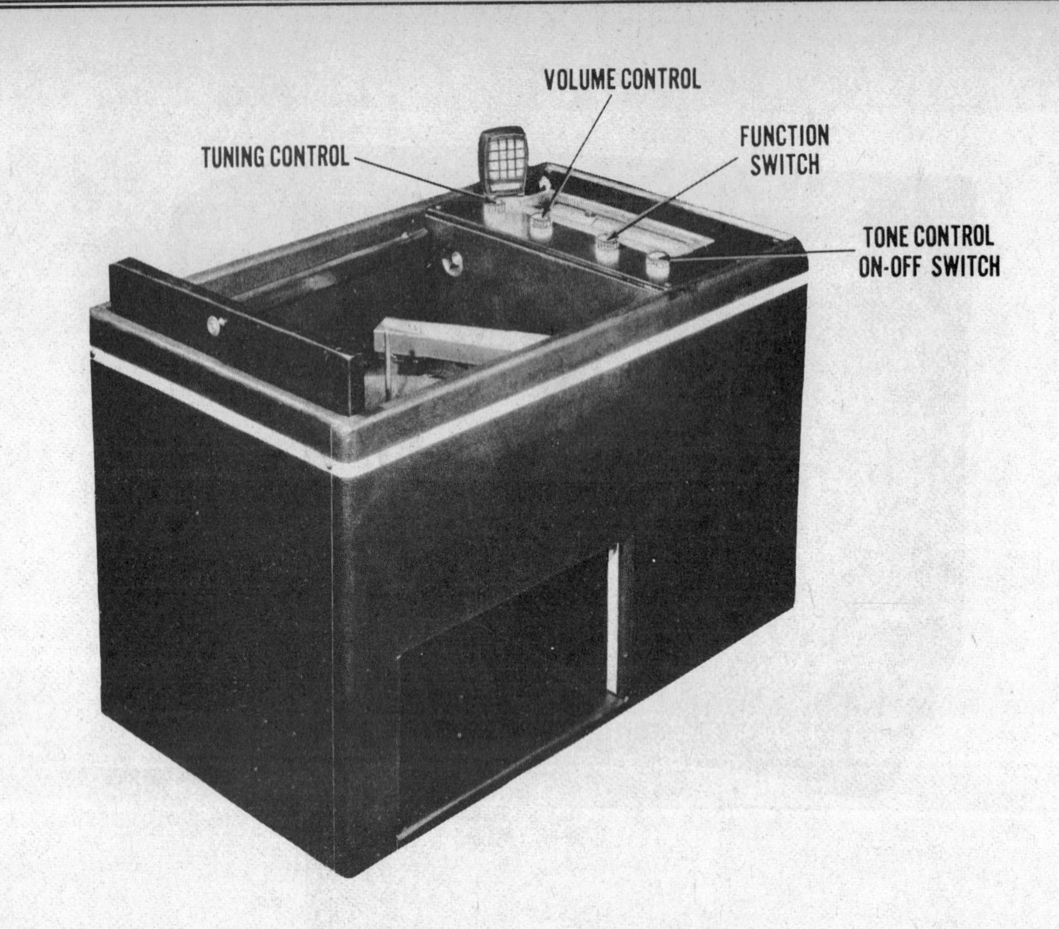

AUDAR

MODEL PICTURED
AV-7T

AC operated AM-phono combo superheterodyne receiver

TUBES
7

POWER SUPPLY
110-120 volts AC

TUNING RANGE
540-1620KC

MFR/SUPPLIER
Audar, Inc.

PHOTOFACT SET
166-6

PUBLISHED
1952

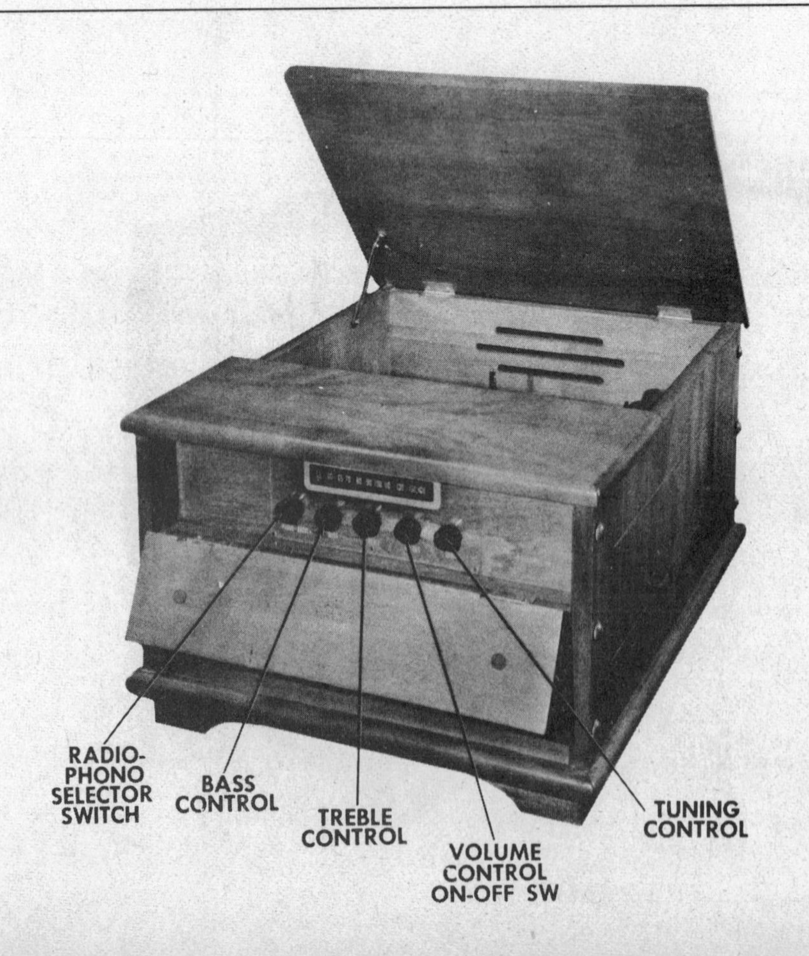

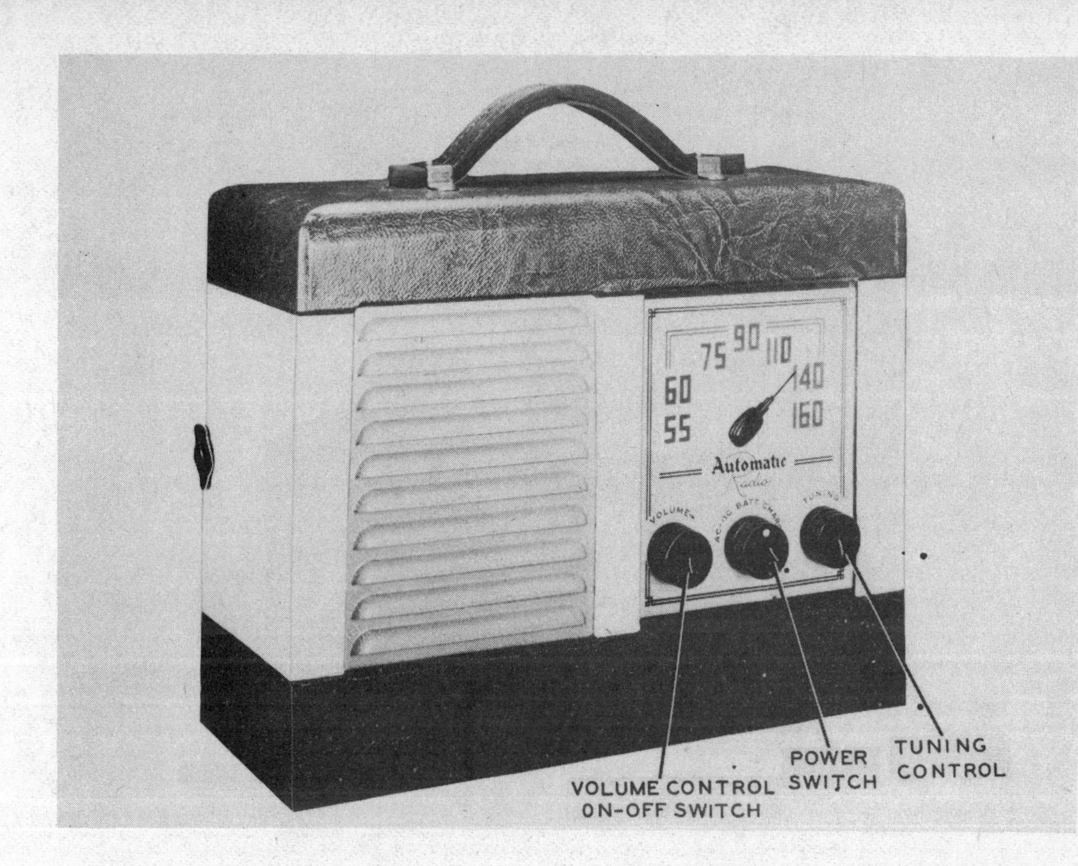

AUTOMATIC

MODEL PICTURED
C60

Three-power AC/DC battery operated portable superheterodyne receiver with self-contained loop antenna

TUBES
5

POWER SUPPLY
110 volts AC/DC or 6 volts A battery & 45 volts B battery

TUNING RANGE
550-1600KC

MFR/SUPPLIER
Automatic Radio Mfg. Co.

PHOTOFACT SET
5-20

PUBLISHED
1946

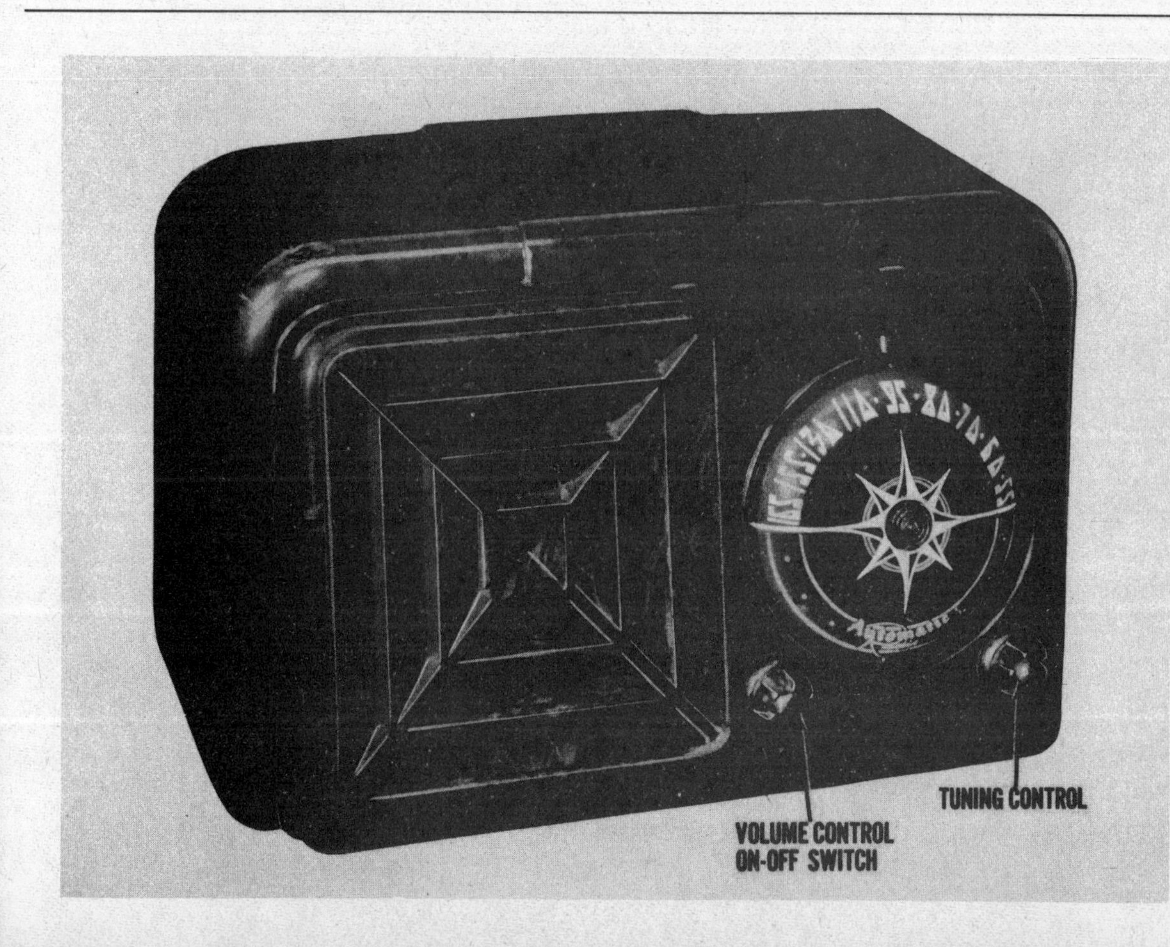

AUTOMATIC

MODEL PICTURED
614X

AC/DC superheterodyne receiver, self-contained loop antenna

TUBES
6

POWER SUPPLY
110-125 volts AC/DC

TUNING RANGE
538-1635KC

MFR/SUPPLIER
Automatic Radio Mfg. Co.

PHOTOFACT SET
8-2

PUBLISHED
1946

AUTOMATIC

MODEL PICTURED
640

AC operated combo phono-superheterodyne receiver with self-contained loop antenna

TUBES
5

POWER SUPPLY
117 volts AC

TUNING RANGE
538-1685KC

MFR/SUPPLIER
Automatic Radio Mfg. Co.

PHOTOFACT SET
10-4

PUBLISHED
1946

AUTOMATIC

MODEL PICTURED
620

AC/DC operated superheterodyne receiver with self-contained loop antenna

TUBES
6

POWER SUPPLY
110-125 volts AC/DC, .235 amp. @ 117 volts AC

TUNING RANGE
538-1630KC

MFR/SUPPLIER
Automatic Radio Mfg. Co.

PHOTOFACT SET
12-3

PUBLISHED
1947

AUTOMATIC

MODEL PICTURED
601 (series A and B)

AC/DC superheterodyne receiver, self-contained loop antenna

TUBES
5

POWER SUPPLY
117 volts AC/DC, .260 amp @ 117 volts AC (A), .250 amp @ 117 volts AC (B)

TUNING RANGE
538-1685KC

MFR/SUPPLIER
Automatic Radio Mfg. Co.

PHOTOFACT SET
13-11 (A)
22-5 (B)

PUBLISHED
1947

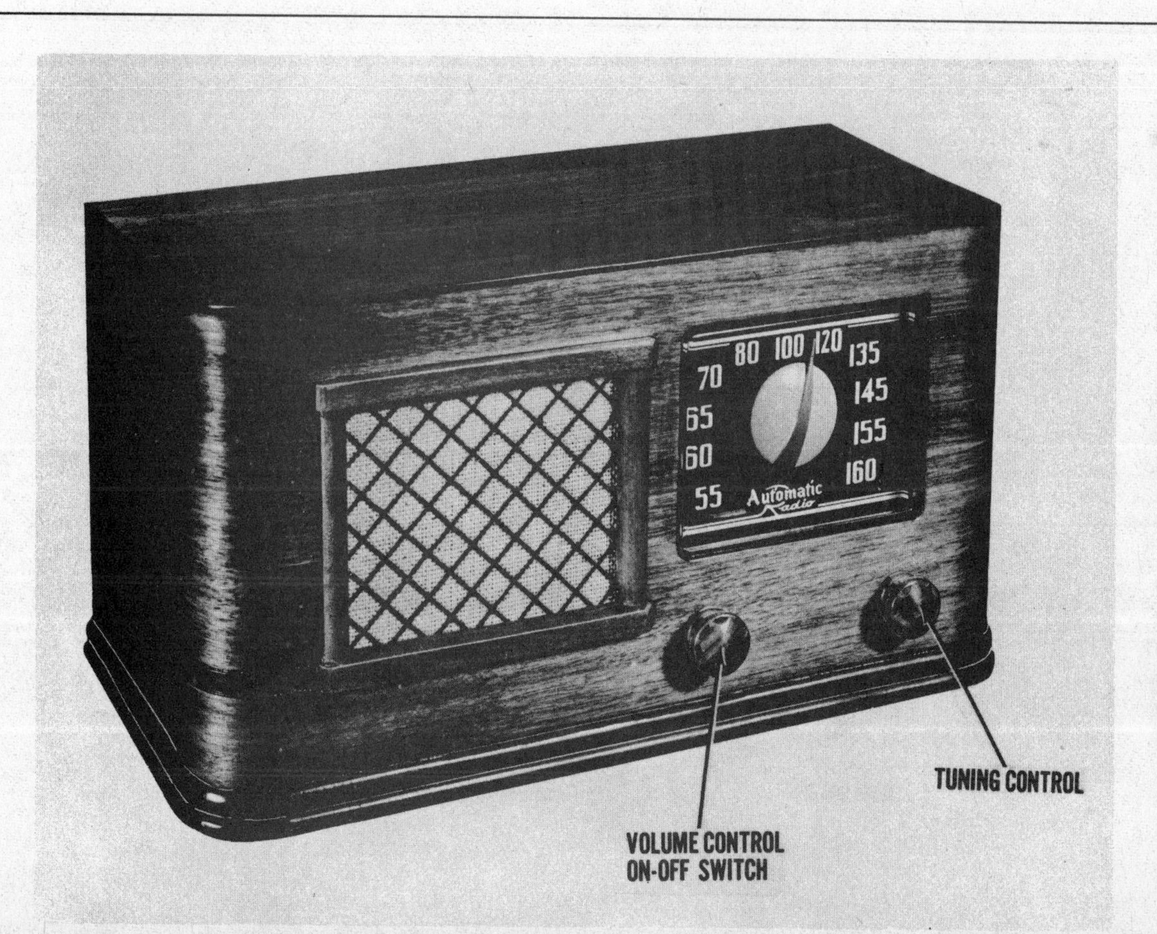

AUTOMATIC

MODEL PICTURED
720

AC/DC operated superheterodyne receiver with loop antenna

TUBES
6

POWER SUPPLY
110-120 volts AC/DC, .260 amp @ 117 volts AC

TUNING RANGE
540-1620KC

MFR/SUPPLIER
Automatic Radio Mfg. Co.

PHOTOFACT SET
21-4

PUBLISHED
1947

AUTOMATIC

MODEL PICTURED
662

AC/DC operated two-band superheterodyne receiver with loop antenna

TUBES
6

POWER SUPPLY
110-120 volts AC/DC, .260 amp @ 117 volts AC

TUNING RANGE
538-1600KC, 5.9-22.5MC

MFR/SUPPLIER
Automatic Radio Mfg. Co.

PHOTOFACT SET
22-6

PUBLISHED
1946

AUTOMATIC

MODEL PICTURED
677

AC operated phono-radio combo superheterodyne receiver with loop antenna

TUBES
7

POWER SUPPLY
110-120 volts AC

TUNING RANGE
538-1685KC

MFR/SUPPLIER
Automatic Radio Mfg. Co.

PHOTOFACT SET
22-7

PUBLISHED
1947

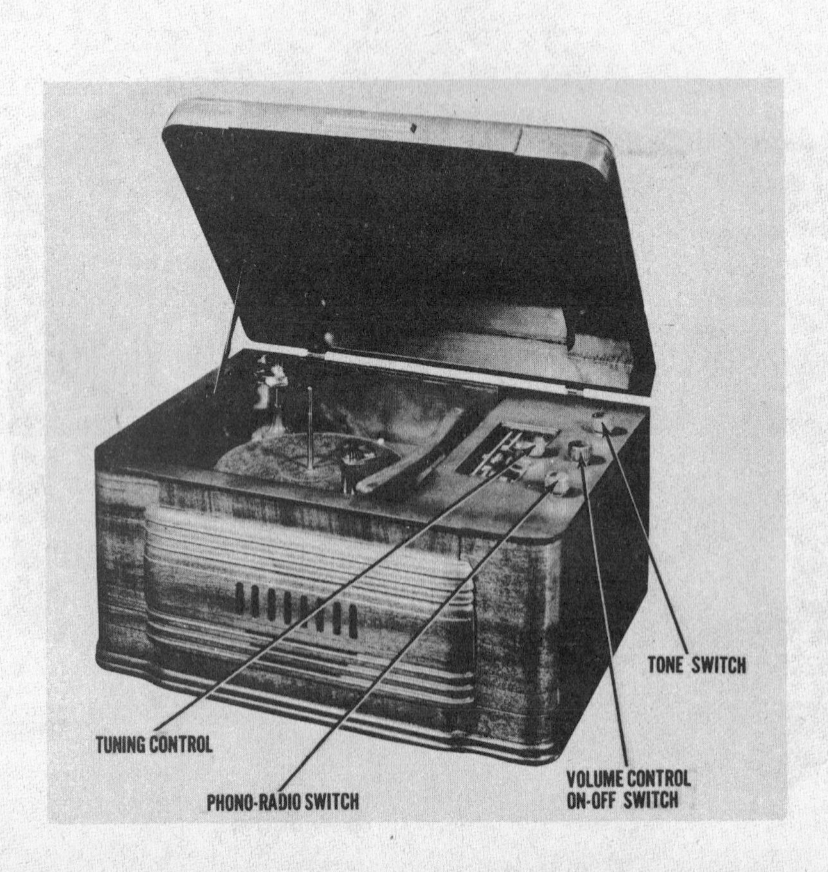

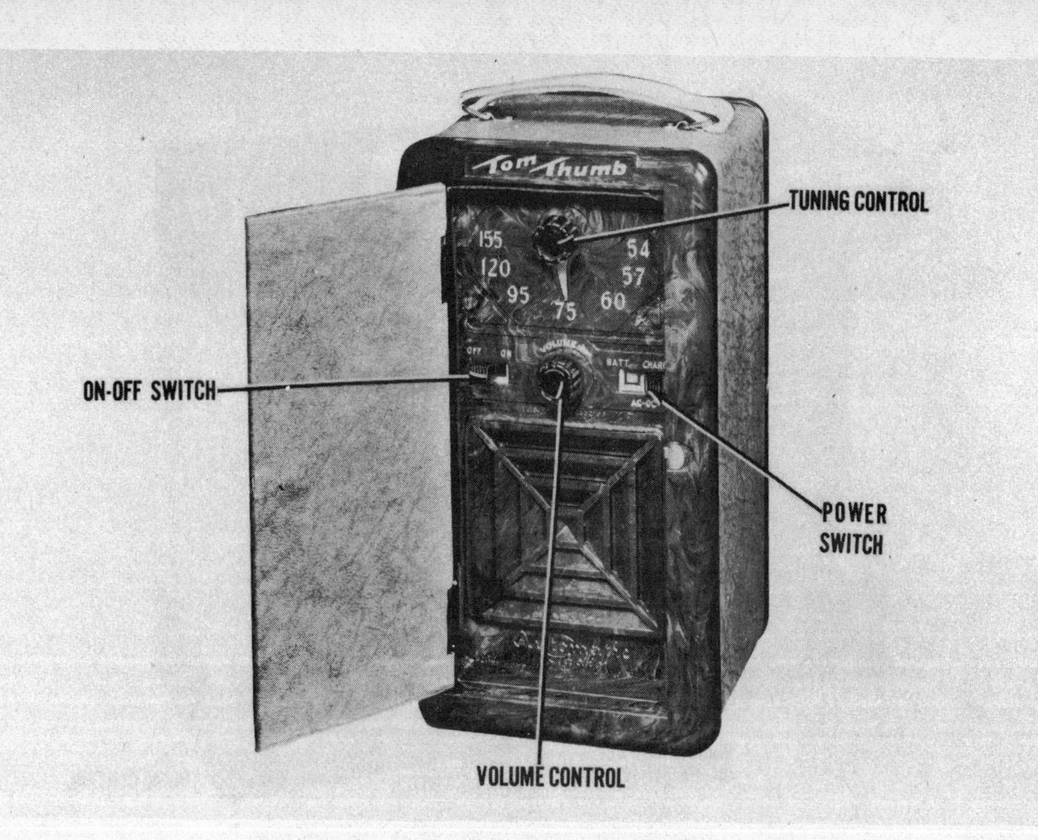

AUTOMATIC

MODEL PICTURED
ATTP
(Tom Thumb)

Three-power portable superheterodyne receiver with loop antenna

TUBES
4

POWER SUPPLY
110-120 volts AC/DC or 7.5 volts A & 67.5 volts B supply

TUNING RANGE
530-1600KC

MFR/SUPPLIER
Automatic Radio Mfg. Co.

PHOTOFACT SET
23-4

PUBLISHED
1947

AUTOMATIC

MODEL PICTURED
F-790

AC operated radio-phono superheterodyne receiver with loop antenna

TUBES
9

POWER SUPPLY
110-120 volts AC

TUNING RANGE
540-1700KC

MFR/SUPPLIER
Automatic Radio Mfg. Co.

PHOTOFACT SET
23-5

PUBLISHED
1947

AUTOMATIC

MODEL PICTURED
C-60X

Three-power portable superheterodyne receiver with loop antenna

TUBES
4

POWER SUPPLY
110-120 volts AC or 6 volts A & 90 volts B supply

TUNING RANGE
538-1600KC

MFR/SUPPLIER
Automatic Radio Mfg. Co.

PHOTOFACT SET
24-10

PUBLISHED
1947

AUTOMATIC

MODEL PICTURED
Tom Thumb Jr.

Battery operated portable superheterodyne receiver with loop antenna

TUBES
5

POWER SUPPLY
1.5 volts A & 67.5 volts B supply

TUNING RANGE
538-1600KC

MFR/SUPPLIER
Automatic Radio Mfg. Co.

PHOTOFACT SET
26-7

PUBLISHED
1947

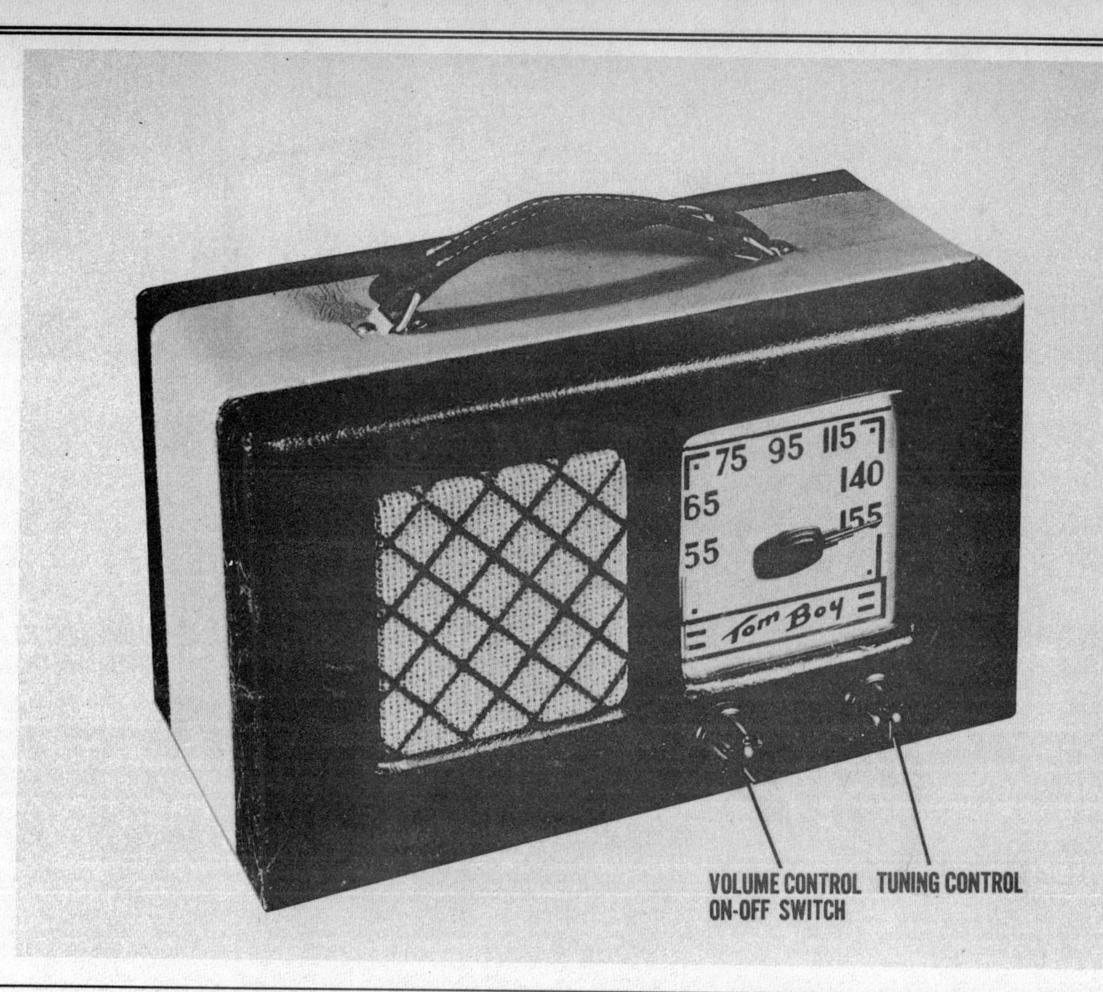

AUTOMATIC

MODEL PICTURED
Tomboy

Battery operated portable superheterodyne receiver with loop antenna

TUBES
4

POWER SUPPLY
1.5 volts A supply & 67.5 volts B supply

TUNING RANGE
538-1600KC

MFR/SUPPLIER
Automatic Radio Mfg. Co.

PHOTOFACT SET
27- 4

PUBLISHED
1947

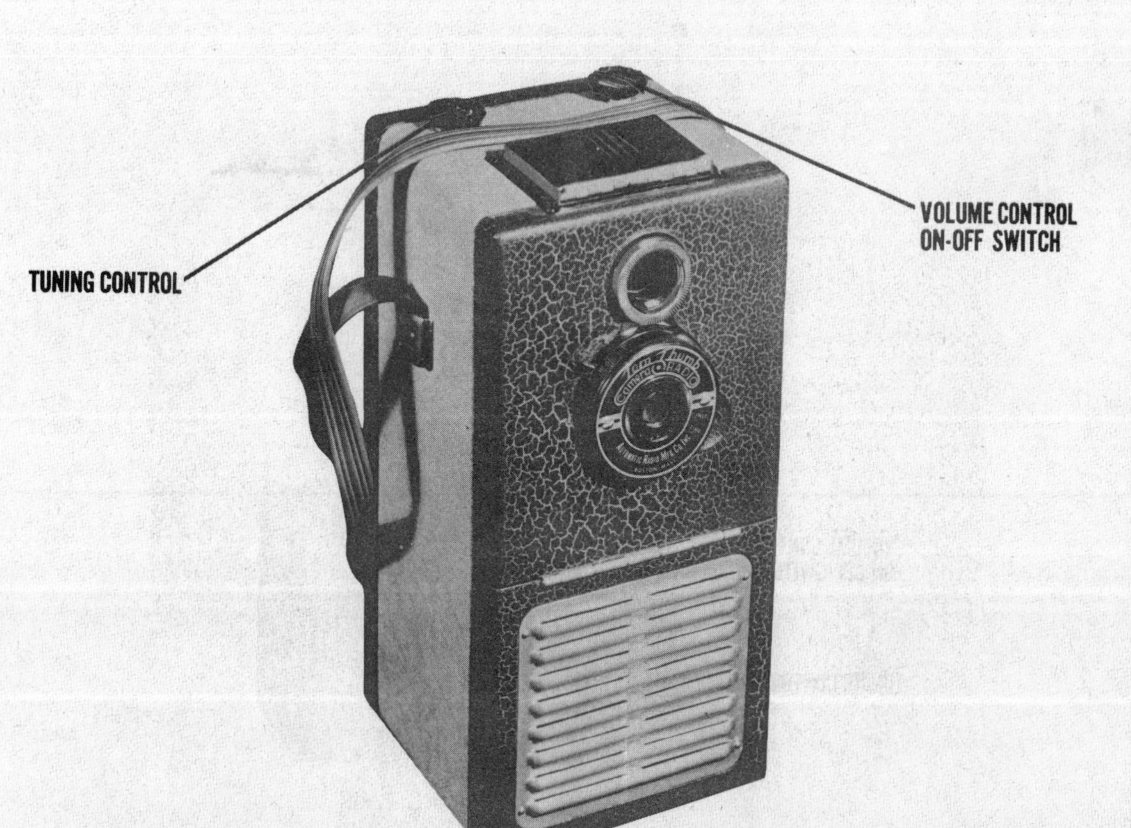

AUTOMATIC

MODEL PICTURED
Tom Thumb Camera/Radio

Battery operated portable superheterodyne receiver with built in camera

TUBES
4

POWER SUPPLY
1.5 volts A supply & 67.5 volts B supply

TUNING RANGE
540-1600KC

MFR/SUPPLIER
Automatic Radio Mfg. Co.

PHOTOFACT SET
49-6

PUBLISHED
1948

AUTOMATIC

MODEL PICTURED
Tom Thumb Buddy

Three-power operated portable superheterodyne radio receiver

TUBES
4

POWER SUPPLY
110-120 volts AC/DC or 3 volts A supply & 67.5 volts B supply

TUNING RANGE
540-1600KC

MFR/SUPPLIER
Automatic Radio Mfg. Co.

PHOTOFACT SET
53-7

PUBLISHED
1949

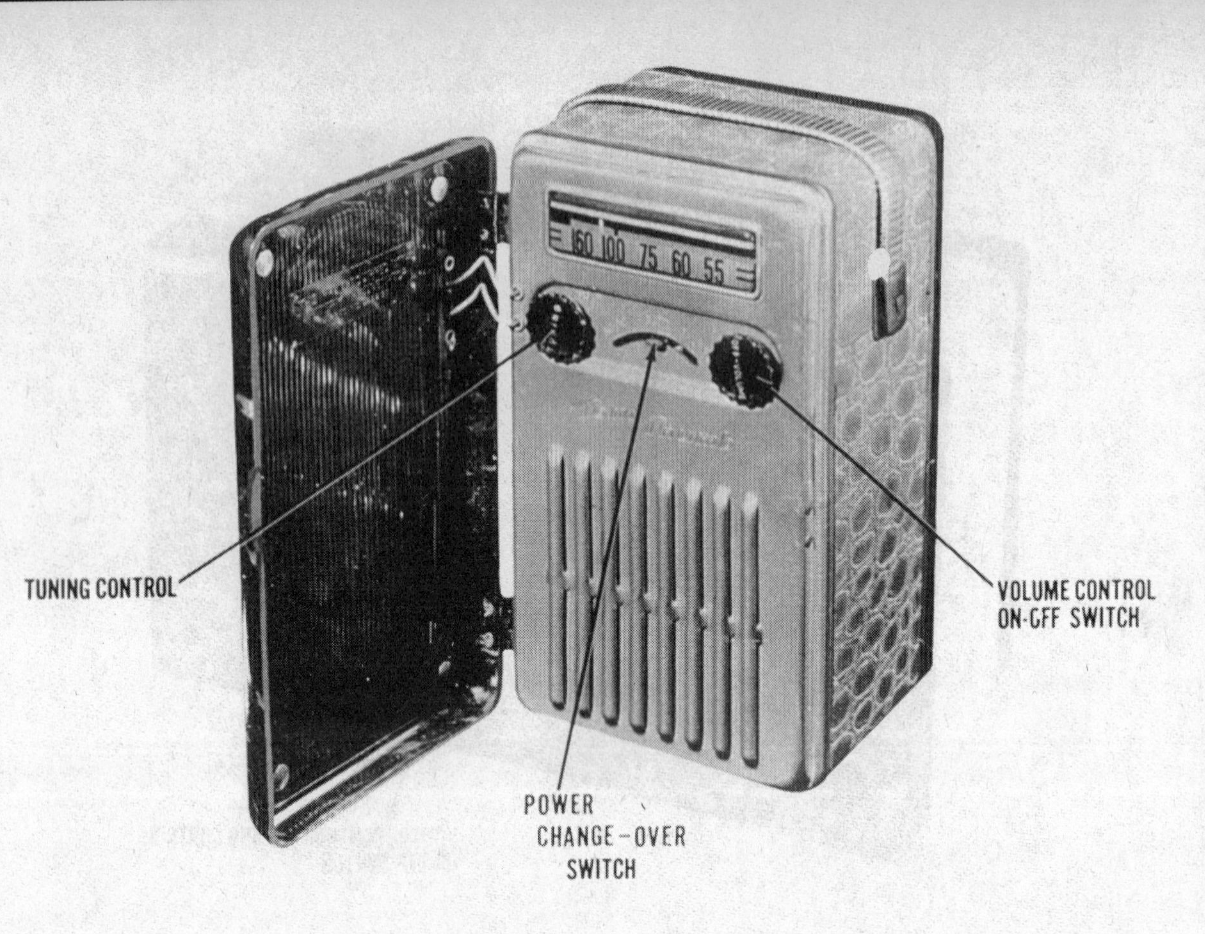

AUTOMATIC

MODEL PICTURED
B-44 (Tom Thumb Bike Radio)

Battery operated portable superheterodyne receiver

TUBES
4

POWER SUPPLY
1.5 volts A & 67.5 volts B supply

TUNING RANGE
530-1610KC

MFR/SUPPLIER
Automatic Radio Mfg. Co.

PHOTOFACT SET
60-5

PUBLISHED
1949

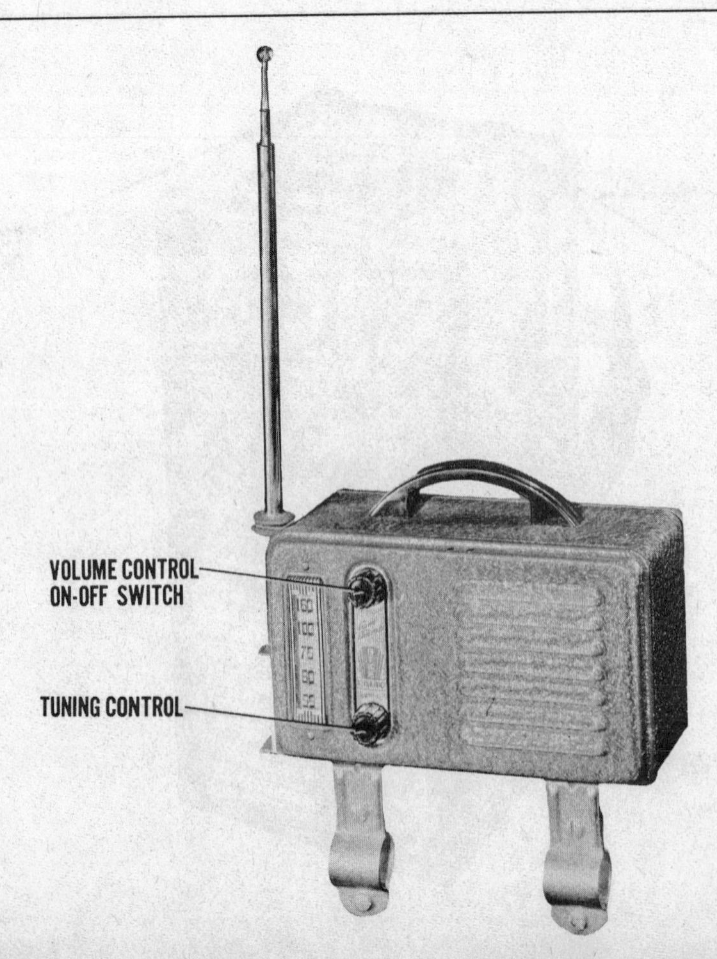

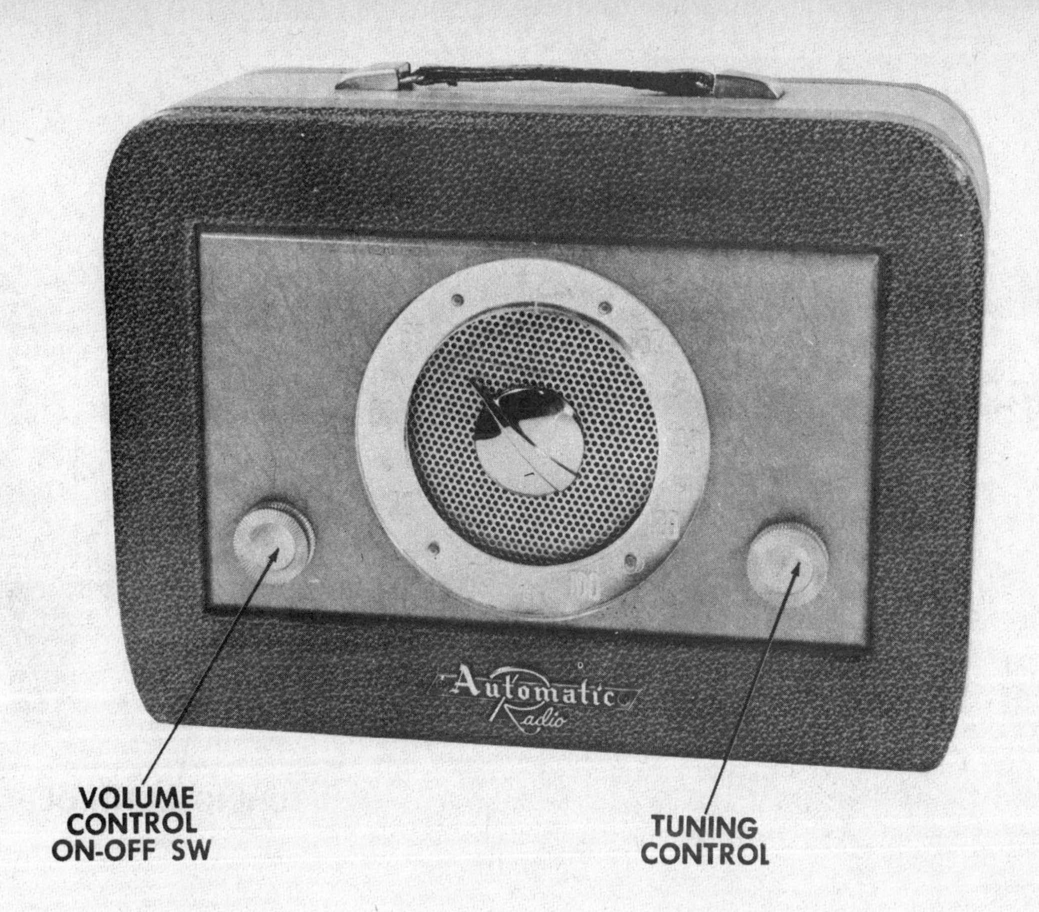

AUTOMATIC

MODEL PICTURED
C51

Three-power portable superheterodyne receiver

TUBES
5

POWER SUPPLY
115-125 volts AC/DC or 9 volts A supply & 90 volts B supply

TUNING RANGE
535-1620KC

MFR/SUPPLIER
Automatic Radio Mfg. Co.

PHOTOFACT SET
178-4

PUBLISHED
1952

AUTOMATIC

MODEL PICTURED
C-54

Three power portable superheterodyne receiver

TUBES
4

POWER SUPPLY
110-120 Volts AC/DC or 9 volts A battery & 90 volts B battery pack

TUNING RANGE
538-1620KC

MFR/SUPPLIER
Automatic Radio Mfg. Co.

PHOTOFACT SET
186-2

PUBLISHED
1952

AUTOMATIC

MODEL PICTURED
CL-152B

AC operated AM superheterodyne receiver with electric clock

TUBES
5

POWER SUPPLY
115-125 volts AC, 60 cycle, .24 amp @ 117 volts AC

TUNING RANGE
538-1650KC

MFR/SUPPLIER
Automatic Radio Mfg. Co.

PHOTOFACT SET
192-3

PUBLISHED
1953

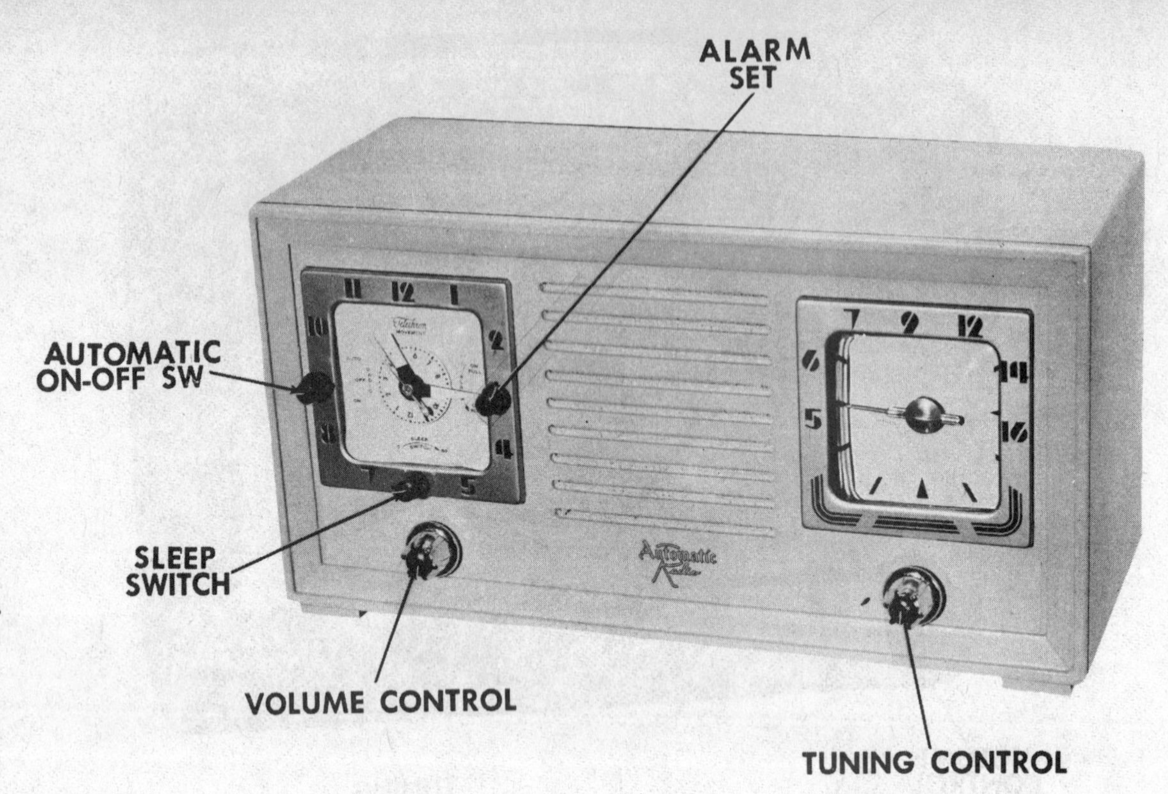

AUTOMATIC

MODEL PICTURED
TT600

Battery operated portable AM superheterodyne receiver with transistor output

TUBES
3

POWER SUPPLY
4 volts A supply & 45 volts B supply

TUNING RANGE
540-1610KC

MFR/SUPPLIER
Automatic Radio Mfg. Co.

PHOTOFACT SET
349-3

PUBLISHED
1957

AUTOMATIC

MODEL PICTURED
CL-100

AC operated AM receiver with electric clock

TUBES
5

POWER SUPPLY
105-125 volts AC, 60 cycles, .29 amp @ 117 volts AC, 28 watts (less clock)

TUNING RANGE
535-1620KC

MFR/SUPPLIER
Automatic Radio Mfg. Co.

PHOTOFACT SET
426-5

PUBLISHED
1959

AUTOMATIC

MODEL PICTURED
P-57

AC/DC battery operated portable AM receiver

TUBES
5

POWER SUPPLY
105-125 volts AC/DC or 3 volts DC A supply & 90 volts DC B supply

TUNING RANGE
538-1610KC

MFR/SUPPLIER
Automatic Radio Mfg. Co.

PHOTOFACT SET
489-4

PUBLISHED
1960

AUTOMATIC

MODEL PICTURED
TT528

Battery operated portable AM receiver

TUBES
4

POWER SUPPLY
1.5 volts DC A supply, 45 volts DC B supply, 150mA @ 1.5 volts DC, 3mA @ 45 volts DC

TUNING RANGE
540-1610KC

MFR/SUPPLIER
Automatic Radio Mfg. Co.

PHOTOFACT SET
491-4

PUBLISHED
1960

AUTOMATIC

MODEL PICTURED
CL-75

AC operated AM receiver with electric clock

TUBES
5

POWER SUPPLY
105-120 volts AC, 60 cycle, 23 watts, .22 amp @ 117 volts AC

TUNING RANGE
538-1610KC

MFR/SUPPLIER
Automatic Radio Mfg. Co.

PHOTOFACT SET
492-5

PUBLISHED
1960

AVIOLA

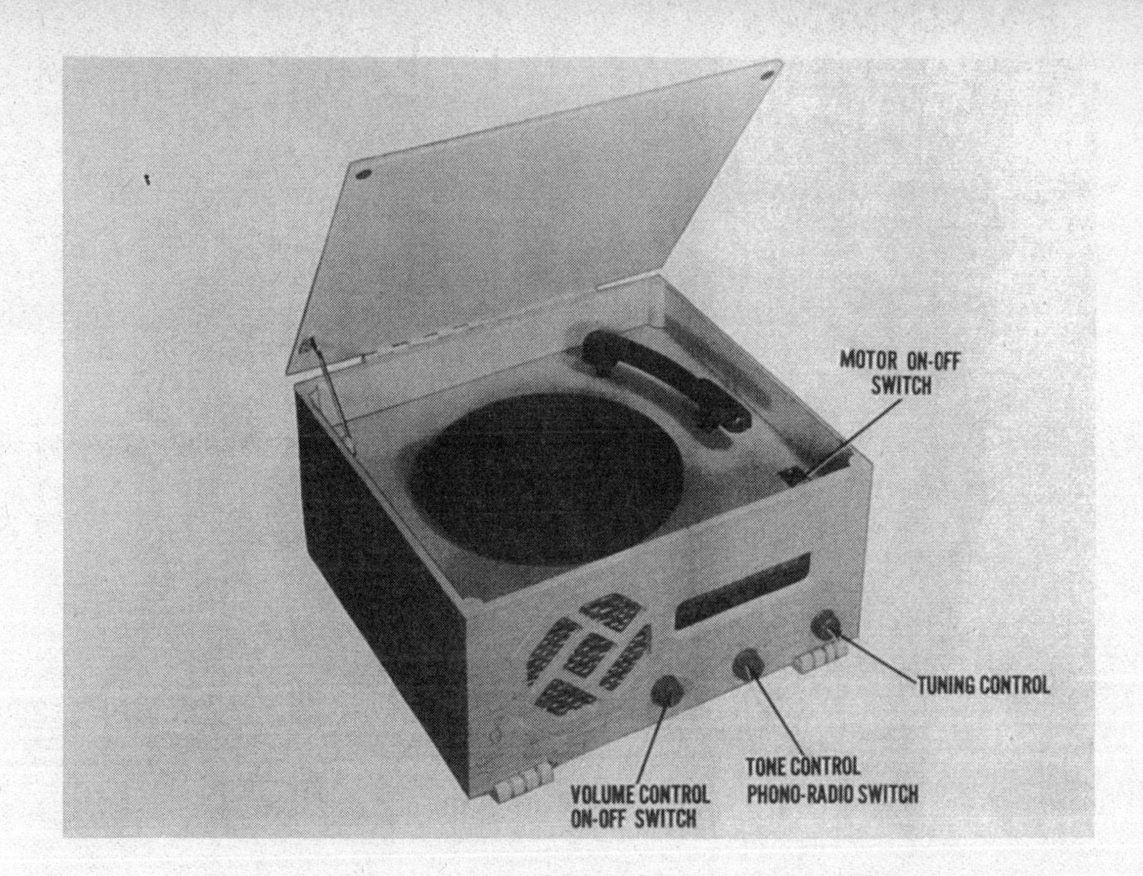

MODEL PICTURED
509

AC operated combo phono-superheterodyne receiver with self-contained loop antenna

TUBES
5

POWER SUPPLY
105-125 volts AC

TUNING RANGE
540-1625KC

MFR/SUPPLIER
Aviola Radio Corp.

PHOTOFACT SET
7-3

PUBLISHED
1946

AVIOLA

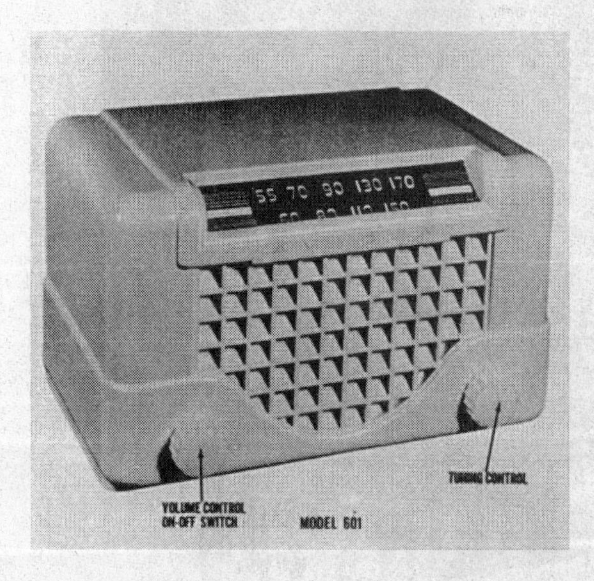

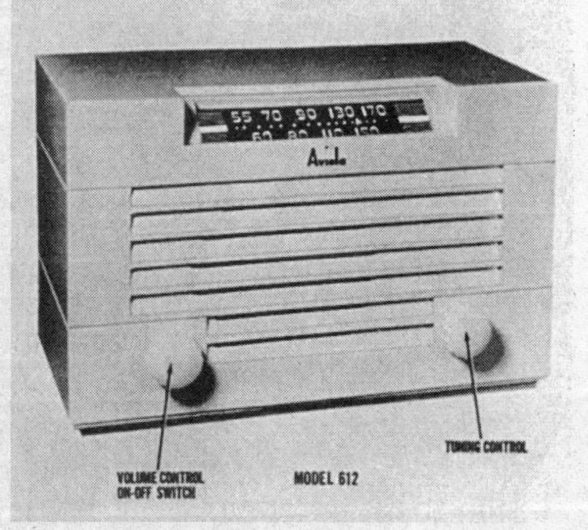

MODELS PICTURED
601 and 612

AC/DC operated superheterodyne receiver with self-contained loop antenna

TUBES
6

POWER SUPPLY
105-125 volts AC/DC, .230 amp @ 117 volts AC

TUNING RANGE
540-1720KC

MFR/SUPPLIER
Aviola Radio Corp.

PHOTOFACT SET
15-3

PUBLISHED
1947

AVIOLA

MODEL PICTURED
608

AC operated combo phono-radio superheterodyne receiver with self-contained loop antenna

TUBES
6

POWER SUPPLY
105-125 volts AC

TUNING RANGE
540-1625KC

MFR/SUPPLIER
Aviola Radio Corp.

PHOTOFACT SET
16-6

PUBLISHED
1947

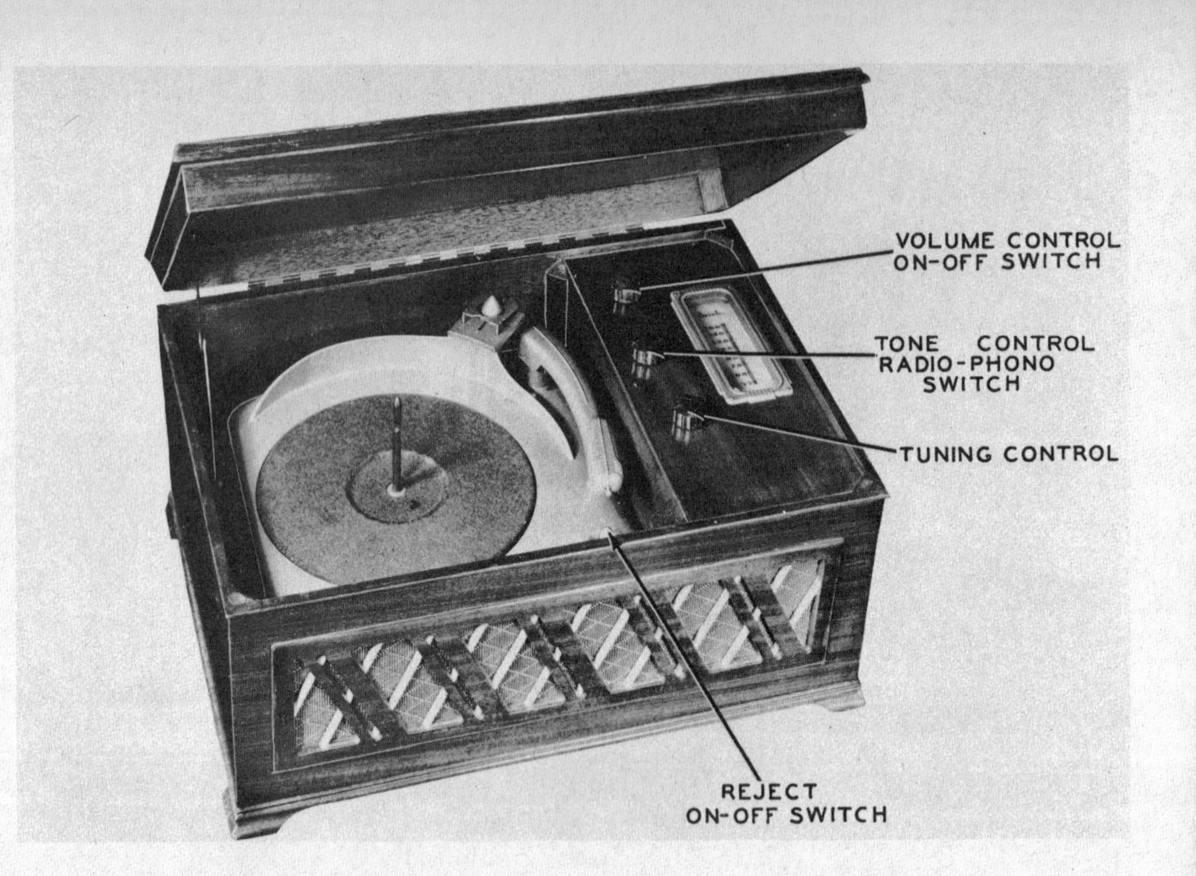

B.F. GOODRICH

MODEL PICTURED
92-523

AC/DC operated superheterodyne receiver with loop antenna

TUBES
5

POWER SUPPLY
110-120 volts AC/DC, .25 amp @ 117 volts AC

TUNING RANGE
535-1620KC

MFR/SUPPLIER
B.F. Goodrich Co.

PHOTOFACT SET
148-7

PUBLISHED
1951

BAGPIPER

MODEL PICTURED
SKR101

Battery operated portable AM superheterodyne receiver

TUBES
4

POWER SUPPLY
1.5 volts A supply & 90 volts B supply

TUNING RANGE
550-1620KC

MFR/SUPPLIER
Lawrence Co.

PHOTOFACT SET
335-2

PUBLISHED
1956

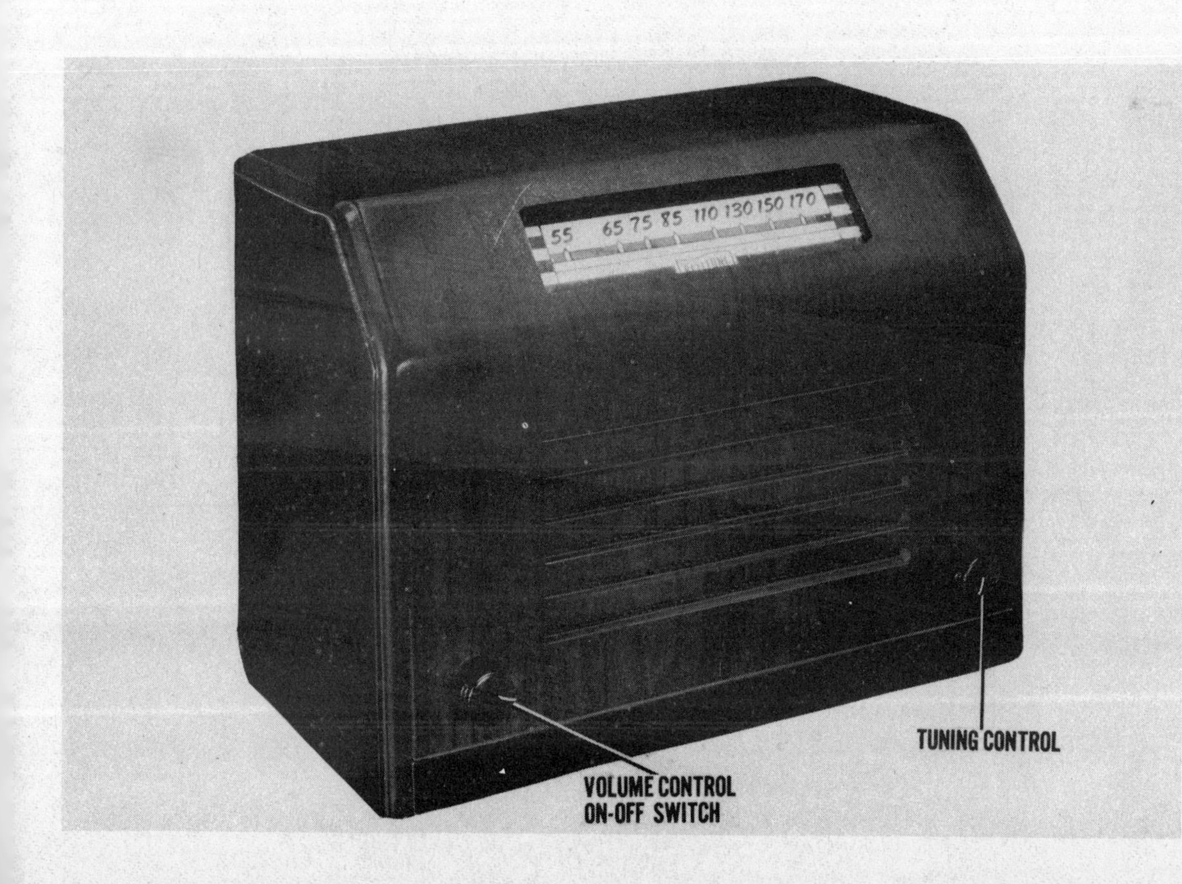

BELLTONE

MODEL PICTURED
500

AC/DC superheterodyne receiver with self-contained loop antenna

TUBES
5

POWER SUPPLY
117 volts AC

TUNING RANGE
540-1700KC

MFR/SUPPLIER
Jewel Radio Corp.

PHOTOFACT SET
5-33

PUBLISHED
1946

BELMONT

MODEL PICTURED
8A59

AC operated combo auto phono, five-band superheterodyne receiver

TUBES
8

POWER SUPPLY
105-125 volts AC, .610 amp @ 117 volts AC

TUNING RANGE
540-1600KC,
5.96-6.19,9.1-10MC,
11.45-12.16MC,
14.94-15.46MC

MFR/SUPPLIER
Belmont Radio Corp.

PHOTOFACT SET
6-4

PUBLISHED
1946

BELMONT

MODEL PICTURED
5D128
(series A)

AC/DC superheterodyne receiver, self-contained loop antenna and pushbutton tuning

TUBES
5

POWER SUPPLY
105-125 volts AC/DC, .230 amp @ 117 volts AC

TUNING RANGE
530-1650KC

MFR/SUPPLIER
Belmont Radio Corp.

PHOTOFACT SET
9-4

PUBLISHED
1946

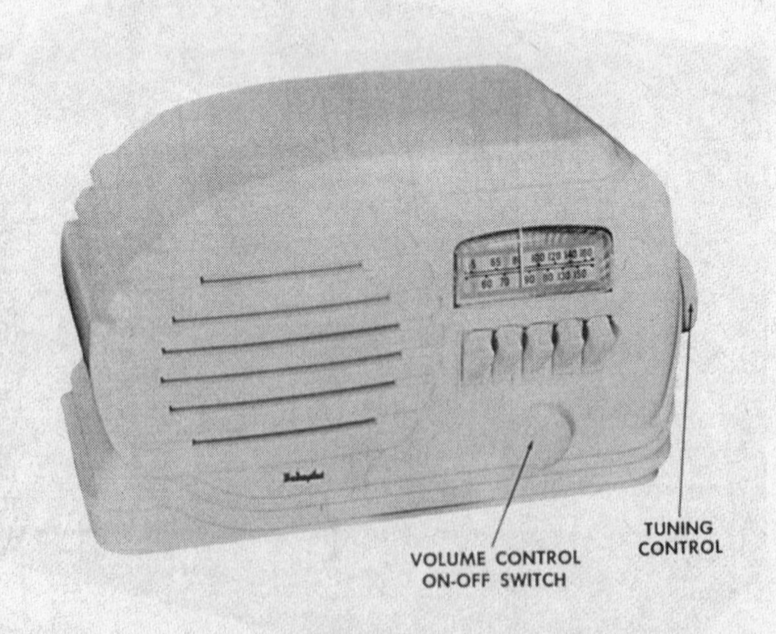

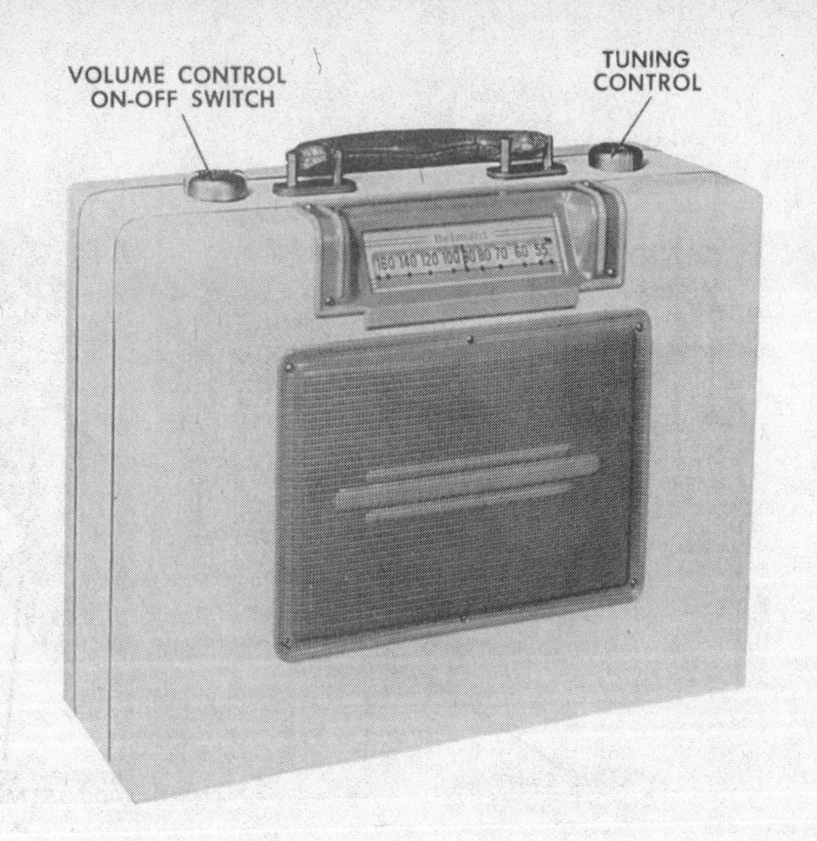

BELMONT

MODEL PICTURED
5P19 (series A)

Three-power portable AC/DC battery superheterodyne receiver

TUBES
5

POWER SUPPLY
105-125 volts AC/DC or 6 volts A battery

TUNING RANGE
530-1650KC

MFR/SUPPLIER
Belmont Radio Corp.

PHOTOFACT SET
9-5

PUBLISHED
1946

BELMONT

MODEL PICTURED
4B112

Battery operated superheterodyne receiver

TUBES
4

POWER SUPPLY
1.5 volts A battery & 90 volts B battery in pack form

TUNING RANGE
540-1700KC

MFR/SUPPLIER
Belmont Radio Corp.

PHOTOFACT SET
10-6

PUBLISHED
1946

BELMONT

MODEL PICTURED
A-6D110

AC/DC superheterodyne receiver, self-contained loop antenna and six-pushbutton tuning

TUBES
6

POWER SUPPLY
105-125 volts AC/DC

TUNING RANGE
535-1650KC

MFR/SUPPLIER
Belmont Radio Corp.

PHOTOFACT SET
17-7

PUBLISHED
1947

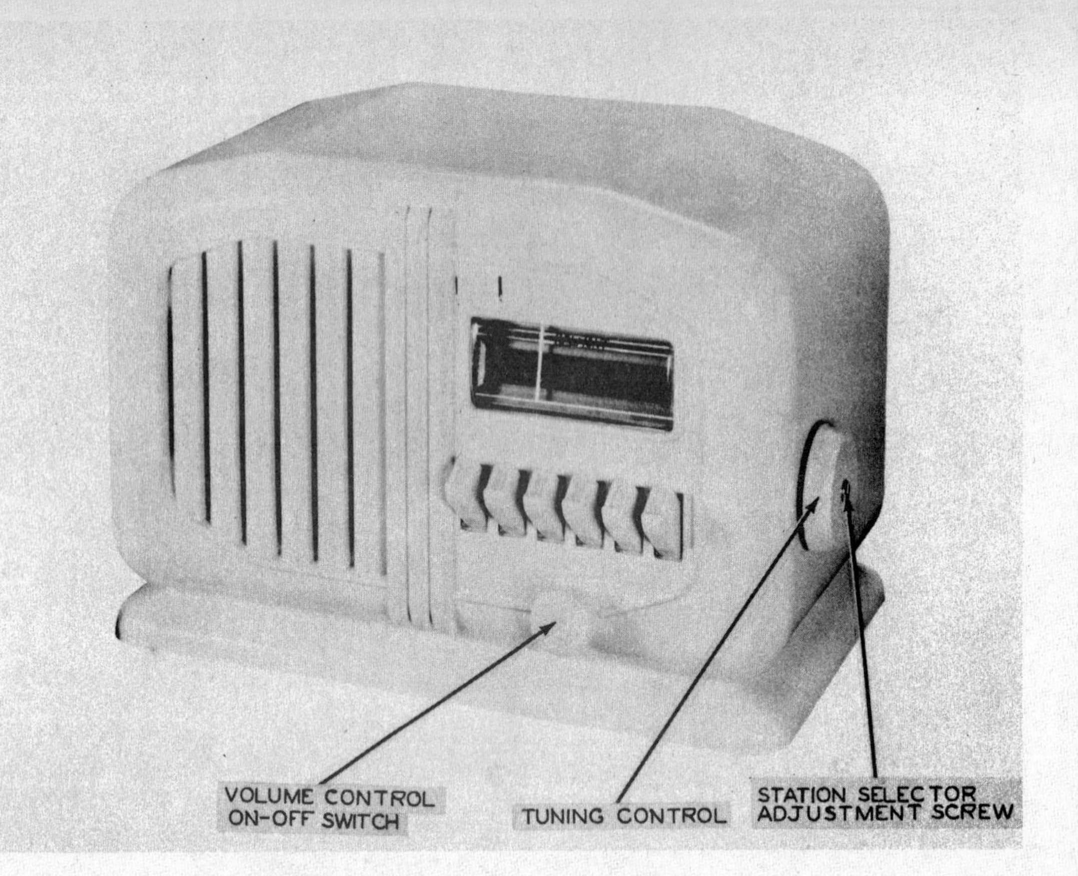

BELMONT

MODEL PICTURED
5D110

AC operated phono-radio combo superheterodyne receiver

TUBES
5

POWER SUPPLY
105-125 volts AC

TUNING RANGE
535-1720KC

MFR/SUPPLIER
Belmont Radio Corp.

PHOTOFACT SET
22-10

PUBLISHED
1947

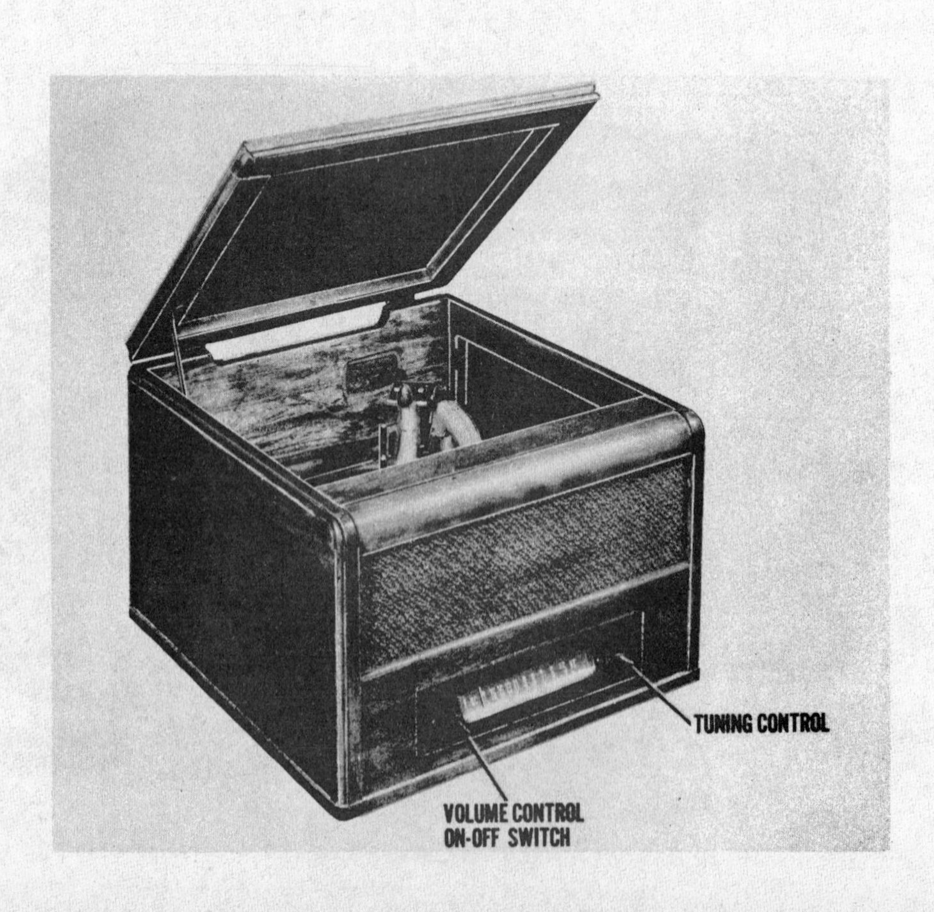

BELMONT

MODEL PICTURED
6D120

AC/DC operated superheterodyne receiver, with self-contained loop antenna and six-pushbutton tuning

TUBES
6

POWER SUPPLY
105-125 volts AC/DC

TUNING RANGE
530-1650KC

MFR/SUPPLIER
Belmont Radio Corp.

PHOTOFACT SET
24-12

PUBLISHED
1947

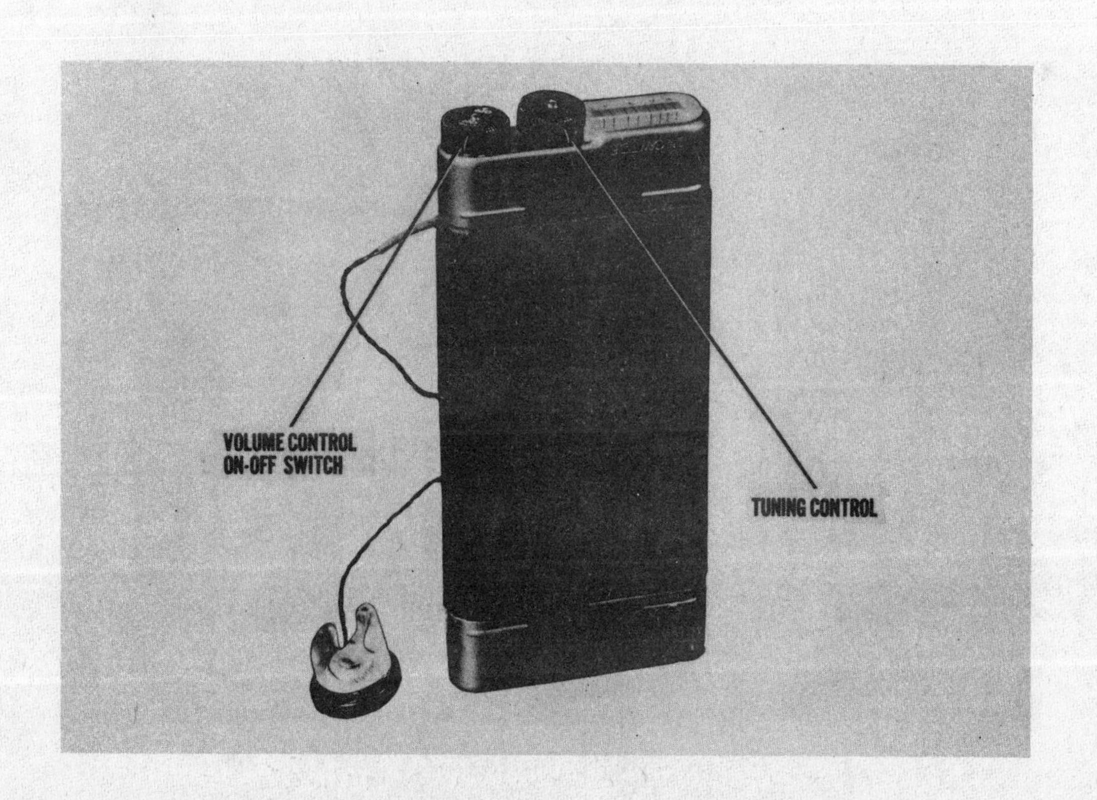

BELMONT

MODEL PICTURED
5P113 (Boulevard)

Battery operated personal portable superheterodyne receiver

TUBES
5

POWER SUPPLY
1.5 volts A supply & 22.5 volts B supply

TUNING RANGE
540-1625KC

MFR/SUPPLIER
Belmont Radio Corp.

PHOTOFACT SET
28-2

PUBLISHED
1947

BENDIX

MODEL PICTURED
676D

AC operated combo automatic phono-superheterodyne receiver, self-contained loop antenna

TUBES
6

POWER SUPPLY
105-125 volts AC

TUNING RANGE
540-1620KC, 6-12MC

MFR/SUPPLIER
Bendix Radio

PHOTOFACT SET
5-23

PUBLISHED
1946

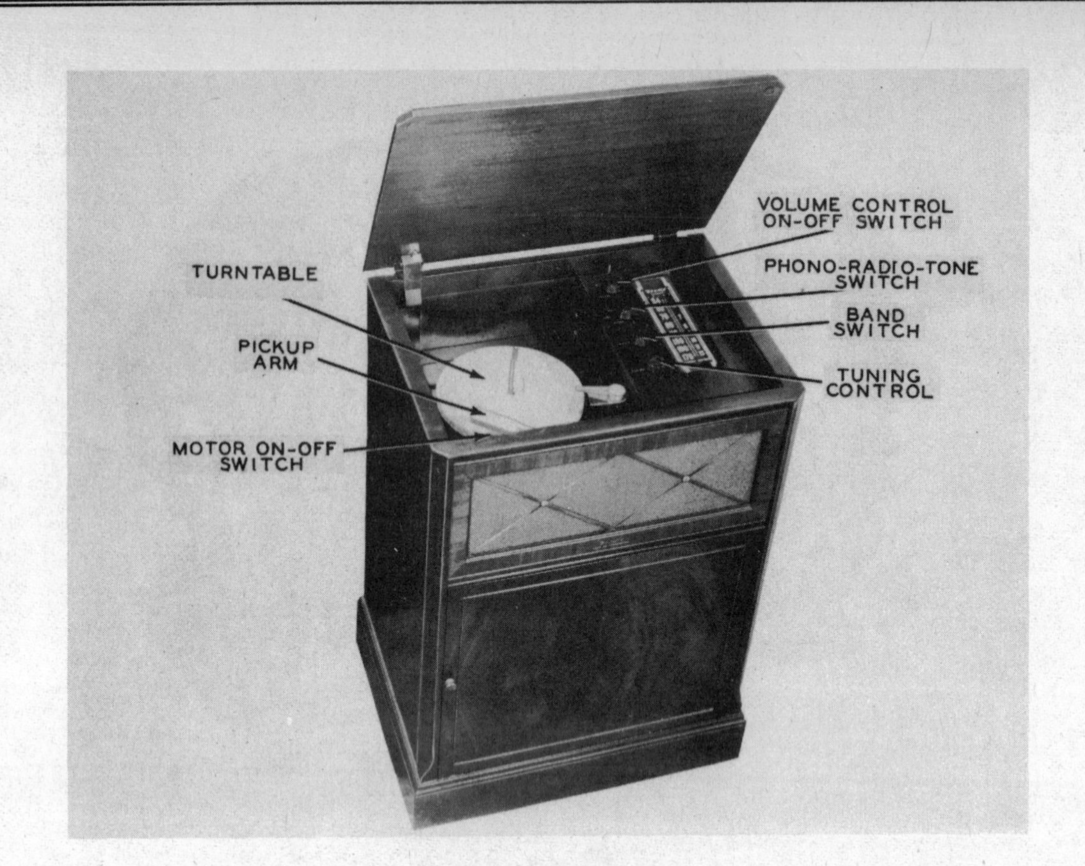

BENDIX

MODEL PICTURED
736-B

AC operated three-band automatic radio-phono superheterodyne receiver, pushbutton tuning

TUBES
7

POWER SUPPLY
105-125 volts AC

TUNING RANGE
540-1620KC, 6-12MC, 11.3-22.0 MC

MFR/SUPPLIER
Bendix Radio

PHOTOFACT SET
10-8

PUBLISHED
1946

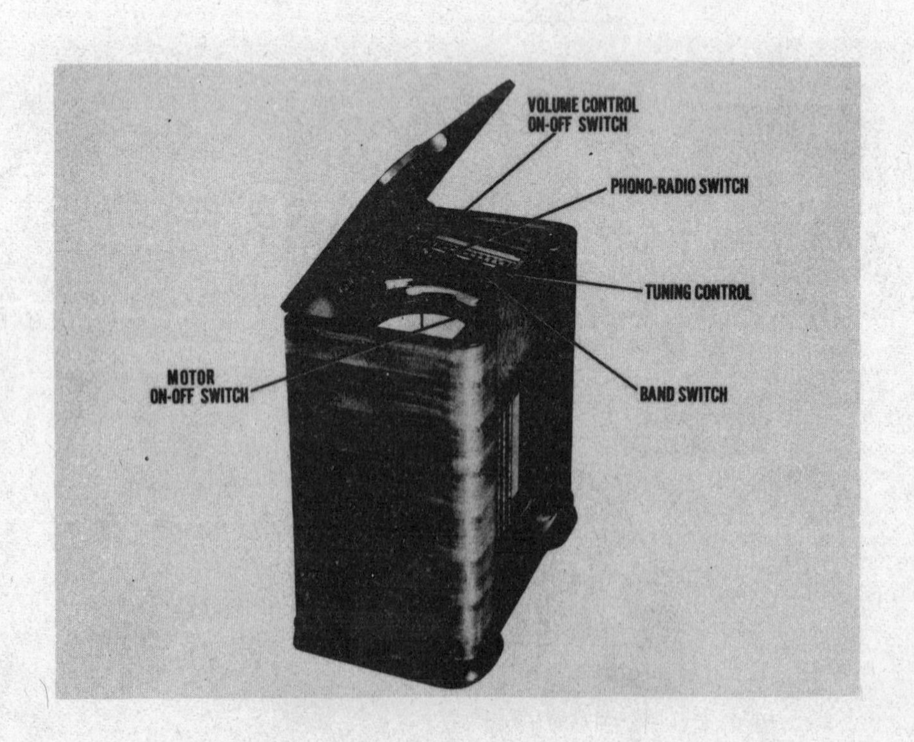

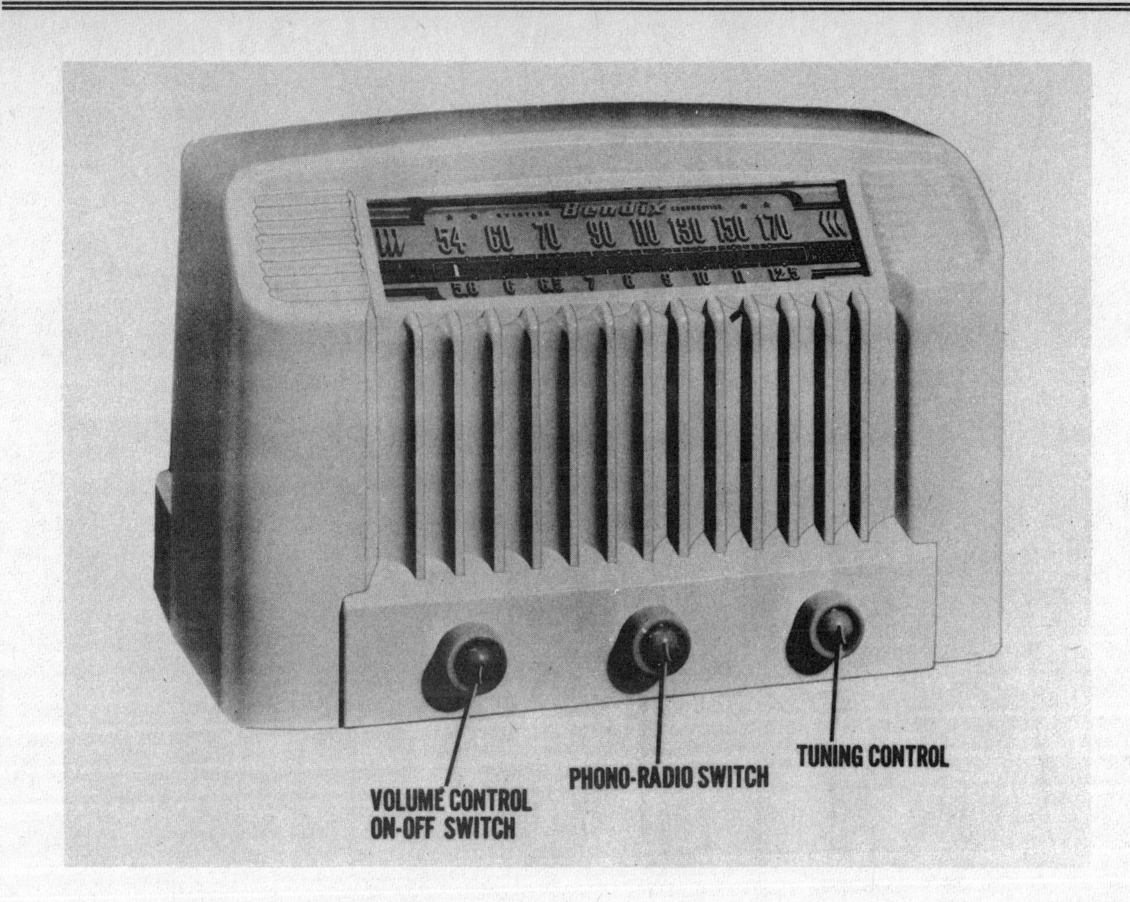

BENDIX

MODEL PICTURED
626-A (0626A)

AC/DC operated two-band superheterodyne receiver, self contained loop antenna.

TUBES
6

POWER SUPPLY
105-125 volts AC/DC, .235 amp. @ 117 volts AC

TUNING RANGE
535-1725KC, 5.7-12.5MC

MFR/SUPPLIER
Bendix Radio

PHOTOFACT SET
12-4

PUBLISHED
1947

MODEL 636A

MODEL 636 C

MODEL 636 B

BENDIX

MODELS PICTURED
636A, 636B, 636C

AC/DC operated superheterodyne receiver with loop antenna and phono provisions

TUBES
6

POWER SUPPLY
105-125 volts AC/DC, .240 amp @ 117 volts AC

TUNING RANGE
535-1725KC

MFR/SUPPLIER
Bendix Radio

PHOTOFACT SET
15-4

PUBLISHED
1947

BENDIX

MODEL PICTURED
697A

AC operated phono-radio combo superheterodyne receiver with loop antenna

TUBES
6

POWER SUPPLY
105-125 volts AC

TUNING RANGE
540-1620KC

MFR/SUPPLIER
Bendix Radio

PHOTOFACT SET
26-8

PUBLISHED
1947

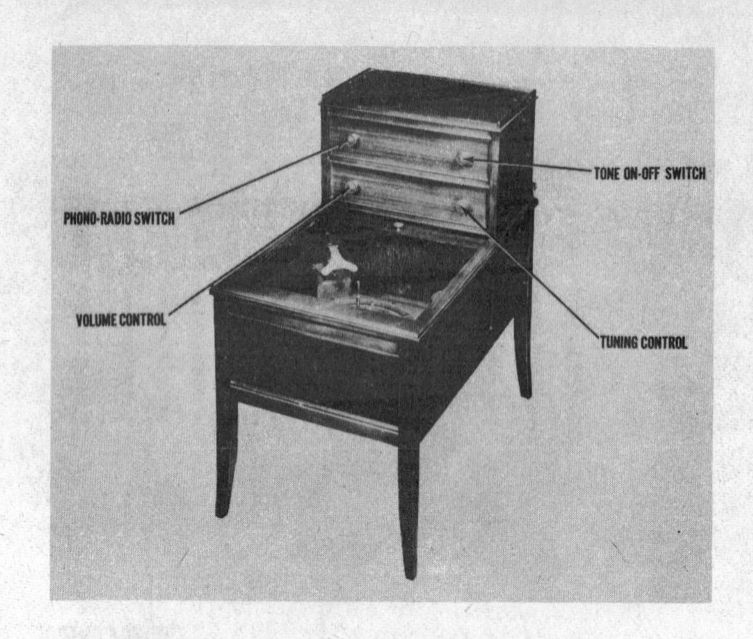

BENDIX

MODEL PICTURED
847-B

AC operated FM/AM radio-phono combo superheterodyne receiver with loop antenna

TUBES
8

POWER SUPPLY
105-125 volts AC

TUNING RANGE
540-1620KC, 88-108MC

MFR/SUPPLIER
Bendix Radio

PHOTOFACT SET
27-5

PUBLISHED
1947

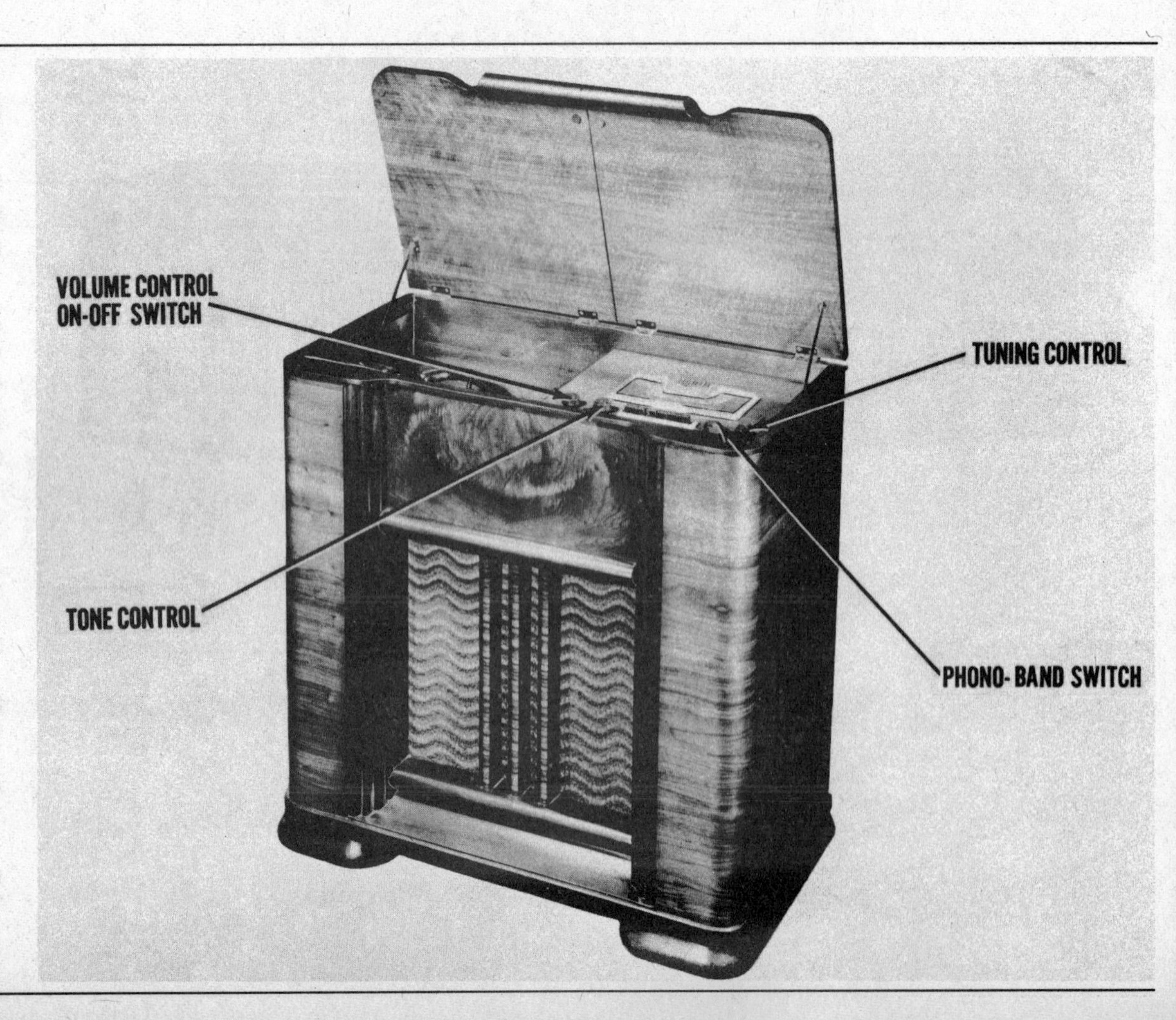

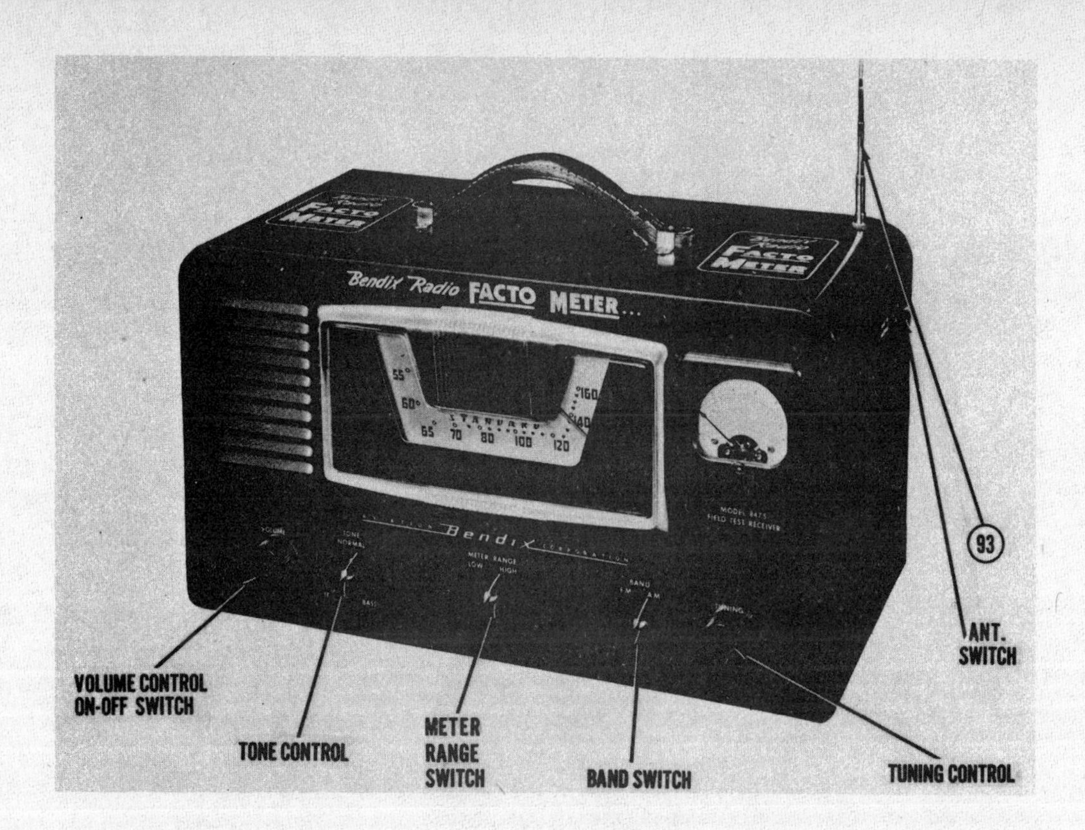

BENDIX

MODEL PICTURED
847S
(Facto Meter)

AC operated AM-FM superheterodyne receiver with telescopic antenna

TUBES
8

POWER SUPPLY
110-120 volts AC

TUNING RANGE
540-1620KC, 88-108MC

MFR/SUPPLIER
Bendix Radio

PHOTOFACT SET
28-3

PUBLISHED
1947

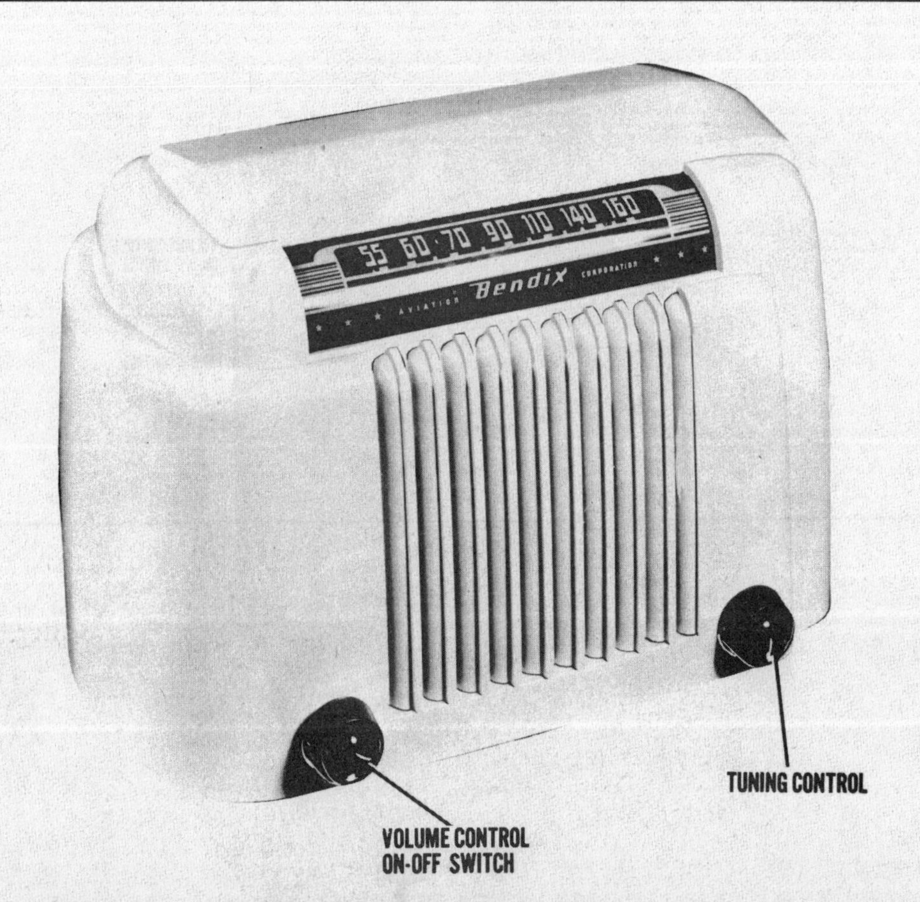

BENDIX

MODEL PICTURED
526MB

AC/DC operated superheterodyne receiver with loop antenna

TUBES
5

POWER SUPPLY
105-125 volts AC/DC

TUNING RANGE
540-1620KC

MFR/SUPPLIER
Bendix Radio

PHOTOFACT SET
29-3

PUBLISHED
1947

BENDIX

MODEL PICTURED
1217B

AC operated phono-radio AM-FM superheterodyne receiver with loop antenna

TUBES
12

POWER SUPPLY
110-120 volts AC

TUNING RANGE
540-1620KC, 88-108MC, 5.7-16MC

MFR/SUPPLIER
Bendix Radio

PHOTOFACT SET
29-4

PUBLISHED
1947

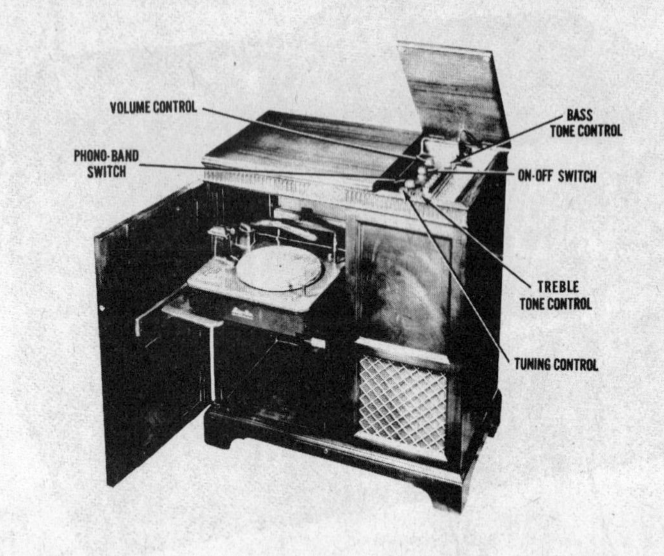

BENDIX

MODEL PICTURED
1524

AC operated phono-radio AM-FM superheterodyne receiver with loop antenna

TUBES
10

POWER SUPPLY
105-125 volts AC

TUNING RANGE
540-1620KC, 88-108MC

MFR/SUPPLIER
Bendix Radio

PHOTOFACT SET
37-3

PUBLISHED
1948

BENDIX

MODEL PICTURED
PAR-80

Three-power operated portable air-marine broadcast superheterodyne receiver

TUBES
6

POWER SUPPLY
105-125 volts AC/DC or 9 volts A & 90 volts B supply in pack form

TUNING RANGE
540-1600KC, 2-5.5MC, Long wave 200-400KC

MFR/SUPPLIER
Bendix Radio

PHOTOFACT SET
39-3

PUBLISHED
1948

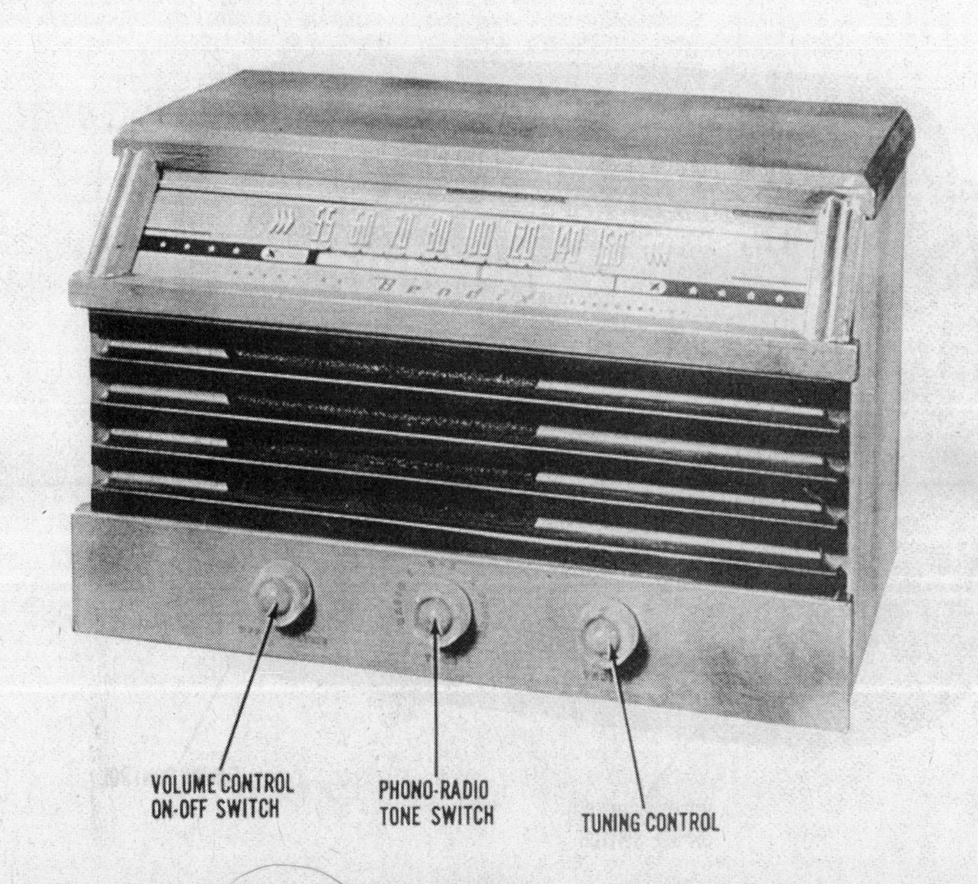

BENDIX

MODEL PICTURED
301

AC/DC operated superheterodyne receiver with loop antenna

TUBES
6

POWER SUPPLY
105-125 volts AC/DC, .25 amp @ 117 volts AC

TUNING RANGE
540-1620KC

MFR/SUPPLIER
Bendix Radio

PHOTOFACT SET
40-2

PUBLISHED
1948

BENDIX

MODEL PICTURED
613

AC operated combo phono-radio superheterodyne receiver with loop antenna

TUBES
5

POWER SUPPLY
110-120 volts AC

TUNING RANGE
540-1620KC

MFR/SUPPLIER
Bendix Radio

PHOTOFACT SET
40-3

PUBLISHED
1948

BENDIX

MODEL PICTURED
110

AC/DC operated superheterodyne receiver with loop antenna

TUBES
5

POWER SUPPLY
105-125 volts AC/DC

TUNING RANGE
540-1620KC

MFR/SUPPLIER
Bendix Radio

PHOTOFACT SET
41-3

PUBLISHED
1948

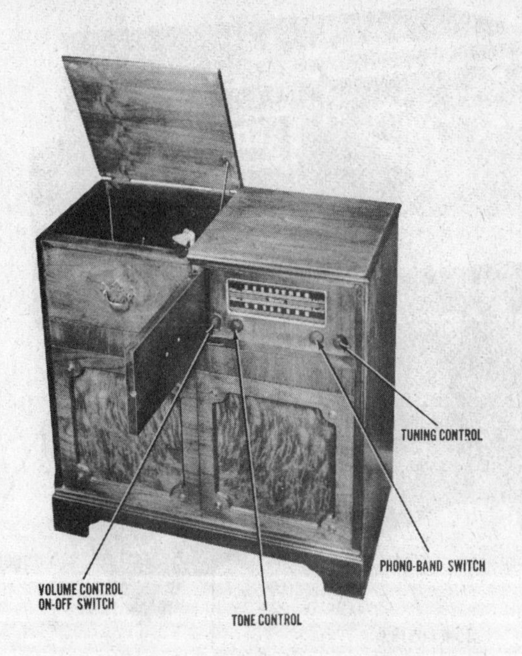

BENDIX

MODEL PICTURED
1521

AC operated combo phono-radio AM-FM superheterodyne receiver with loop antenna

TUBES
8

POWER SUPPLY
110-120 volts AC

TUNING RANGE
540-1620KC

MFR/SUPPLIER
Bendix Radio

PHOTOFACT SET
42-4

PUBLISHED
1948

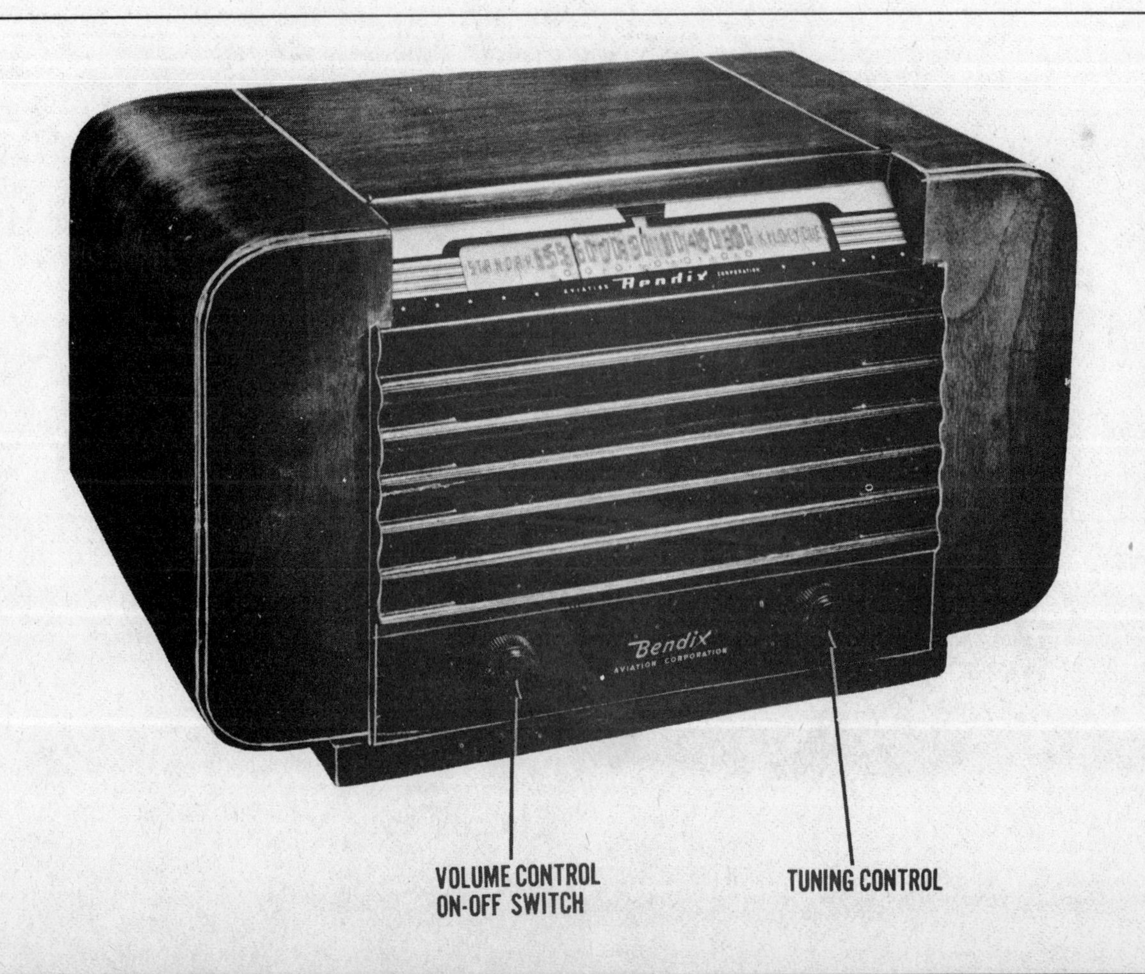

BENDIX

MODEL PICTURED
416A

Battery operated superheterodyne receiver with loop antenna

TUBES
4

POWER SUPPLY
1.5 volts A supply & 90 volts B supply in pack form

TUNING RANGE
540-1620KC

MFR/SUPPLIER
Bendix Radio

PHOTOFACT SET
43-5

PUBLISHED
1948

BENDIX

MODEL PICTURED
1531

AC operated combo phono-radio two-band superheterodyne receiver

TUBES
6

POWER SUPPLY
105-125 volts AC

TUNING RANGE
540-1620KC, 6-12MC

MFR/SUPPLIER
Bendix Radio

PHOTOFACT SET
43-6

PUBLISHED
1948

BENDIX

MODEL PICTURED
1217D

AC operated combo phono-radio two-band AM-FM superheterodyne receiver

TUBES
14

POWER SUPPLY
110-120 volts AC

TUNING RANGE
540-1620KC, 88-108MC, 5.7-16MC

MFR/SUPPLIER
Bendix Radio

PHOTOFACT SET
46-5

PUBLISHED
1948

BENDIX

MODEL PICTURED
55P2

AC/DC operated superheterodyne receiver with loop antenna

TUBES
5

POWER SUPPLY
105-120 volts AC/DC

TUNING RANGE
540-1620KC

MFR/SUPPLIER
Bendix Radio

PHOTOFACT SET
51-4

PUBLISHED
1948

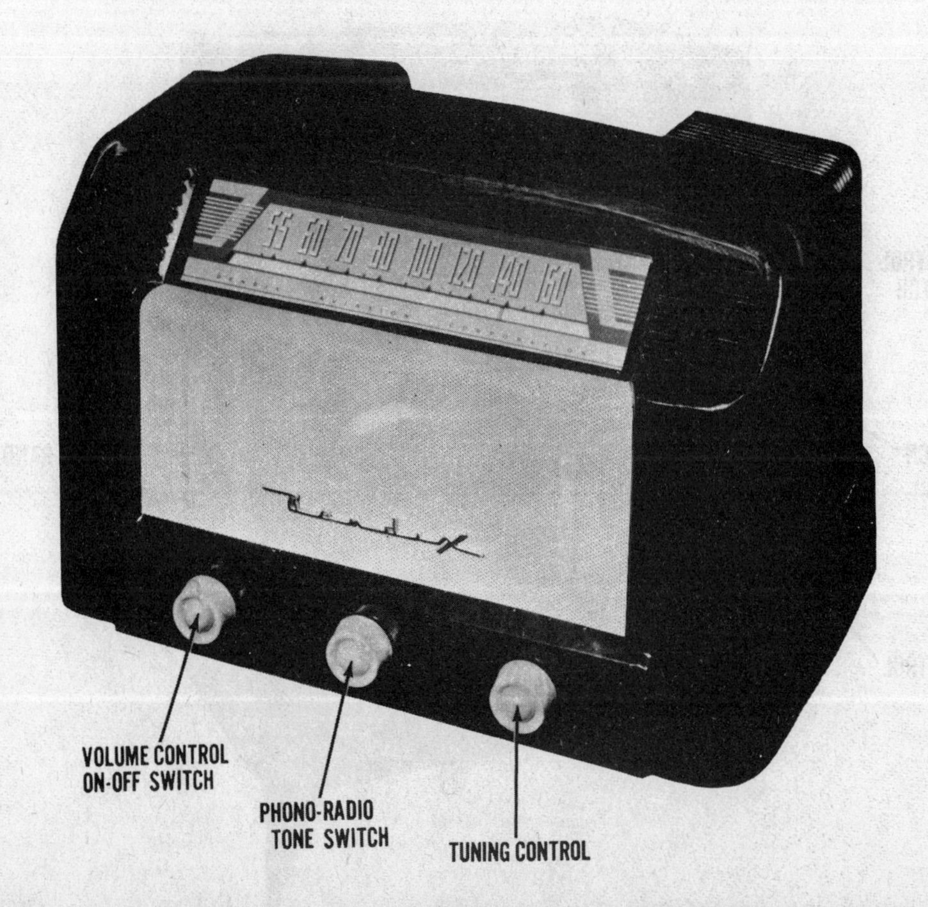

BENDIX

MODEL PICTURED
65P4

AC/DC operated superheterodyne reciever with loop antenna

TUBES
6

POWER SUPPLY
105-120 volts AC/DC

TUNING RANGE
540-1620KC

MFR/SUPPLIER
Bendix Radio

PHOTOFACT SET
52-4

PUBLISHED
1948

BENDIX

MODEL PICTURED
55X4

Three-power operated portable superheterodyne receiver with loop antenna

TUBES
5

POWER SUPPLY
110-120 volts AC/DC or 6 volts A supply & 67.5 volts B supply

TUNING RANGE
530 KC-1630KC

MFR/SUPPLIER
Bendix Radio

PHOTOFACT SET
58-6

PUBLISHED
1949

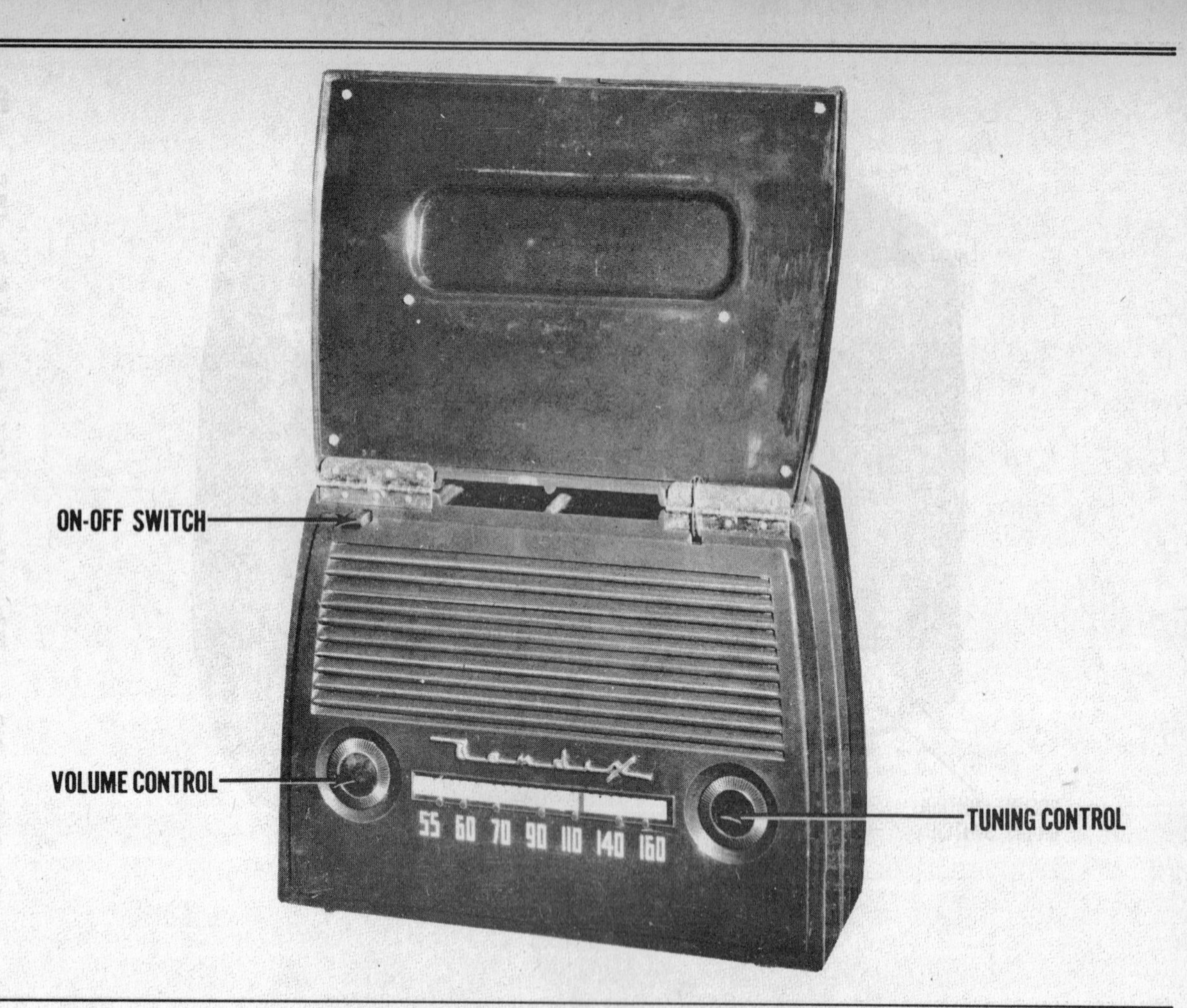

BENDIX

MODEL PICTURED
75M8

AC operated combo phono-radio, AM-FM superheterodyne receiver with loop antenna

TUBES
6

POWER SUPPLY
110-120 volts AC

TUNING RANGE
540-1620KC, 88-108MC

MFR/SUPPLIER
Bendix Radio

PHOTOFACT SET
59-5

PUBLISHED
1949

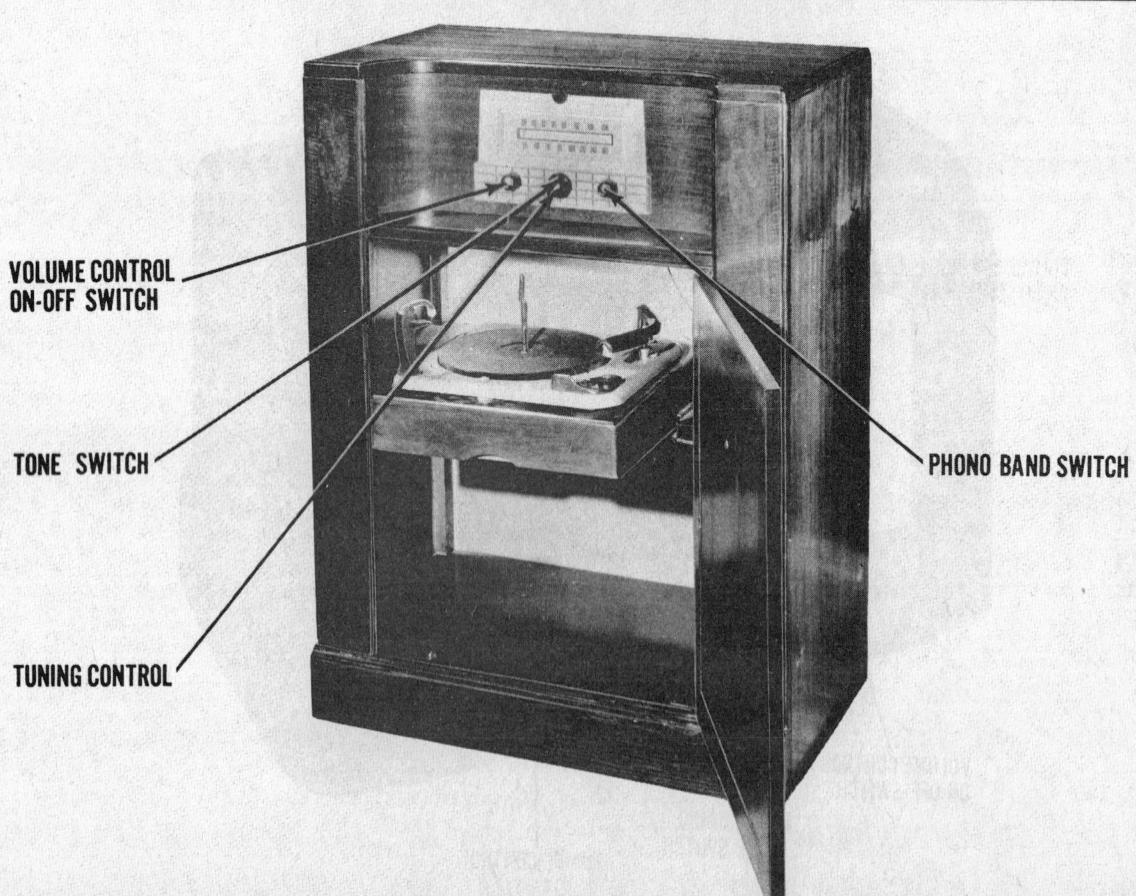

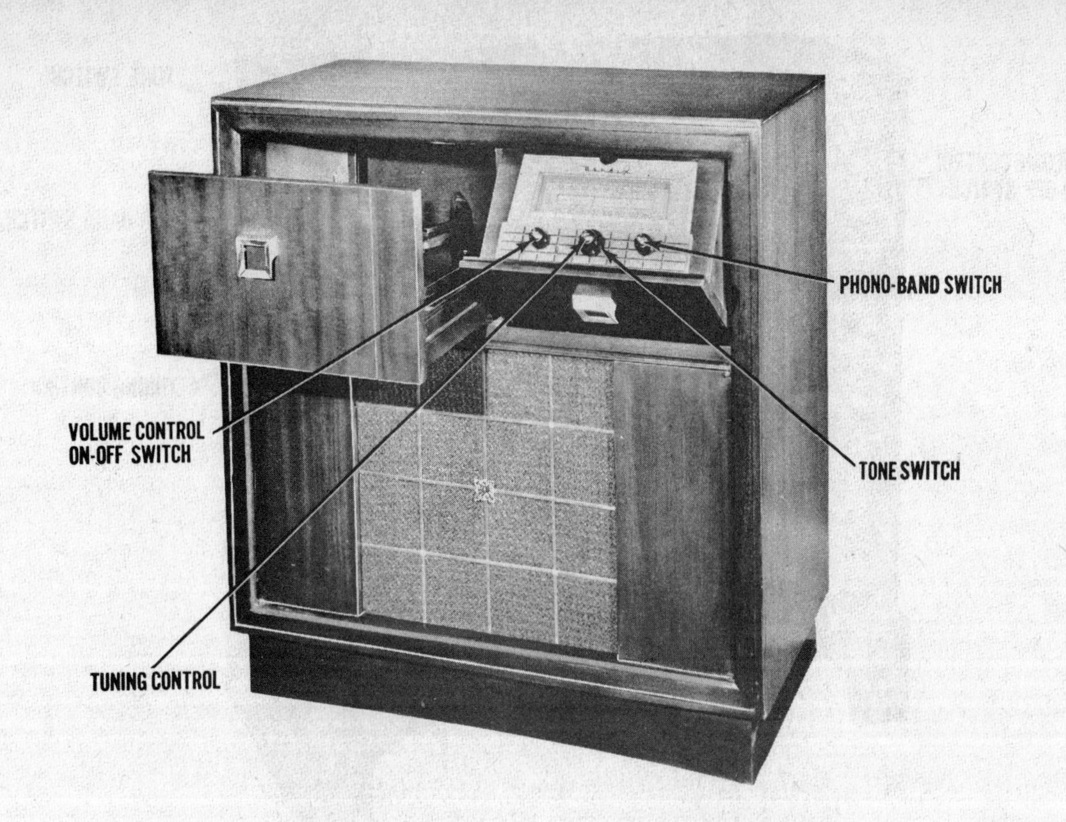

BENDIX

MODEL PICTURED
95M3

AC operated combo phono-radio AM-FM superheterodyne receiver with loop antenna

TUBES
9

POWER SUPPLY
105-120 volts AC

TUNING RANGE
540-1620KC, 88-108MC

MFR/SUPPLIER
Bendix Radio

PHOTOFACT SET
60-7

PUBLISHED
1949

BENDIX

MODEL PICTURED
69M9

AC operated combo phono-radio AM-FM superheterodyne receiver with loop antenna

TUBES
6

POWER SUPPLY
105-120 volts

TUNING RANGE
540-1620KC, 88-108MC

MFR/SUPPLIER
Bendix Radio

PHOTOFACT SET
63-3

PUBLISHED
1949

BENDIX

MODEL PICTURED
79M7

AC operated combo phono-radio AM-FM superheterodyne receiver with loop antenna

TUBES
7

POWER SUPPLY
105-120 volts AC

TUNING RANGE
540-1620KC, 88-108MC

MFR/SUPPLIER
Bendix Radio

PHOTOFACT SET
66-3

PUBLISHED
1949

BENDIX

MODEL PICTURED
951

AC operated combo phono-radio AM-FM superheterodyne receiver

TUBES
7

POWER SUPPLY
110-120 volts AC

TUNING RANGE
540-1620 KC, 88-108MC

MFR/SUPPLIER
Bendix Radio

PHOTOFACT SET
136-6

PUBLISHED
1951

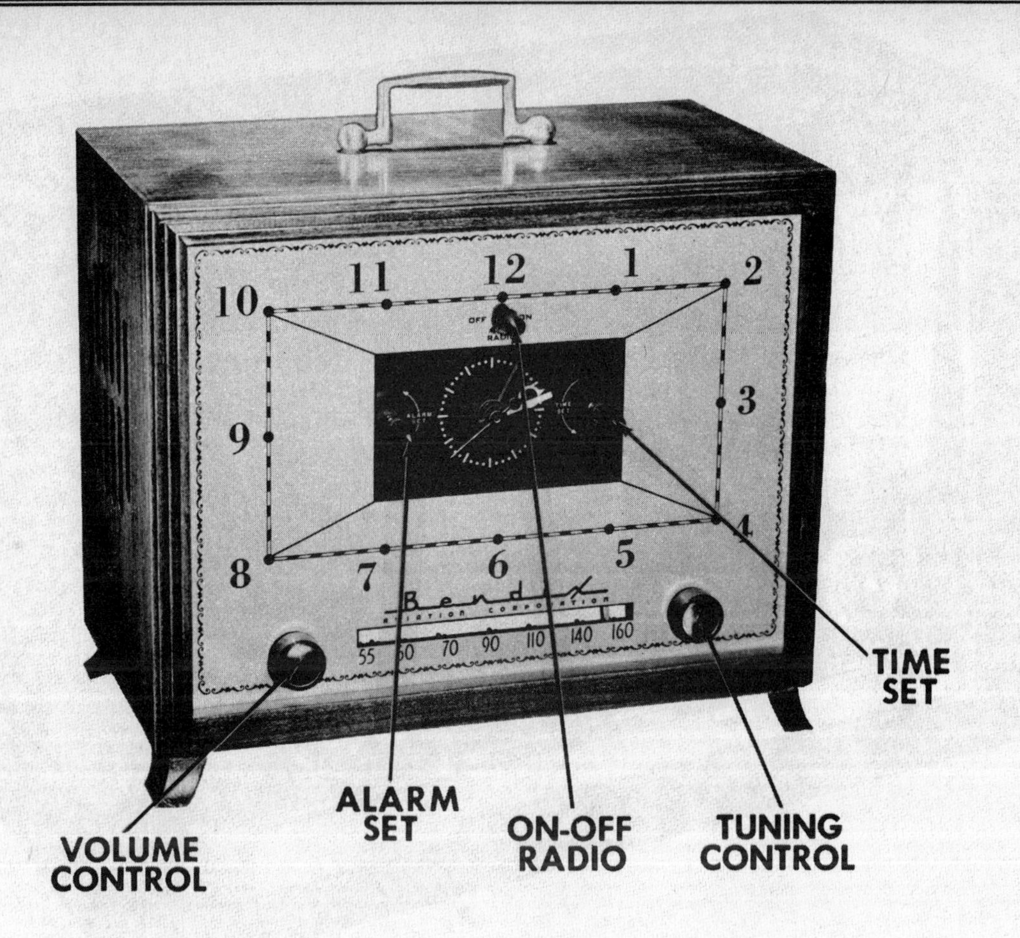

BENDIX

MODEL PICTURED
753F

AC operated AM superheterodyne receiver with electric clock

TUBES
5

POWER SUPPLY
105-120 volts AC, 60 cycle, .25 amp @ 117 volts AC

TUNING RANGE
540-1620KC

MFR/SUPPLIER
Bendix Radio

PHOTOFACT SET
199-3

PUBLISHED
1953

BENRUS

MODEL PICTURED
10B01B15B

AC operated AM superheterodyne receiver with electric clock

TUBES
5

POWER SUPPLY
110-120 volts AC, 60 cycle

TUNING RANGE
535-1620KC

MFR/SUPPLIER
Benrus Watch Co.

PHOTOFACT SET
299-3

PUBLISHED
1955

BLAUPUNKT

MODEL PICTURED
Granada 2330

AC operated
FM-BC-SW-LW receiver

TUBES
7

POWER SUPPLY
110/127/155/220/240 volts AC, 60 cycles, .63 amp @ 117 volts AC, 65 watts

TUNING RANGE
145-375KC,
515-1620KC,
87.4-108MC,
5.95-18.2MC

MFR/SUPPLIER
N. Pickens Import Co.

PHOTOFACT SET
418-4

PUBLISHED
1958

BLAUPUNKT

MODEL PICTURED
Barcelona

AC operated
FM-BC-SW-LW receiver

TUBES
7

POWER SUPPLY
110/127/155/220/240 volts AC, 60 cycles, .63 amp @ 117 volts AC, 65 watts

TUNING RANGE
145-375KC,
515-1620KC,
87.4-108MC,
5.95-18.2MC

MFR/SUPPLIER
N. Pickens Import Co.

PHOTOFACT SET
422-5

PUBLISHED
1958

BLAUPUNKT

MODEL PICTURED
Sultan 2320

AC operated
FM-BC-SW-LW receiver

TUBES
6

POWER SUPPLY
240/220/155/127/110 volts AC, 60 cycles, .62 AM @ 117 volts AC, 61 watts

TUNING RANGE
148-375KC,
515-1620KC,
87.4-108MC,
5.95-18.2MC

MFR/SUPPLIER
N. Pickens Import Co.

PHOTOFACT SET
424-4

PUBLISHED
1958

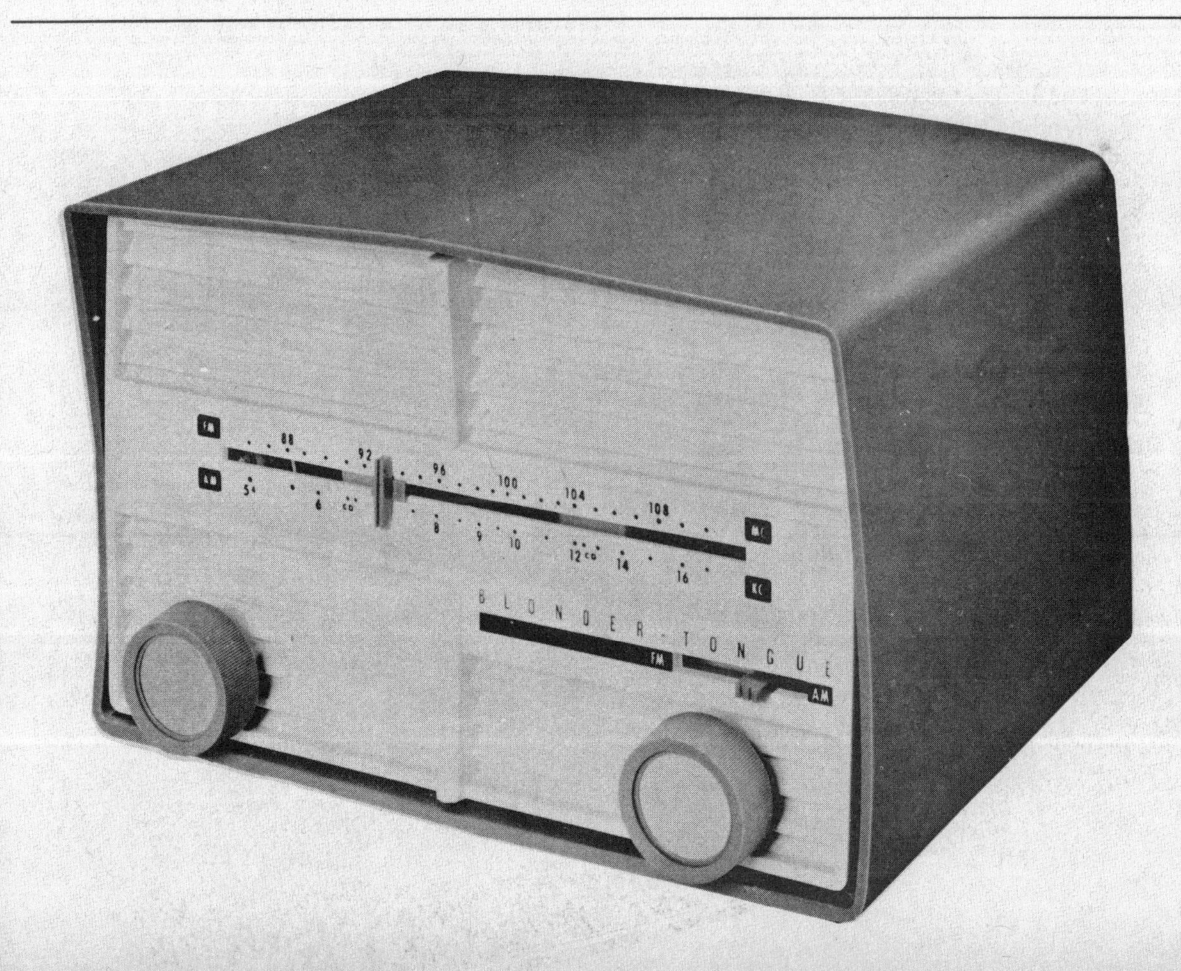

BLONDER-TONGUE

MODEL PICTURED
R-98

AC/DC operated FM-AM receiver

TUBES
6

POWER SUPPLY
110-120 volts AC/DC, 32 watts, .375 amp @ 117 volts AC

TUNING RANGE
535-1620KC, 88-108MC

MFR/SUPPLIER
Blonder-Tongue Laboratories Inc.

PHOTOFACT SET
457-4

PUBLISHED
1959

BLONDER-TONGUE

MODEL PICTURED
R-20

AC/DC operated FM receiver

TUBES
6

POWER SUPPLY
110-120 volts AC/DC, 30 watts, .32 amp @ 117volts AC

TUNING RANGE
535-1620KC, 88-108MC

MFR/SUPPLIER
Blonder-Tongue Laboratories Inc.

PHOTOFACT SET
503-6

PUBLISHED
1960

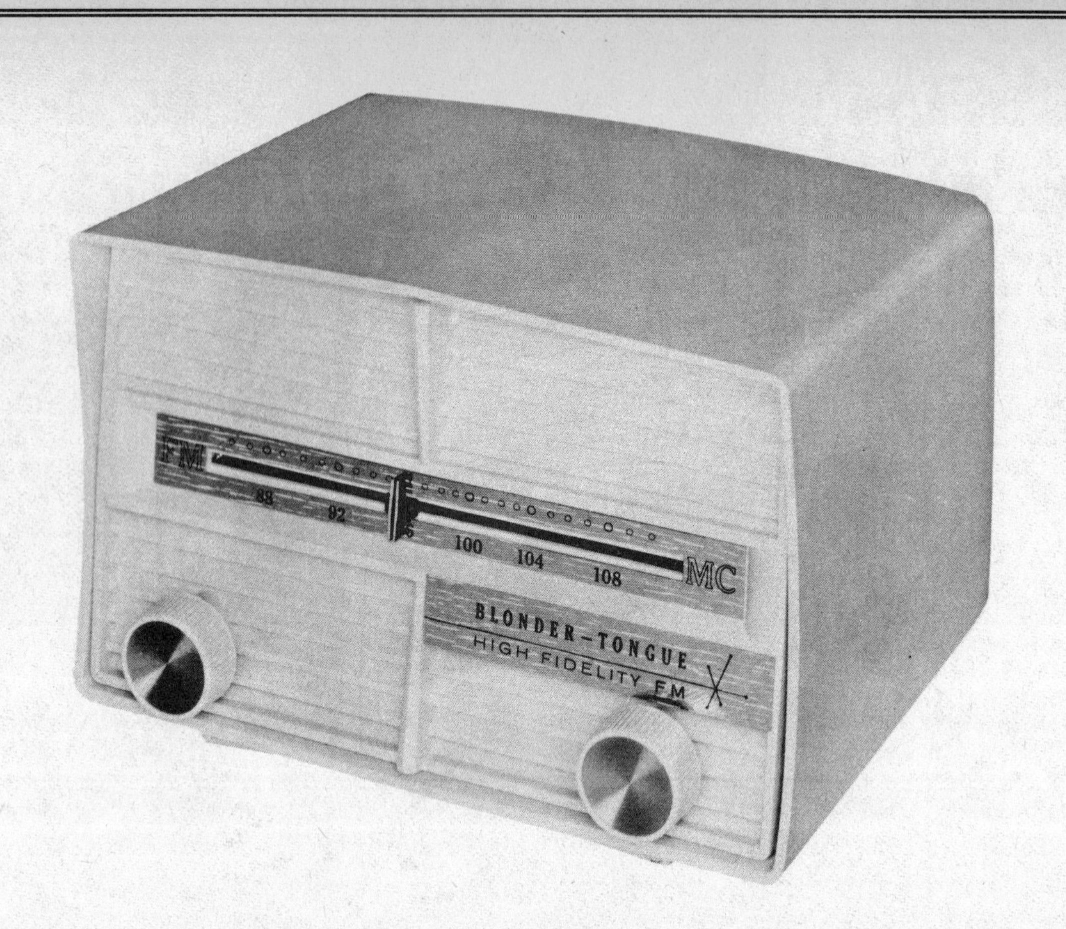

BRADFORD

MODEL PICTURED
WGEC-95190C

AC operated AM-FM receiver, FM stereo, four-speed automatic record changer

TUBES
12

POWER SUPPLY
110-120 volts AC, 60 cycle

TUNING RANGE
530-1650KC, 87-108.5MC

MFR/SUPPLIER
Bradford Electronics

PHOTOFACT SET
774-6

PUBLISHED
1965

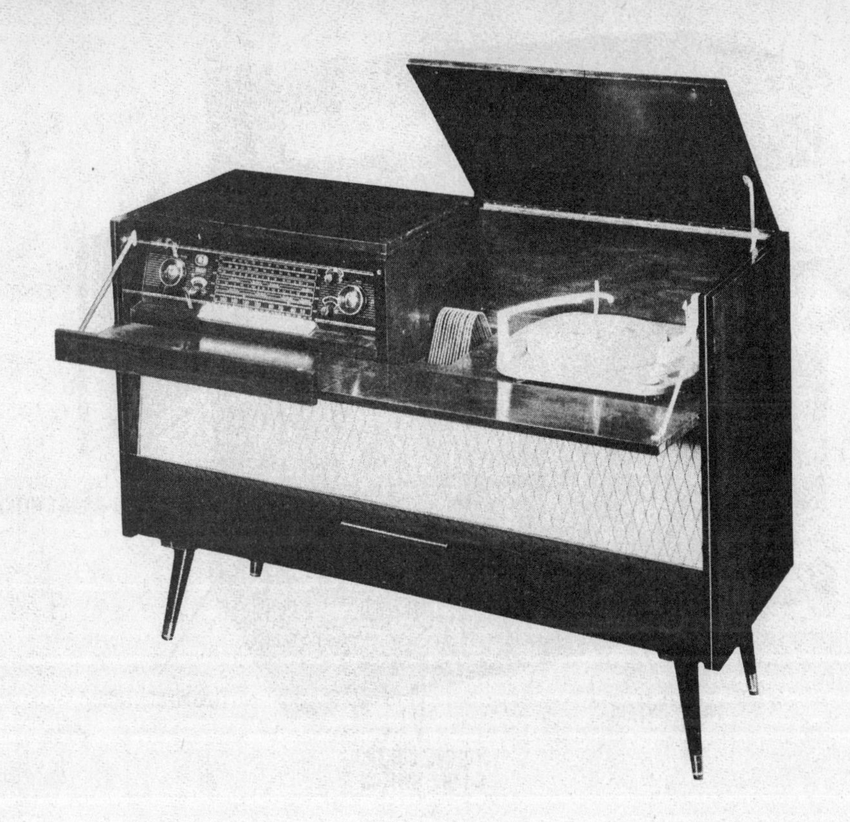

BRAUN

MODEL PICTURED
MM4D

AC operated FM-BC-SW receiver with four-speed auto record changer

TUBES
9

POWER SUPPLY
110-125-220-240 volts AC, 60 cycle

TUNING RANGE
535-1620KC, 87-108MC, 12-23MC, 6-12.5MC

MFR/SUPPLIER
Electronic Utilities Corp.

PHOTOFACT SET
409-6

PUBLISHED
1958

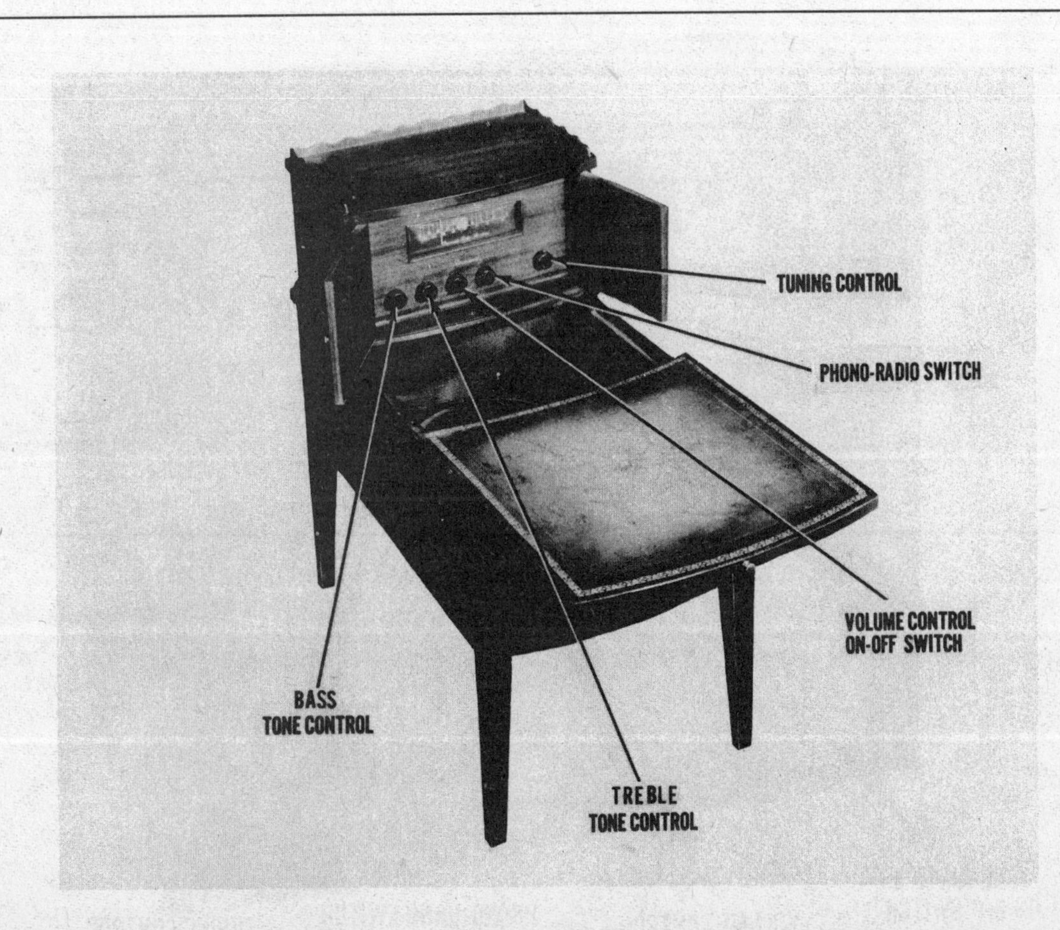

BRUNSWICK

MODEL PICTURED
BJ-6836 (Tuscany)

AC operated radio-phono combo superheterodyne receiver with loop antenna

TUBES
9

POWER SUPPLY
110-120 volts AC

TUNING RANGE
530-1750KC

MFR/SUPPLIER
Radio & TV Inc.

PHOTOFACT SET
28-4

PUBLISHED
1947

BRUNSWICK

MODEL PICTURED
T-4000 (Buckingham)

AC operated radio-phono AM-FM superheterodyne receiver, self-contained loop antenna

TUBES
10

POWER SUPPLY
105-125 volts AC

TUNING RANGE
540-1620KC, 87.5-108MC

MFR/SUPPLIER
Radio & TV Inc.

PHOTOFACT SET
29-5

PUBLISHED
1947

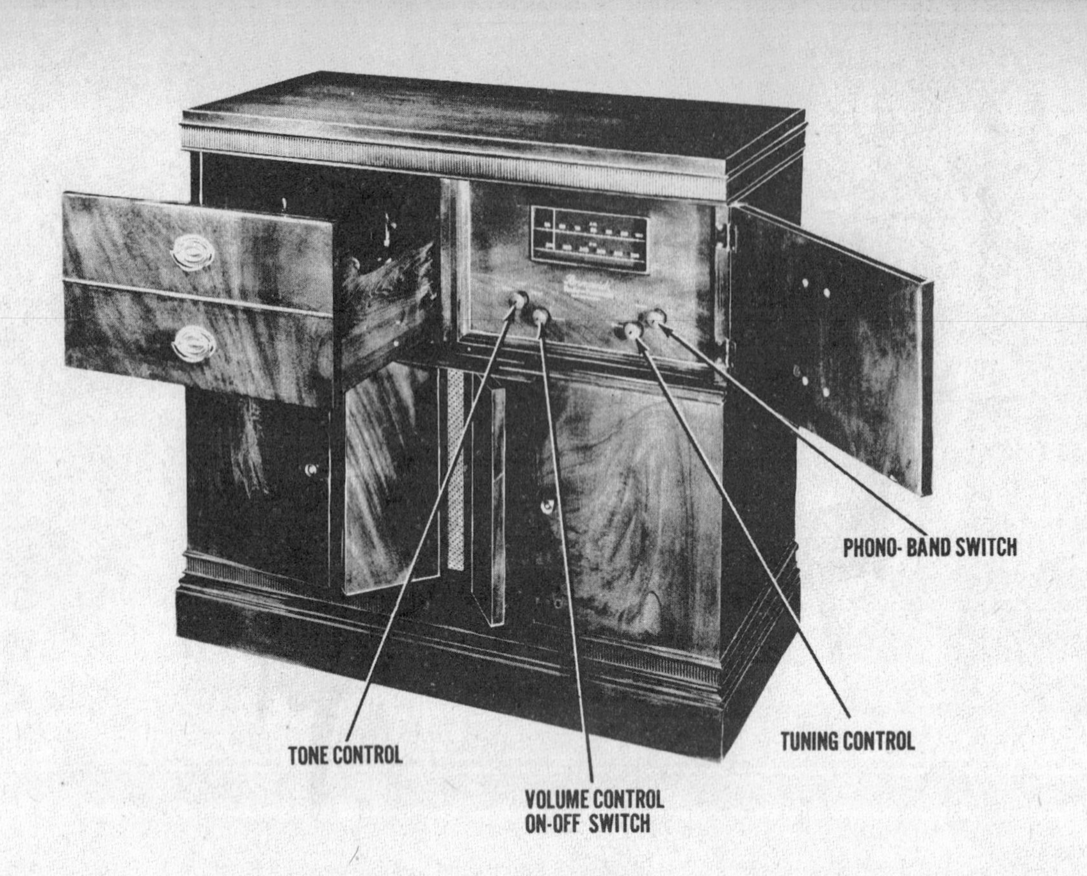

BRUNSWICK

MODEL PICTURED
5000

AC operated combo phono-radio AM-FM superheterodyne receiver

TUBES
11

POWER SUPPLY
105-125 volts AC

TUNING RANGE
540-1700KC, 88-108MC

MFR/SUPPLIER
Radio & TV Inc.

PHOTOFACT SET
42-5

PUBLISHED
1948

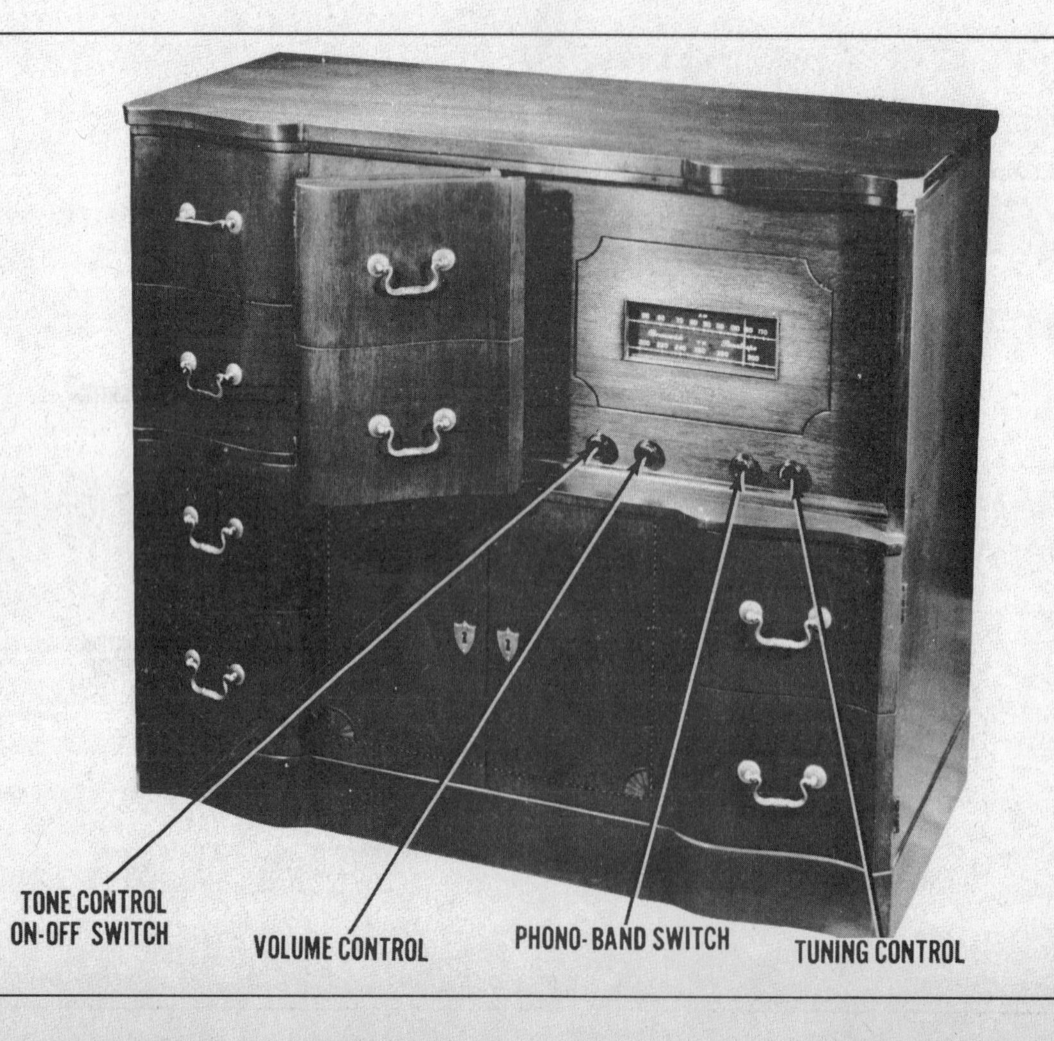

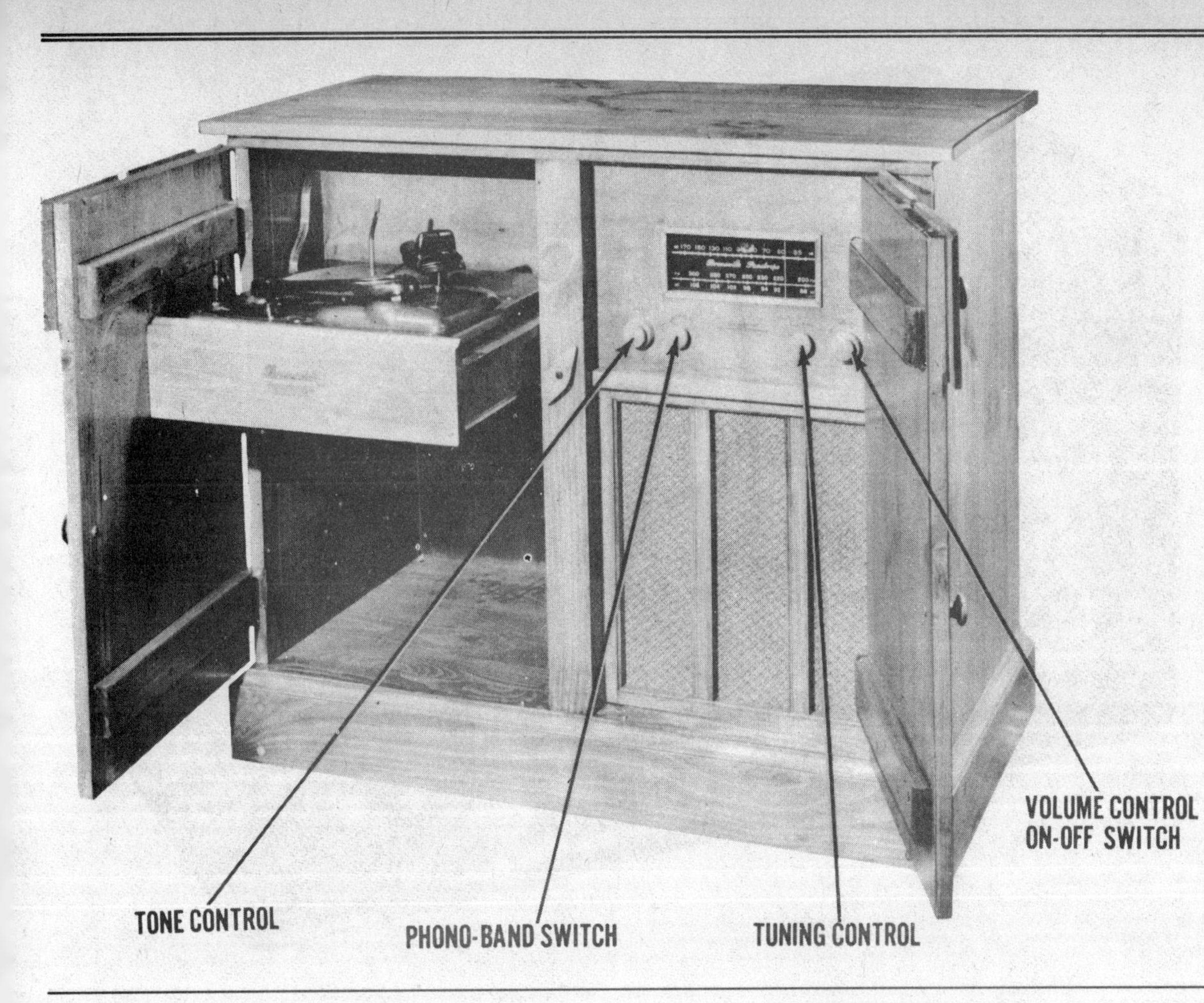

BRUNSWICK

MODEL PICTURED
D-1000

AC operated phono-radio AM-FM superheterodyne receiver with loop antenna

TUBES
14

POWER SUPPLY
110-120 volts AC

TUNING RANGE
540-1620KC, 88-108MC

MFR/SUPPLIER
Radio & TV Inc.

PHOTOFACT SET
56-7

PUBLISHED
1949

BULOVA

MODEL PICTURED
278

Battery operated portable AM transistorized receiver

TUBES
0

POWER SUPPLY
9 volts DC

TUNING RANGE
540-1620KC

MFR/SUPPLIER
Electronics Guild

PHOTOFACT SET
369-5

PUBLISHED
1957

BULOVA

MODEL PICTURED
304

AC/DC operated AM receiver

TUBES
5

POWER SUPPLY
105-120 volts AC/DC

TUNING RANGE
535-1620KC

MFR/SUPPLIER
Electronics Guild

PHOTOFACT SET
369-6

PUBLISHED
1957

BULOVA

MODEL PICTURED
310

AC/DC operated AM receiver

TUBES
5

POWER SUPPLY
105-120 volts AC/DC

TUNING RANGE
535-1620KC

MFR/SUPPLIER
Electronics Guild

PHOTOFACT SET
370-2

PUBLISHED
1957

BULOVA

MODEL PICTURED
100

AC operated AM receiver with electric clock

TUBES
5

POWER SUPPLY
105-125 volts AC, 60 cycle

TUNING RANGE
535-1640KC

MFR/SUPPLIER
Electronics Guild

PHOTOFACT SET
371-2

PUBLISHED
1957

BULOVA

MODEL PICTURED
220

Battery operated portable AM receiver

TUBES
4

POWER SUPPLY
1.5 volts A supply & 67.5 volts B supply

TUNING RANGE
540-1640KC

MFR/SUPPLIER
Electronics Guild

PHOTOFACT SET
372-2

PUBLISHED
1957

BULOVA

MODEL PICTURED
200

Three-power portable AM receiver

TUBES
4

POWER SUPPLY
110-120 volts AC/DC or 3 volts A & 67.5 volts B supply

TUNING RANGE
535-1640KC

MFR/SUPPLIER
Electronics Guild

PHOTOFACT SET
382-6

PUBLISHED
1957

CALRAD

MODEL PICTURED
60A183

Battery operated transistorized portable AM receiver

TUBES
0

POWER SUPPLY
9 volts DC, 6mA @ 9 volts DC (no signal, minimum volume), 10mA @ 9 volts DC (signal, normal volume)

TUNING RANGE
535-1650KC

MFR/SUPPLIER
Burstein-Applebee Co.

PHOTOFACT SET
497-5

PUBLISHED
1960

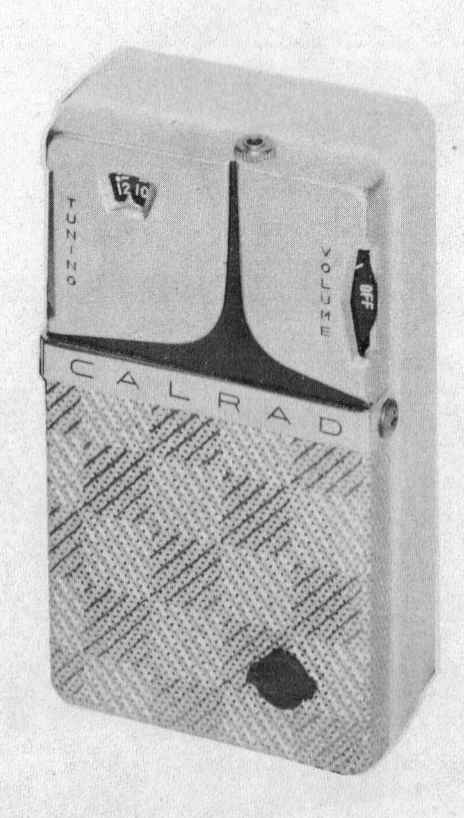

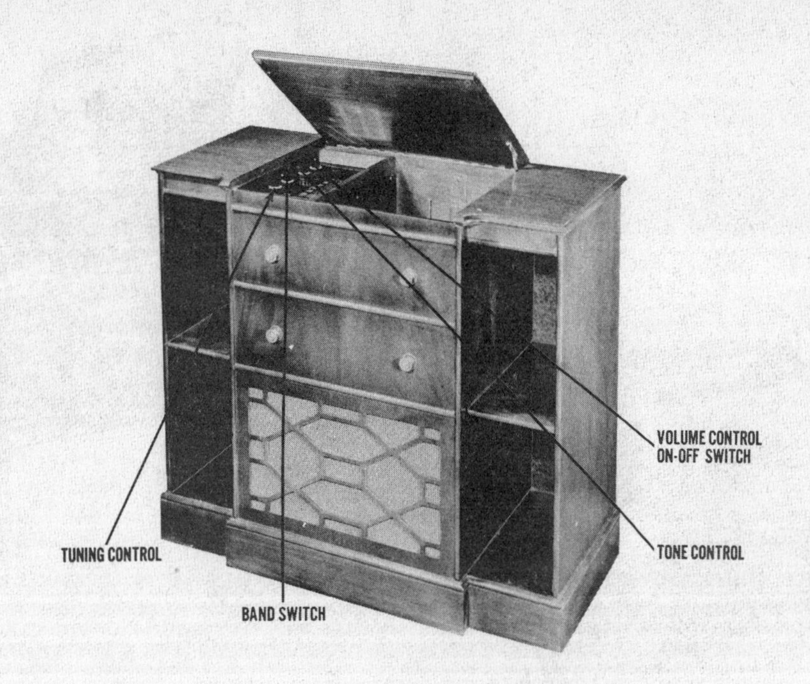

CAPEHART

MODEL PICTURED
33P9

AC operated combo phono-radio AM-FM superheterodyne receiver with loop antenna

TUBES
10

POWER SUPPLY
105-125 volts AC

TUNING RANGE
540-1620KC,
87.5-108.5MC

MFR/SUPPLIER
Farnsworth Television & Radio Corp.

PHOTOFACT SET
64-3

PUBLISHED
1949

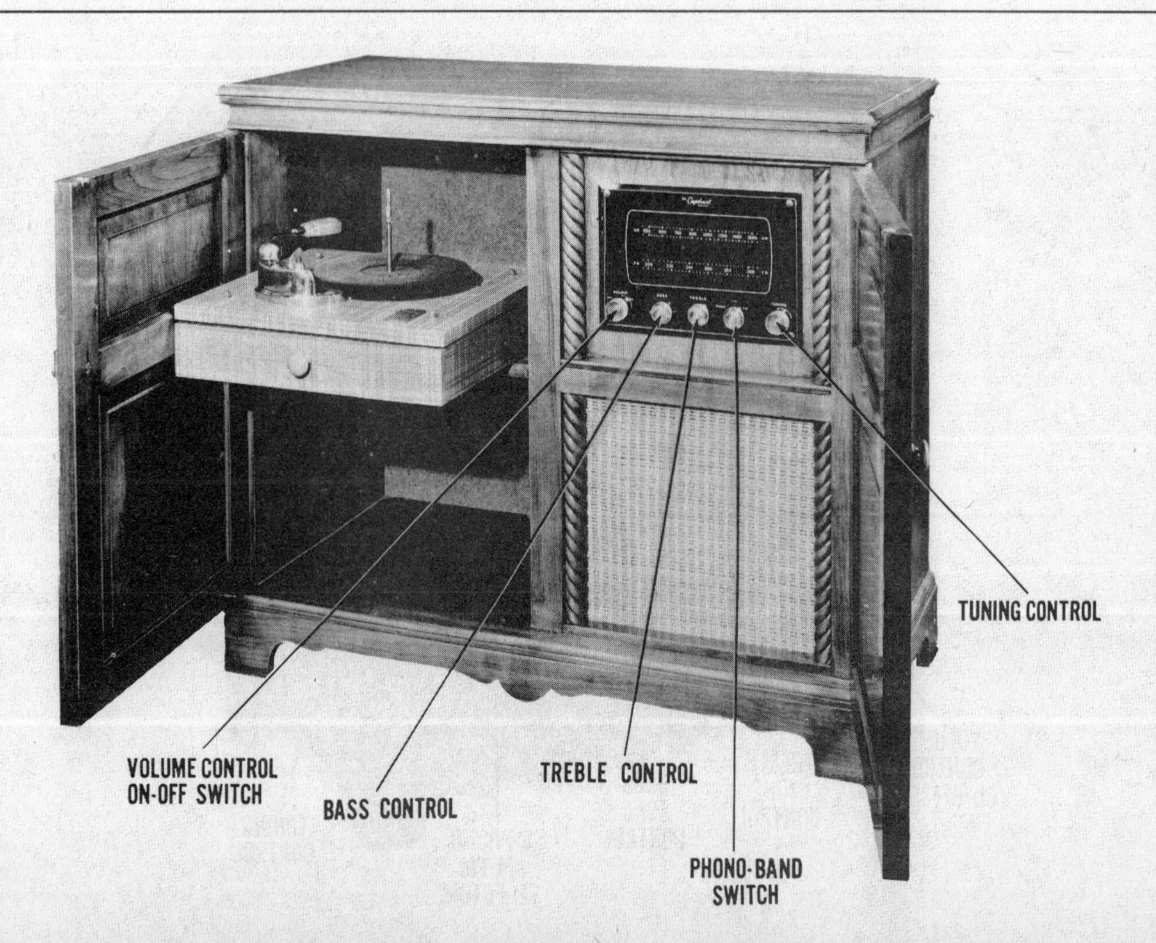

CAPEHART

MODEL PICTURED
29P4

AC operated combo phono-radio AM-FM superheterodyne receiver with loop antenna

TUBES
11

POWER SUPPLY
105-125 volts AC

TUNING RANGE
540-1600KC,
87.5-108.5MC

MFR/SUPPLIER
Farnsworth Television & Radio Corp.

PHOTOFACT SET
65-3

PUBLISHED
1949

CAPEHART

MODEL PICTURED
413P, 115P2

AC operated phono-radio two-band AM-FM superheterodyne receiver with loop antenna

TUBES
36

POWER SUPPLY
105-125 volts AC

TUNING RANGE
540-1600KC, 88-108MC, 9.4-15.4MC

MFR/SUPPLIER
Farnsworth Television & Radio Corp.

PHOTOFACT SET
67-6

PUBLISHED
1949

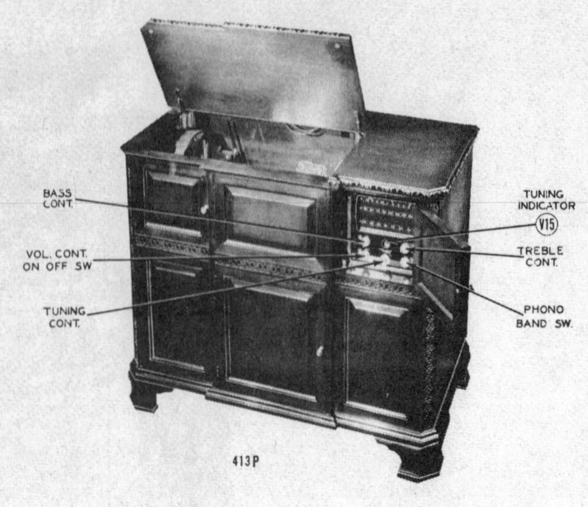

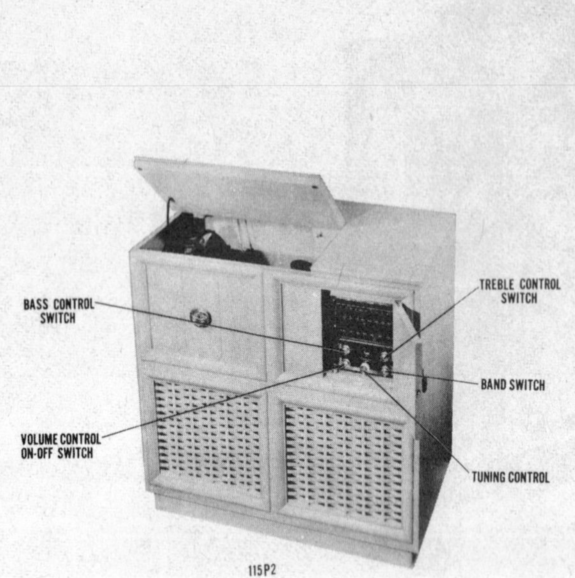

CAPEHART

MODEL PICTURED
1002F

AC operated phono-radio AM-FM superheterodyne receiver with loop antenna

TUBES
11

POWER SUPPLY
105-125 volts AC

TUNING RANGE
540-1620KC, 88-108MC

MFR/SUPPLIER
Capehart-Farnsworth Corp.

PHOTOFACT SET
135-4

PUBLISHED
1951

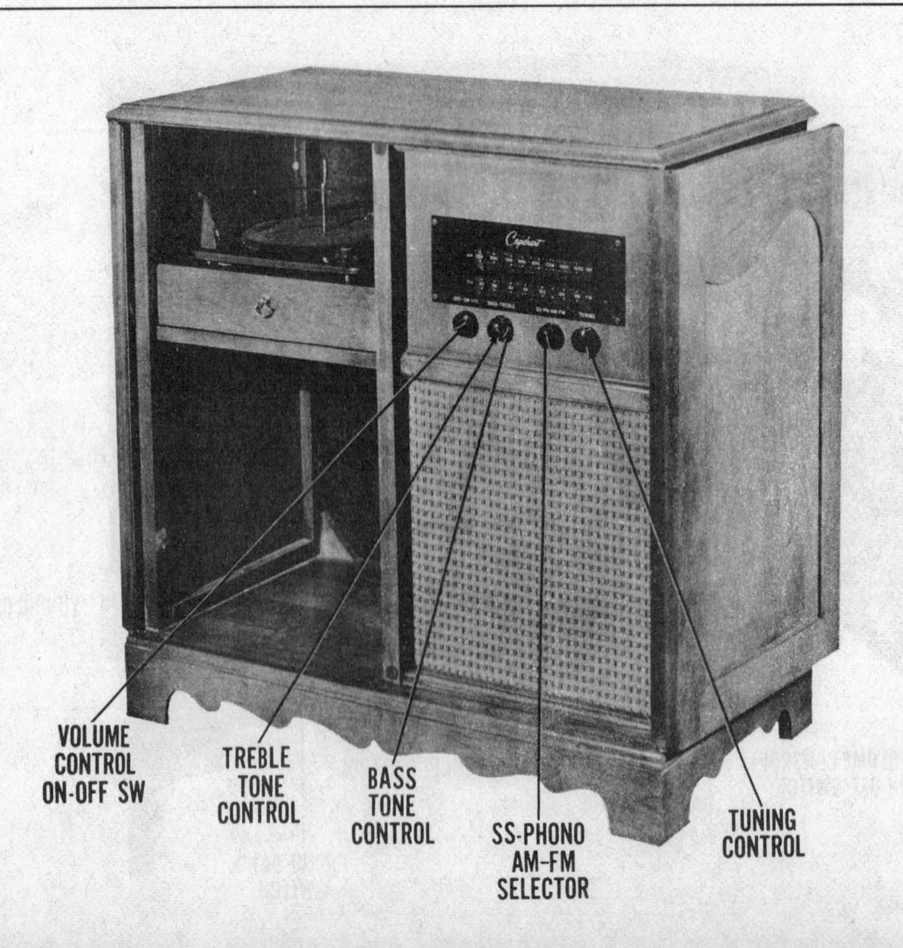

CAPEHART

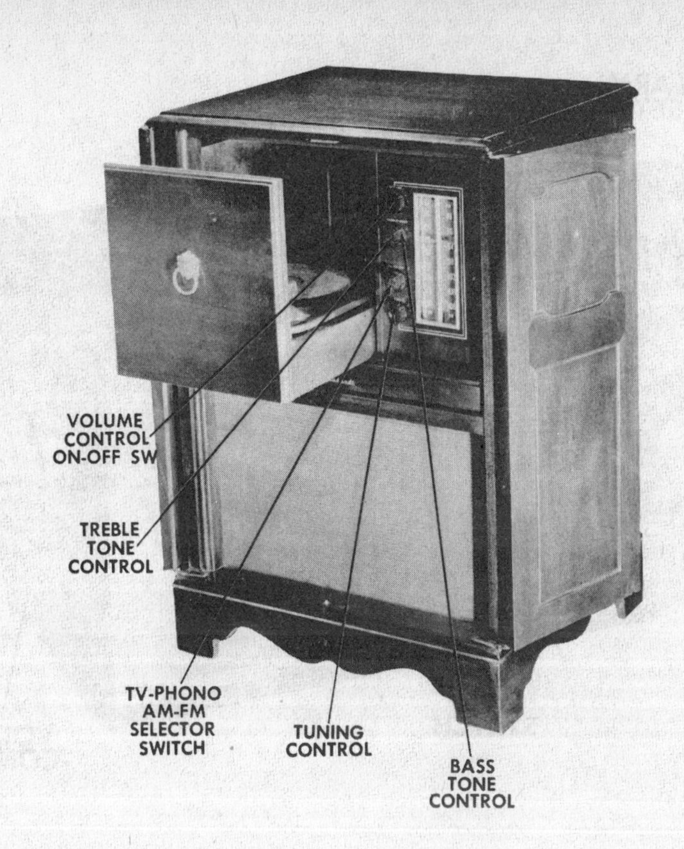

MODEL PICTURED
1007AM

AC operated phono-radio AM-FM superheterodyne receiver with loop antenna

TUBES
11

POWER SUPPLY
110-120 volts AC

TUNING RANGE
540-1620KC, 88-108MC

MFR/SUPPLIER
Capehart-Farnsworth Corp.

PHOTOFACT SET
150-5

PUBLISHED
1951

CAPEHART

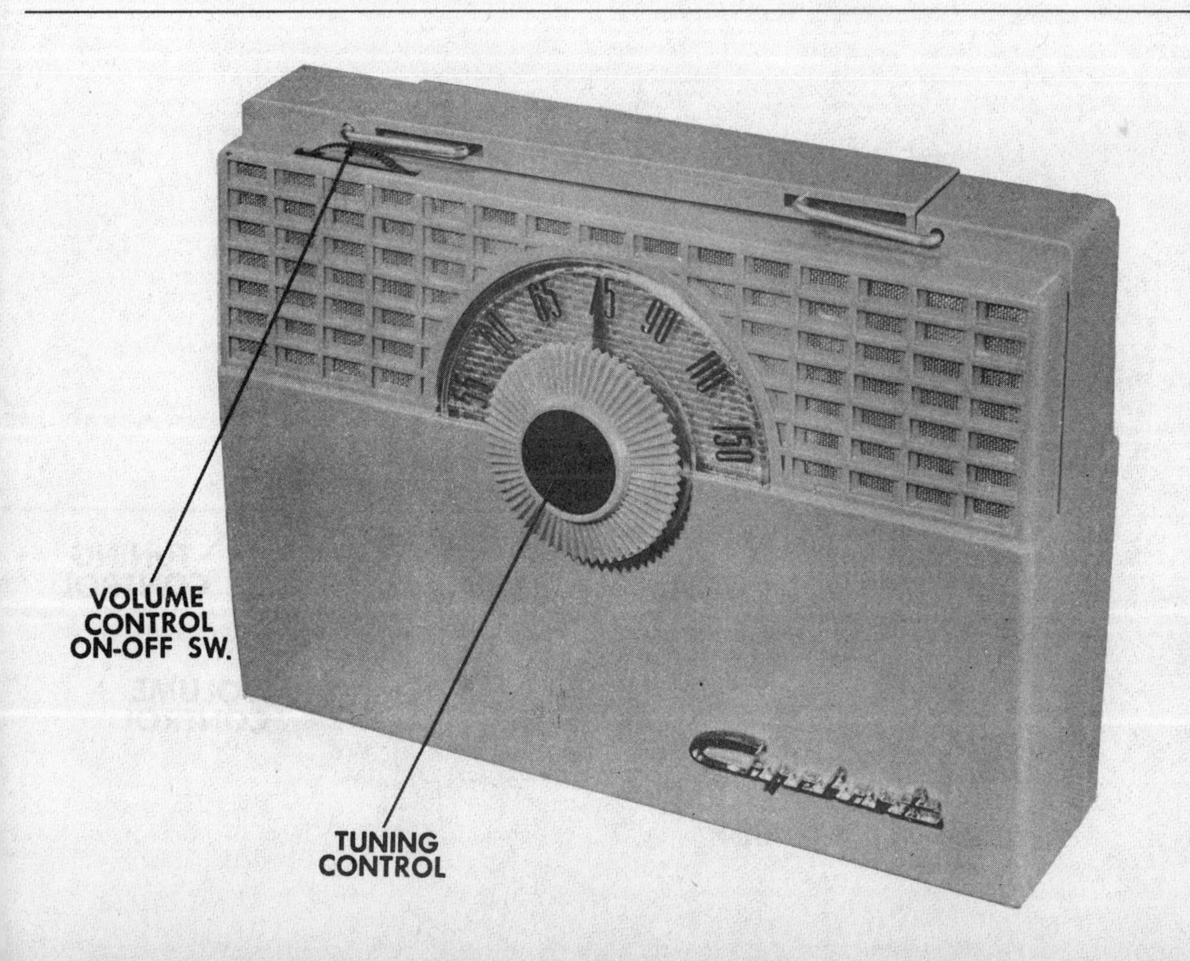

MODEL PICTURED
10

Battery operated portable AM superheterodyne receiver

TUBES
4

POWER SUPPLY
1.5 volts DC A supply & 67.5 volts DC B supply

TUNING RANGE
532-1620KC

MFR/SUPPLIER
Capehart-Farnsworth Corp.

PHOTOFACT SET
166-7

PUBLISHED
1952

CAPEHART

MODEL PICTURED
TC-62

AC operated AM superheterodyne receiver with electric clock

TUBES
6

POWER SUPPLY
105-125 volts AC, 60 Cycles

TUNING RANGE
535-1620KC

MFR/SUPPLIER
Capehart-Farnsworth Corp.

PHOTOFACT SET
192-4

PUBLISHED
1953

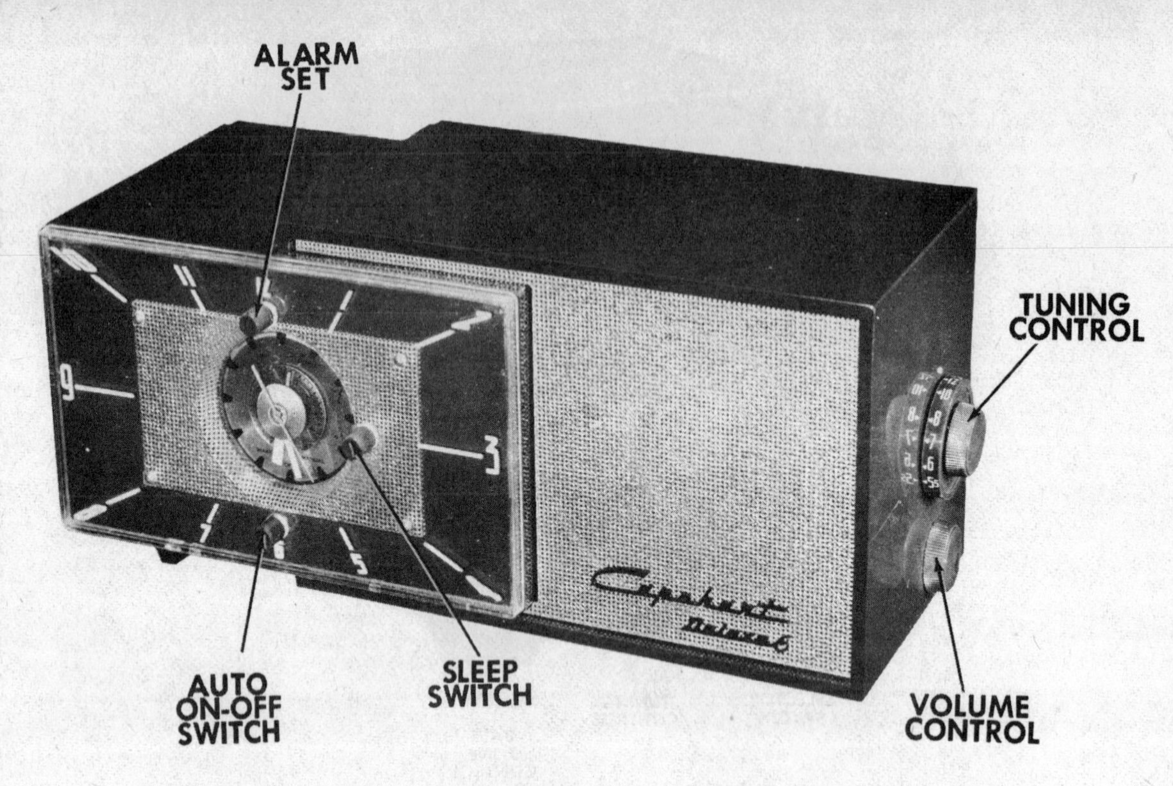

CAPEHART

MODEL PICTURED
TC-100

AC operated AM superheterodyne receiver with electric clock

TUBES
5

POWER SUPPLY
110-120 volts AC, 60 cycle, .24 amp @ 117 volts AC

TUNING RANGE
540-1620KC

MFR/SUPPLIER
Capehart-Farnsworth Corp.

PHOTOFACT SET
203-5

PUBLISHED
1953

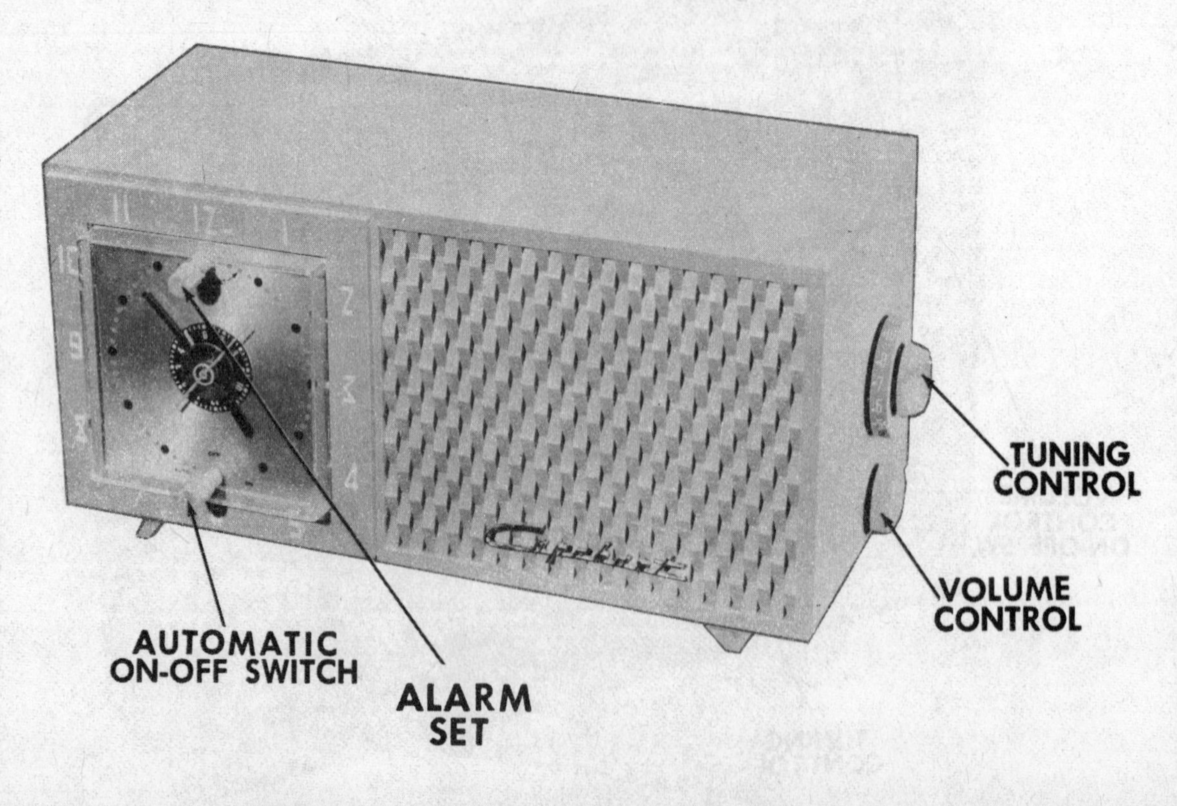

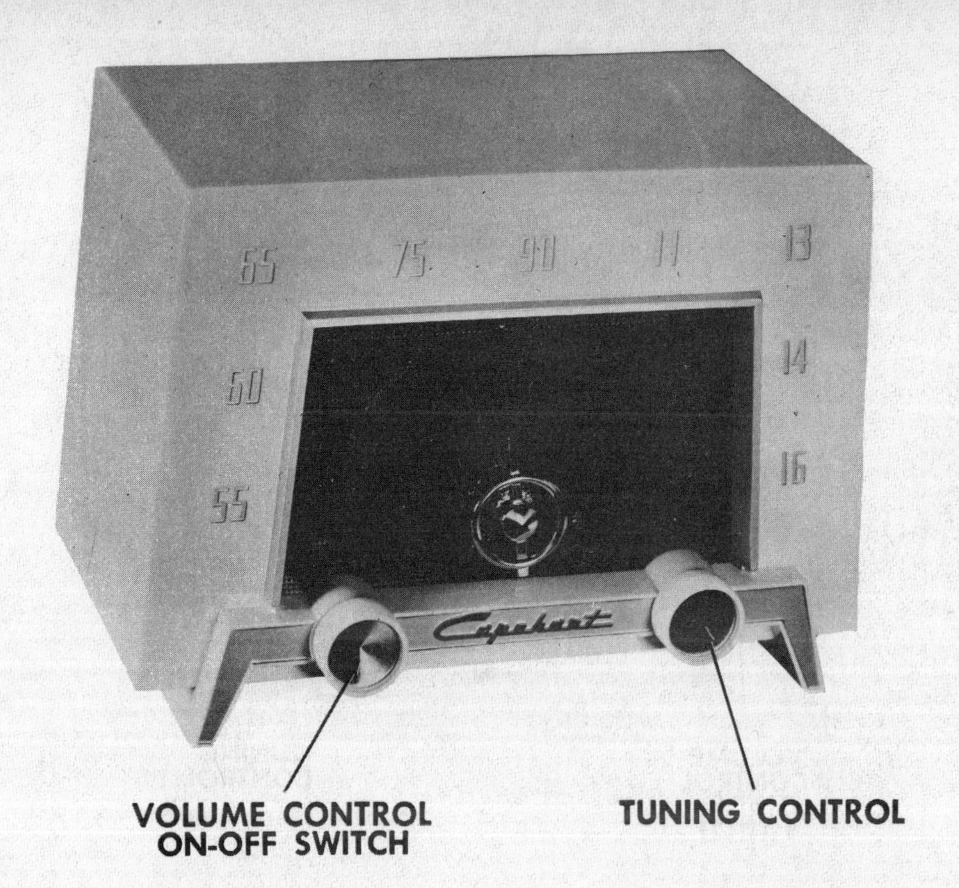

CAPEHART

MODEL PICTURED
T-522

AC/DC operated AM superheterodyne receiver

TUBES
5

POWER SUPPLY
105-125 volts AC/DC

TUNING RANGE
540-1620KC

MFR/SUPPLIER
Capehart-Farnsworth Corp.

PHOTOFACT SET
209-1

PUBLISHED
1953

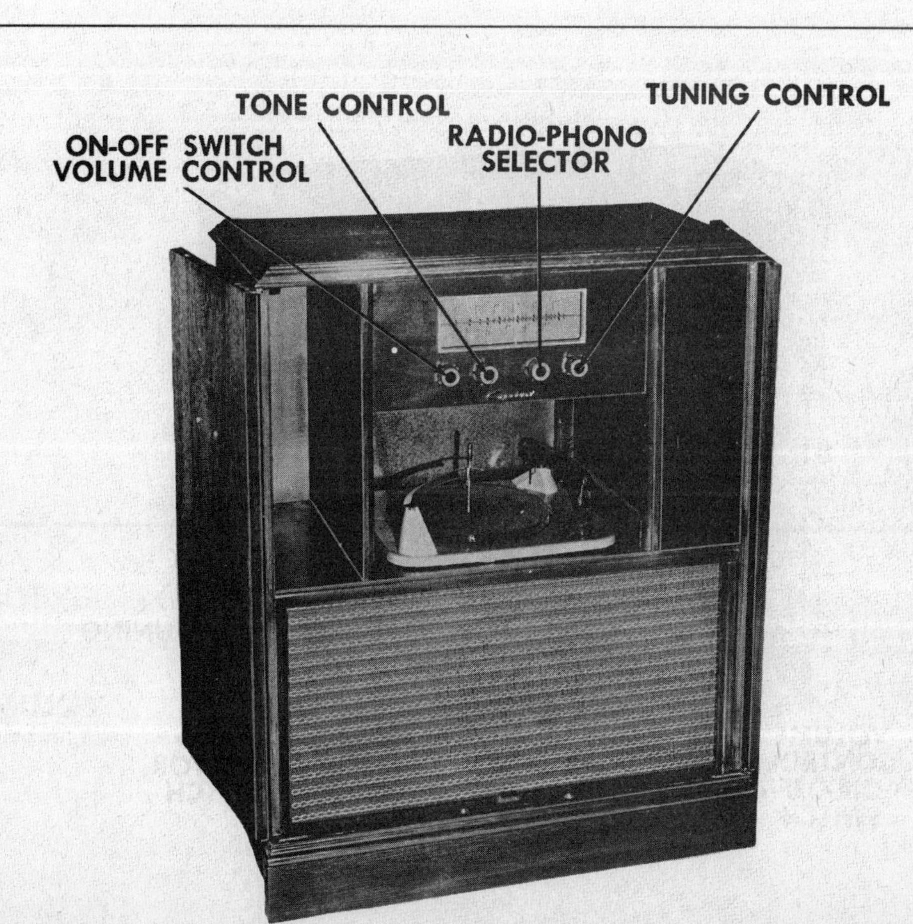

CAPEHART

MODEL PICTURED
RP-152

AC operated AM-phono combo superheterodyne receiver with three-speed auto record changer

TUBES
7

POWER SUPPLY
110-120 volts AC, 60 cycle

TUNING RANGE
537-1620KC

MFR/SUPPLIER
Capehart-Farnsworth Corp.

PHOTOFACT SET
215-4

PUBLISHED
1953

CAPEHART

MODEL PICTURED
P-213

Three-power portable superheterodyne receiver

TUBES
4

POWER SUPPLY
105-125 volts AC/DC or 3 volts A supply & 67.5 volts B supply

TUNING RANGE
540-1620KC

MFR/SUPPLIER
Capehart-Farnsworth Corp.

PHOTOFACT SET
234-3

PUBLISHED
1954

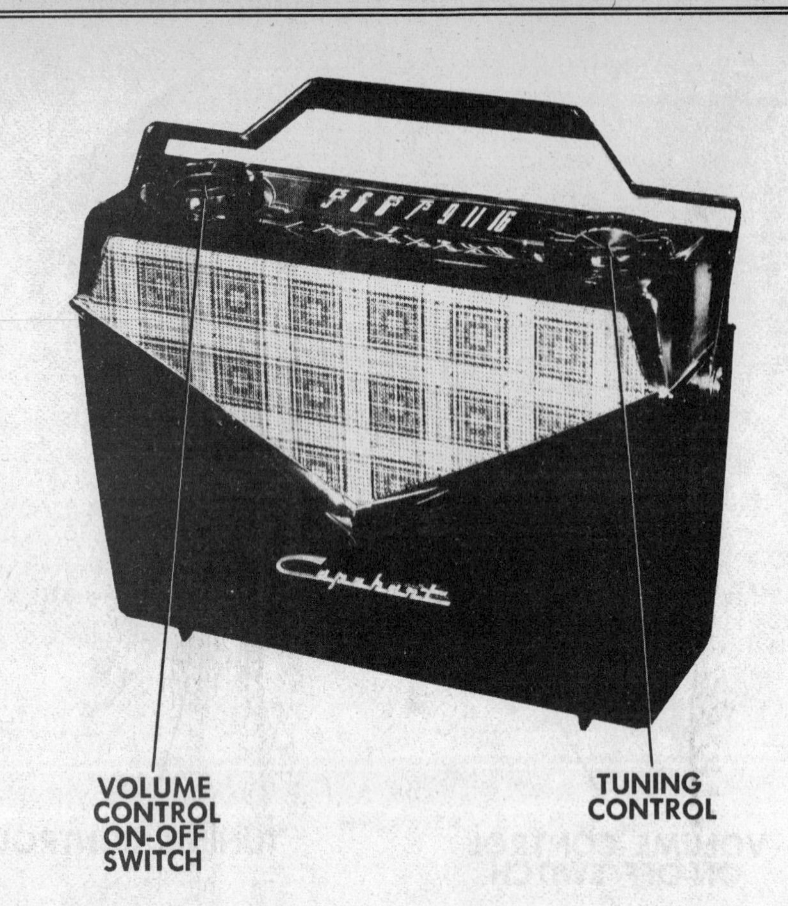

CAPEHART

MODEL PICTURED
RP153

AC operated AM-FM-phono superheterodyne receiver with three-speed auto record changer

TUBES
8

POWER SUPPLY
105-125 volts AC, 60 cycle

TUNING RANGE
537-1620KC, 88-108MC

MFR/SUPPLIER
Capehart-Farnsworth Corp.

PHOTOFACT SET
258-4

PUBLISHED
1954

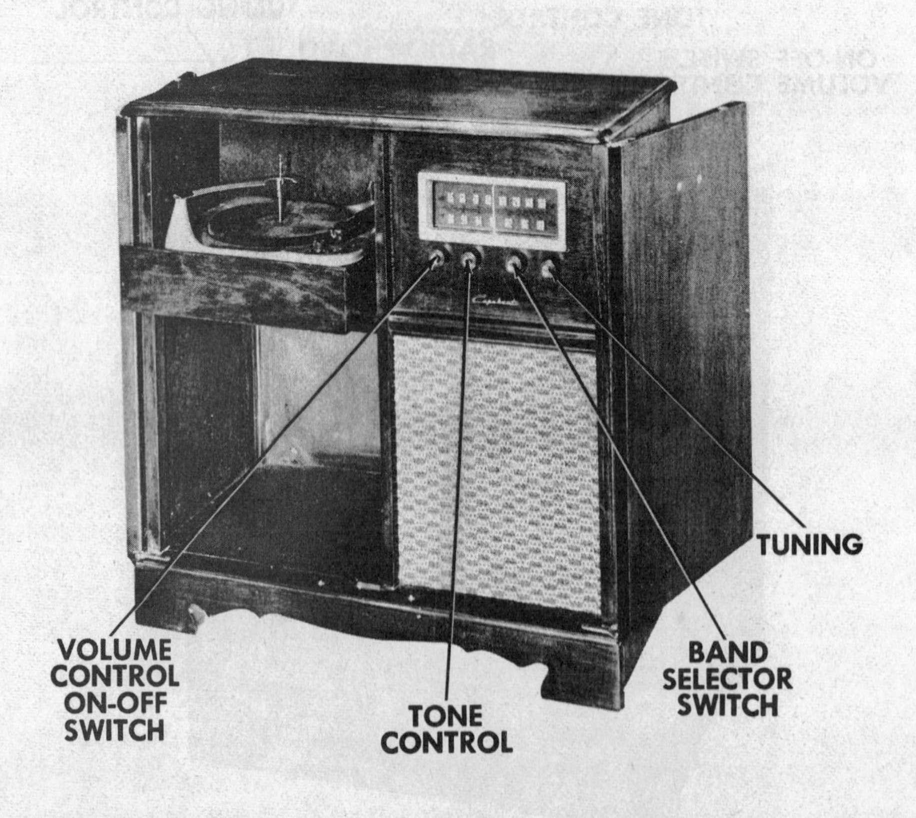

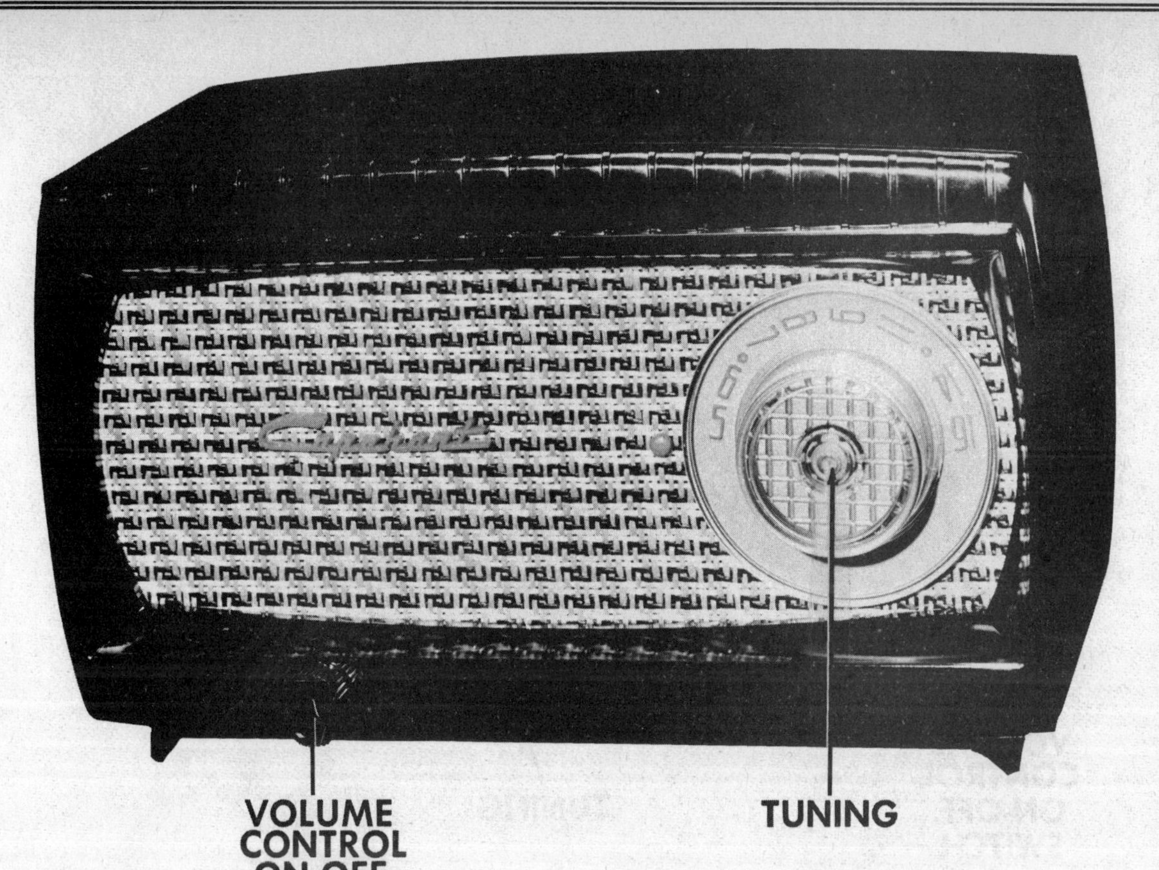

CAPEHART

MODEL PICTURED
2T55

AC/DC operated AM superheterodyne receiver

TUBES
5

POWER SUPPLY
105-125 volts AC/DC

TUNING RANGE
540-1620KC

MFR/SUPPLIER
Capehart-Farnsworth Corp.

PHOTOFACT SET
261-4

PUBLISHED
1954

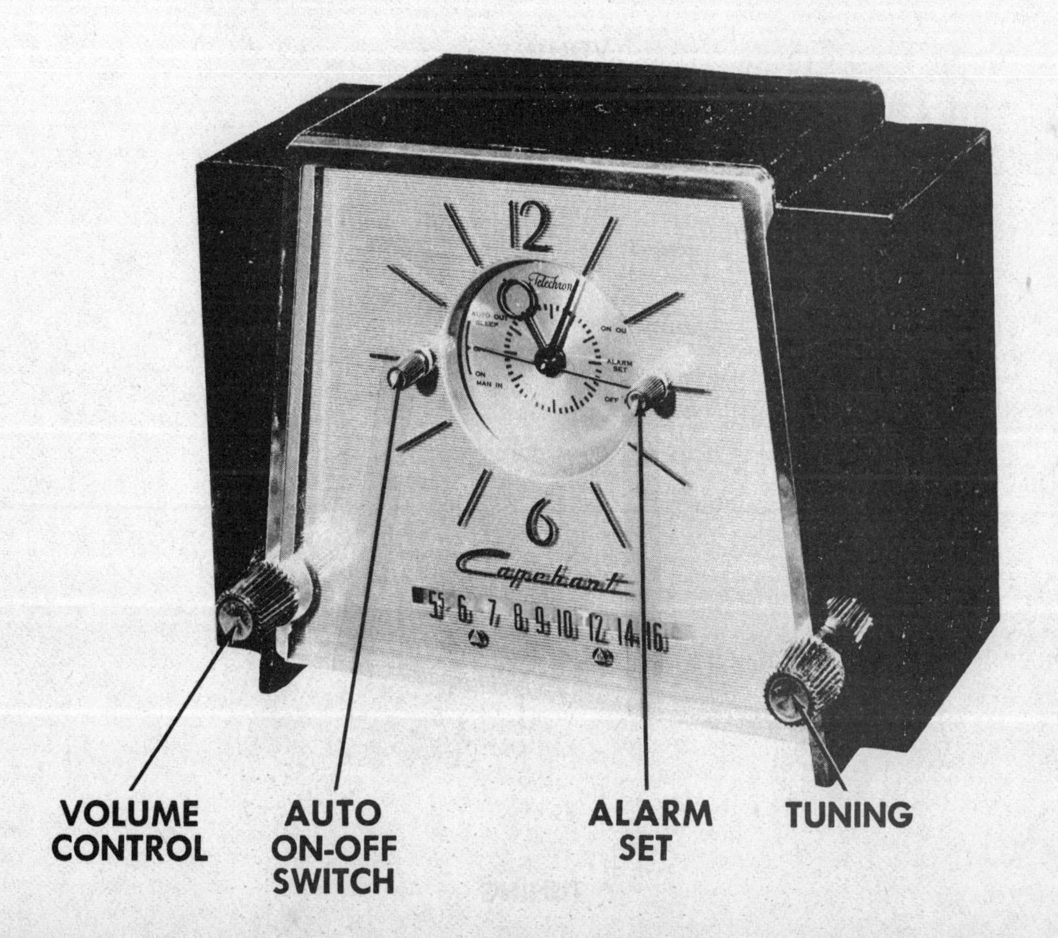

CAPEHART

MODEL PICTURED
C-14

AC operated AM superheterodyne receiver with electric clock

TUBES
4

POWER SUPPLY
105-120 volts AC, 60 cycle

TUNING RANGE
540-1620KC

MFR/SUPPLIER
Capehart-Farnsworth Corp.

PHOTOFACT SET
263-4

PUBLISHED
1955

CAPEHART

MODEL PICTURED
T-54

AC/DC operated AM superheterodyne receiver

TUBES
5

POWER SUPPLY
105-125 volts AC/DC

TUNING RANGE
540-1620KC

MFR/SUPPLIER
Capehart-Farnsworth Corp.

PHOTOFACT SET
265-3

PUBLISHED
1955

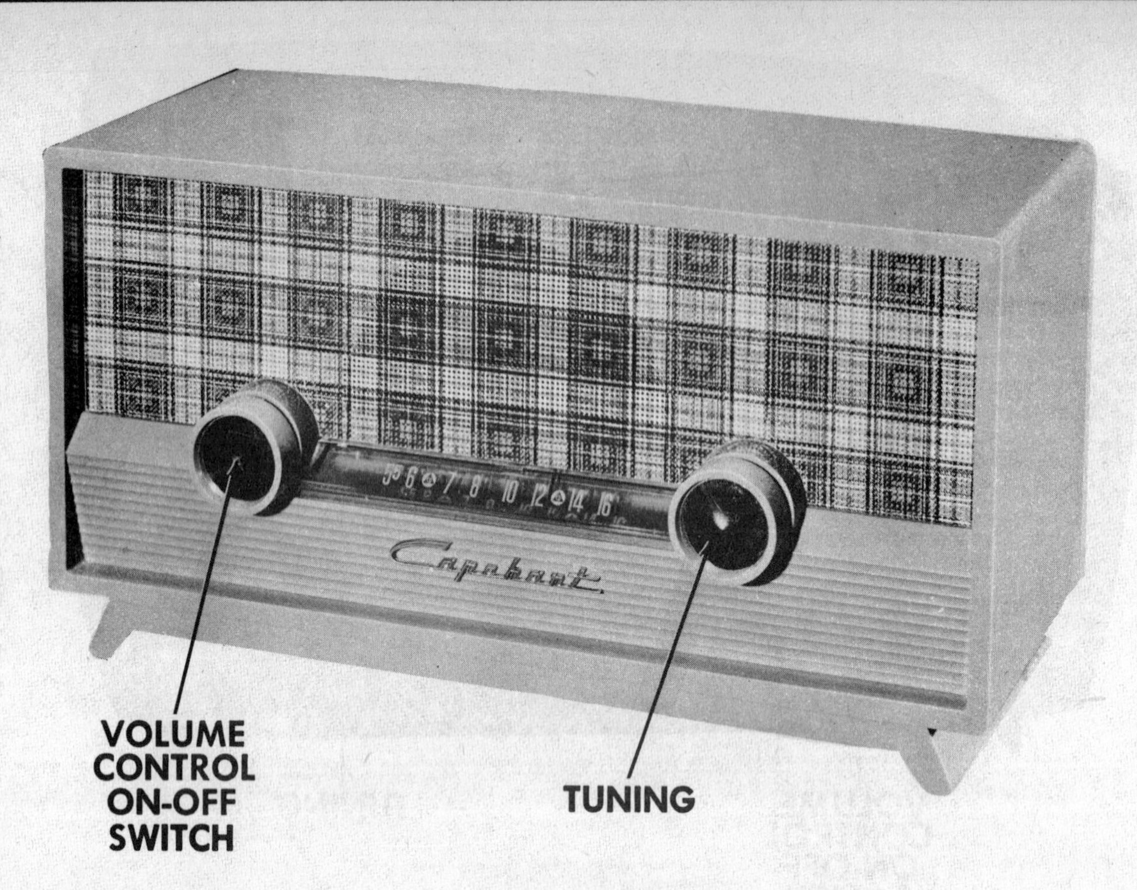

CAPEHART

MODEL PICTURED
1P55

Three-power portable AM superheterodyne receiver

TUBES
4

POWER SUPPLY
105-125 volts AC/DC or 3 volts A & 67.5 volts B supply

TUNING RANGE
540-1620KC

MFR/SUPPLIER
Capehart-Farnsworth Corp.

PHOTOFACT SET
268-4

PUBLISHED
1955

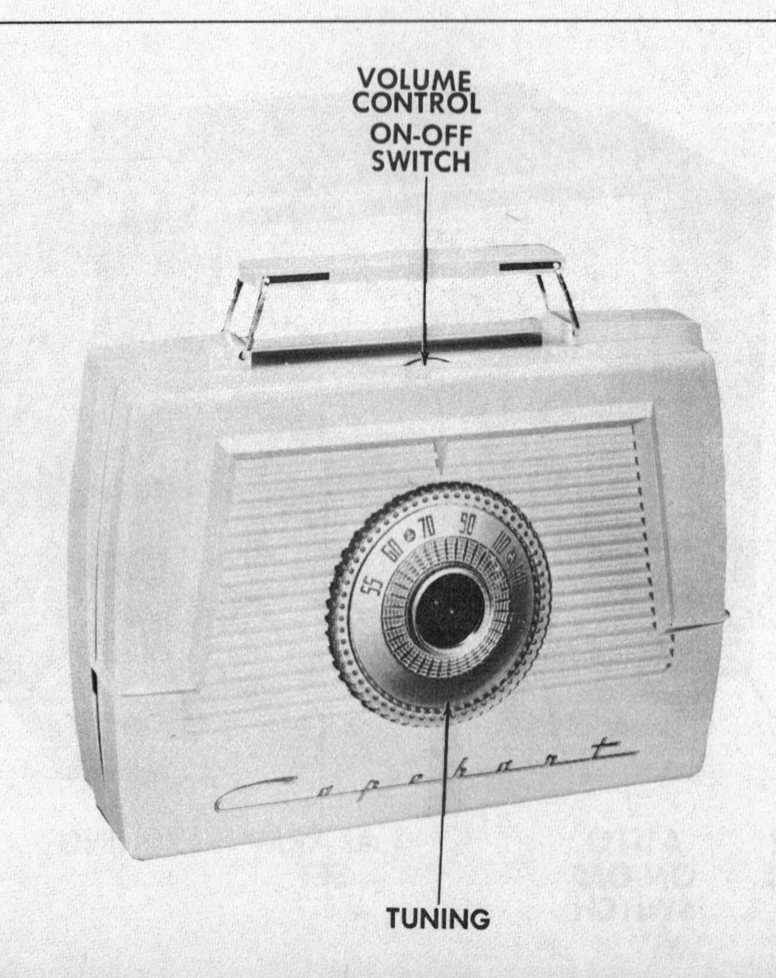

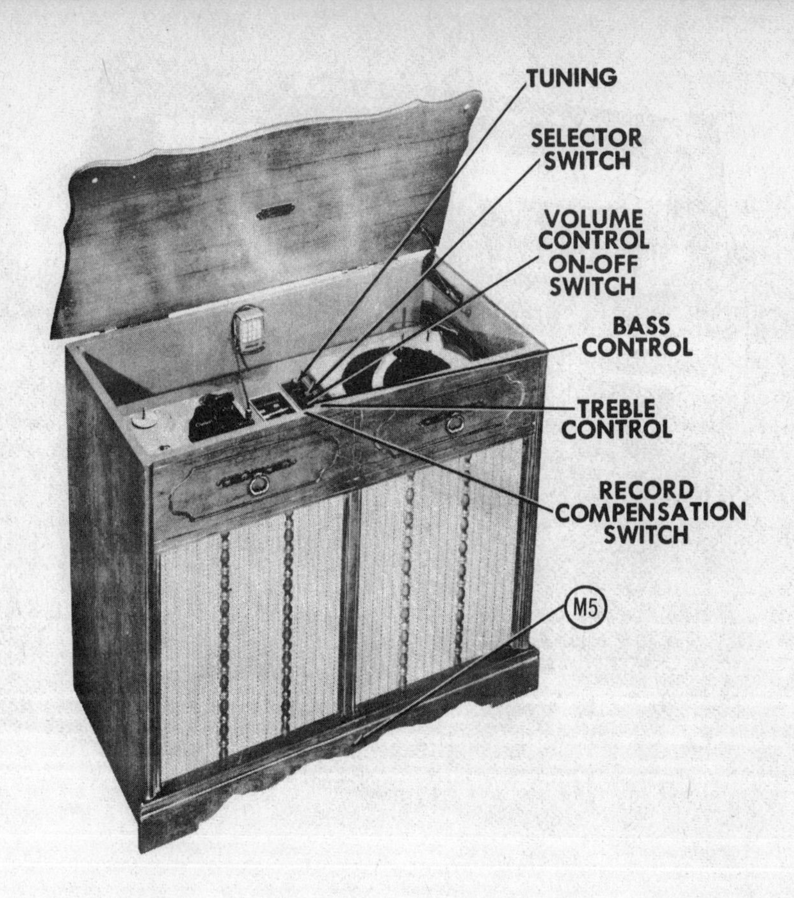

CAPEHART

MODEL PICTURED
17RPQ155F

AC operated AM-FM radio-phono and tape recorder

TUBES
11

POWER SUPPLY
105-120 volts AC, 60 cycle

TUNING RANGE
540-1620KC, 88-108MC

MFR/SUPPLIER
Capehart-Farnsworth Corp.

PHOTOFACT SET
294-4

PUBLISHED
1955

CAPEHART

MODEL PICTURED
2C56

AC operated AM superheterodyne receiver with electric clock

TUBES
5

POWER SUPPLY
105-125 volts AC, 60 cycle

TUNING RANGE
540-1620KC

MFR/SUPPLIER
Capehart-Farnsworth Corp.

PHOTOFACT SET
316-5

PUBLISHED
1956

CAPEHART

MODEL PICTURED
2P56

Three-power portable AM superheterodyne receiver

TUBES
4

POWER SUPPLY
105-125 volts AC/DC or 3 volts A supply & 67.5 volts B supply

TUNING RANGE
540KC-1620KC

MFR/SUPPLIER
Capehart-Farnsworth Corp.

PHOTOFACT SET
330-2

PUBLISHED
1956

CAPEHART

MODEL PICTURED
88P66BNL

Three-power portable multi-band superheterodyne receiver

TUBES
5

POWER SUPPLY
105-120 volts AC/DC or 9 volts A & 90 volts B supply in pack form

TUNING RANGE
Eight tuning bands

MFR/SUPPLIER
Capehart-Farnsworth Corp.

PHOTOFACT SET
331-3

PUBLISHED
1956

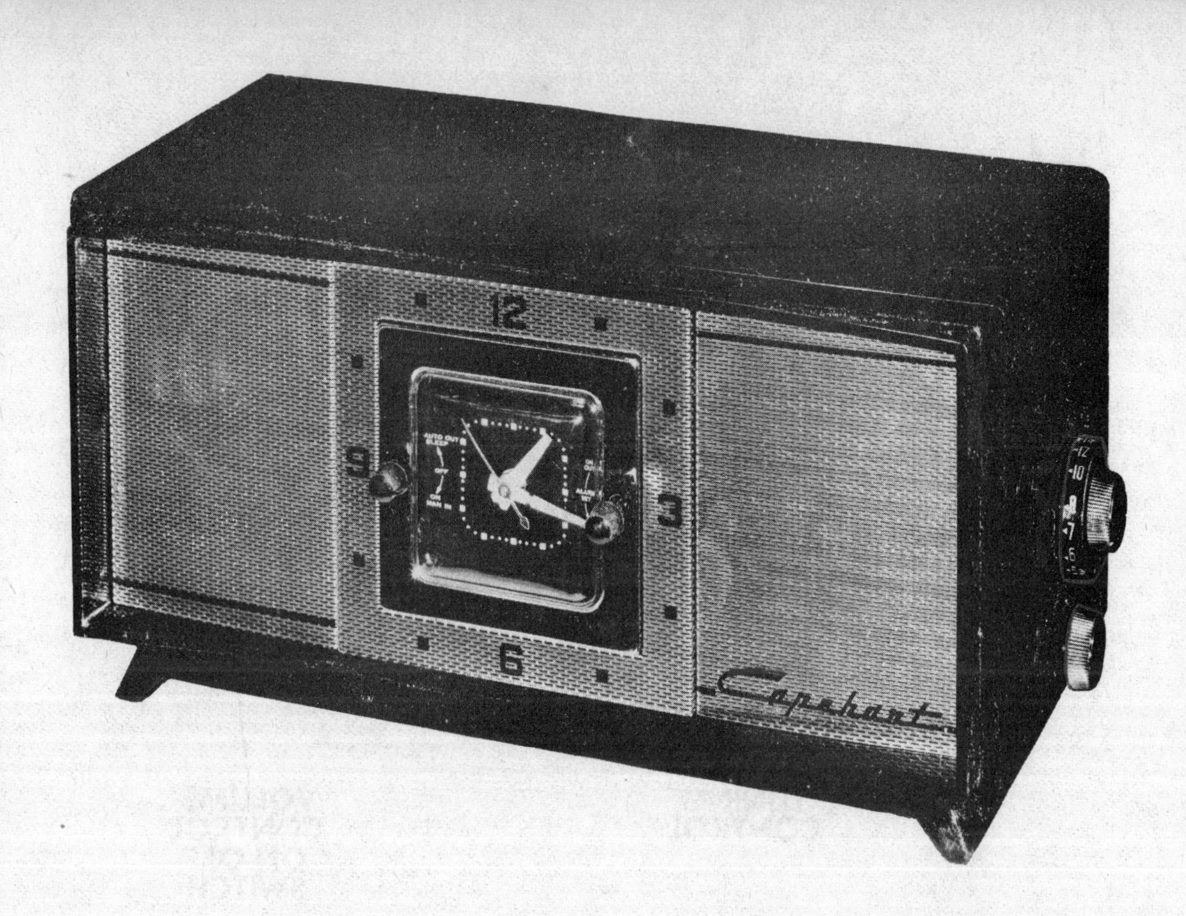

CAPEHART

MODEL PICTURED
75C56

AC operated AM superheterodyne receiver with electric clock

TUBES
5

POWER SUPPLY
105-125 volts AC, 60 cycle

TUNING RANGE
540-1620KC

MFR/SUPPLIER
Capehart-Farnsworth Corp.

PHOTOFACT SET
340-3

PUBLISHED
1956

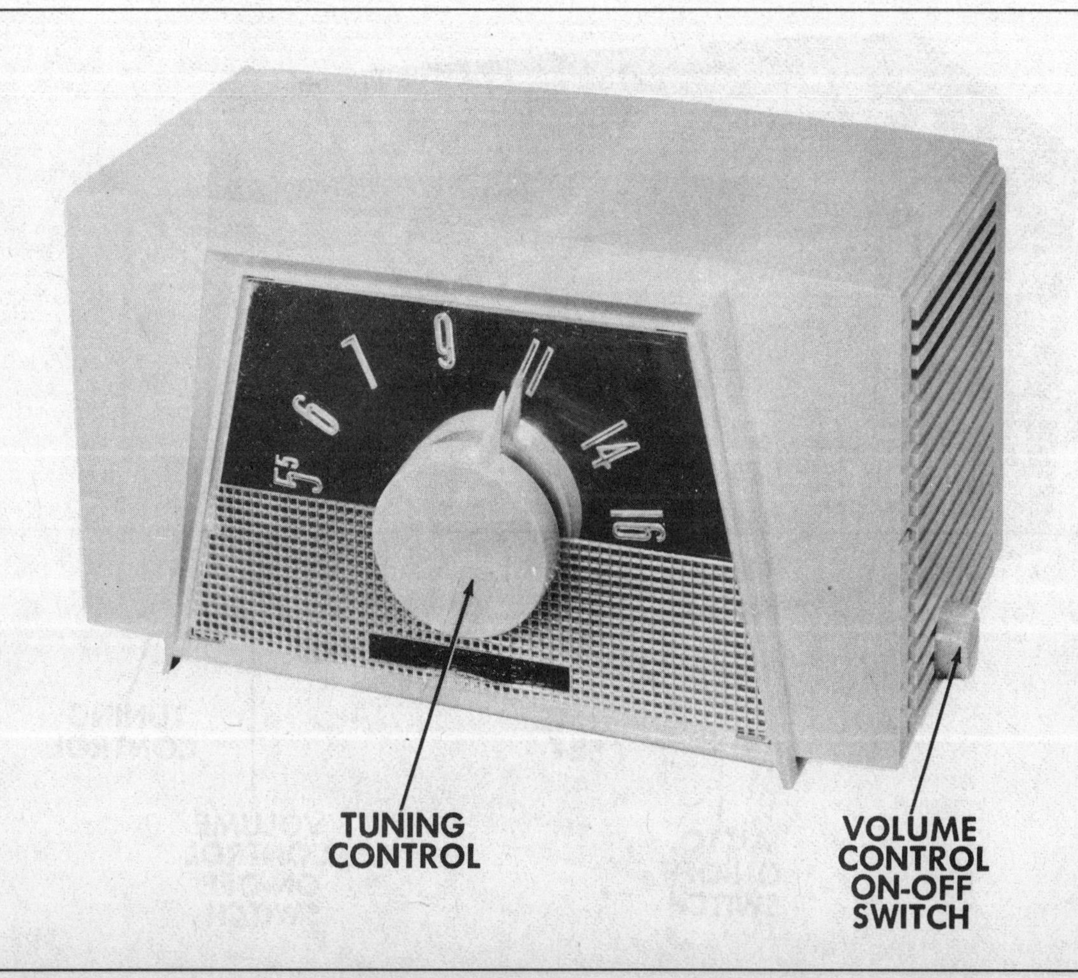

CAVALIER

MODEL PICTURED
5B1

AC/DC operated AM superheterodyne receiver

TUBES
5

POWER SUPPLY
110-120 volts AC/DC, .2 amp. @ 117 volts AC

TUNING RANGE
535-1620KC

MFR/SUPPLIER
Hinners-Galanek Radio Corp.

PHOTOFACT SET
238-6

PUBLISHED
1954

CAVALIER

MODEL PICTURED
5AT1

AC/DC operated AM superheterodyne receiver

TUBES
5

POWER SUPPLY
110-120 volts AC/DC, .26 amp @ 117 volts AC

TUNING RANGE
535-1620KC

MFR/SUPPLIER
Hinners-Galanek Radio Corp.

PHOTOFACT SET
241-4

PUBLISHED
1954

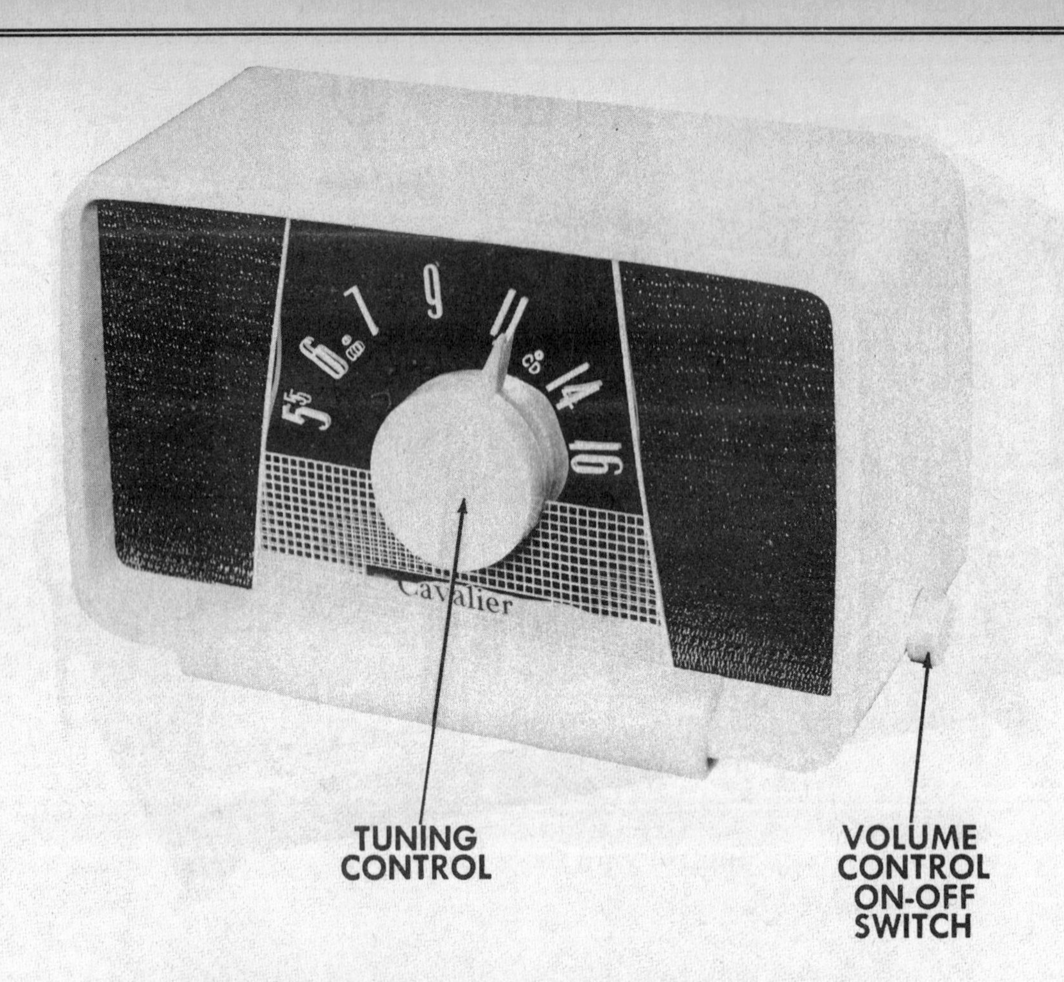

CAVALIER

MODEL PICTURED
5C1

AC operated AM superheterodyne receiver with electric clock

TUBES
5

POWER SUPPLY
110-120 volts AC, 60 cycles, .25 amp @ 117 volts AC

TUNING RANGE
535-1620KC

MFR/SUPPLIER
Hinners-Galanek Radio Corp.

PHOTOFACT SET
242-4

PUBLISHED
1954

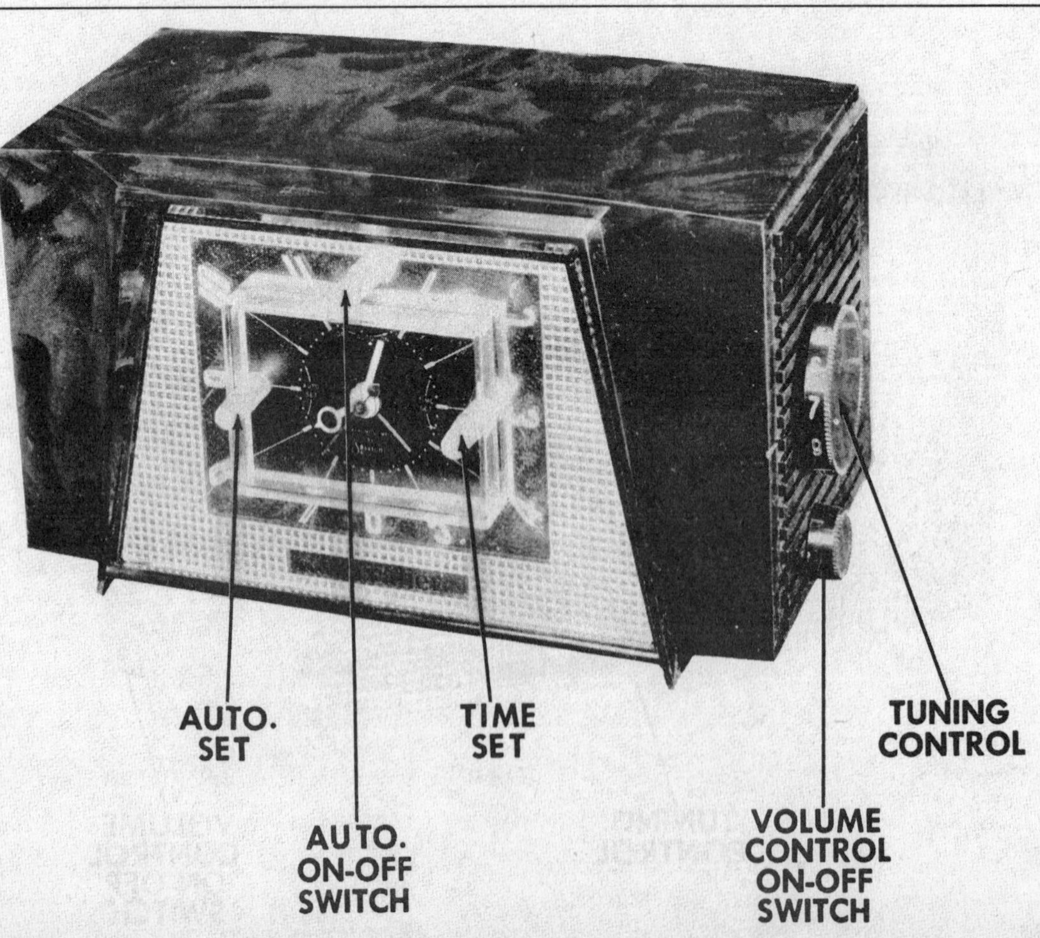

CAVALIER

MODEL PICTURED
5R1

AC/DC operated AM superheterodyne receiver

TUBES
5

POWER SUPPLY
110-120 volts AC/DC

TUNING RANGE
535-1640KC

MFR/SUPPLIER
Hinners-Galanek Radio Corp.

PHOTOFACT SET
265-4

PUBLISHED
1955

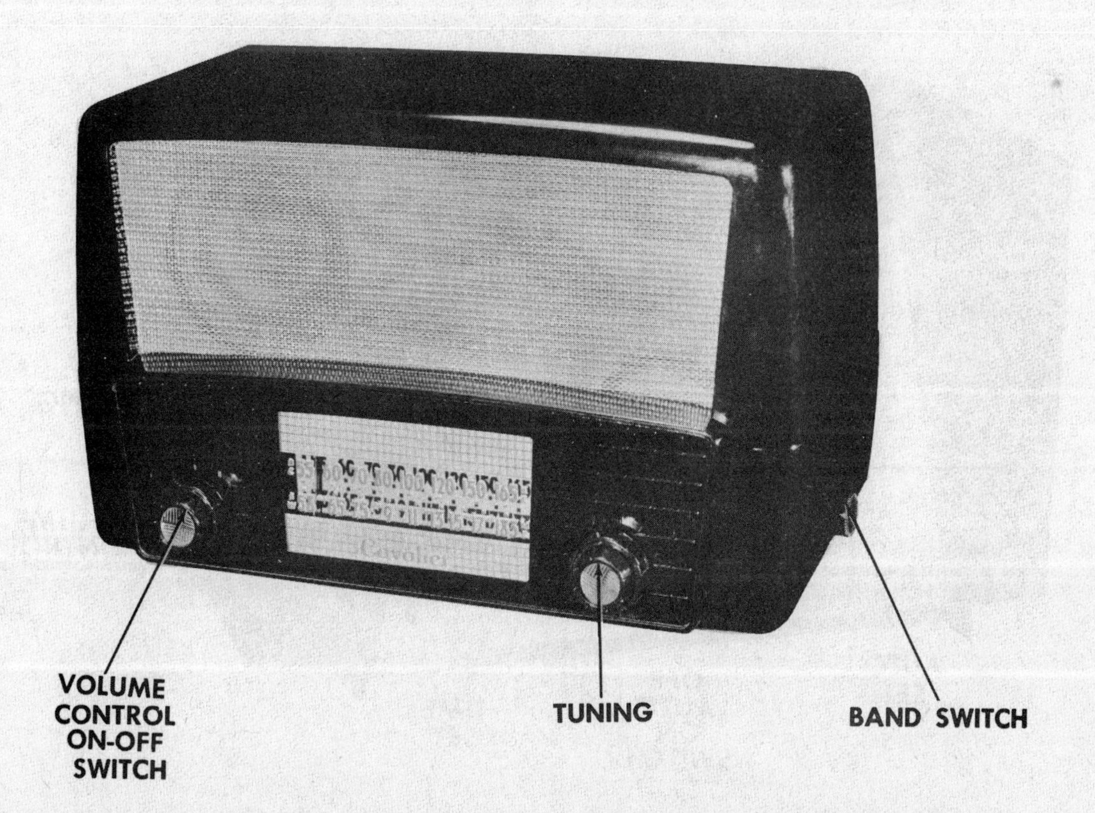

CAVALIER

MODEL PICTURED
6A2

AC/DC operated two-band AM superheterodyne receiver

TUBES
6

POWER SUPPLY
110-120 volts AC/DC

TUNING RANGE
540-1650KC, 5.8-18.3MC

MFR/SUPPLIER
Hinners-Galanek Radio Corp.

PHOTOFACT SET
265-5

PUBLISHED
1955

CAVALIER

MODEL PICTURED
4P3

Three-power portable AM superheterodyne receiver

TUBES
4

POWER SUPPLY
110-120 volts AC/DC or 7.5 volts A & 90 volts B supply in pack form

TUNING RANGE
535-1630KC

MFR/SUPPLIER
Hinners-Galanek Radio Corp.

PHOTOFACT SET
266-4

PUBLISHED
1955

CAVALIER

MODEL PICTURED
4CL4

AC operated AM superheterodyne receiver with electric clock

TUBES
4

POWER SUPPLY
110-120 volts AC, 60 cycle

TUNING RANGE
535-1620KC

MFR/SUPPLIER
Hinners-Galanek Radio Corp.

PHOTOFACT SET
273-4

PUBLISHED
1955

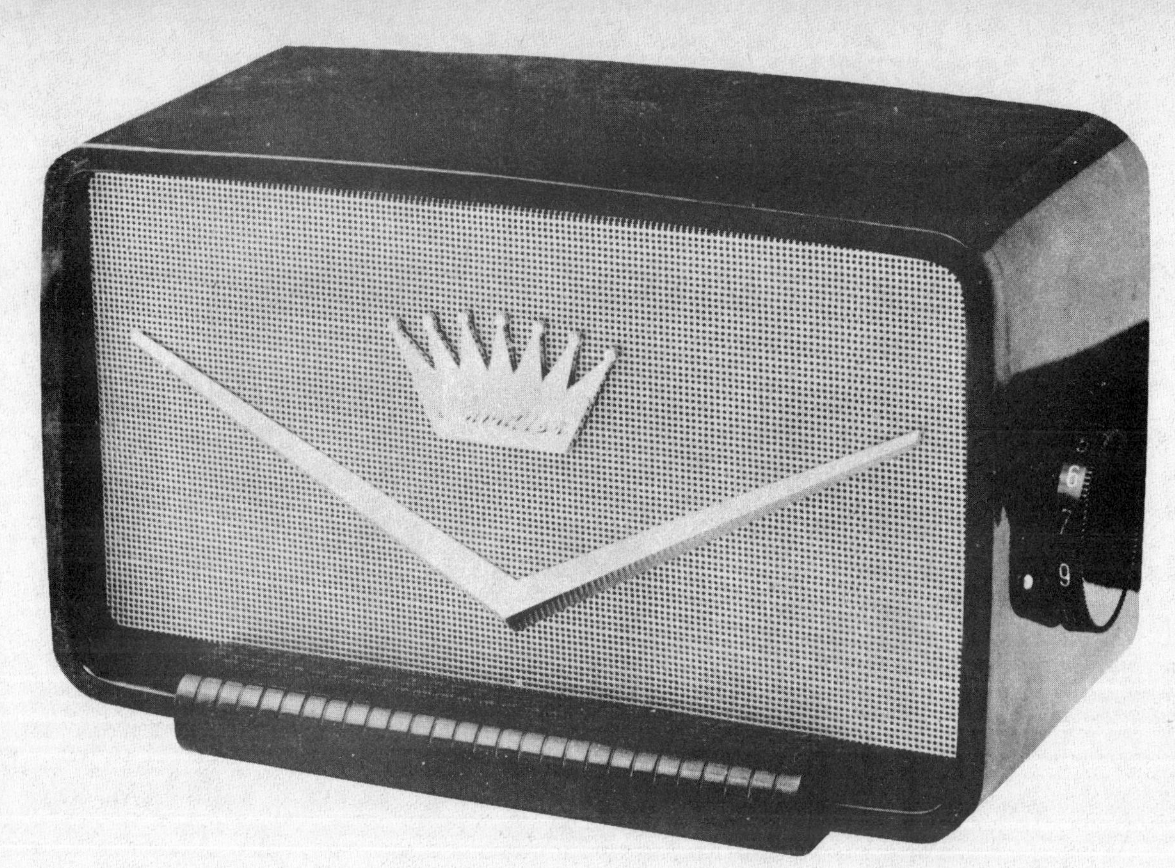

CAVALIER

MODEL PICTURED
603

AC/DC operated AM receiver

TUBES
6

POWER SUPPLY
105-125 volts AC/DC

TUNING RANGE
540-1650KC

MFR/SUPPLIER
Hinners-Galanek Radio Corp.

PHOTOFACT SET
353-2

PUBLISHED
1957

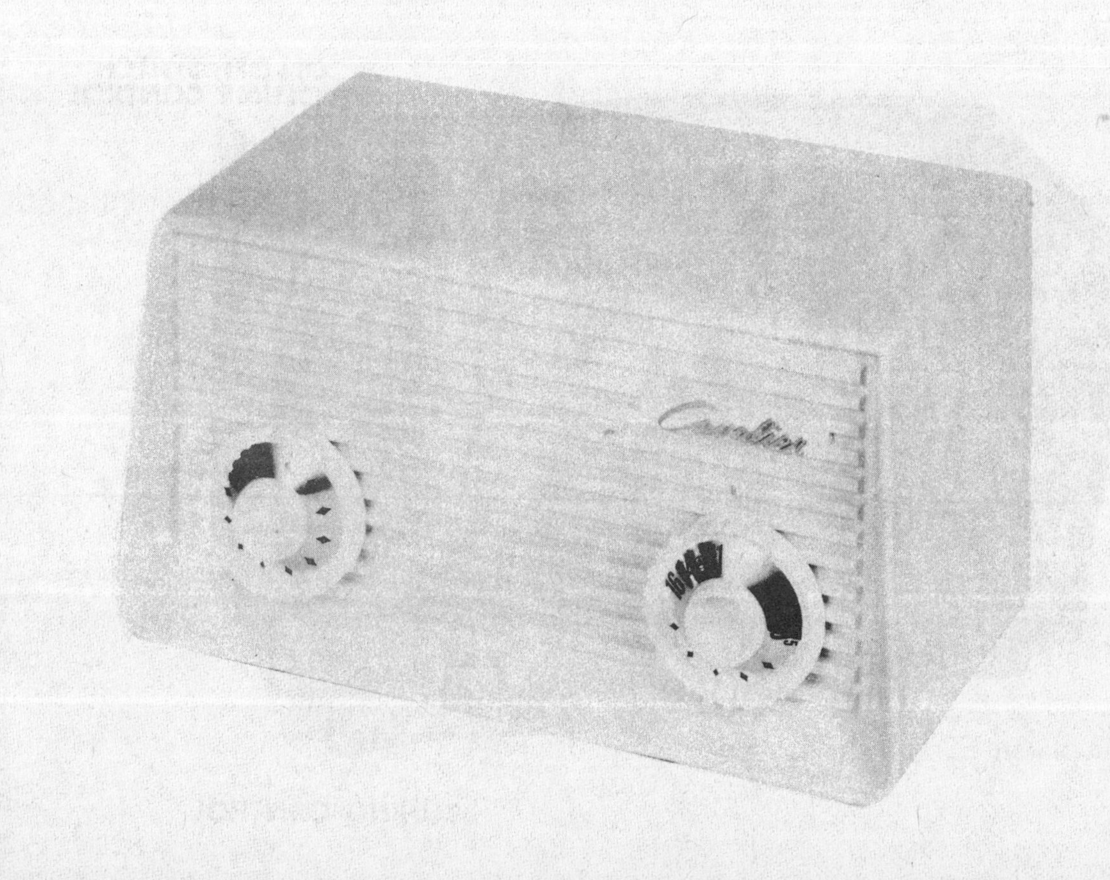

CAVALIER

MODEL PICTURED
562

AC/DC operated AM receiver

TUBES
5

POWER SUPPLY
105-125 volts AC/DC, 25 watts, .27 amp @ 117 volts AC

TUNING RANGE
540-1650KC

MFR/SUPPLIER
Hinners-Galanek Radio Corp.

PHOTOFACT SET
473-6

PUBLISHED
1960

CBS COLUMBIA

MODEL PICTURED
541

AC operated AM superheterodyne receiver with electric clock

TUBES
5

POWER SUPPLY
110-120 volts AC, 60 cycle, .27 amp @ 117 volts AC

TUNING RANGE
540-1620KC

MFR/SUPPLIER
CBS-Columbia, Inc.

PHOTOFACT SET
211-4

PUBLISHED
1953

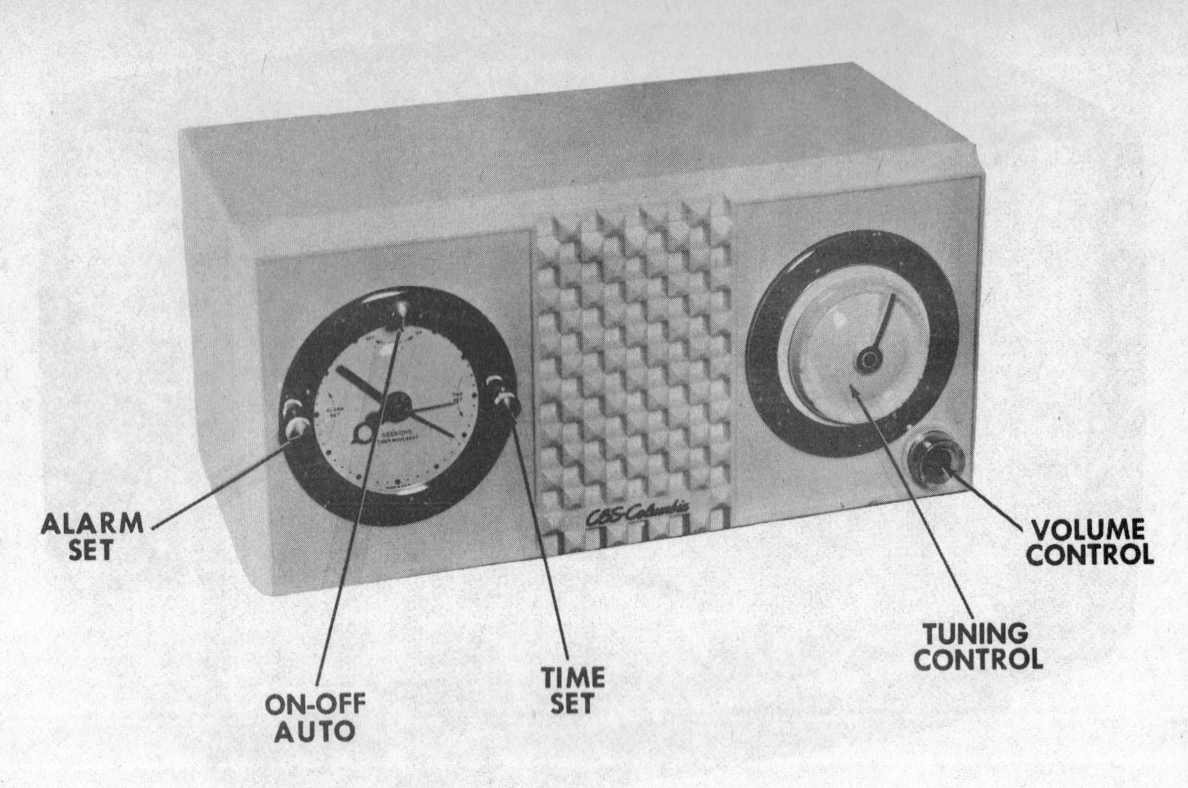

CBS COLUMBIA

MODEL PICTURED
525

Three-power portable superheterodyne receiver

TUBES
4

POWER SUPPLY
110-120 volts AC/DC or 7.5 volts A supply & 67.5 volts B supply

TUNING RANGE
535-1620KC

MFR/SUPPLIER
CBS-Columbia, Inc.

PHOTOFACT SET
222-4

PUBLISHED
1953

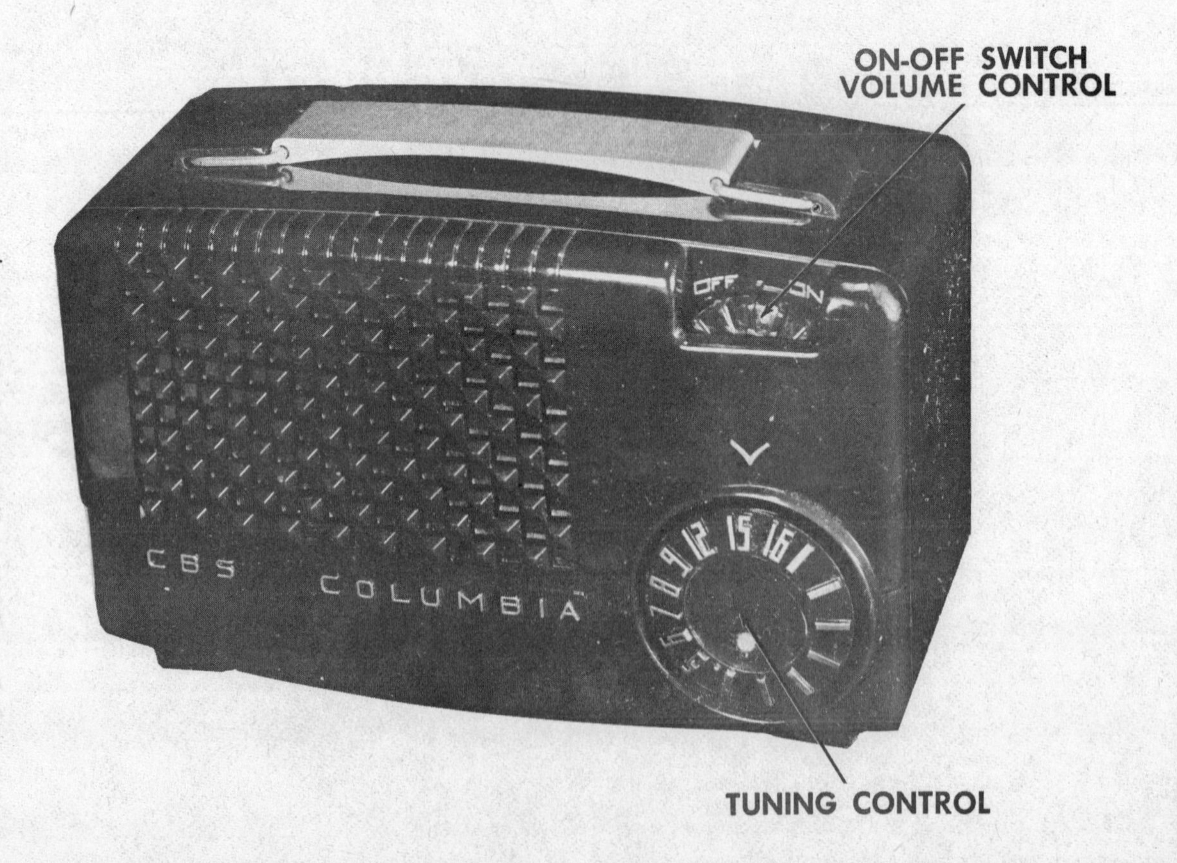

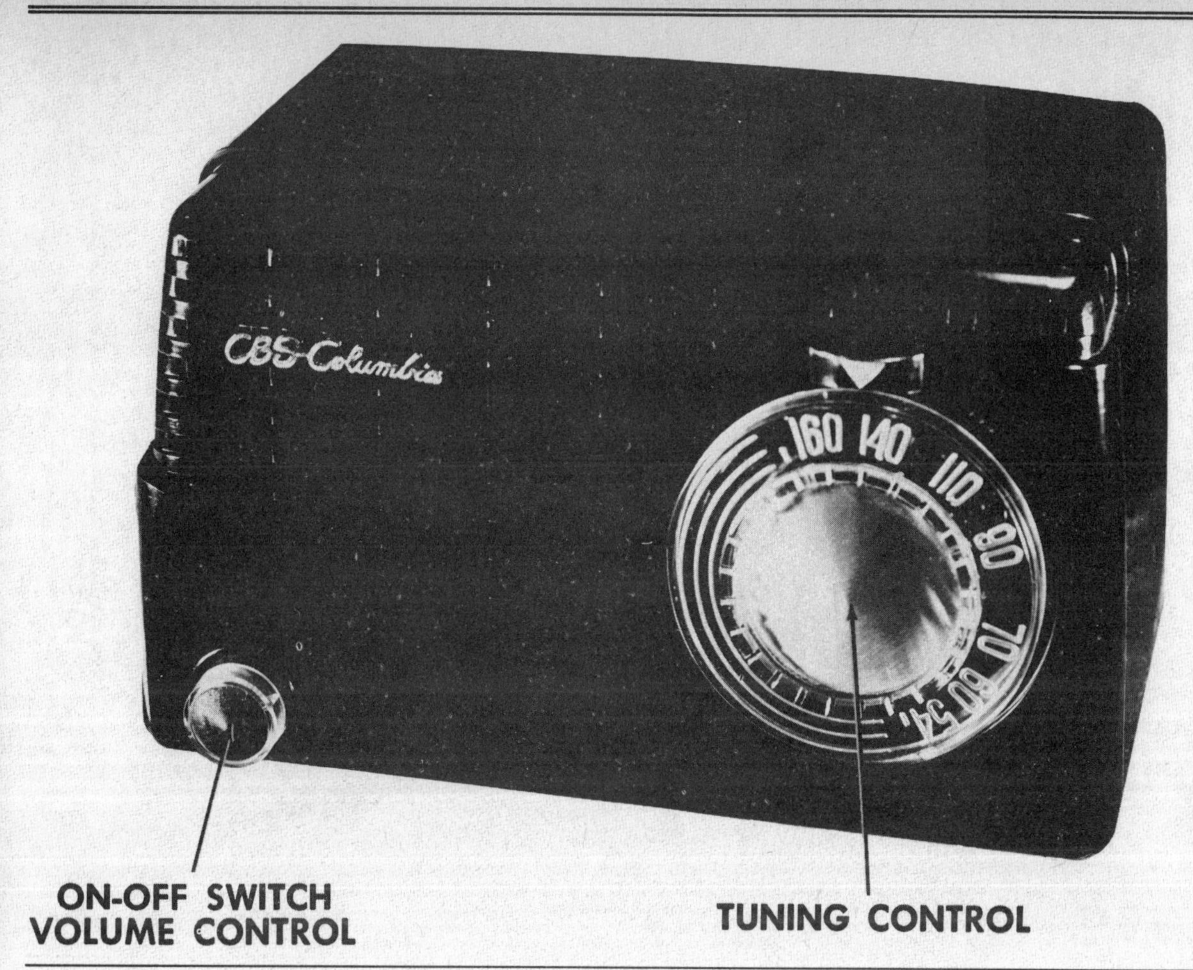

CBS COLUMBIA

MODEL PICTURED
515A

AC/DC operated AM superheterodyne receiver

TUBES
5

POWER SUPPLY
110-120 volts AC/DC, .22 amp @ 117 volts AC

TUNING RANGE
535-1620KC

MFR/SUPPLIER
CBS-Columbia, Inc.

PHOTOFACT SET
223-4

PUBLISHED
1953

CBS COLUMBIA

MODEL PICTURED
5220

Three-power portable AM superheterodyne receiver

TUBES
4

POWER SUPPLY
110-120 volts AC/DC or 7.5 volts A supply & 67.5 volts B supply

TUNING RANGE
535-1620KC

MFR/SUPPLIER
CBS-Columbia, Inc.

PHOTOFACT SET
261-5

PUBLISHED
1954

CBS COLUMBIA

MODEL PICTURED
C220

AC operated AM superheterodyne receiver with electric clock

TUBES
5

POWER SUPPLY
110-120 volts AC, 60 cycles

TUNING RANGE
535KC-1620KC

MFR/SUPPLIER
CBS-Columbia, Inc.

PHOTOFACT SET
337-1

PUBLISHED
1956

CBS COLUMBIA

MODEL PICTURED
T200

AC/DC operated AM superheterodyne receiver

TUBES
5

POWER SUPPLY
105-120 volts AC/DC, .27 amp @ 117 volts AC

TUNING RANGE
535KC-1620KC

MFR/SUPPLIER
CBS-Columbia, Inc.

PHOTOFACT SET
338-1

PUBLISHED
1956

CBS COLUMBIA

MODEL PICTURED
5440

AC operated AM superheterodyne receiver with electric clock

TUBES
5

POWER SUPPLY
110-120 volts AC, 60 cycle

TUNING RANGE
535-1620KC

MFR/SUPPLIER
CBS-Columbia, Inc.

PHOTOFACT SET
340-4

PUBLISHED
1956

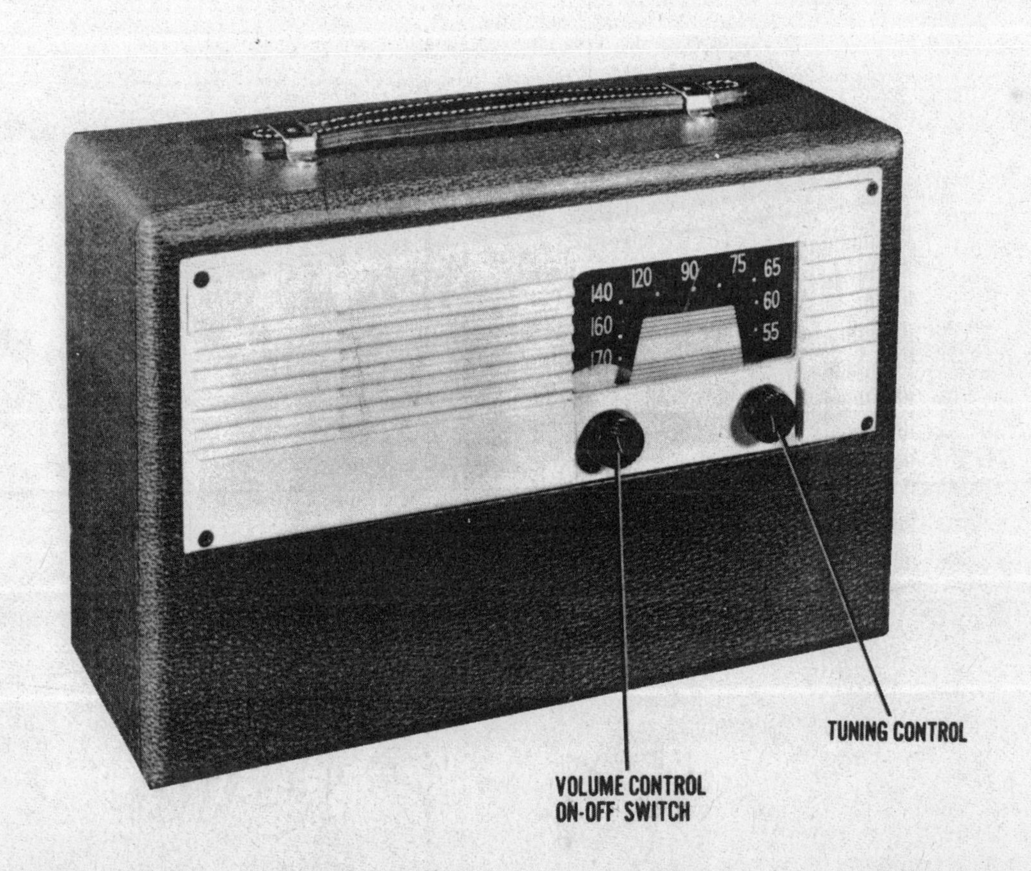

CHANCELLOR

MODEL PICTURED
35P

Three-power portable superheterodyne receiver with loop antenna

TUBES
4

POWER SUPPLY
110-120 volts AC/DC or 7.5 volts A & 90 volts B supply

TUNING RANGE
540-1600KC

MFR/SUPPLIER
Radionic Equipment Corp.

PHOTOFACT SET
30-25

PUBLISHED
1947

CHANNEL MASTER

MODEL PICTURED
6501

Battery operated portable transistorized AM receiver

TUBES
0

POWER SUPPLY
9 Volts DC

TUNING RANGE
540-1620KC

MFR/SUPPLIER
Channel Master Corp.

PHOTOFACT SET
462-5

PUBLISHED
1959

CHANNEL MASTER

MODEL PICTURED
6506

Battery operated portable transistorized AM receiver

TUBES
0

POWER SUPPLY
6 volts DC

TUNING RANGE
535-1620KC

MFR/SUPPLIER
Channel Master Corp.

PHOTOFACT SET
470-6

PUBLISHED
1960

CHANNEL MASTER

MODEL PICTURED
6512

Battery operated transistorized two-band portable receiver

TUBES
0

POWER SUPPLY
6 volts DC

TUNING RANGE
530-1650KC,
3.8MC-12.5MC

MFR/SUPPLIER
Channel Master Corp.

PHOTOFACT SET
485-6

PUBLISHED
1960

CHANNEL MASTER

MODEL PICTURED
6503

Battery operated transistorized portable AM receiver

TUBES
0

POWER SUPPLY
9 volts DC

TUNING RANGE
530-1650KC

MFR/SUPPLIER
Channel Master Corp.

PHOTOFACT SET
495-6

PUBLISHED
1960

CISCO

MODEL PICTURED
9A5

AC/DC operated superheterodyne receiver with self-contained loop antenna

TUBES
5

POWER SUPPLY
110-120 volts AC/DC

TUNING RANGE
540-1720KC

MFR/SUPPLIER
Cities Service Oil Co.

PHOTOFACT SET
20-3

PUBLISHED
1947

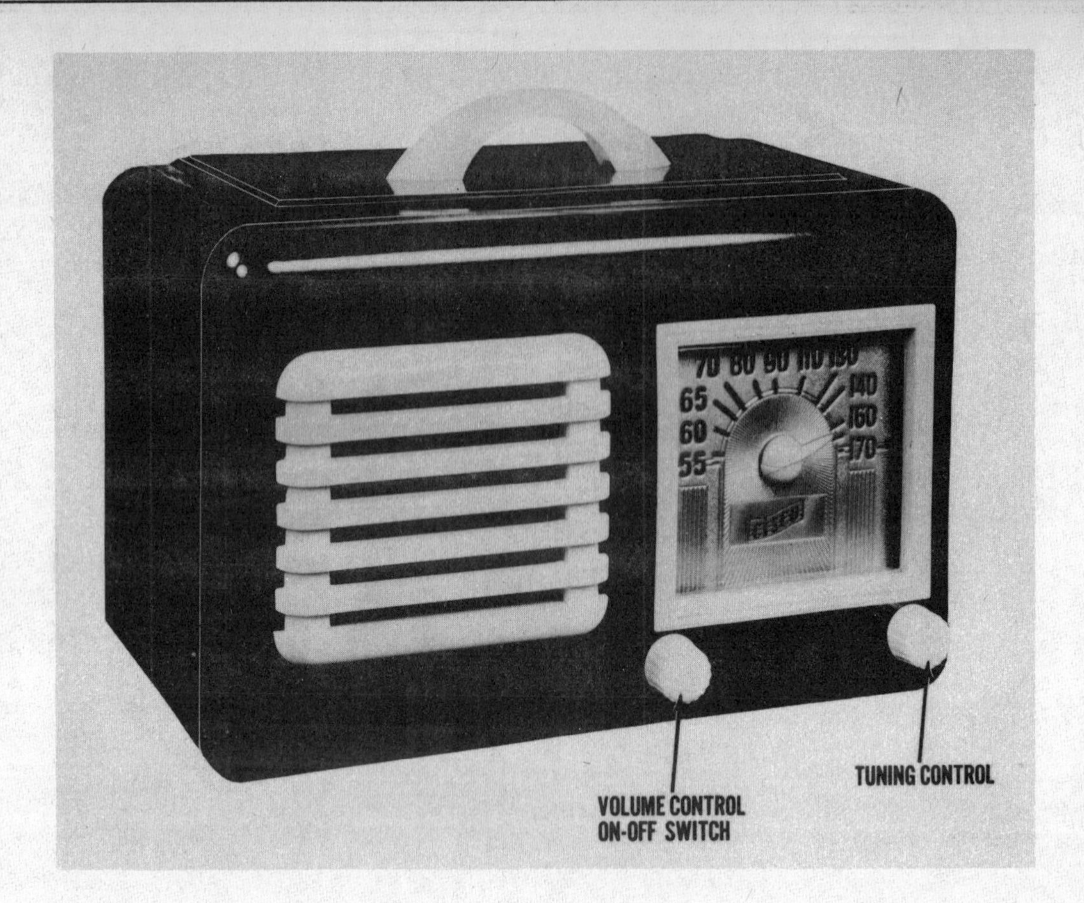

CISCO

MODEL PICTURED
1A5

AC/DC operated superheterodyne receiver with loop antenna

TUBES
5

POWER SUPPLY
117 volts AC/DC

TUNING RANGE
540-1700KC

MFR/SUPPLIER
Cities Service Oil Co.

PHOTOFACT SET
37-4

PUBLISHED
1948

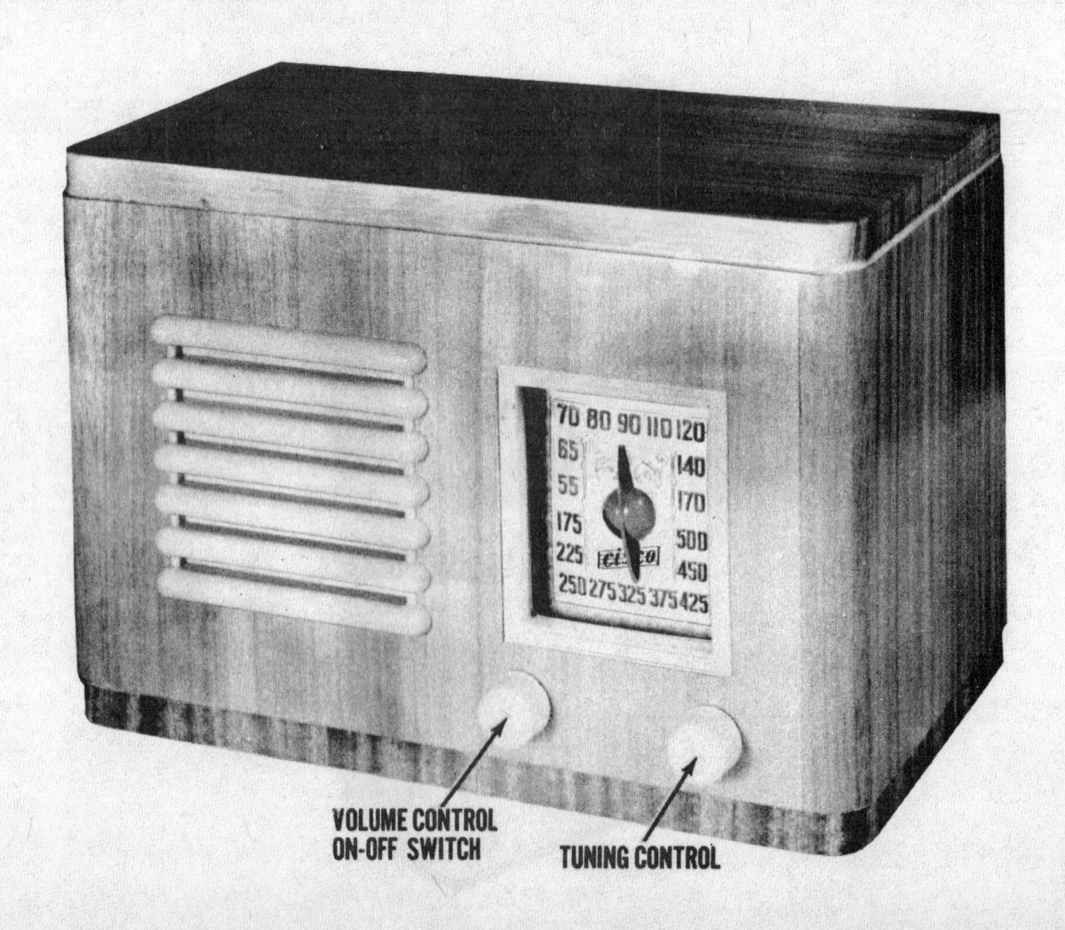

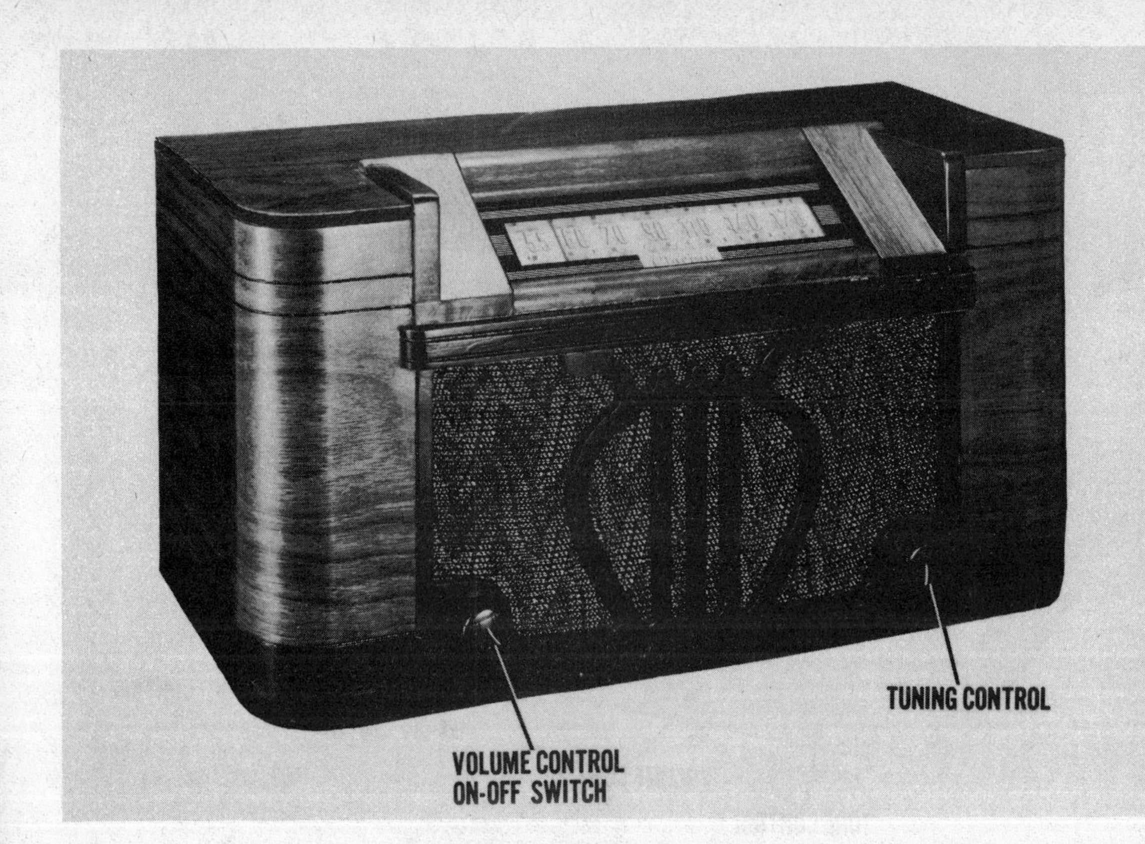

CLARION

MODEL PICTURED
C108

Battery operated superheterodyne receiver

TUBES
4

POWER SUPPLY
1.5 volts A battery & 90 volts B battery

TUNING RANGE
535-1725KC

MFR/SUPPLIER
Warwick Mfg. Corp.

PHOTOFACT SET
5-8

PUBLISHED
1946

CLARION

MODEL PICTURED
C101

AC operated combo auto phono-superheterodyne receiver, self-contained loop antenna

TUBES
5

POWER SUPPLY
117 volts AC

TUNING RANGE
540-1630KC

MFR/SUPPLIER
Warwick Mfg. Corp.

PHOTOFACT SET
5-9

PUBLISHED
1946

CLARION

MODEL PICTURED
C-103

AC operated superheterodyne receiver with self-contained loop antenna

TUBES
6

POWER SUPPLY
117 volts AC, .47 amp @ 117 volts AC

TUNING RANGE
535-1725KC

MFR/SUPPLIER
Warwick Mfg. Corp.

PHOTOFACT SET
6-6

PUBLISHED
1946

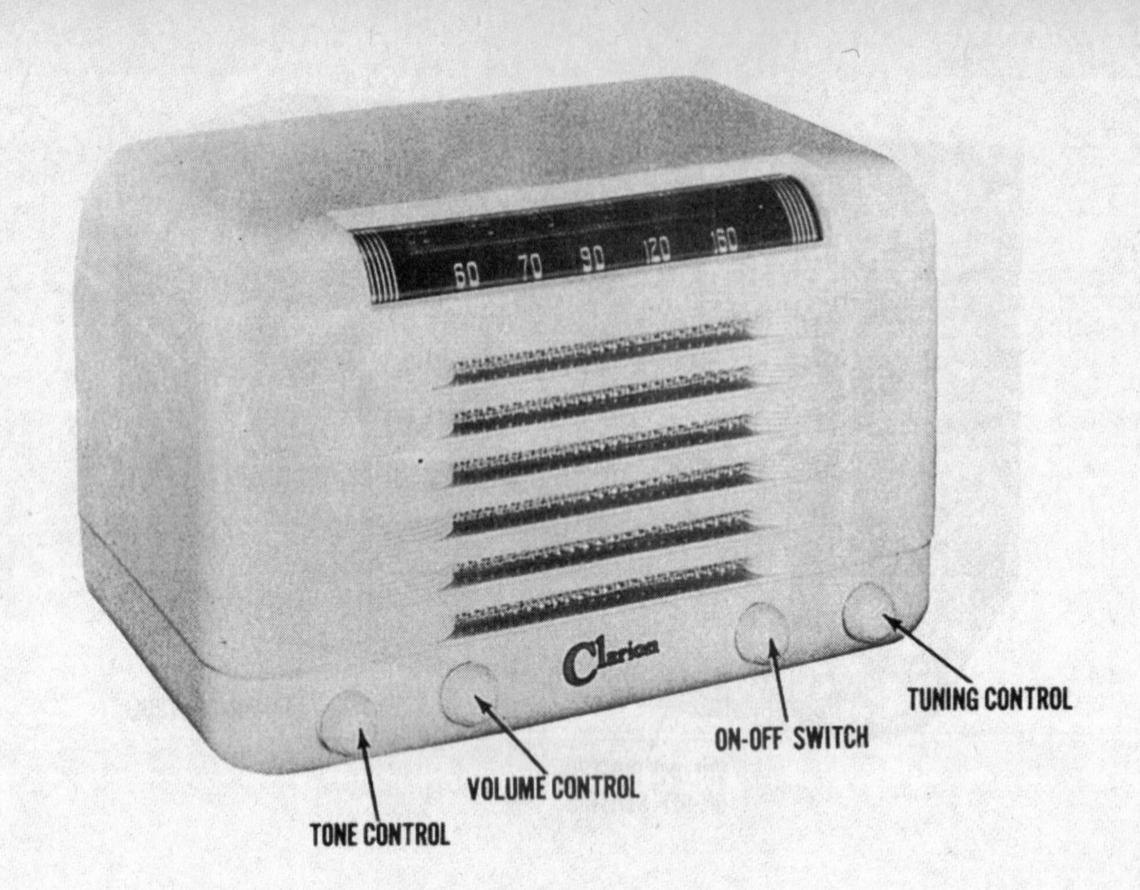

CLARION

MODEL PICTURED
C105-A

AC operated combo auto phono-superheterodyne receiver, self-contained loop antenna

TUBES
6

POWER SUPPLY
117 volts AC

TUNING RANGE
535-1725KC

MFR/SUPPLIER
Warwick Mfg. Corp.

PHOTOFACT SET
6-7

PUBLISHED
1946

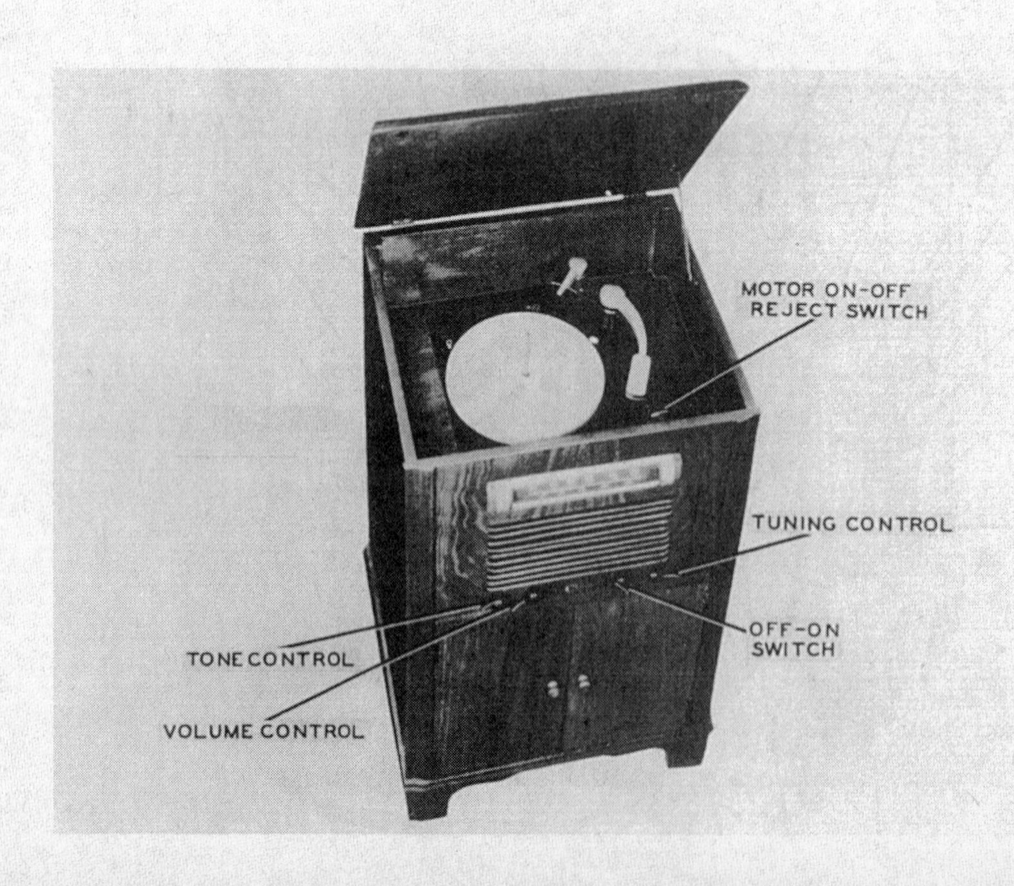

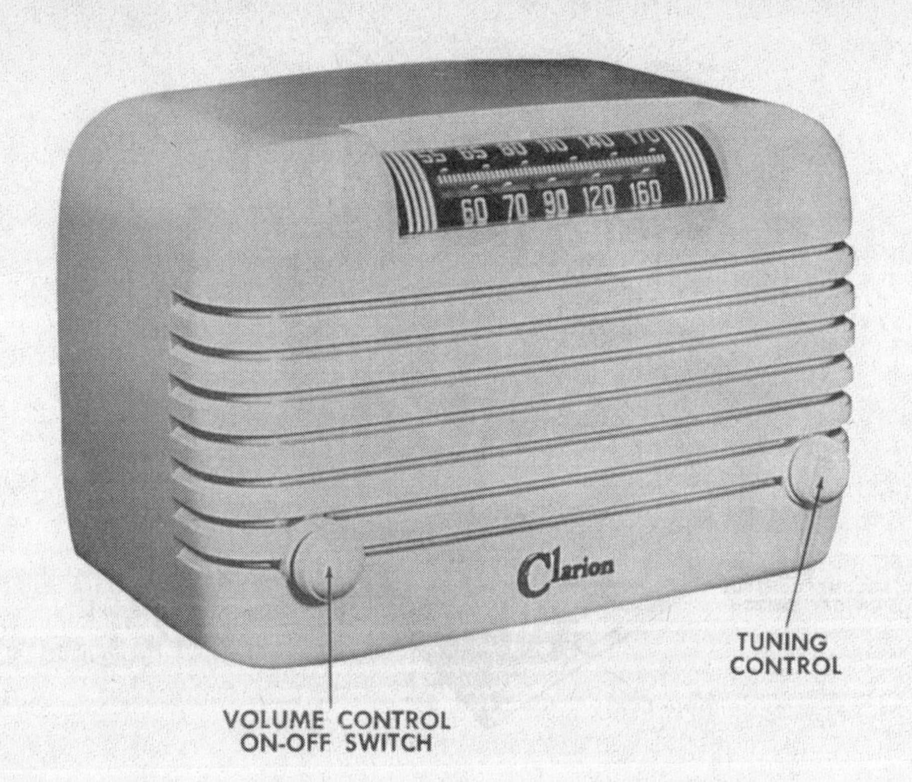

CLARION

MODEL PICTURED
C102

AC/DC operated superheterodyne receiver, self-contained loop antenna

TUBES
5

POWER SUPPLY
117 volts AC/DC, .260 amp @ 117 volts AC

TUNING RANGE
535-1725K

MFR/SUPPLIER
Warwick Mfg. Corp.

PHOTOFACT SET
9-6

PUBLISHED
1946

CLARION

MODEL PICTURED
11011

Three-power portable superheterodyne receiver, self-contained loop antenna

TUBES
4

POWER SUPPLY
110-120 volts AC/DC or 7.5 volts A battery & 90 volts B battery

TUNING RANGE
535-1600KC

MFR/SUPPLIER
Warwick Mfg. Corp.

PHOTOFACT SET
17-8

PUBLISHED
1947

CLARION

MODEL PICTURED
11305

AC operated phono-radio combo superheterodyne receiver with self-contained loop antenna

TUBES
4

POWER SUPPLY
110-120 volts AC

TUNING RANGE
535-1725KC

MFR/SUPPLIER
Warwick Mfg. Corp.

PHOTOFACT SET
18-11

PUBLISHED
1947

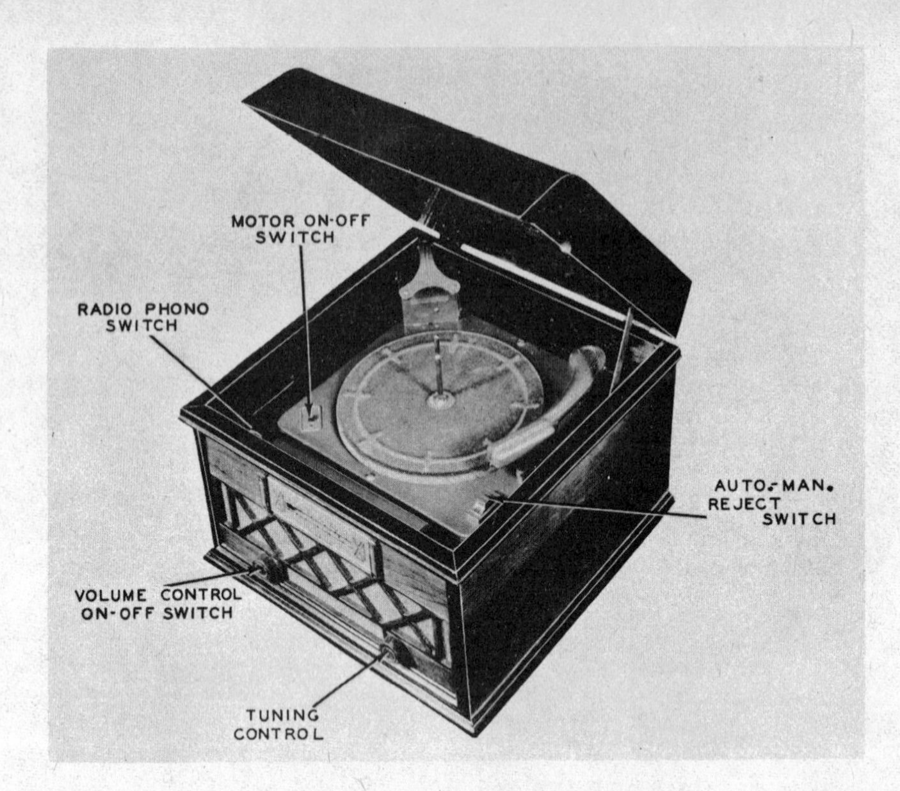

CLARION

MODEL PICTURED
11801

AC/DC superheterodyne receiver with loop antenna

TUBES
5

POWER SUPPLY
110-120 volts AC/DC

TUNING RANGE
540-1630KC

MFR/SUPPLIER
Warwick Mfg. Corp.

PHOTOFACT SET
23-6

PUBLISHED
1947

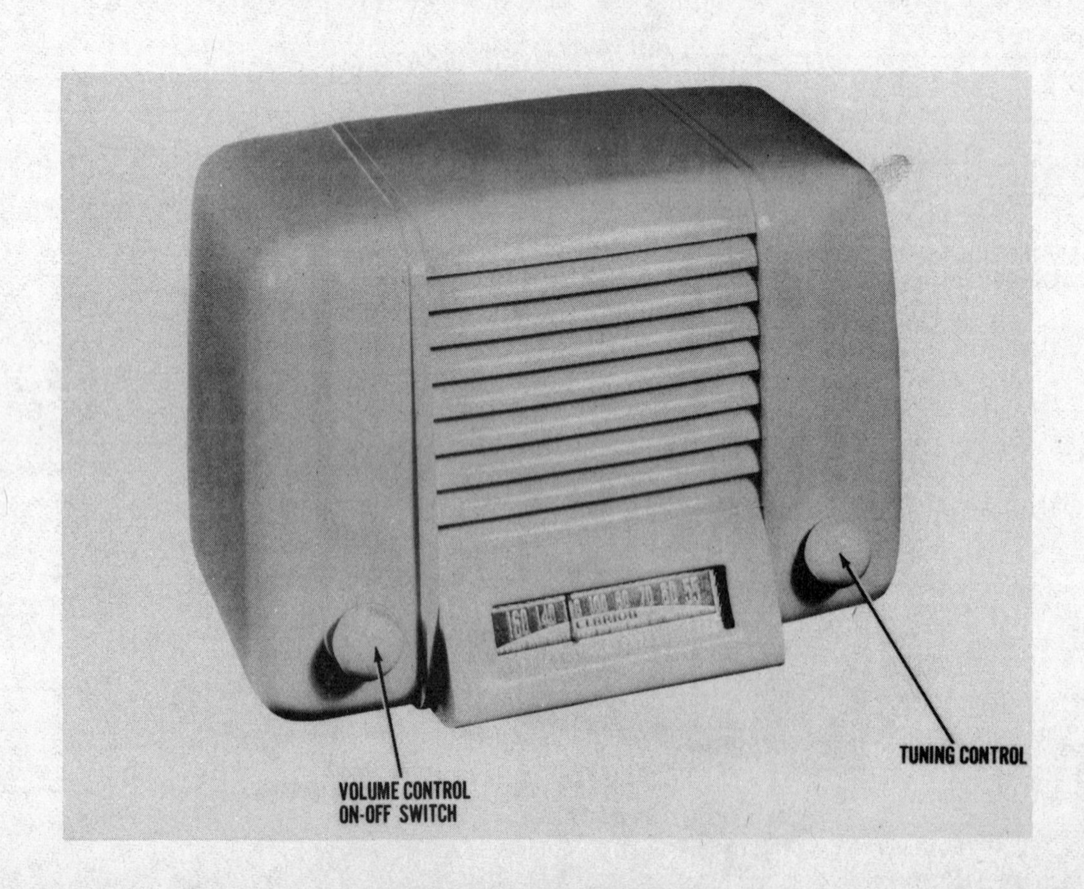

CLARION

MODEL PICTURED
11411-N

Three-power portable superheterodyne receiver with loop antenna

TUBES
4

POWER SUPPLY
110-120 volts AC

TUNING RANGE
545-1600KC

MFR/SUPPLIER
Warwick Mfg. Corp.

PHOTOFACT SET
30-5

PUBLISHED
1947

CLARION

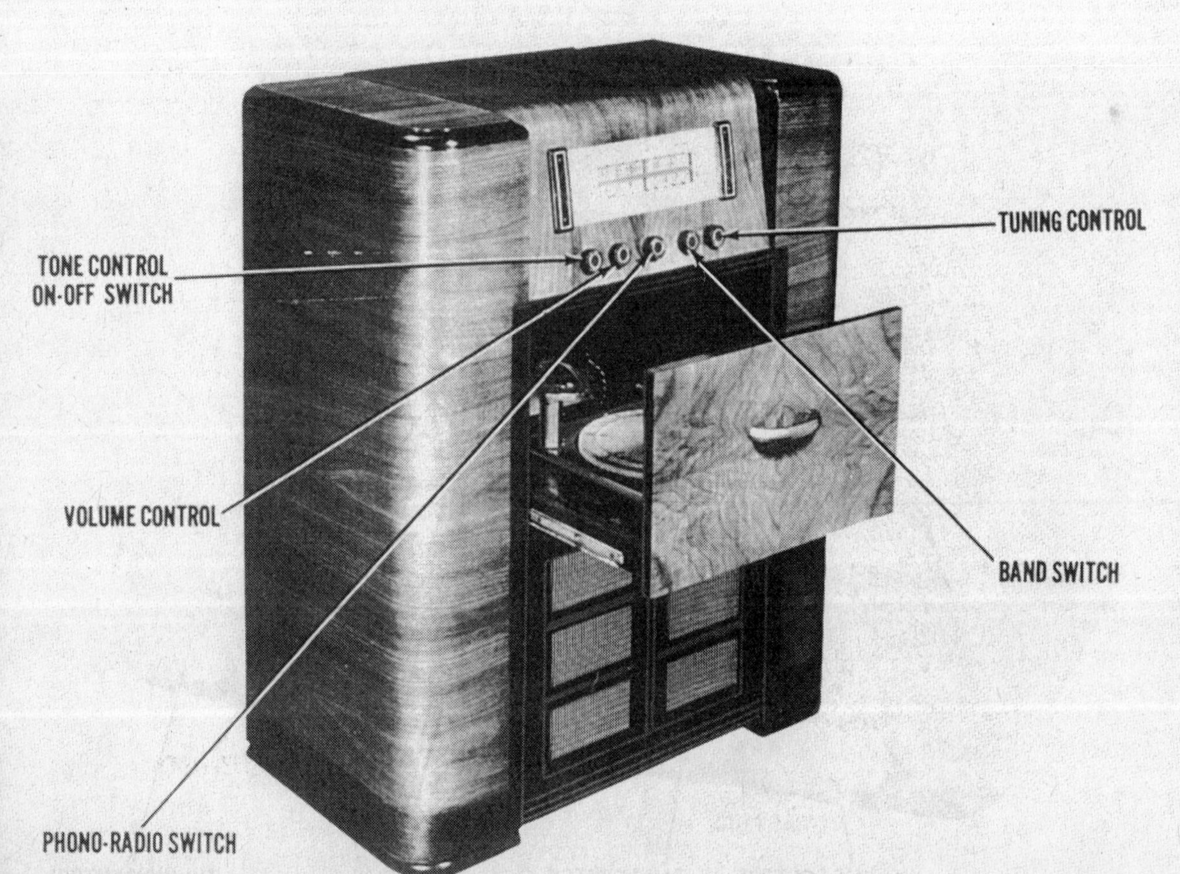

MODEL PICTURED
1231OW

AC operated combination phono-radio two-band superheterodyne receiver with loop antenna.

TUBES
6

POWER SUPPLY
110-120 volts AC

TUNING RANGE
535-1725KC, 6-18.2MC

MFR/SUPPLIER
Warwick Mfg. Corp.

PHOTOFACT SET
31-6

PUBLISHED
1948

CLARION

MODEL PICTURED
12708

AC operated phono-radio superheterodyne receiver with loop antenna

TUBES
4

POWER SUPPLY
110-120 volts AC

TUNING RANGE
535-1725KC

MFR/SUPPLIER
Warwick Mfg. Corp.

PHOTOFACT SET
41-5

PUBLISHED
1948

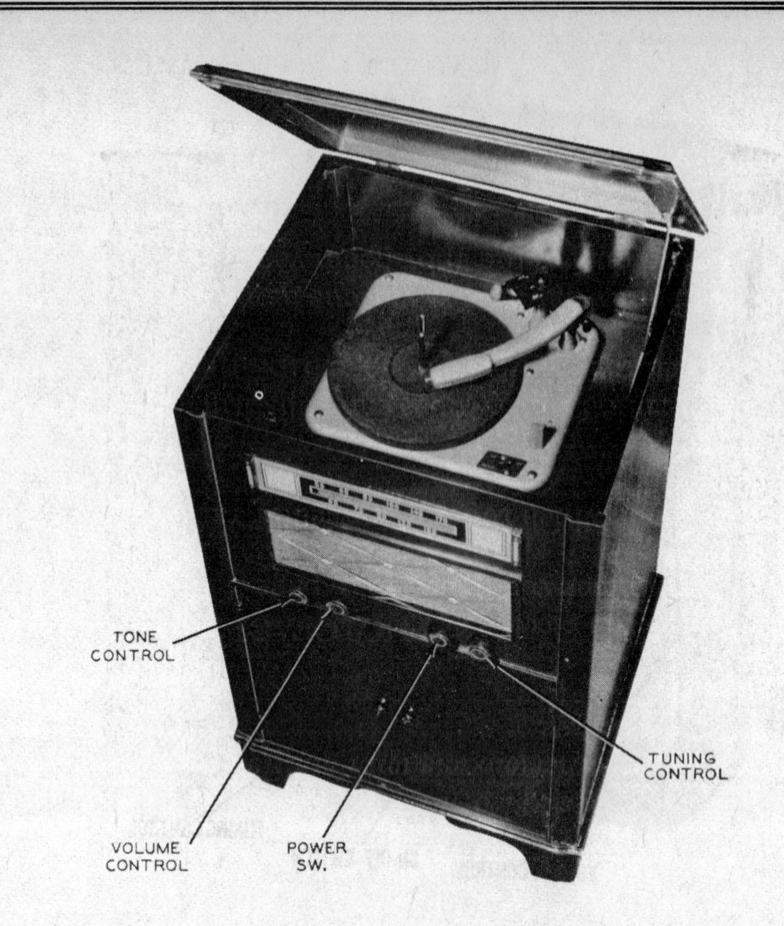

CLARION

MODEL PICTURED
13101

AC/DC operated AM-FM superheterodyne receiver

TUBES
10

POWER SUPPLY
110-120 volts AC/DC,
.453 amp @ 117 volts AC

TUNING RANGE
540-1600KC, 88-108MC

MFR/SUPPLIER
Warwick Mfg. Corp.

PHOTOFACT SET
46-7

PUBLISHED
1948

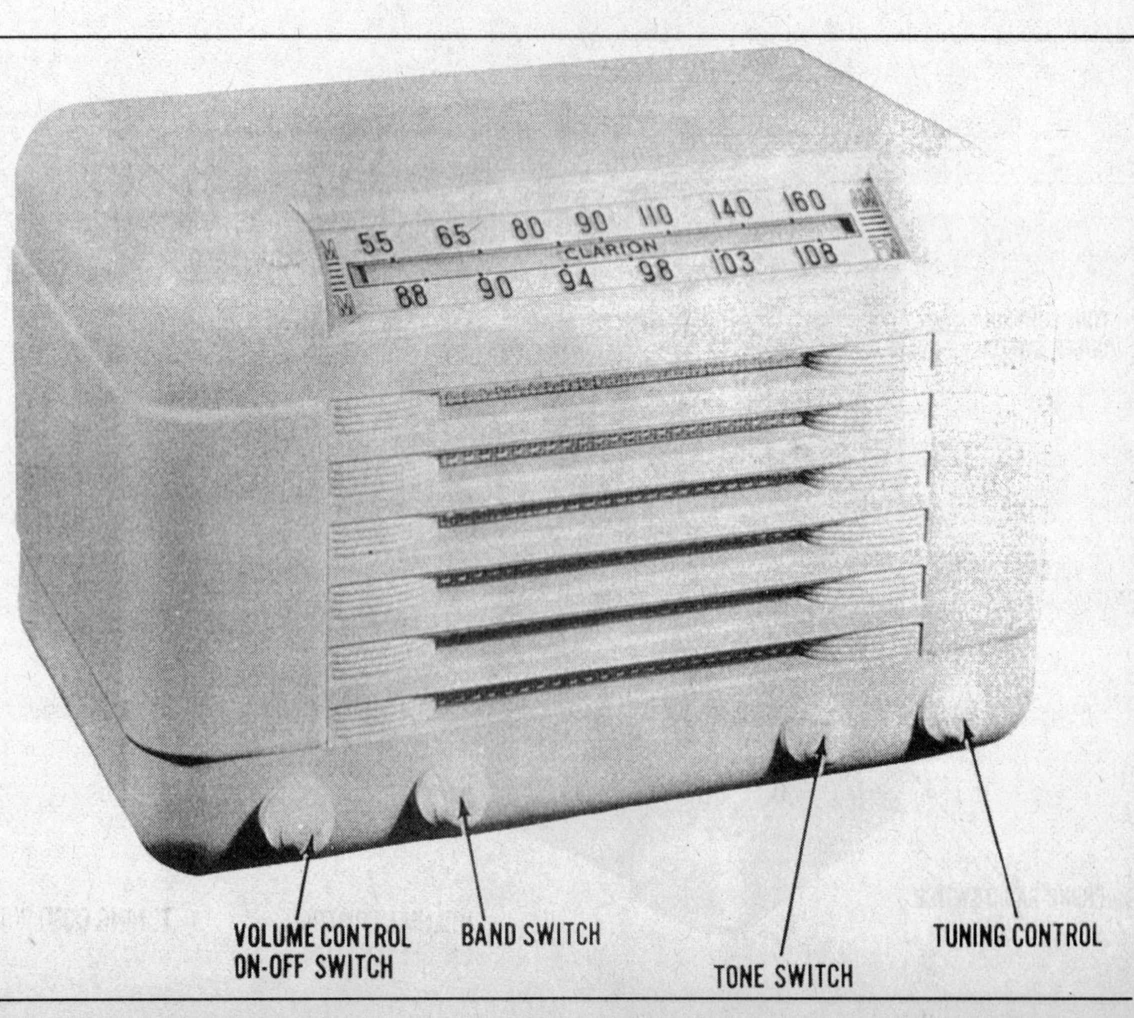

CLARION

MODEL PICTURED
12110M

AC operated combo phono-radio AM-FM superheterodyne receiver with loop antenna

TUBES
10

POWER SUPPLY
110-120 volts AC

TUNING RANGE
540-1600KC, 88-108MC

MFR/SUPPLIER
Warwick Mfg. Corp.

PHOTOFACT SET
54-5

PUBLISHED
1949

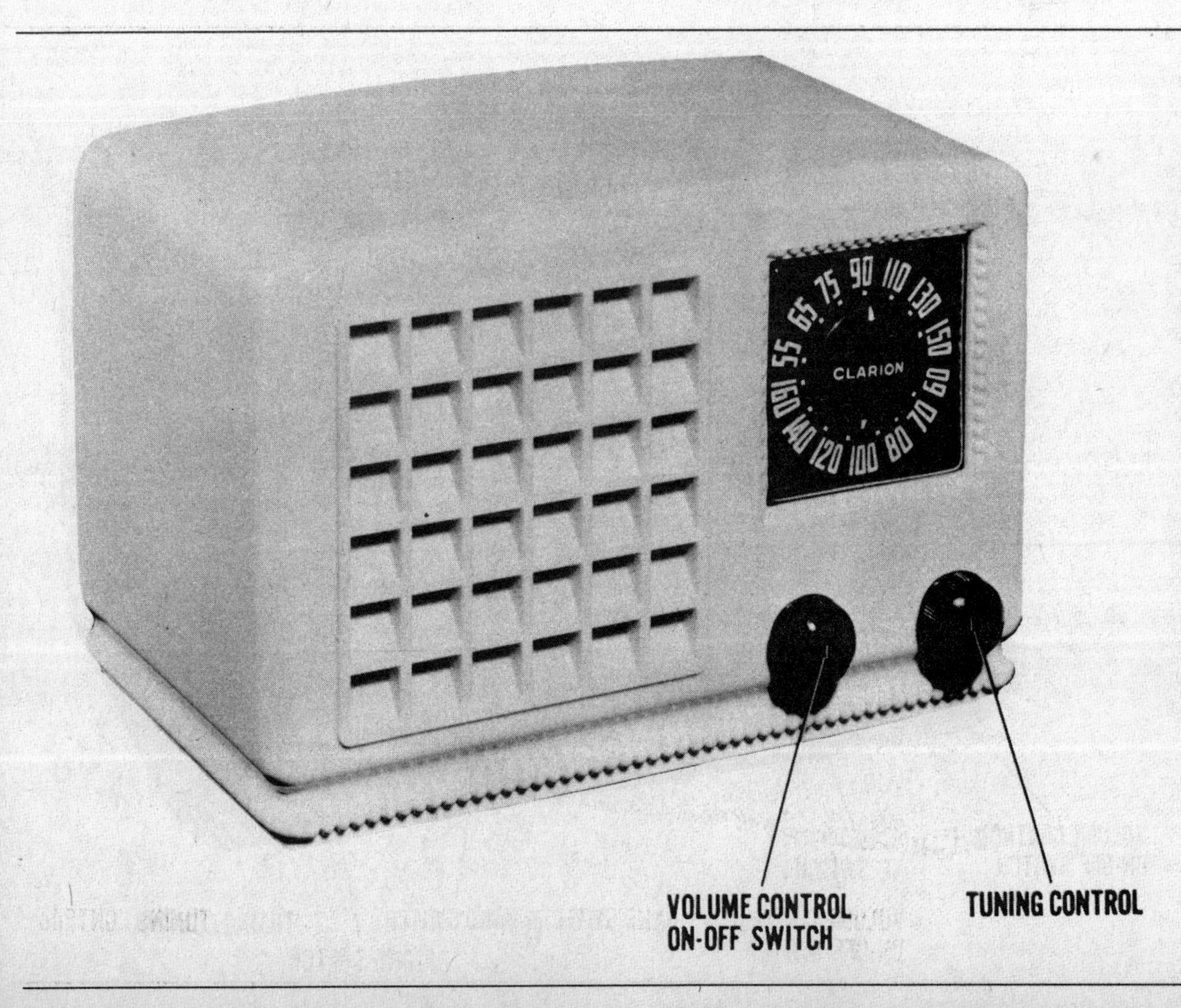

CLARION

MODEL PICTURED
14601

AC/DC operated superheterodyne receiver with loop antenna

TUBES
5

POWER SUPPLY
110-120 volts AC/DC

TUNING RANGE
540-1630KC

MFR/SUPPLIER
Warwick Mfg. Corp.

PHOTOFACT SET
60-9

PUBLISHED
1949

CLARION

MODEL PICTURED
13201

Battery operated portable superheterodyne receiver

TUBES
4

POWER SUPPLY
1.5 volts A supply & 90 volts B supply in pack form

TUNING RANGE
530-1725KC

MFR/SUPPLIER
Warwick Mfg. Corp.

PHOTOFACT SET
62-8

PUBLISHED
1949

CLARION

MODEL PICTURED
14965

AC operated AM-FM superheterodyne receiver with loop antenna

TUBES
8

POWER SUPPLY
105-125 volts AC, .54 amp @ 117 volts AC

TUNING RANGE
540-1600KC, 88-108MC

MFR/SUPPLIER
Warwick Mfg. Corp.

PHOTOFACT SET
66-5

PUBLISHED
1949

CLEARSONIC

MODEL PICTURED
5C66

AC/DC operated superheterodyne receiver with self-contained loop antenna

TUBES
5

POWER SUPPLY
110-120 volts AC/DC

TUNING RANGE
535-1620KC

MFR/SUPPLIER
U.S. TV Mfg. Co.

PHOTOFACT SET
17-9

PUBLISHED
1947

INDEXES

Use these indexes to learn more about the radios pictured in this book and other models with similar configurations and similar tube usage.

- To locate the picture or PHOTOFACT set number of a model number you know, go to page 296.
- To find out which tubes a particular radio holds, go to page 305.
- To determine which tubes can substitute for the tubes you need to replace, go to page 315.
- To find out what brands various manufacturers produced during the baby boom years, go to page 317.

Pictured Radios and Similar Models, by Model Number

BRAND/MODEL	PF SET	SIMILAR TO	PAGE
ADMIRAL			
HIFI6	258-2	-	25
HIFI7	258-2	HIFI6	25
HIFI8	258-2	HIFI6	25
SS1081	509-3	Y1149	41
SS1092	509-3	Y1149	41
SS1093	509-3	Y1149	41
SS1139	509-3	Y1149	41
SS632	447-4	SS642	34
SS633	447-4	SS642	34
SS634	447-4	SS642	34
SS642	447-4	-	34
SS643	447-4	SS642	34
SS644	447-4	SS642	34
SS649	447-4	SS642	34
SS654	447-4	SS642	34
SS662	447-4	SS642	34
SS663	447-4	SS642	34
SS664	447-4	SS642	34
SS671	447-4	SS642	34
Y1081	509-3	Y1149	41
Y1102	509-3	Y1149	41
Y1103	509-3	Y1149	41
Y1139	509-3	Y1149	41
Y1149	509-3	-	41
Y1171	509-3	Y1149	41
Y1172	509-3	Y1149	41
Y1173	509-3	Y1149	41
Y1191	509-3	Y1149	41
Y1192	509-3	Y1149	41
Y2008	510-4	Y2027	42
Y2011	510-4	Y2027	42
Y2012	510-4	Y2027	42
Y2013	510-4	Y2027	42
Y2023	510-4	Y2027	42
Y2027	510-4	-	42
Y2028	510-4	Y2027	42
Y793	478-4	692	38
Y797	478-4	692	38
Y798	478-4	692	38
Y821	479-3	717	39
Y822	479-3	717	39
Y833	476-3	-	38
Y837	476-3	Y833	38
Y8385L5	476-3	Y833	38
Y839	476-3	Y833	38
Y846	476-3	Y833	38
Y847	476-3	Y833	38
Y848	476-3	Y833	38
Y853	483-3	Y858	40
Y858	483-3	-	40
Y865	483-3	Y858	40
Y866	483-3	Y858	40
Y873	476-3	Y833	38
Y875	476-3	Y833	38
Y878	476-3	Y833	38
202	411-4	-	31
215	411-4	202	31
217	411-4	202	31
218	411-4	202	31
221	413-4	227	32
227	413-4	-	32
228	413-4	227	32
231	431-3	531	33
237	431-3	531	33
242	410-4	-	30
244	410-4	242	30
245	410-4	242	30
248	410-4	242	30
251	446-4	-	33
275	410-4	242	30
278	410-4	242	30
279	410-4	242	30
292	410-5	-	31
296	410-5	292	31
298	410-5	292	31
299	410-5	292	31
303	426-4	-	32
304	426-4	303	32
4D1	49-1	-	9
4E21	354-2	-	28
4F22	354-2	4E21	28
4F24	354-2	4E21	28
4F26	354-2	4E21	28
4F28	354-2	4E21	28
4H22	354-2	4E21	28
4H24	354-2	4E21	28
4H26	354-2	4E21	28
4H28	354-2	4E21	28
4L21	446-4	251	33
4L24	446-4	251	33
4L25	446-4	251	33
4L26	446-4	251	33
4L28	446-4	251	33
4M22	446-4	251	33
4M23	446-4	251	33
4M25	446-4	251	33
4M28	446-4	251	33
4P21	374-1	4P24	29
4P22	374-1	4P24	29
4P24	374-1	-	29
4P28	374-1	4P24	29
4R11	108-3	-	18
4R12	108-3	4R11	18
4T11	143-2	4W19	20
4W18	143-2	4W19	20
4W19	143-2	-	20
4X11	261-1	-	25
4X12	261-1	4X11	25
4X18	261-1	4X11	25
4X19	261-1	4X11	25
4Z11	274-2	-	26
4Z12	274-2	4Z11	26
4Z14	274-2	4Z11	26
4Z18	274-2	4Z11	26
4Z19	274-2	4Z11	26
521	448-4	-	35
528	448-4	521	35
531	431-3	-	33
537	431-3	531	33
561	452-3	566	35
566	452-3	-	35
581	446-5	-	34
582	446-5	581	34
5A32/12	191-2	5A32/16	22
5A32/15	191-2	5A32/16	22
5A32/16	191-2	-	22
5A33/12	191-2	5A32/16	22
5A33/15	191-2	5A32/16	22
5A33/16	191-2	5A32/16	22
5B42	341-2	5W32	27
5B43	341-2	5W32	27
5B48	341-2	5W32	27
5D31	256-3	-	24
5D32	256-3	5D31	24
5D33	256-3	5D31	24
5E21	139-2	5E22	19
5E22	139-2	-	19

BRAND/MODEL	PF SET	SIMILAR TO	PAGE
5E23	139-2	5E22	19
5E31	224-2	-	23
5E32	224-2	5E31	23
5E33	224-2	5E31	23
5E38	224-2	5E31	23
5E39	224-2	5E31	23
5F11	57-1	-	11
5F12	57-1	115F	11
5G21	137-2	5G22	19
5G21/15	137-2	5G22	19
5G22	137-2	-	19
5G22/15	137-2	5G22	19
5G23	137-2	5G22	19
5G23/15	137-2	G22	19
5J21	136-2	-	18
5J22	136-2	5J21	18
5J23	136-2	5J21	18
5L21	160-1	-	21
5L22	160-1	5L21	21
5L23	160-1	5L21	21
5M21	157-2	-	21
5M22	157-2	5M21	21
5R11	59-1	-	12
5R12	59-1	5R11	12
5R13	59-1	5R11	12
5R14	59-1	5R11	12
5R32	272-1	-	26
5R33	272-1	5R32	26
5R35	272-1	5R32	26
5R36	272-1	5R32	26
5R37	272-1	5R32	26
5R38	272-1	5R32	26
5RP41	387-6	-	30
5RP42	387-6	5RP41	30
5S21AN	197-2	5S22AN	23
5S22AN	197-2	-	23
5S23AN	197-2	5S22AN	23
5S32	272-1	5R32	26
5S33	272-1	5R32	26
5S34	272-1	5R32	26
5S35	272-1	5R32	26
5S38	272-1	5R32	26
5T12	68-1	-	13
5T31	279-1	-	27
5T32	279-1	5T31	27
5T33	279-1	5T31	27
5T34	279-1	5T31	27
5T38	279-1	5T31	27
5W11	79-2	-	16
5W12	79-2	5W11	16
5W32	341-2	-	27
5W33	341-2	5W32	27
5W34	341-2	5W32	27
5W38	341-2	5W32	27
5W39	341-2	5W32	27
5X11	76-3	-	15
5X12	76-3	5X11	15
5X13	76-3	5X11	15
691	478-4	-	38
692	478-4	-	38
6A21	103-1	-	17
6A22	103-1	6A21	17
6A23	103-1	6A21	17
6C11	53-1	-	10
6C22	252-3	6C23A	24
6C22A	252-3	6C23A	24
6C23	252-3	6C23A	24
6C23A	252-3	-	24
6J21	140-2	-	20
6J22	140-2	6J21	20
6N25	165-3	6N26	22
6N26	165-3	-	22
6N27	165-3	6N26	22
6P32	6-1	-	3
6Q11	78-1	6Q12	16
6Q12	78-1	-	16
6Q13	78-1	6Q12	16
6Q14	78-1	6Q12	16
6R11	54-1	-	10
6RT42A	18-1	-	4
6RT43	4-24	-	3
6RT44	18-2	-	4
6S11	107-1	6S12	17
6S12	107-1	-	17
6V11	62-1	6V12	12
6V12	62-1	-	12
6W11	71-1	6W12	14
6W12	71-1	-	14
6Y18	75-1	-	15
6Y19	75-1	6Y18	15
703	478-5	-	39
708	478-5	703	39
711	479-3	717	39
717	479-3	-	39
739	471-3	-	36
742	474-4	-	37
743	474-4	742	37
751	473-3	-	37
757	473-3	751	37
7849	476-3	Y83	383
7C60B	48-2	7C60M	8
7C60M	48-2	-	8
7C60W	48-2	7C60M	8
7C65B	36-1	7C65W	8
7C65M	36-1	7C65W	8
7C65W	36-1	-	8
7C73	32-1	-	7
7G11	54-2	7G14	11
7G12	54-2	7G14	11
7G14	54-2	-	11
7G15	54-2	7G14	11
7G16	54-2	7G14	11
7L12	375-6	7L16	29
7L14	375-6	7L16	29
7L16	375-6	-	29
7L18	375-6	7L16	29
7M12	369-1	7M14	28
7M14	369-1	-	28
7M16	369-1	7M14	28
7M18	369-1	7M14	28
7P325H1	26-1	7P33	5
7P33	26-1	-	5
7P34	26-1	7P33	5
7RT42	26-2	-	6
7T01	31-1	-	7
7T04	31-1	-	7
7T06	24-1	7T12	5
7T10	30-1	-	6
7T12	24-1	-	5
7T14	30-1	7T10	6
7T15	30-1	7T10	6
801	454-4	-	36
802	454-4	801	36
808	454-4	801	36
811	452-3	566	35
811B	499-3	816B	40
816	452-3	566	35
816B	499-3	-	40
8D15	67-1	-	13
8D16	67-1	8D15	13
909	502-4	-	41
9B14	49-2	-	9
9B15	49-2	9B14	9
9B16	49-2	9B14	9

BRAND/MODEL	PF SET	SIMILAR TO	PAGE
652.5T5E	260-4	-	102
652.5T5V	260-4	652.5T5E	102
652.5X5	286-2	-	102
652.6T1E	205-2	-	99
652.6T1V	205-2	652.6T1E	99
6541	17-2	-	66
6547	17-2	6541	66
659.511	167-2	-	97
659.513	167-2	659.511	97
659.520E	185-4	-	98
659.520I	185-4	659.520E	98
6611	15-2	6634	65
6612	15-2	6634	65
6613	15-2	6634	65
6630	15-2	6634	65
6631	15-2	6634	65
6632	15-2	6634	65
6634	15-2	-	65
6635	15-2	6634	65
7553	45-3	-	70
782.5C1	287-2	-	103
782.5R1	287-2	782.5C1	103
782.FM-99-AC	290-2	-	103
7B	52-1	-	74
9	50-2	-	73
9008I	99-2	9008W	90
9008W	99-2	-	90
9009I	97-2	9009W	89
9009W	97-2	-	89
9012I	94-1	9012W	87
9012W	94-1	-	87
915I,W	129-2	-	95
935	128-2	-	94

AIRLINE

BRAND/MODEL	PF SET	SIMILAR TO	PAGE
GAA-1003A	404-4	-	160
GAA-1009A	404-4	GAA-1003A	160
GAA-2620A	384-6	-	159
GAA-2621A	384-6	GAA-2620A	159
GAA-2640A	496-5	-	165
GAA-2641A	496-5	GAA-2640A	165
GAA-2642A	384-6	GAA-2620A	159
GAA-2643A	384-6	GAA-2620A	159
GAA-2689B	496-5	GAA-2640A	165
GAA-990A	320-3	-	154
GEN-1090A	350-2	-	156
GEN-1103A	349-2	-	156
GEN-1120A	458-4	GEN-1120C	163

BRAND/MODEL	PF SET	SIMILAR TO	PAGE
GEN-1120B	458-4	GEN-1120C	163
GEN-1120C	458-4	-	163
GEN-1628A	428-3	GEN-1670A	161
GEN-1655A	358-1	-	157
GEN-1656A	358-1	GEN-1655A	157
GEN-1657A	358-1	GEN-1655A	157
GEN-1658A	358-1	GEN-1655A	157
GEN-1660A	358-1	GEN-1655A	157
GEN-1661A	358-1	GEN-1655A	157
GEN-1670A	428-3	-	161
GEN-2645A	452-4	-	162
GEN-2645B	452-4	GEN-2645A	162
GEN-2645C	452-4	GEN-2645A	162
GEN-2646A	452-4	GEN-2645A	162
GEN-2646B	452-4	GEN-2645A	162
GEN-2646C	452-4	GEN-2645A	162
GEN-2653A	452-4	GEN-2645A	162
GRX-1650A	359-2	GSL-1650A	158
GRX-1651A	359-2	GSL-1650A	158
GSE-1077A	250-3	-	148
GSE-1078A	250-3	GSE-1077A	148
GSE-1606A	292-2	-	151
GSE-1607A	292-2	GSE-1606A	151
GSE-1620A	317-2	-	153
GSE-1621A	317-2	GSE-1620A	153
GSE-1622A	317-2	GSE-1620A	153
GSE-1625A	325-3	-	155
GSE-1626A	325-3	GSE-1625A	155
GSL-1079-A	294-2	-	152
GSL-1575A	323-3	-	154
GSL-1581A	280-3	-	150
GSL-1582A	280-3	GSL-1581A	150
GSL-1614A	289-2	-	151
GSL-1615A	289-2	GSL-1614A	151
GSL-1616A	289-2	GSL-1614A	151
GSL-1617A	289-2	GSL-1614A	151
GSL-1650A	359-2	-	158
GSL-1651A	359-2	GSL-1650A	158
GTC-1085A	356-2	-	157
GTC-1086A	356-2	GTC-1085A	157
GTM-1108A	379-5	-	159
GTM-1109A	392-6	-	160
GTM-1117A	473-4	-	164
GTM-1201A	478-6	-	165
GTM-1638A	411-5	GTM-1639B	161
GTM-1639A	411-5	GTM-1639B	161
GTM-1639B	411-5	-	161
GTM-1666A	467-3	-	164

BRAND/MODEL	PF SET	SIMILAR TO	PAGE
WG-1572C	251-1	-	149
WG-1635A	306-2	-	152
WG-1636A	306-2	WG-1635A	152
WG-1637A	363-3	-	158
WG-2602A	307-3	-	153
WG-2603A	307-3	WG-2602A	153
WG-2673A	453-3	-	162
WG-2674A	453-3	WG-2673A	162
WG-2683A	466-4	WG-2684A	163
WG-2684A	466-4	-	163
WG-3503A	453-3	WG-2673A	162
WG-3503B	453-3	WG-2673A	162
05GAA-992A	125-2	-	132
05GCB-1540A	131-2	05GCB-1541A	134
05GCB-1541A	131-2	-	134
05GHM-1061A	133-3	-	134
05GHM-934A	167-3	15GHM-934A	139
05WG-1813A	127-4	-	132
05WG-2748F	139-4	-	135
05WG-2749D	129-3	-	133
05WG-2752	100-3	-	131
15BR-1536B	146-2	-	137
15BR-1537B	146-2	15BR-1536B	137
15BR-1543A	145-2	15BR-1544A	136
15BR-1544A	145-2	-	136
15BR-1547A	143-3	-	135
15BR-1548A	191-3	25BR-1549B	143
15BR-1549A	191-3	25BR-1549B	143
15BR-2756B	148-3	-	137
15BR-2757A	148-3	15BR-2756B	137
15GAA-995A	168-3	-	139
15GHM-1070A	184-3	-	142
15GHM-934A	167-3	-	139
15GSE-2764A	165-4	-	138
15GSL-1564A	169-3	-	140
15GSL-1564B	169-3	15GSL-1564A	140
15GSL-1565A	169-3	15GSL-1564A	140
15GSL-1565B	169-3	15GSL-1564A	140
15GSL-1566A	169-3	15GSL-1564A	140
15GSL-1566B	169-3	15GSL-1564A	140
15GSL-1567A	169-3	15GSL-1564A	140
15GSL-1567B	169-3	15GSL-1564A	140
15WG-2745C	130-2	-	133
15WG-2749E	151-4	15WG-2749F	138
15WG-2749F	151-4	-	138
15WG-2752D	151-4	15WG-2749F	138
15WG-2752E	151-4	15WG-2749F	138
15WG-2758A	144-2	-	136

BRAND/MODEL	PF SET	SIMILAR TO	PAGE
AIRLINE *cont'd*			
2509	328-2	-	155
25BR-1542A	203-3	-	145
25BR-1548B	191-3	25BR-1549B	143
25BR-1549B	191-3	-	143
25GAA-996A	182-2	-	141
25GHM-1073A	242-2	-	147
25GHM-2012A	256-4	-	150
25GSE-1555A	174-3	-	140
25GSE-1556A	174-3	25GSE-1555A	140
25GSL-1560A	189-2	-	142
25GSL-1561A	189-2	25GSL-1560A	142
25GSL-1814A	198-1	-	144
25WG-1507C	177-4	25WG-1570A	141
25WG-1570A	177-4	-	141
25WG-1570B	177-4	25WG-1570A	141
25WG-1571A	177-4	25WG-1570A	141
25WG-1571B	177-4	25WG-1570A	141
25WG-1572A	177-4	25WG-1570A	141
25WG-1572B	177-4	25WG-1570A	141
25WG-1573A	196-2	-	144
25WG-2758C	195-3	-	143
25WG-2766A	195-3	25WG-2758C	143
35BR-1557A	251-2	-	149
35BR-1558A	251-2	35BR-1557A	149
35BR-1559A	251-2	35BR-1557A	149
35GAA-3969A	227-1	-	145
35GHM-1073B	242-2	25GHM-1073A	147
35GHM-1073C	242-2	25GHM-1073A	147
35GHM-1074A	243-2	-	147
35GHM-2012A	256-4	25GHM-2012A	150
35GHM-2020A	256-4	25GHM-2012A	150
35GSL-2770A	249-3	-	148
35WG-1573B	228-1	-	146
35WG-2767A	241-2	-	146
54KP-1209A	8-1	54KP-1209B	107
54KP-1209B	8-1	-	107
54WG-1801A	4-33	64WG-1801C	105
54WG-1801B	4-33	64WG-1801C	105
54WG-2500A	4-15	-	104
54WG-2700A	4-15	54WG-2500A	104
64BR-1205A	10-3	-	109
64BR-1206A	10-3	64BR-1205A	109
64BR-1208A	16-4	-	109
64BR-1513A	24-4	74BR-1514B	112
64BR-1513B	24-4	74BR-1514B	112
64BR-1514A	24-4	74BR-1514B	112
64BR-1514B	24-4	74BR-1514B	112
64BR-1808A	16-5	-	110
64BR-2200A	16-4	64BR-1208A	109
64WG-1050A	10-2	-	108
64WG-1052A	9-2	-	108
64WG-1207B	18-5	-	110
64WG-1511A	5-5	-	106
64WG-1511B	5-5	64WG-1511A	106
64WG-1512A	5-5	64WG-1511A	106
64WG-1512B	5-5	64WG-1511A	106
64WG-1801C	4-33	-	105
64WG-1804A	4-27	64WG-1804B	104
64WG-1804B	4-27	-	104
64WG-1807A	5-4	-	105
64WG-1807B	5-4	64WG-1807A	105
64WG-1809A	5-5	64WG-1511A	106
64WG-1809B	5-5	64WG-1511A	106
64WG-2007A	5-6	64WG-2007B	106
64WG-2007B	5-6	-	106
64WG-2009	6-2	64WG-2009A	107
64WG-2009A	6-2	-	107
64WG-2009B	6-2	64WG-2009A	107
64WG-2010B	18-6	74WG-2010B	111
74BR-1513B	24-4	74BR-1514B	112
74BR-1514B	24-4	-	112
74BR-2001B	23-2	-	112
74BR-2701A	24-5	-	113
74KR-1210A	41-1	-	118
74KR-2713A	43-2	-	119
74KR2706B	35-1	-	116
74WG-1056A	29-1	-	115
74WG-1057A	32-2	-	116
74WG-1207B	18-5	64WG-1207B	110
74WG-1509A	27-1	74WG-1510A	114
74WG-1510A	27-1	-	114
74WG-2002A	26-4	-	113
74WG-2004A	47-2	-	121
74WG-2010B	18-6	-	111
74WG-2504A	28-1	-	115
74WG-2504B	28-1	74WG-2504A	115
74WG-2504C	28-1	74WG-2504A	115
74WG-2505A	18-7	-	111
74WG-2704A	28-1	74WG-2504A	115
74WG-2704B	28-1	74WG-2504A	115
74WG-2704C	28-1	74WG-2504A	115
74WG-2709A	26-5	-	114
84GAA3967A	91-3	-	130
84GCB-1062A	52-26	-	121
84GSE2730A	70-1	84GSE2731A	125
84GSE2731A	70-1	-	125
84HA-1810C	69-2	-	124
84HA1529A	85-2	94HA1529A	127
84HA1530A	85-2	94HA1529A	127
84KR-1520A	56-4	-	122
84KR2511A	68-4	-	124
84WG-1060A	42-1	-	118
84WG-2015A	38-1	-	117
84WG-2506B	58-5	-	123
84WG-2712A	43-3	-	119
84WG-2714A	36-2	-	117
84WG-2714F	56-5	-	122
84WG-2714G	56-5	84WG-2714F	122
84WG-2714H	56-5	84WG-2714F	122
84WG-2714J	56-5	84WG-2714F	122
84WG-2718A	45-5	84WG-2720A	120
84WG-2718B	45-5	84WG-2720A	120
84WG-2720A	45-5	-	120
84WG-2721A	46-3	-	120
84WG-2721B	46-3	84WG-2721A	120
84WG-2724A	45-5	84WG-2720A	120
92BR-2740A	89-1	94BR-2740A	129
94BR-1533A	88-1	-	128
94BR-2741A	89-1	94BR-2740A	129
94BR-2741B	89-1	94BR-2740A	129
94GCB-1064A	96-2	-	130
94GHM-934A	167-3	15GHM-934A	139
94GSE-2735A	72-3	-	126
94GSE-2736A	72-3	94GSE-2735A	126
94HA1527C	67-3	94HA1528C	123
94HA1528C	67-3	-	123
94HA1529A	85-2	-	127
94HA1530A	85-2	94HA1529A	127
94WG-1059A	75-3	-	126
94WG-1804D	86-2	-	128
94WG-1811A	99-4	-	131
94WG-246A	71-5	94WG-2742A	125
94WG-2742A	71-5	-	125
94WG-2742C	71-5	94WG-2742A	125
94WG-2742D	71-5	94WG-2742A	125
94WG-2745A	76-4	-	127
94WG-2746B	71-5	94WG-2742A	125
94WG-2747A	71-5	94WG-2742A	125
94WG-2748A	90-1	-	129
94WG-2749A	90-1	94WG-2748A	129
ALGENE			
AR5U	22-3	-	166
AR-6U	22-4	-	166

BRAND/MODEL	PF SET	SIMILAR TO	PAGE
AMC			
126	16-1	-	167
AMI			
PBA	361-3	PBA (Mark 1)	167
ANDREA			
2CR-P12WB	416-4	CRP-24WA	171
CO-U15	27-3	-	170
CRP-24W	367-4	-	170
CRP-24WA	416-4	-	171
P-I63	18-8	-	168
T-16	21-2	-	168
T-U15	24-7	-	169
T-U16	21-3	-	169
W69P Spacemaster Deluxe	371-1	-	171
ANSLEY			
32	5-27	-	172
41 Paneltone	4-38	-	172
53	24-8	-	173
APEX INDUSTRIES			
4B5	37-2	-	173
ARCADIA			
37D14-600	9-3	-	174
ARIA			
554-1-61A	7-2	-	174
ARTHUR ANSLEY			
R-1	200-2	-	175
ARTONE			
524	76-6	-	175
ARVIN			
150TC	39-2	151TC	179
151TC	39-2	-	179
152T	33-1	-	178
1581	411-6	-	209
160T	49-5	-	181
161T	49-5	160T	181
182TFM	32-3	-	177
240-P	42-2	-	179
2410P	47-3	-	181
241P	47-3	2410P	181
242T	52-3	-	182
243T	52-3	242T	182
244P	47-3	2410P	181
250-P	43-4	-	180
253T	53-5	255T	182
254T	53-5	255T	182
255T	53-5	-	182
2563	356-3	-	204
2564	356-3	2563	201
256T	53-5	255T	182
2572	386-6	-	206
2573	386-6	2572	206
2581	427-4	-	210
2584	430-4	5583	211
2585	428-4	-	210
2598	485-3	-	215
264T	64-2	-	183
265T	64-2	264T	183
341T	84-3	-	185
350-PB	100-4	350-PL	186
350-PL	100-4	-	186
350P	69-3	-	183
351-PB	100-4	350-PL	186
351-PL	100-4	350-PL	186
351P	69-3	350P	183
352-PL	100-4	350-PL	186
353-PL	100-4	350-PL	186
3561	358-2	-	205
356T	78-2	-	184
357T	78-2	356T	184
3582	444-5	-	213
3586	454-5	-	213
3588	433-5	-	212
360TFM	70-2	-	184
361TFM	70-2	360TFM	184
440T	96-3	-	185
442	34-2	-	178
446P	106-2	-	186
450T	110-3	-	188
451T	110-3	450T	188
460T	107-3	-	187
461T	107-3	460T	187
462-CB	116-3	462-CM	188
462-CM	116-3	-	188
480TFM	107-4	-	187
481TFM	107-4	480TFM	187
482CFB	117-4	-	189
482CFM	117-4	482CFB	189
540T	143-4	-	189
547	42-3	-	180
547A	42-3	547	180
551T	154-2	-	190
552AN	13-9	555	176
552N	13-9	555	176
553	159-4	-	192
554CCB	155-3	554CCM	191
554CCM	155-3	-	191
555	13-9	-	176
555A	13-9	555	176
5561	351-3	-	204
5571	387-7	-	207
5572	387-7	5571	207
5578	393-6	-	208
5583	430-4	-	211
5591	473-5	-	214
5592	474-5	5594	215
5594	474-5	-	215
580TFM	152-2	-	190
581TFM	227-2	-	196
582CFB	156-4	-	191
582CFM	156-4	582CFB	191
60R23	508-5	-	218
60R28	508-5	60R23	218
60R29	508-5	60R23	218
60R33	507-4	-	217
60R35	507-4	60R33	217
60R38	507-4	60R33	217
60R47	506-6	60R49	217
60R49	506-6	-	217
60R58	507-4	60R33	217
650-P	175-6	-	193
651T	251-3	-	197
655SWT	187-2	-	193
657-T	168-5	-	192
664	29-2	-	177
6640	29-2	664	177
664A	29-2	664	177
665	18-10	-	176
741T	225-4	-	195
746P	225-5	-	196
747P	225-5	746P	196
753T	220-2	-	194
758T	221-3	-	194
760T	223-3	-	195
780TFM	227-2	581TFM	196
840T	263-3	-	199
848T	259-2	-	198
849T	259-2	848T	198

BRAND/MODEL	PF SET	SIMILAR TO	PAGE
850T	262-3	-	199
851T	266-3	-	200
852P	258-3	-	197
853T	262-3	850T	199
854P	258-3	852P	197
855T	266-3	851T	200
8571	394-7	-	208
8572	394-7	8571	208
8573	394-7	8571	208
8576	395-5	-	209
857T	275-4	-	200
8581	432-4	-	212
8583	432-4	8581	212
8584	429-3	-	211
858T	261-2	-	198
859T	261-2	858T	198
950T	295-3	-	201
950T1	340-2	-	203
950T2	384-7	-	206
951T	295-3	950T	201
951T1	340-2	950T1	203
952P	300-2	-	201
954P	300-2	952P	201
955T	304-1	-	202
9562	348-1	-	203
956T	304-1	955T	202
956T1	391-5	-	207
9574P	377-6	-	205
957T	321-2	-	202
9594	486-5	-	216
9595	470-4	-	214
9598	485-4	-	216

ASTRA-SONIC

BRAND/MODEL	PF SET	SIMILAR TO	PAGE
748	53-6	-	218

ATLAS

BRAND/MODEL	PF SET	SIMILAR TO	PAGE
AB-45	14-5	-	219

AUDAR

BRAND/MODEL	PF SET	SIMILAR TO	PAGE
AV-7T	166-6	-	220
PR-6	13-10	-	219
(Telvar) RER.9	65-2	-	220
WC-7T	166-6	AV-7T	220

AUTOMATIC

BRAND/MODEL	PF SET	SIMILAR TO	PAGE
ATTP (Tom Thumb)	23-4	-	225
B-44 (Tom Thumb Bike Radio)	60-5	-	228
C-54	186-2	-	229
C-60X	24-10	-	226
C51	178-4	-	229
C60	5-20	-	221
CL-100	426-5	-	231
CL-152B	192-3	-	230
CL-152M	192-3	CL-152B	230
CL-164B	192-3	CL-152B	230
CL-75	492-5	-	232
F-790	23-5	-	225
P-57	489-4	-	231
Series B	10-4	640	222
Tom Thumb Camera/Radio	49-6	-	227
Tom Thumb Jr	26-7	-	226
Tom Thumb Buddy	53-7	-	228
Tom Thumb	49-6	Tom Thumb Camera	227
Tomboy	27-4	-	227
TT528	491-4	-	232
TT600	349-3	-	230
601 (A and B)	13-11	-	223
602 (A and B)	13-11	601	223
614X	8-2	-	221
616X	8-2	614X	221
620	12-3	-	222
640	10-4	-	222
660	22-6	662	224
662	22-6	-	224
666	22-6	662	224
677	22-7	-	224
720	21-4	601	223

AVIOLA

BRAND/MODEL	PF SET	SIMILAR TO	PAGE
509	7-3	-	233
601	15-3	-	233
608	16-6	-	234
612	15-3	601	233
618	16-6	608	234

B.F.GOODRICH

BRAND/MODEL	PF SET	SIMILAR TO	PAGE
92-523	148-7	-	234
92-524	148-7	92-523	234
92-525	148-7	92-523	234
92-526	148-7	92-523	234
92-527	148-7	92-523	234
92-528	148-7	92-523	234

BAGPIPER

BRAND/MODEL	PF SET	SIMILAR TO	PAGE
SKR101	335-2	-	235

BELLTONE

BRAND/MODEL	PF SET	SIMILAR TO	PAGE
500	5-33	-	235

BELMONT

BRAND/MODEL	PF SET	SIMILAR TO	PAGE
4B112 (Series A)	10-6	-	237
4B113	10-6	4B112	237
5D110	22-10	-	238
5D128 (Series A)	9-4	-	236
5P113 Boulevard	28-2	-	239
5P19 (Series A)	9-5	-	237
6D120	24-12	-	239
8A59	6-4	-	236
A-6D110	17-7	-	238

BENDIX

BRAND/MODEL	PF SET	SIMILAR TO	PAGE
110	41-3	-	246
110W	41-3	110	246
111	41-3	110	246
111W	41-3	110	246
112	41-3	110	246
114	41-3	110	246
115	41-3	110	246
1217	29-4	1217B	244
1217B	29-4	-	244
1217D	29-4	1217B	244
1217D	46-5	-	248
1518	37-3	1524	244
1519	37-3	1524	244
1521	42-4	-	247
1524	37-3	-	244
1525	37-3	1524	244
1531	43-6	-	248
1533	43-6	1531	248
300	40-2	301	245
300W	40-2	301	245
301	40-2	-	245
302	40-2	301	245
416A	43-5	-	247
526MA	29-3	526MB	243
526MB	29-3	-	243
526MC	29-3	526MB	243
55L2	51-4	55P2	249
55L3	51-4	55P2	249
55P2	51-4	-	249
55P3	51-4	55P2	249
55X4	58-6	-	250
613	40-3	-	246
626-A (0626A)	12-4	-	241
636A	15-4	-	241

BRAND/MODEL	PF SET	SIMILAR TO	PAGE
636B	15-4	636A	241
636C	15-4	636A	241
65P4	52-4	-	249
676B	5-23	676D	240
676C	5-23	676D	240
676D	5-23	-	240
697A	26-8	-	242
69B8	63-3	69M9	251
69M8	63-3	69M9	251
69M9	63-3	-	251
736-B	10-8	-	240
753F	199-3	-	253
753M	199-3	753F	253
753W	199-3	753F	253
75B5	59-5	75MA	250
75M5	59-5	75MA	250
75M8	59-5	75MA	250
75P6	59-5	75MA	250
75W5	59-5	75MA	250
79M7	66-3	-	252
847-B	27-5	-	242
847S	28-3	-	243
951	136-6	-	252
951W	136-6	951	252
95B3	60-7	95M9	251
95M3	60-7	95M9	251
95M9	60-7	-	251
PAR-80	39-3	-	245
PAR-80A	39-3	PAR-80	245

BENRUS

BRAND/MODEL	PF SET	SIMILAR TO	PAGE
10B01B15B	299-3	-	253

BLAUPUNKT

BRAND/MODEL	PF SET	SIMILAR TO	PAGE
Americano	422-5	Barcelona	254
Arizona 57	418-4	Granada 2330	254
Arizona	424-4	Sultan 2320	255
Arkansas	422-5	Barcelona	254
Barcelona	422-5	-	254
Elvira	422-5	Barcelona	254
Gina	422-5	Barcelona	254
Granada 2330	418-4	-	254
Granada 2330 Monaco	418-4	Granada 2330	254
Hawaiian	422-5	Barcelona	254
Jewel	422-5	Barcelona	254
Kongo	422-5	Barcelona	254
Rio	424-4	Sultan 2320	254
Romeo	422-5	Barcelona	254
Sultan 2320	424-4	-	255

BLONDER-TONGUE

BRAND/MODEL	PF SET	SIMILAR TO	PAGE
R-98	457-4	-	255
R-20	503-6	-	256

BRADFORD

BRAND/MODEL	PF SET	SIMILAR TO	PAGE
WGEC-95190C	774-6	-	256

BRAUN

BRAND/MODEL	PF SET	SIMILAR TO	PAGE
MM4D	409-6	-	257
MM4LO	409-6	MM4D	257
MM4W	409-6	MM4D	257

BRUNSWICK

BRAND/MODEL	PF SET	SIMILAR TO	PAGE
5000	42-5	-	258
BJ-6836 Tuscany	28-4	-	257
C-3300 Darby	28-4	BJ-6836	257
D-1000	56-7	-	259
D-1100	56-7	D-1000	259
D-68T6	29-5	T-4000	258
T-4000	29-5	-	258
T-4000-1/2	29-5	T-4000	258
T-9000	56-7	D-1000	259

BULOVA

BRAND/MODEL	PF SET	SIMILAR TO	PAGE
100	371-2	-	261
110	371-2	100	261
200 Series	382-6	200	262
220 Series	372-2	220	261
230 Series	372-2	220	261
240 Series	382-6	200	262
270 Series	369-5	278	259
300 Series	369-6	304	260
310 Series	370-2	310	260
320 Series	370-2	310	260

CALRAD

BRAND/MODEL	PF SET	SIMILAR TO	PAGE
60A183	497-5	-	262

CAPEHART

BRAND/MODEL	PF SET	SIMILAR TO	PAGE
C-14	263-4	-	269
P-213	234-3	-	268
RP-152	215-4	-	267
RP153	258-4	-	268
RP254	258-4	RP153	268
T-522	209-1	-	267
T-54	265-3	-	270
TC-100	203-5	-	266
TC-101	203-5	TC-100	266
TC-62	192-4	-	266
10	166-7	-	265
1002F	135-4	-	264
1003M	135-4	1002F	264
1004B	135-4	1002F	264
1007AM	150-5	-	265
114N4	65-3	29P4	263
115P2	67-6	413P	264
116N4	65-3	29P4	263
116P4	65-3	29P4	263
118P4	65-3	29P4	263
17RPQ155F	294-4	263-4	271
19N4	65-3	29P4	263
1P55	268-4	-	270
21P4	65-3	29P4	263
24N4	65-3	29P4	263
24P4	65-3	29P4	263
26N4	65-3	29P4	263
29P4	65-3	-	263
2C56	316-5	-	271
2P56	330-2	-	272
2T55	261-4	-	269
30P4	65-3	29P4	263
31N4	65-3	29P4	263
31P4	65-3	29P4	263
32P9	64-3	33P9	263
33P9	64-3	-	263
34P10	64-3	33P9	263
35P7	135-4	1002F	264
3T55	261-4	2T55	269
413P	67-6	115P2	264
414P	67-6	413P	264
75C56	340-3	-	273
88P66BNL	331-3	-	272

CAVALIER

BRAND/MODEL	PF SET	SIMILAR TO	PAGE
4CL4	273-4	-	276
4P3	266-4	-	276
562	473-6	-	277
5AT1	241-4	-	274
5B1	238-6	-	273
5C1	242-4	-	274
5R1	265-4	-	275
603	353-2	-	277
6A2	265-5	-	275

CBS COLUMBIA

BRAND/MODEL	PF SET	SIMILAR TO	PAGE
C220	337-1	-	280
C230	337-1	C220	280

BRAND/MODEL	PF SET	SIMILAR TO	PAGE
C231	337-1	C220	280
C232	337-1	C220	280
C240	337-1	C220	280
T200	338-1	-	280
T201	338-1	T200	280
T202	338-1	T200	280
T203	338-1	T200	280
T204	338-1	T200	280
515A	223-4	-	279
516A	223-4	515A	279
517A	223-4	515A	279
5220	261-5	-	279
525	222-4	-	278
526	222-4	525	278
540	211-4	541	278
541	211-4	-	278
5440	340-4	-	281

CHANCELLOR

BRAND/MODEL	PF SET	SIMILAR TO	PAGE
35P	30-25	-	281

CHANNEL MASTER

BRAND/MODEL	PF SET	SIMILAR TO	PAGE
6501	462-5	-	282
6502	462-5	6501	282
6503	495-6	-	283
6504	495-6	6503	283
6506	470-6	-	282
6512	485-6	-	283
6514	485-6	6512	283

CISCO

BRAND/MODEL	PF SET	SIMILAR TO	PAGE
1A5	37-4	-	284
9A5	20-3	-	284

CLARION

BRAND/MODEL	PF SET	SIMILAR TO	PAGE
C-103	6-6	-	286
C101	5-9	-	285
C102	9-6	-	287
C105-A	6-7	C-105-A	286
C108	5-8	-	285
11011	17-8	-	287
11305	18-11	-	288
11411-N	30-5	-	289
11801	23-6	-	288
12110M	54-5	-	291
1231OW	31-6	-	289
12708	41-5	-	290
13101	46-7	-	290
13201	62-8	-	292
13203	62-8	13201	292
14601	60-9	-	291
14965	66-5	-	292

CLEARSONIC

BRAND/MODEL	PF SET	SIMILAR TO	PAGE
5C66	17-9	-	293

Pictured Radios and Associated Tubes

BRAND/MODEL	PAGE	USES THESE TUBES
ADMIRAL		
202	31	1R5, 1U4, 1U5, 3V4
227	32	Transistorized
242	30	12BE6, 12BA6, 12AV6, 50C5, 35W4
251	33	12AU6, 12AV6, 50C5, 35W4
292	31	12BE6, 12BA6, 12AV6, 50C5, 35W4
303	32	12DT8, 12AU6, 12BA6, 12AU6, 19T8, 50C5
4D1	9	1R5, 1U4, 1U5, 3S5
4E21	28	1R5, 1U4, 1U5, 3V4
4P24	29	Transistorized
4R11	18	1R5, 1U4, 1U5, 3V4
4W19	20	1R5, 1U4, 1U5, 3V4
4X11	25	1R5, 1U4, 1U5, 3V4
4Z11	26	1R5, 1U4, 1U5, 3V4
521	35	Transistorized
531	33	Transistorized
566	35	Transistorized
581	34	Transistorized
5A32/16	22	12BE6, 12BA6, 12AV6, 50C5, 35W4
5D31	24	12BE6, 12BA6, 12AV6, 50C5, 35W4
5E22	19	12SA7, 12SK7, 12SQ7, 50L6GT, 35Z5GT
5E31	23	12BE6, 12BA6, 12AV6, 50C5, 35W4
5F11	11	1R5, 1U4, 1U5, 3V4
5G22	19	12BE6, 12BA6, 12AV6, 50C5, 35W4
5J21	18	12SA7, 12SK7, 12SQ7, 50L6GT, 35Z5GT
5L21	21	12BE6, 12BA6, 12AV6, 50C5, 35W4
5M21	21	12SA7, 12SK7, 12SQ7, 50L6GT, 35Z5GT
5R11	12	12SA7, 12SK7, 12SQ7, 50L6GT, 35Z5GT
5R32	26	12BE6, 12BA6, 12AV6, 50C5, 35W4
5RP41	30	12BE6, 12BA6, 12AV6, 50C5, 35W4
5S22AN	23	12BE6, 12BA6, 12AV6, 50C5, 35W4
5T12	13	12SA7, 12SK7, 12SQ7, 50L6GT, 35Z5GT
5T31	27	12BE6, 12BA6, 12AV6, 50C5, 35W4
5W12	16	12BE6, 12BA6, 12AV6, 50C5, 35W4
5W32	27	12BE6, 12BA6, 12AV6, 50C5, 35W4
5X11	15	12SA7GT, 12SK7, 12SQ7GT, 50L6GT, 35Z5GT
692	38	Transistorized
6A21	17	12SK7, 12SA7, 12SQ7, 35L6GT, 35Z5GT, 12SK7
6C11	10	1U4, 1R5, 1U4, 1U5, 3U4
6C23A	24	12SK7, 12SA7, 12SK7, 12SQ7, 35L6GT, 35Z5GT
6J21	20	12BA6, 12J5, 12SK7, 35L6GT, 35Z5GT, 12SQ7
6N26	22	6BA6, 6BE6, 6BA6, 6AV6, 6AQ5, 5Y3GT
6P32	3	1N5GT, 1A7GT, 1N5GT, 1H5GT, 3Q5GT, 117Z6GT
6Q12	16	12AT7, 12BA6, 12BA6, 12AL5, 12AV6, 50C5
6R11	10	12BA7, 12BA6, 12BA6, 12AL5, 12SJ7, 50L6GT
6RT42A	4	12SA7, 12SK7, 12SQ7, 50L6GT, 35Z5GT/G
6RT43	3	12SA7, 12SK7, 12SQ7, 50L6GT, 35Z5GT
6RT44	4	6SK7, 6SA7, 6SK7, 6SQ7GT/G, 6SQ7GT/G, 6V6GT/G, 5Y3GT/G
6S12	17	12BA6, 12J5GT, 12SK7, 12SQ7, 35L6GT, 35Z5GT
6V12	12	12SA7, 12SK7, 12SQ7, 12SQ7, 35L6GT, 35Z5GT
6W12	14	12BA7, 12BA6, 12BA6, 12AL5, 12SJ7GT, 50L6GT
6Y18	15	1U4, 1R5, 1U4, 1U5, 3V4
703	39	Transistorized
717	39	Transistorized
739	36	Transistorized
742	37	Transistorized
751	37	Transistorized
7C60M	8	12SA7, 12SK7GT, 12SQ7, 12SJ7GT, 35L6GT, 35Z5GT
7C65W	8	6SA7, 6SK7, 6SQ7, 6SQ7, 6K6GT, 6K6GT, 5Y3GT
7C73	7	6BA6, 6BA6, 6SB7Y, 6BA6, 6BA6, 6AL5, 6SJ7, 6V6GT, 5Y3GT
7G14	11	6SA7, 6SK7, 6SQ7, 6SQ7, 6K6GT, 6K6GT, 5Y3GT
7L16	29	Transistorized
7M14	28	Transistorized
7P33	5	1U4, 1R5, 1U4, 1S5, 3V4
7RT42	6	12SA7, 12SK7, 12SQ7, 12SJ7, 35L6GT, 35Z5GT
7T01	7	12SA7, 12SK7GT, 12SQ7, 50L6GT, 35Z5GT
7T04	7	12SA7, 12SK7GT, 12SQ7, 50L6GT, 35Z5GT
7T10	6	12SA7, 12SK7, 12SQ7 OR 14B6, 35Z5GT OR 35Y4, 50L6GT OR 50A5
7T12	5	1A7GT, 1N5GT, 1H5GT, 3Q5
801	36	Transistorized
816B	40	Transistorized

BRAND/MODEL	PAGE	USES THESE TUBES
8D15	13	6BA6, 6SB7Y, 6BA6, 6BA6, 6AL5, 6SQ7, 6V6GT, 5Y3GT
909	41	Transistorized
9B14	9	6BA6, 6BA, 6SB7Y, 6BA6, 6BA6, 6AL5, 6SQ7, 6V6GT, 5U4G
9E15	14	12AT7, 6BA6, 6BA6, 6AL5, 6SQ7, 6SJ7, 6K6GT, 6K6GT, 5U4G
HIFI6	25	6AU6, 6AU6, 6AU6, 6BJ6, 12AT7, 6CB6, 6BJ6, 6AV6, 12AX7, 12AT7, 5Y3GT, 6AL7GT, 6AV6, 6AL5, 6AU6, 6SN7GT, 6SN7GT, 5881, 5881, 5U4G
SS642	34	6DJ8, 12AT7, 6AU6, 6BE6, 6BA6, 6AU6, 6AL5, 12AX7, 12AX7, 6BQ5, 6BQ5, 5AR4, 6AV6, 12AX7, 6BQ5, 6BQ5, 5Y3GT
Y1149	41	6DJ8, 12AT7, 6BE6, 6BA6, 6AU6A, 6AU6A, 6AL5, 12AU7, 12AU7, 12AU7A, 6BQ5, 6BQ5, 12AU7, 6BQ5, 6BQ5, 5U4GB
Y2027	42	Transistorized
Y833	38	12BE6, 12BA6, 12AV6, 50C5, 35W4
Y858	40	12BE6, 12BA6, 12AV6, 50C5, 35W4

AERMOTIVE

BRAND/MODEL	PAGE	USES THESE TUBES
181-AD	42	14A7, 14Q7, 14A7, 14B6, 50A5, 50A5, 35Y4, 35Y4

AIR CHIEF

BRAND/MODEL	PAGE	USES THESE TUBES
4-A-10	46	12SA7, 12SK7, 12SQ7, 35L6GT, 35Z5GT
4-A-11	50	12BA6, 12BE6, 12SK7, 12SQ7, 35L6GT, 35Z5GT
4-A-12	51	12AT7, 12BE6, 12BA6, 12BA6, 19T8, 50L6GT
4-A-15	49	6AG5, 6SK7, 6AG5, 6C4, 6SA7, 6SK7, 6SK7, 6SK7, 6H6, 6SQ7, 6SQ7, 6SJ7, 6V6GT, 6V6GT, 5A4G
4-A-17	49	7A8, 7B7, 7C6, 7C6, 35L6GT, 35L6GT, 35Z5GT
4-A-2	44	12BE6, 12BA6, 12AT6, 50B5, 35W4
4-A-20	45	12SK7, 12SA7, 12SF7, 12SK7, 35L6GT, 35Z5GT
4-A-24	43	1A7GT, 1N5GT, 1H5GT, 1A5GT
4-A-25	43	12SA7GT, 12SK7, 12SQ7, 50L6GT, 35Z5GT
4-A-26	48	12SA7, 12SK7, 12SQ7, 50L6GT, 35Z5GT
4-A-27	46	12SA7, 12SK7, 12SQ7, 50L6GT, 35Z5GT
4-A-3	47	12BD6, 12BE6, 12BD6, 12AT6, 35L6GT, 35W4
4-A-37	44	6SK7, 6SA7, 6SK7, 6SQ7, 6SQ7, 6V6GT, 6V6GT, 5Y3GT
4-A-42	47	6AG5, 6SB7Y, 6SG7, 6SA7, 6SF7, 6H6, 6SL7GT, 6V6GT, 6V6GT, 5Y3GT,
4-A-60	50	6BE6, 6SB7Y, 6BA6, 6BA6, 6AL5, 6SQ7, 6SJ7, 6V6GT, 6Y3GT
4-A-61	51	12SA7, 12SK7, 12SQ7, 50L6GT, 35Z5GT
4-C-3	45	1N5GT, 1A7GT, 1N5GT, 1H5GT, 3Q5GT, 117Z6GT
4-C-5	48	1R5, 1T4, 1S5, 3Q4

AIR KING

BRAND/MODEL	PAGE	USES THESE TUBES
4604	52	7H7, 7A4, 6SK7GT, 6V6GT, 6X5GT, 7B6
4700	57	7Q7, 7A7, 7B6, 7C5, 1273, 7C5, 7Z4
4704	53	12SK7GT, 12SA7GT, 12SK7GT, 12SQ7GT, 35L6GT, 35Z5GT
4705	52	12SK7GT, 12SA7GT, 12SK7GT, 12SQ7GT, 35L6GT, 35Z5GT
800	60	6BA6, 6BE6, 6BA6, 6BA6, 6H6, 6AT6, 6V6GT, 5Y3GT
A-400	54	12SA7GT, 12SQ7GT, 50L6GT, 35Z5GT
A-403	53	12SA7GT, 12SQ7GT, 50L6GT, 35Z5GT
A-410	56	1R5, 1T4, 1S5, 3Q4
A-410 (revised)	57	1R5, 1T4, 1S5, 3Q4
A-426	58	1R5, 1T4, 1S5, 3S4
A-502	56	12SA7GT, 12SK7GT, 12SQ7GT, 50L6GT, 35Z5GT
A-510	54	1R5, 1U4, 1S5, 3Q4
A-511	55	12SA7GT, 12SK7GT, 12SQ7GT, 50L6GT, 35W4 OR 35Z5GT
A-520	59	1R5, 1T4, 1S5, 3V4
A-600	55	12SK7GT, 12SA7GT, 12SK7GT, 12SQ7GT, 35L6GT, 35W4
A-604	60	6SK7GT, 6SA7GT, 6SK7GT, 6SQ7GT, 6V6GT, 6X5GT
A-625	59	12SK7GT, 12SA7GT, 12SK7GT, 12SQ7GT, 35L6GT, 35W4
A-650	58	14F8, 12SA7GT, 12SK7GT, 12SQ7GT, 35L5GT, 35W4

AIR KNIGHT

BRAND/MODEL	PAGE	USES THESE TUBES
CA-500	61	12BE6, 12BA6, 12AT6, 50B5, 35W4
N5-RD291	61	12BE6, 12BA6, 12AT6, 50B5, 35W4

AIRADIO

BRAND/MODEL	PAGE	USES THESE TUBES
3100	62	12BA6, 12BE6, 6BJ6, 6BJ6, 12BA6, 12AL5, 6AQ6, 50B5

AIRCASTLE

BRAND/MODEL	PAGE	USES THESE TUBES
10002	76	12SK7GT, 12SA7GT, 12SK7GT, 12SQ7GT, 35L6GT, 35Z5GT
10003-I	77	12SA7GT, 12SK7GT, 12SQ7GT, 50L6GT, 35Z5GT
10005	80	12AT7, 12BA6, 12BE6, 12BA6, 12BA6, 12AL5, 6AQ6, 35B5
10023	78	12BE6, 12BA6, 12AT6, 50B5, 35W4
102B	89	1R5, 1U4, 1U5, 3V4, 117Z3
106B	62	12SA7, 12SK7, 12SQ7, 50L6GT, 35Z5GT

BRAND/MODEL	PAGE	USES THESE TUBES
108014	78	6SK7, 6SA7, 6SK7, 6SQ7, 6K6GT, 5Y3GT
121104	84	12BA6, 12BE6, 12BA6, 12BA6, 12BA6, 6AL5, 12BE6, 12BA6, 12AT6, 50L6GT,
131504	79	12BA6, 12BE6, 12BA6, 12BA6, 12BA6, 6AL5, 12BE6, 12BA6, 12AT6, 50L6GT,
132564	82	1A7GT, 1N5GT, 1H5GT, 3Q5GT
138104	76	6SK7, 6SA7GT, 6SK7, 6SQ7GT, 6K6GT, 5Y3GT
138124	81	6SK7, 6SA7, 6SK7GT, 6SQ7, 6K6GT, 5Y3GT
147114	77	1R5, 1U4, 1U5, 3V4
150084	83	6BE6, 12AT7, 6BA6, 6BA6, 6AL5, 6AT6, 6AQ5, 5Y3GT
153	93	12BE6, 12BA6, 12AT6, 12AT6, 50C5, 50C5, 35W4
171	88	12SA7, 12SK7, 12SQ7, 50L6GT, 35Z5GT
198	85	6BJ6, 12AT7, 12BE6, 6BJ6, 12BA6, 19T8, 35C5, 35W4
201	85	12BA6, 12BE6, 12BA6, 12AT6, 35C5, 35W4
211	81	12BE6, 12AT6, 50B5, 35W4
212	82	6BJ6, 12AT7, 12BE6, 6BJ6, 12BA6, 19T8, 35C5, 35W4
213	80	1R5, 1U4, 1U5, 3S4, 117Z3
227I	86	12BE6, 12AT6 OR 12AV6, 50B5, 35W5
350	95	12AT7, 6BE6, 6BA6, 6BA6, 6AL5, 6AT6, 6AQ5, 6X4
472-053VM	96	12BE6, 12BA6, 12AT6, 50C5, 35W4
472.254	100	12BE6, 12BA6, 12AT6, 12AT6, 50C5, 50C5, 35W4
5000	65	12SA7GT, 12BA6 OR 12SK7GT, 12AT6 OR 12SQ7GT, 50L6GT, 35W4 OR 35Z5GT
5001	65	12SA7GT, 12BA6 OR 12SK7GT, 12AT6 OR 12SQ7GT, 50L6GT, 35W4 OR 35Z5GT
5002	67	12SK7 OR 12BA6, 12SA7, 12SK7 OR 12BA6, 12SQ7 OR 12AT6, 35L6GT, 35Z5GT
5003	68	12K7GT, 12SA7, 12SK7GT, 12SQ7GT, 50L6GT
5008	70	12BA6, 12BE6, 12BA6, 12AT6, 35L6GT, 35W4
5011	63	12SK7GT, 12SA7GT, 12SK7GT, 12AT6, 35L6GT, 35W4
5015.1	91	12SA7, 12SK7, 12SQ7, 50L6GT, 35Z5GT
5020	66	1A7GT, 1N5GT, 1LH4, 3Q5GT
5022	93	1R5, 1U4, 1U5, 3V4
5024	69	1A7GT, 1N5GT, 1H5GT, 3Q5GT
5025	68	1A7GT, 1N5GT, 1H5GT, 3Q5GT
5027	72	1A7GT, 1N5GT, 1H5GT, 3Q5GT
5029	74	1R5, 1U4, 1S5, 3S4 OR 3V4

BRAND/MODEL	PAGE	USES THESE TUBES
5035	71	12BE6, 12BA6, 12AT6, 50B5, 35W4
5036	83	12SA7GT, 12SQ7GT, 50L6GT, 35Z5GT
5050	72	12BE6, 12AT6, 50B5, 35W4, 12BA6
5052	69	12BE6, 12BA6, 12AT6, 50B5, 35W4
5056-A	92	12SA7, 12SQ7, 50L6, 35Z5
568	64	12SA7, 12SK7GT, 12SQ7, 50L6GT, 35Z5GT
603-880	100	6BE6, 6BA6, 6AV6, 6AQ5, 6AV6, 6X4
6050	84	12SA7GT, 12SK7GT, 12SQ7GT, 50L6GT, 35Z5GT
6053	88	12SA7GT, 12SK7GT, 12SQ7GT, 50L6GT, 35Z5GT
606-400WB	91	1A7GT, 1N5GT, 1H5GT, 3Q5GT
607.299	98	12AU6, 12AV6, 50C5, 35W4
607-314	92	12AU6, 6C4, 12BA6, 12AV6, 50C5, 35W4
607-316-1	96	12BE6, 12BA6, 12AV6, 50C5 OR 35B5, 35Z5GT OR 35Z4GT OR 35W4
610.CL152B,M	99	12BE6, 12BA6, 12AT6, 50C5, 35W4
651	64	14Q7, 14A7/12B7, 14B6, 50A5, 35Y4
6514	67	14Q7, 14A7/12B7, 14B6, 50A5, 35Y4
652.505	97	12BE6, 12BA6, 12AV6 OR 12AT6, 50C5, 35W4
652.5C1M	101	12BE6, 12BA6, 12AT6, 50C5, 35W4
652.5C1M,V	101	12BE6, 12BA6, 12AT6, 50C5, 35W4
652.5T5E	102	12BE6, 12BA6, 12AT6, 50C5, 35W4
652.5X5	102	12BE6, 12BA6, 12AT6, 12AX7, 35L6GT, 35L6GT
652.6T1E	99	6BJ6, 12BE6, 6BJ6, 12AT6, 50C5, 35W4
6541, 6547	66	14Q7, 14A7, 14B6, 50A5, 35Y4
659.511	97	12BE6, 12AT6 OR 12AV6, 50C5, 35W4
659.520E	98	12BE6, 12AV6 or 12AT6, 50C5, 35W4
6634	65	14Q7, 14A7, 14B6, 35A5, 35A5, 35Y4
7553	70	14Q7, 14A7, 14B6, 50A5, 35Y4
782.5C1	103	12BE6, 12BA6, 12AT6, 50C5, 35W4
782.FM-99-AC	103	6CB6, 12AT7, 6BE6, 6BA6, 6AU6, 6AL5, 6AT6, 6AS5, 5Y3GT
7B	74	6BA6, 7Q7, 7F8, 7AH7, 6SH7, 7A6, 6SQ7, 7F7, 7C5, 7C5, 5Y3GT
9	73	14F8, 12BE6, 12BA6, 12AT6, 35B5, 35W4
9008W	90	12BE6, 12AV6 OR 12AT6, 50B5, 35W4
9009W	89	12BE6 OR 12SA7, 35W4 OR 35Z5, 12BA6 OR 12SK7, 12AT6 OR 12SQ7, 50B5
9012W	87	12BE6, 12AV6 OR 12AT6, 50B5, 35W4
915I,W	95	12BA6, 12AT6, 50B5, 35W4
935	94	12BE6, 12BA6, 12AT6, 50C5, 35W4
DM700	86	1R5, 1L4, 1U5, 3S4
G-516	71	12SA7GT, 12SK7GT, 12SQ7GT, 50L6GT, 35Z5GT
G-518	71	12SA7GT, 12SK7GT, 12SQ7GT, 50L6GT, 35Z5GT

BRAND/MODEL	PAGE	USES THESE TUBES
G-724	75	6SB7Y, 12SG7, 12SG7, 7A6, 12SQ7GT, 35L6GT, 35Z5GT
G-725	73	6SB7Y, 12SG7, 12SG7, 7A6, 12SQ7GT, 35L6GT, 35Z5GT
G521	75	1LA6, 1LN5, 1LN5, 1LH4, 3LF4
PC-8	90	12SA7, 12SK7, 12SQ7, 50L6GT, 35Z5GT
PX	63	1A7GT, 1N5GT, 1H5GT, 3Q5GT
REV248	94	12SK7, 12SA7, 12SK7, 12SQ7, 35L6GT, 35Z5GT
SC-448	79	6BE6, 6BA6, 6BA6, 6AL5
WEU-262	87	6BA6, 6BE6, 6BA6, 6BA6, 6AL5, 6SQ7, 25L6GT, 25Z6GT

AIRLINE

BRAND/MODEL	PAGE	USES THESE TUBES
05GAA-992A	132	12SA7GT, 12SK7, 12SQ7, 50L6, 35Z5GT
05GCB-1541A	134	12BE6r, 12AV6 or 12AT6, 50C5, 35W4
05GHM-1061A	134	1R5, 1T4, 1U5, 3Q5
05WG-1813A	132	6AB4, 6AB4, 6BE6, 6BA6, 6BA6, 6AL5, 6AV6, 6V6GT, 6X5GT
05WG-2748F	135	6AB4, 6AB4, 6BE6, 6BA6, 6BA6, 6AL5, 6AV6, 6V6GT, 6X5GT
05WG-2749D	133	6AB4, 6AB4, 6BE6, 6BA6, 6BA6, 6AL5, 6AV6, 6V6GT, 5Y3GT
05WG-2752	131	12AT7, 6BE6, 6BA6, 6BA6, 6AL5, 6AV6, 6V6GT, 5Y3GT
15BR-1536B	137	12BE6, 12BA6, 12AV6, 50C5, 35Z5GT
15BR-1544A	136	12BE6, 12BA6, 12AV6 OR 12AT6, 50C5, 35Z5GT
15BR-1547A	135	12BE6, 12BA6, 12AV6, 50L6GT
15BR-2756B	137	12AT7, 6BA7, 6BA6, 6AU6, 6AL5, 6AV6, 6V6GT, 5Y3GT
15GAA-995A	139	12SA7, 12SK7, 12SQ7, 50L6, 35Z5GT
15GHM-1070A	142	1R5, 1L4 OR 1T4, 1U5, 3V4 OR 3Q5
15GHM-934A	139	12BE6, 12BA6, 12AV6, 50B5, 35W4
15GSE-2764A	138	12SK7, 12SA7, 12SK7, 12SQ7, 35L6GT, 35Z5GT
15GSL-1564A	140	12AU6, 12AV6, 50C5, 35Z5GT
15WG-2745C	133	6BA6, 12AT7, 6BA6, 6BA6, 6AL5, 6AV6, 6AV6, 6K6GT, 6K6GT, 5Y3GT,
15WG-2749F	138	12AT7, 6BE6, 6BA6, 6BA6, 6AL5, 6AV6, 6V6GT, 5Y3GT
15WG-2758A	136	12AT7, 6BE6, 6BA6, 6BA6, 6AL5, 6AV6, 6V6GT, 6X5
2509	155	6BE6, 6BA6, 6AV6, 12AU7, 6BL7GT, 5Y3GT
25BR-1542A	145	12BE6, 12BA6, 12AV6, 50L6GT
25BR-1549B	143	12BE6, 12BA6, 12AV6 or 12AT6, 50C5
25GAA-996A	141	12BE6, 12BA6, 12AT6, 50C5, 35W4
25GHM-1073A	147	1T4, 1R5, 1U4, 1U5, 3V4
25GHM-2012A	150	12BE6, 12BA6, 12AT6, 50C5, 35W4
25GSE-1555A	140	12BE6, 12BA6, 12AV6, 50C5, 35W4
25GSL-1560A	142	12AU6, 12AV6, 50C5, 35W4
25GSL-1814A	144	12AU6, 6C4, 12BA6, 12AV6, 50C5, 35W4
25WG-1570A	141	12AT7, 6BE6, 6BA6, 6BA6, 6AL5, 6AU6, 6V6GT, 6X5GT
25WG-1573A	144	6BA6., 6BE6, 6BA6, 6AV6, 6V6GT, 6X5GT
25WG-2758C	143	12AT7, 6BE6, 6BA6, 6BA6, 6AL5, 6AV6, 6V6GT, 6X5GT
35BR-1557A	149	12BE6, 12BA6, 12AV6 OR 12AT6, 50C5, 35W4
35GAA-3969A	145	6BE6, 6BA6, 6AV6, 6AQ5, 6AV6, 6X4
35GHM-1074A	147	1R5, 1U4, 1U5, 3V4
35GSL-2770A	148	12BA6, 12BE6, 12BA6, 12AV6, 35C5, 35W4
35WG-1573B	146	6BA6, 6BE6, 6BA6, 6AV6, 6V6GT, 6X5GT
35WG-2767A	146	6BA6, 12AT7, 6BA6, 6BA6, 6AL5, 6AV6, 12AX7, 6V6GT, 6V6GT, 12AX7, 6X4, 6X4
54KP-1209B	107	1N5GT, 1A7GT, 1N5GT, 1H5GT, 3Q5GT
54WG-2500A	104	6SJ7, 6J5, 6SK7, 6SK7, 6SQ7, 6V6GT, 5Y3GT
54WG-2700A	104	6SJ7, 6J5, 6SK7, 6SK7, 6SQ7, 6V6GT, 5Y3GT
64BR-1205A	109	1R5, 1T4, 1S5, 3S4
64BR-1208A	109	1A7GT/G, 1N5GT/G, 1N5GT/G, 1H5GT/G, 3Q5GT/G
64BR-1808A	110	6SK7GT, 6SA7, 6SK7GT, 6SQ7GT/G, 6J5GT, 6K6GT/G, 6K6GT/G, 5Y3GT/G
64WG-1050A	108	1R5, 1T4, 1S5, 3S4
64WG-1052A	108	1R5, 1U4, 1U4, 1S5, 3Q4
64WG-1207B	110	1R5, 1U4, 144, 1S5, 3Q4
64WG-1511A	106	12SK7, 12SA7, 12SF7, 12SJ7, 35L6GT, 35Z5GT
64WG-1801C	105	12SA7, 12SK7, 12SQ7, 50L6GT, 35Z5GT
64WG-1804B	104	12SK7, 12SA7, 12SF7, 12SJ7, 35L6GT, 35Z5GT
64WG-1807A	105	6SA7, 6SK7, 6SF7, 6SJ7, 6V6GT, 6X5GT
64WG-1809A	106	12SK7, 12SA7, 12SF7, 12SJ7, 35L6GT, 35Z5GT
64WG-2007B	106	12SA7, 12SK7, 12SQ7, 50L6GT, 35Z5GT
64WG-2009A	107	12SK7, 12SA7, 12SF7, 12SJ7, 35L6GT, 35Z5GT
74BR-1514B	112	12SG7, 12SA7, 12SF7, 12SJ7, 35L6GT, 35Z5GT
74BR-2001B	112	12BE6, 12BD6, 12AT6, 50B5, 35W4
74BR-2701A	113	6SK7GT, 6K6GT/G, 6K6GT/G, 5Y3GT/G, 6SA7, 6SK7GT, 6SQ7GT/G, 6J5GT
74KR-1210A	118	1A7GT, 1N5GT, 1H5GT, 3Q5GT
74KR-2713A	119	6SK7, 6SA7, 6SK7, 6SQ7, 6V6GT, 5Y3GT

BRAND/MODEL	PAGE	USES THESE TUBES
74KR2706B	116	6SK7, 6SA7, 6SK7, 6SQ7, 6V6GT, 5Y3GT
74WG-1056A	115	1R5, 1U4, 1U4, 1S5, 3V4
74WG-1057A	116	1R5, 1U4, 1S5, 3S4
74WG-1510A	114	12SK7, 12SA7, 12SF7, 12SJ7, 35L6GT, 35Z5GT
74WG-2002A	113	12SK7, 12SA7, 12SF7, 12SJ7, 35L6GT, 35Z5GT
74WG-2004A	121	12SA7, 12SK7, 12SQ7, 50L6GT, 35Z5GT
74WG-2010B	111	6SA7, 6SK7, 6SF7, 6SJ7, 6V6GT, 6X5GT
74WG-2504A	115	6SA7, 6SK7, 6SF7, 6SJ7, 6V6GT, 6X5GT
74WG-2505A	111	6BA6, 6J6, 6BE6, 6BA6, 6BA6, 6AL5, 6AT6, 6U5GT, 6V6GT, 5Y3GT,
74WG-2704A	115	6SA7, 6SK7, 6SF7, 6SJ7, 6V6GT, 6X5GT
74WG-2709A	114	6SJ7, 6J5, 6SK7, 6SK7, 6SQ7, 6V6GT, 5Y3GT
84GAA3967A	130	6SA7GT, 6SK7GT, 6SQ7GT, 6SJ7GT, 6V6GT, 6X5GT
84GCB-1062A	121	1R5, 1T4, 1S5, 3S4
84GSE2731A	125	12BE6, 12BA6, 12AT6, 50C5, 35W4
84HA-1810C	124	12BE6, 6BJ6, 6BJ6, 12AL5, 12SQ7, 35L6GT, 35Z5GT
84KR-1520A	122	12SG7, 12SQ7, 50L6GT, 35Z5GT
84KR2511A	124	12SA7, 12SK7, 12SQ7, 50L6GT, 35Z5GT
84WG-1060A	118	1R5, 1U4, 1U5, 3V4
84WG-2015A	117	6BE6, 6BA6, 6BA6, 6AL5, 6AV6, 6V6GT, 6X5GT
84WG-2506B	123	6BE6, 6BA6, 6BA6, 6AL5, 6AV6, 6V6GT, 5Y3GT
84WG-2712A	119	6BA6, 12AT7, 6BE6, 6BA6, 6BA6, 6AL5, 6U5, 6AV6, 6AV6, 6V6GT,6V6GT, 5U4GT
84WG-2714A	117	6BE6, 6BA6, 6BA6, 6AL5, 6AT6, 6V6GT, 5Y3GT
84WG-2714F	122	6BA7, 6BA6, 6BA6, 6AL5, 6AV6, 6V6GT, 5Y3GT
84WG-2720A	120	6BA6, 12AT7, 6BE6, 6BA6, 6BA6, 6AL5, 6AV6, 6U5/6G5, 6V6GT, 5Y3GT,
84WG-2721A	120	6BE6, 6BA6, 6BA6, 6AL5, 6AV6, 6V6GT, 5Y3GT
94BR-1533A	128	12AT7, 12BA7, 12BA6, 12AU7, 12AL5, 12AV6, 50C5
94BR-2740A	129	12AT7, 12BA7, 12BA6, 12AU6, 12AL5, 12AV6, 50L6GT
94GCB-1064A	130	1R5, 1T4, 1S5, 3S4
94GSE-2735A	126	6BA6, 6BE6, 6BA6, 6AU6, 6AL5, 6AV6, 6V6GT, 5Y3GT
94HA1528C	123	12SA7, 12SK7, 12SQ7, 50L6GT, 35Z5GT
94HA1529A	127	12BE6, 6BJ6, 6BJ6, 12AL5, 12SQ7, 35L6GT, 35Z5GT

BRAND/MODEL	PAGE	USES THESE TUBES
94WG-1059A	126	1T4, 1R5, 1U4, 1U5, 3V4
94WG-1804D	128	12SK7, 12SA7, 12SJ7, 35L6GT, 35Z5GT, 12SF7
94WG-1811A	131	12AT7, 6BE6, 6BA6, 6BA6, 6AL5, 6AV6, 6V6GT, 6X5GT
94WG-2742A	125	6BA7, 6BA6, 6BA6, 6AL5, 6AV6, 6V6GT, 5Y3GT
94WG-2745A	127	6BA6, 12AT7, 6BA6, 6BA6, 6AL5, 6AV6, 6VA6, 6K6GT, 6K6GT, 5Y3GT,
94WG-2748A	129	12AT7, 6BE6, 6BA6, 6BA6, 6AL5, 6AV6, 6V6GT, 6X5GT OR 5Y3GT
GAA-1003A	160	12BE6, 12BA6, 12AT6, 50C5, 35W4
GAA-2620A	159	5UHGB, 6V6GT, 6SN7GTB, 6V6GT, 12AU7, 6BA6, 6BE6, 6AL5, 6AU6, 6AU6, 6AU6, 12AT7, 6AB4, 6CB6
GAA-2640A	165	6CB6A, 6AB4, 12AT7, 6AU6A, 6AU6A, 6AL5, 6BA6, 6BE6, 6BA6, 12AX7, 12AX7, 6BQ5, 6BQ5, 6BQ5, 6BQ5
GAA-990A	154	12BE6, 12BA6, 12AT6, 50C5, 35W4
GEN-1090A	156	1R5, 1U4, 1U5, 3V4
GEN-1103A	156	1R5, 1U4, 1U5, 3V4
GEN-1120C	163	Transistorized
GEN-1655A	157	12BE6, 12BA6, 12AV6, 50C5, 35W4
GEN-1670A	161	12AU6, 12AV6, 50C5, 35W4
GEN-2645A	162	6CB6, 12AT7, 6BA6, 6BA6, 6BE6, 6BA6, 6BA6, 6AL5, 6E6, 12AU7, 12AU7, EL84/6BQ5, EL84/6BQ5, EL84/6BQ5, EL84/6BQ5, 5U4GB
GSE-1077A	148	1R5, 1U4, 1U5, 3V4
GSE-1606A	151	12BE6, 12BA6, 12AV6, 50C5, 35W4
GSE-1620A	153	12BE6, 12BA6, 12AV6, 50C5, 35W4
GSE-1625A	155	12BE6, 12BA6, 12AV6, 50C5, 35W4
GSL-1079-A	152	1U4, 1U4, 1U5, 3V5, 1L6
GSL-1575A	154	12AU6, 12AV6, 50C5, 35W4
GSL-1581A	150	12BE6, 12BA6, 12AV6, 50C5, 35W4
GSL-1614A	151	12AU6, 12AV6, 50C5, 35W4
GSL-1650A	158	12BE6, 12BA6, 12AV6, 50C5, 35W4
GTC-1085A	157	1R5, 1U4, 1U5, 3V4
GTM-1108A	159	Transistorized
GTM-1109A	160	Transistorized
GTM-1117A	164	1T4, 1R5, 1U4, 1U5, 3V4
GTM-1201A	165	Transistorized
GTM-1639B	161	12BE6, 12BA6, 12BA6, 12AV6, 35C5, 35W4
GTM-1666A	164	12BA6, 12AT7, 12AU6, 12BA6, 12AL5, 12AV6, 50C5
WG-1572C	149	6BJ6, 6CB6, 6AU6, 6BA6, 6AL5, 6AV6, 6AQ5, 6X5GT
WG-1635A	152	12BA6, 12BF6, 12BA6, 12AV6, 35C5, 35W4
WG-1637A	158	12BA6, 12BF6, 12BA6, 12AV6, 35C5, 35W4
WG-2602A	153	6BA6, 6BE6, 6BA6, 6AV6, 12AX7, 6AQ5, 6AQ5, 5Y3GT

BRAND/MODEL	PAGE	USES THESE TUBES
WG-2673A	162	6CB6, 12AT7, 6AB4, 6AU6, 6BA6, 6BE6, 6BA6, 6AU6, 6AL5, 7025,12AX7, EL84/6BQ5, EL84/6BQ5, EL84/6BQ5, EL84/6BQ5, 5Y3GT, 5U4GB
WG-2684A	163	6CB6, 12AT7, 6AB4, 6AU6A, 6BA6, 6BE6, 6BA6, 6AU6, 6AL5

ALGENE

AR5U	166	14Q7, 14A7, 14B6, 50A5, 35Y4
AR-6U	166	1R5, 1T4, 1T4, 1S5, 3Q4, 117Z3

AMC

126	167	6SS7, 12SA7, 6SS7, 12SQ7, 50L6GT, 35Z5GT

AMI

PBA (Mark 1)	167	6AN4, 6AB4, 12AT7, 6BA6, 6BE6, 6BA6, 6BA6, 6AU6, 6AU6, 6AL5, 12AU7, 6X4, 12AD7, 6CG7, 12AX7, 6L6GB, 6L6GB, 5U4GB

ANDREA

CO-U15	170	12SA7, 12SK7, 12SQ7, 50L6GT, 35Z5GT
CRP-24W	170	6BA6, 12AT7, 6BE6, 6BA6, 6AU6, 6AL5, 12AD7, 6AV6, 6X5GT, 12AX7, 6V6GT, 6V6GT, 6V6GT, 6V6GT, 5Y3GT, 5Y3GT
CRP-24WA	171	6CB6, 12AX7, 6U8, 6V6GT, 6AB4, 6V6GT, 6BH6, 6V6GT, 6BE6, 6V6GT, 6BH6, 5Y3GT, 6BH6, 5Y3GT, 6BH6, 12AX7, 6AL5, 6V6GT, EM81, 6V6GT
P-I63	168	1R5, 1N5GT, 1N5GT, 1H5GT, 3Q5GT, 35Z5GT
T-16	168	6K8, 6SK7, 6H6, 6SF5, 6V6GT, 5Y3GT
T-U15	169	12SA7, 12SK7, 12SQ7, 50L6GT, 35Z5GT
T-U16	169	6K8, 6SK7, 6H6, 6SF5, 25L6G, 25Z6G
W69P Spacemaster Deluxe	171	1U4, 1R5, 1U5, 3V4, 35Z5GT, 1U4

ANSLEY

32	172	12SK7GT, 12SA7, 6SK7GT, 6SQ7GT, 6J5, 50L6GT, 50L6GT, 25Z5GT, 25Z5GT
41 Paneltone	172	6SK7, 6SA7, 6SK7, 6SQ7GT, 6U5/6G5, 6V6GT, 5Y3GT
53	173	6BA6, 6BE6, 6C5, 6AC7, 6AC7, 6SJ7, 6SJ7, 6H6GT, 6SK7, S6SA7, 6SK7, 6SQ7GT, 6E5, 6J5GT, 6V6GT, 6V6GT, 5Y3GT, 6C4

APEX INDUSTRIES

4B5	173	12BE6, 12BA6, 12AT6, 50B5, 35W4

ARCADIA

37D14-600	174	12SA7, 12SK7, 6SS7, 12SQ7, 12SJ7, 35L6GT, 35Z5GT

ARIA

554-1-61A	174	6SK7, 6SA7GT, 6SK7, 6SQ7GT, 6K6GT, 5Y3GT

ARTHUR ANSLEY

R-1	175	6CB6, 6U8, 6BJ6, 6BE6, 6BA6, 6AU6, 6AU6, 6T8, 6V6GT, 5Y3GT,

ARTONE

524	175	14Q7, 14A7, 14B6, 50A5, 35Y4

ARVIN

151TC	179	12BD6, 12BE6, 12BD6, 12AT6, 50L6GT
152T	178	12SA7, 12SK7, 12SQ7, 50L6GT, 35Z5GT
1581	209	12BE6, 12AV6, 50C5, 35W4
160T	181	12SK7, 12SA7, 12SK7, 12SQ7, 35L6GT, 35Z5GT
182TFM	177	6C4, 12BA6, 12BE6, 12BA6, 12BA6, 6AQ6, 12H6, 50L6GT
240-P	179	1R5, 1U4, 1S5, 1LB4
2410P	181	1R5, 1U4, 1S5, 1LB4
242T	182	12SA7, 12SQ7, 50L6GT, 35Z5GT
250-P	180	1U4, 1R5, 1U4, 1S5, 3V4
255T	182	12SK7, 12SA7, 12SA7, 50L6GT, 35Z5GT
2563	204	12BE6, 12BA6, 12AV6, 50C5, 35W4
2572	206	12BE6, 12BA6, 12AV6, 50C5, 35W4
2581	210	12BE6, 12BA6, 12AV6, 50C5, 35W4
2585	210	12BE6, 12BA6, 12AV6, 50C5, 35W4
2598	215	Transistorized
264T	183	12SK7GT, 12SA7GT, 12SK7GT, 12SQ7, 35L6GT, 35Z5GT
341T	185	12SA7, 12SQ7, 50L6GT, 35Z5GT
350-PL	186	1U4, 1R5, 1U4, 1U5, 3V4
350P	183	1U4, 1R5, 1U4, 1U5, 3V4
3561	205	6BJ6, 12BE6, 6BJ6, 12AV6, 50C5, 35W4
356T	184	12SA7GT, 12SK7, 12SQ7GT, 50L6GT, 35Z5GT
3582	213	12BA6, 12BE6, 12BA6, 12AV6, 12AV6, 25F5, 25F5
3586	213	6BQ7A, 6BA6, 6BA6, 6AL5, 6BE6, 6BA6, 12AX7, 6AQ5A, 6AQ5A
3588	212	Transistorized
360TFM	184	12AT7, 12BE6, 12BA6, 12BA6FM, 19T8, 50L6GT
440T	185	12SA7GT, 12SQ7GT, 50L6GT, 35Z5GT
442	178	12SA7, 12SQ7, 50L6GT, 35Z5GT
446P	186	1R5, 1T4, 1U5, 3S4

BRAND/MODEL	PAGE	USES THESE TUBES
450T	188	12BE6, 12BA6, 12AT6, 50C5, 35W4
460T	187	12SK7GT, 12SA7GT, 12SK7GT, 12SQ7GT, 35L6GT, 35Z5GT
462-CM	188	12SK7GT, 12SA7, 12SK7GT, 12SQ7GT, 35L6GT, 35Z5GT
480TFM	187	6BA6, 12AT7, 6BE6, 6BA6, 6BA6, 6T8, 6V6GT, 6X4
482CFB	189	6BA6, 12AT7, 6BE6, 6BA6, 6BA6, 6T8, 6C6GT, 6X4
540T	189	12SA7, 12SQ7GT, 50L6GT, 35Z5GT
547	180	12SA7, 12SK7, 12SQ7, 50L6GT, 35Z5GT
551T	190	6BE6, 6BA6, 6AV6, 6V6GT, 5Y3GT
553	192	12BE6, 12BA6, 12AT6 OR 12AV6, 50C5, 35W4
554CCM	191	6BE6, 6BA6, 6AV6, 6V6GT, 5Y3GT
555	176	12SA7GT, 12SK7GT, 12SQ7GT, 50L6GT, 35Z5GT
5561	204	12BE6, 12BA6, 12AV6, 50C5, 35W4
5571	207	12BE6, 12BA6, 12AV6, 50C5, 35W4
5578	208	12BE6, 12BA6, 12AV6, 50C5, 35W4
5583	211	12BE6, 12BA6, 12AV6, 50C5, 35W4
5591	214	18FX6, 18FW6, 18FY6, 32ET5, 36AM3
5594	215	18FX6, 18FW6, 18FY6, 32ET5, 36AM3
580TFM	190	6BA6, 12AT7, 6BE6, 6BA6, 6BA6, 6T8, 6V6GT, 6X4
581TFM	196	6CB6, 12AT7, 6BE6, 6BA6, 6BA6, 6T8, 6V6GT, 6X4
582CFB	191	6BA6, 12AT7, 6BE6, 6BA6, 6BA6, 6T8, 6V6GT, 6X4
60R23	218	Transistorized
60R33	217	Transistorized
60R49	217	Transistorized
650-P	193	1T4, 1R5, 1T4, 1U5, 3V4
651T	197	12BE6, 12BA6, 12AT6, 50C5, 35W4
655SWT	193	12BE6, 12BA6, 12AT6 OR 12AV6, 50C5, 35W4
657-T	192	12BE6, 12BA6, 12AT6 OR 12AV6, 50C5, 35W4
664	177	12SK7, 12SA7GT, 12SK7, 12SQ7GT, 35L6GT, 35Z5GT
665	176	6SS7, 12SA7, 6SS7, 12SQ7, 50L6GT, 35Z5GT
741T	195	12BE6, 12AV6, 50C5, 35W4
746P	196	1R5, 1T4, 1U5, 3V4
753T	194	12BE6, 12BA6, 12AV6, 50C5, 35W4
758T	194	12BE6, 12BA6, 12AV6, 50C5, 35W4
760T	195	12SK7, 12SA7, 12SK7, 12SQ7, 35L6GT, 35Z5GT
840T	199	12SA7, 12SQ7, 50L6GT, 35Z5GT
848T	198	12BE6, 12AV6, 50C5, 35W4
850T	199	12BE6, 12BA6, 12AV6, 50C5, 35W4
851T	200	12BE6, 12BA6, 12AV6, 50C5, 35W4
852P	197	1R5, 1T4, 1U5, 3V4
8571	208	1R5, 1U4, 1U5, 3V4
8576	209	Transistorized
857T	200	12BE6, 12BA6, 12AV6, 50C5, 35W4
8581	212	1R5, 1U4, 1U5, 3V4
8584	211	Transistorized
858T	198	12BE6, 12BA6, 12AV6, 50C5, 35W4
950T	201	12BE6, 12BA6, 12AV6, 50C5, 35W4
950T1	203	12BE6, 12BA6, 12AT6, 50C5, 35W4
950T2	206	12BE6, 12BA6, 12AV6, 50C5, 35W4
952P	201	1R5, 1T4, 1U5, 3V4
955T	202	12BE6, 12BA6, 12AT6, 50C5, 35W4
9562	203	Transistorized
956T1	207	12BE6, 12BA6, 12AV6, 50C5, 35W4
9574P	205	Transistorized
957T	202	12BE6, 12BA6, 12AV6, 50C5, 35W4
9594	216	Transistorized
9595	214	Transistorized
9598	216	Transistorized

ASTRA-SONIC

BRAND/MODEL	PAGE	USES THESE TUBES
748	218	6BE6, 6BA6, 6AT6, 6BJ7, 6C4, 6V6GT, 6X4

ATLAS

BRAND/MODEL	PAGE	USES THESE TUBES
AB-45	219	12SK7, 12SA7, 12SQ7, 35L6GT, 35Z5GT, 12SK7

AUDAR

BRAND/MODEL	PAGE	USES THESE TUBES
(Telvar) RER-9	220	6SK7, 6SA7, 6SF7, 6SQ7, 6U5/6G5, 6SN7GT, 6V6GT, 5Y3GT
AV-7T	220	6BE6, 6BA6, 6AT6, 12AX7, 6AK6, 6AK6, 5Y3GT
PR-6	219	14Q7 OR 12SA7, 14B6 OR 12SQ7, 50A5 RO 50L6GT, 35Y4 OR 35Z5GT

AUTOMATIC

BRAND/MODEL	PAGE	USES THESE TUBES
ATT (Tom Thumb)	225	1R5, 1T4, 1S5, 3S4
B-44	228	1R5, 1U4 OR 1T4, 1S5, 3S4
C-54	229	1R5, 1U4 or 1T4, 1U5, 3V4
C-60X	226	1R5, 1T4, 1S5, 3W4
C51	229	1U4 OR 1T4, 1R5, 1U4 OR 1T4, 1U5, 3V4
C60	221	1R5, 1T4, 1S5, 3Q4, 0Y4
CL-100	231	12BE6, 12BA6, 12AV6 OR 12AT6, 50C5, 35W4
CL-152B	230	12BE6, 12BA6, 12AT6, 50C5, 35W4
CL-75	232	12BE6, 12BA6, 12AV6 OR 12AT6, 50C5, 35W4
F-790	225	12SK7, 12SA7, 12SK7, 12SQ7, 12SQ7, 50L6GT, 50L6GT, 35Z5GT, 35Z5GT

BRAND/MODEL	PAGE	USES THESE TUBES
P-57	231	1U4 OR 1T4, 1R5, 1U4 OR 1T4, 1U5, 3V4
Tom Thumb Buddy	228	1R5, 1L4 OR 1T4, 1S5, 3S4
Tom Thumb Jr	226	1R5, 1T4, 1T4, 1S5, 3S4
Tom Thumb Camera/Radio	227	1R5, 1L4 OR 1T4, 1S4, 3S4
Tomboy	227	1R5, 1T4; 1S5, 3S4
TT528	232	1V6, 1AH4, 1AJ5, 1AG4
TT600	230	1V6, 1AH4, 1AJ5
601 (series A and B)	223	12SA7GT, 14A7/12B7, 12SQ7GT, 50L6GT, 35Z5GT
614X	221	12SK7GT, 12SA7GT, 12SK7GT, 12SQ7GT, 35L6GT, 35Z5GT
620	222	12SK7, 12SA7, 12SK7, 12SQ7, 35L6GT, 35Z5GT
640	222	14Q7, 14A7, 14B6, 50A5, 35Y4 OR TYPES:, 12SA7, 12SK7, 12SQ7, 50L6GT, 3
662	224	14A7, 14Q7, 14A7, 14B6, 35L6GT, 35Y4
677	224	14A7, 14Q7, 14A7, 14B6, 50L6GT, 35Y4, 35Y4
720	223	12SK7GT, 12SA7GT, 12SK7GT, 12SQ7GT, 35L6GT, 35Z5GT

AVIOLA

BRAND/MODEL	PAGE	USES THESE TUBES
509	233	12SA7, 12SK7, 12SQ7, 50L6GT/G, 35Z5GT/G
601	233	12SK7, 12SA7, 12SK7, 12SQ7, 35L6GT, 35Z5GT
608	234	12BA6, 12BE6, 12BA6, 12AT6, 35L6GT, 35W4

B.F.GOODRICH

BRAND/MODEL	PAGE	USES THESE TUBES
92-523	234	12BE6, 12BA6, 12AT6, 50C5, 35W4

BAGPIPER

BRAND/MODEL	PAGE	USES THESE TUBES
SKR101	235	1R5, 1U4, 1U5, 3V4

BELLTONE

BRAND/MODEL	PAGE	USES THESE TUBES
500	235	12SA7GT, 12SK7GT, 12SQ7GT, 50L6GT, 35Z5GT

BELMONT

BRAND/MODEL	PAGE	USES THESE TUBES
4B112 (series A)	237	1A7GT, 1N5GT, 1H5GT, 3Q5GT
5D110	238	12BE6, 12BD6, 12AT6, 50B5, 35W4
5D128 (series A)	236	12SA7GT, 12SK7, 12SQ7, 50L6GT, 35Z5GT
5P113 (Boulevard)	239	2E32, 2G22, 2E32, 2E42, 2E36
5P19 (series A)	237	1A7GT, 1N5GT, 1H5GT, 1A5GT, 35Z5GT
6D120	239	12SK7GT, 12SA7GT, 12SK7GT, 12SQ7GT, 35L6GT, 35Z5GT
8A59	236	6SK7, 6SA7, 6SK7, 6SQ7, 6J5GT, 6K6GT, 6K6GT, 5Y3GT
A-6D110	238	12SK7GT, 12SA7GT, 12SK7GT, 12SQ7GT, 35L6GT, 35Z5GT

BENDIX

BRAND/MODEL	PAGE	USES THESE TUBES
110	246	12SA7, 12SK7, 12SQ7, 50L6GT, 35Z5GT
1217B	244	6AG5, 7F8, 6BE6, 7AH7, 7AH7, 6H6, 6H6, 6SN7GT, 6SN7GT, 6V6GT,6V6GT, 5U4G
1217D	248	6AG5, 7F8, 6BE6, 7AH7, 7AH7, 6H6, 6H6, 6SN7GT, 6AQ6, 6BA6, 6SN7GT, 6V6GT, 6V6GT, 5U4G
1521	247	6AG5, 7F8, 7AH7, 7AG7, 6H6, 6SQ7, 6V6GT, 5Y3GT
1524	244	6AG5, 7F8, 7AH7, 7AG7, 6H6, 6AQ6, 6BA6, 6SQ7, 6V6GT, 5Y3GT,
1531	248	7B7, 7A8, 7B7, 7C6, 7C5, 7Y4
301	245	14A7, 14Q7, 14A7, 14B6, 35A5, 35Y4
416A	247	1LA6, 1LN5, 1LB4, 1LD5
526MB	243	12BE6, 12BA6, 12AT6, 50B5, 35W4
55P2	249	12SA7GT, 12SK7, 12SQ7, 50L6GT, 35Z5GT
55X4	250	1R5, 1T4, 1U5, 1LB4, 117Z3
613	246	12BE6, 12BA6, 12AT6, 50B5, 35W4
626-A (0626A)	241	14A7, 14Q7, 14A7, 14B6, 35A5, 35Y4
636A	241	14A7, 14Q7, 14A7, 14B6, 35A5, 35Y4
65P4	249	14A7, 35Y4, 14Q7, 14A7, 14B6, 35A5
676D	240	6SK7, 6SA7, 6SK7, 6SQ7, 6V6GT, 5Y3GT
697A	242	12BE6, 12BA6, 12AT6, 12BA6, 35B5, 35B5
69M9	251	12BA6, 12AT7, 12BA6, 12BA6, 19T8, 50L6GT
736-B	240	6SK7, 6SA7, 6SF7, 6SC7, 6V6GT, 6V6GT, 5Y3GT
753F	253	12BE6, 12BA6, 12AT6, 50C5, 35W4
75MA	250	12BA6, 12AT7, 12BA6, 12BA6, 19T8, 50L6GT
79M7	252	6BA6, 12AT7, 6BA6, 6BA6, 6T8, 6V6GT, 5Y3GT
847-B	242	6AG5, 7F8, 7AH7, 7AG7, 6H6, 6SQ7, 6V6GT, 5Y3GT
847S	243	6AG5, 7F8, 7AH7, 7AG7, 6H6, 6SQ7, 6V6GT, 5Y3GT
951	252	12AT7, 6BE6, 6BA6, 6AU6, 6T8 OR 6V8, 6V6GT OR 6K6GT, 5X3GT
95M9	251	6BA6, 12AT7, 6BA6, 6BA6, 6T8, 6SN7GT, 6K6GT, 6K6GT, 5Y3GT
PAR-80	245	1T4, 1R5, 1T4, 1S5, 3Q4, 117Z3

BRAND/MODEL	PAGE	USES THESE TUBES
BENRUS		
10B01B15B	253	12BE6, 12BA6, 12AV6, 50C5, 35W4
BLAUPUNKT		
Barcelona	254	ECC85/6AQ8, EC92/6AB4, ECH81/6AJ8, EF89, EABC80/6T8, EM80/6BR5, EL84/6
Granada 2330	254	6AQ8/ECC85, 6AB4/EC92, 6AJ8/ECH81, 6DA6/EF89, 6T8/EABC80, 6BR5/EM80, 6
Sultan 2320	255	ECC85/6AQ8, ECH81/6AJ8, EF89/6DA6, EABC80/6T8, EM80/6BR5, EL84/6BQ5
BLONDER-TONGUE		
R-20	256	12BA6, 12AT7, 12BA6, 12AU6, 50C5, 19T8
R-98	255	12BA6, 12AT7, 12BA6, 12BA6, 12AV6, 50C5
BRADFORD		
WGEC-95190C	256	6CB6, 6AQ8, 6BA6, 6BA6, 6EQ7, 6AL5, 12AX7A, 12AX7A, 6BQ5, 6BQ5, 6CA4, 6EA8
BRAUN		
MM4D	257	6AQ8/ECC85, 6DA6/EF89, 6AJ8/ECH81, 6DA6/EF89, 6T8/EABC80, 6BR5/EM80, 6
BRUNSWICK		
BJ-6836	257	6SK7, 6SA7, 6SF7, 6J5GT, 6U5, 6SN7GT, 6V6GT, 6V6GT, 5Y3GT
T-4000	258	6AG5, 6SB7Y, 6SG7, 6SA7, 6SF7, 6H6, 6SL7GT, 6V6GT, 6V6GT, 5Y3GT,
5000	258	6BA6, 7Q7, 7F8, 7AH7, 6SH7 OR 7AG7, 7A6, 6SQ7GT, 7F7, 7C5, 7C5, 5Y3GT
D-1000	259	6AG5, 6SB7Y, 6SG7 OR 6SK7, 6SA7, 6SK7, 6SK7, 6SK7, 6H6, 6SQ7, 6U5/6G5, 6SN7GT, 6K6GT, 6K6GT, 5Y3GT
BULOVA		
100	261	12BE6, 12BA6, 12AV6, 50C5, 35W4
200	262	1R5, 1U4, 1U5, 3V4
220	261	1R5, 1U4, 1U5, 3V4
278	259	Transistorized
304	260	12BE6, 12BA6, 12AV6, 50C5, 35W4
310	260	12BE6, 12BA6, 12AV6, 50C5, 35W4
CALRAD		
60A183	262	Transistorized
CAPEHART		
10	265	1R5, 1U4, 1U5, 3V4
1002F	264	6AG5, 12AT7, 6BE6, 6SK7, 6SK7, 6SK7, 6T8, 6SQ7, 6V6GT, 6V6GT, 5Y3GT
1007AM	265	6BA6, 12AT7, 6BE6, 6BA6, 6BA6, 6AL5, 6SQ7, 6SQ7, 6V6GT, 6V6GT,5Y3GT
115P2	264	6AG5, 6SK7, 6AG5, 6SA7, 6C4, 6J5, 6SK7, 6SK7, 6SK7, 6SK7,6SK7, 6H6, 6SR7, 6SN7GT, 6AF6, 6J7, 6SN7GT, 6J5, 6J5, 6J5
17RPQ155F	271	6BA6, 6X6, 6BA6, 6BA6, 6AL5, 6AV6, 12AX7, 12AX7, 6V6GT, 6V6GT, 5Y3GT
1P55	270	1R5, 1U4, 1U5, 3V4
29P4	263	6AG5, 6SB7Y, 6SG7, 6SA7, 6SF7, 6H6, 6SC7 OR 6J7, 6SL7GT, 6V6GT, 6V6GT, 5Y3GT
2C56	271	12BE6, 12BA6, 12AV6, 50C5, 35W4
2P56	272	1R5, 1U4, 1U5, 3V4
2T55	269	12BE6, 12BA6, 12AV6, 50C5, 35W4
33P9	263	6AG5, 12AT7, 6BE6, 6SK7, 6SK7, 6SK7, 6T8, 6SC7, 6V6GT, 5Y3GT,
413P	264	6AG5, 6SK7, 6AG5, 6SA7, 6C4, 6J5, 6SK7, 6SK7, 6SK7, 6SK7,6SK7, 6H6, 6SR7, 6SN7GT, 6AF6, 6J7, 6SN7GT, 6J5, 6J5, 6J5
75C56	273	12BE6, 12BA6, 12AV6, 50C5, 35W4
88P66BNL	272	1U4, 1L6, 1U4, 1U5, 3V4
C-14	269	12BE6, 12BA6, 12AV6, 6AK6
P-213	268	1R5, 1U4, 1U5, 3V4
RP-152	267	6SK7, 6SA7, 6J5, 6SK7, 6SQ7, 6V6GT, 6X5GT
RP153	268	6BA6, 6X8, 6BA6, 6BA6, 6AL5, 6AV6, 6V6GT, 6X5GT
T-522	267	12BE6, 12BA6, 12AV6, 50C5, 35W4
T-54	270	12BE6, 12BA6, 12AV6, 50L6GT, 35Z5GT
TC-100	266	12BE6, 12BA6, 12AV6, 50C5, 35W4
TC-62	266	12BA6, 12BE6, 12BA6, 12AV6, 35C5, 35W4
CAVALIER		
4CL4	276	12AU6, 12AV6, 50C5, 35W4
4P3	276	1R5, 1U5, 3V4, 1U4
562	277	12BE6, 12BA6, 12AV6, 50C5, 35W4
5AT1	274	12BE6, 12BA6, 12AT6, 50C5, 35W4
5B1	273	12BE6, 12BA6, 12AT6, 50C5, 35W4
5C1	274	12BE6, 12BA6, 12AT6, 50B5, 35W4
5R1	275	12BE6, 12BA6, 12AT6, 50C5, 35W4
603	277	12BA6, 12BE6, 12BA6, 12AT6, 35C5, 35W4
6A2	275	12BA6, 12BE6, 12BA6, 12AT6, 35C5, 35W4
CBS COLUMBIA		
515A	279	12SA7GT, 12SK7GT, 12SQ7GT, 50L6GT, 35Z5GT
5220	279	1R5, 1U4, 1U5, 3V4

BRAND/MODEL	PAGE	USES THESE TUBES
525	278	1R5, 1U4, 1U5, 3V4
541	278	12SA7GT, 12SK7GT, 12SQ7GT, 50L6GT, 35Z5GT
5440	281	12BE6, 12BA6, 12AV6, 50C5, 35W4
C220	280	12BE6, 12BA6, 12AV6, 50C5, 35W4
T200	280	12BE6, 12BA6, 12AV6, 50C5, 35W4

CHANCELLOR

BRAND/MODEL	PAGE	USES THESE TUBES
35P	281	1R5, 1T4, 1S5, 3Q4

CHANNEL MASTER

BRAND/MODEL	PAGE	USES THESE TUBES
6501	282	Transistorized
6503	283	Transistorized
6506	282	Transistorized
6512	283	Transistorized

CISCO

BRAND/MODEL	PAGE	USES THESE TUBES
1A5	284	12SA7GT, 12SK7GT, 12SQ7GT, 50L6GT, 35Z5GT
9A5	284	12SA7GT, 12SK7GT, 12SQ7GT, 50L6GT, 35Z5GT

CLARION

BRAND/MODEL	PAGE	USES THESE TUBES
11011	287	1R5, 1U4, 1S5, 3Q4
11305	288	12SA7, 12SK7, 12SQ7, 50L6GT
11411-N	289	1R5, 1U4, 1S5, 3Q4
11801	288	12SA7, 12SK7GT, 12SQ7GT, 50L6GT, 35Z5GT
12110M	291	12BA6, 12BE6, 12BE6, 12BA6, 12BA6, 12BA6, 12BA6, 6AL5, 12AT6, 50L6GT,
1231OW	289	6SK7, 6SA7, 6SK7, 6SQ7, 6K6GT, 5Y3GT
12708	290	12SA7, 12SK7, 12SQ7, 50L6GT
13101	290	12BA6, 12BE6, 12BA6, 12BA6, 12BA6, 6AL5, 12BE6, 12BA6, 12AT6, 50L6GT,
13201	292	1A7GT, 1N5GT, 1H5GT, 3Q5GT
14601	291	12SA7GT, 35Z5GT, 50L6GT, 12SK7GT, 12SQ7GT
14965	292	12AT7, 6BE6, 6BA6, 6BA6, 6AL5, 6AT6, 6AQ5, 5Y3GT
C-103	286	6SK7GT, 6SA7GT, 6SK7GT, 6SQ7GT, 6K6GT, 5Y3GT
C-105-A	286	6SK7GT, 6SA7GT, 6SK7GT, 6SQ7GT, 6K6GT, 5Y3GT
C101	285	12SA7GT, 12SK7GT, 12SQ7GT, 50L6GT, 35Z5GT
C102	287	12SA7, 12SK7, 12SQ7, 50L6GT, 35Z5GT
C108	285	1A7GT, 1N5GT, 1H5GT, 3Q5GT

CLEARSONIC

BRAND/MODEL	PAGE	USES THESE TUBES
5C66	293	12SA7GT, 12SK7GT, 12SQ7GT, 50L6GT, 35Z5GT

Tube Substitutions

TUBE	MAY BE REPLACED BY
0Y4	0Y4G
1A5G	1A5GT, 1A5GT/G, 1T5GT
1A5GT	1A5G, 1A5GT/G, 1T5GT
1A7GT	1A7G, 1B7G*, GT*
1A7GT/G	1A7G, GT, 1B7G*, GT
1AH5	1AF5, 1AR5, 1S5*
1AJ4	1AF4, 1AM4, 1T4SF, 1AE4*, 1T4*
1H5GT	1H5, G, GT/G
1H5GT/G	1H5, G, GT
1L4	1AE4*
1L6	1U6*
1LA6	1LC6
1LB4	-
1LD5	-
1LH4	-
1LN5	1LC5
1N5GT	1N5G, 1N5GT/G, 1P5G, 1P5GT
1N5GT/G	1N5G, GT, 1P5G, GT
1R5	1R5SF*, 1AQ5*
1S4	-
1S5	1AF5*, 1AH5*, 1AR5*
1T4	1AE4*, 1AF4*, 1AJ4*, 1AM4*, 1T4SF*
1U4	1AF4*
1U5	1DN5, 1AS5*, 1U5SF*
1U6	1L6*
1V	6Z3
2A3	2A3H
3LF4	3LE4#, 3D6*
3Q4	3S4#, 3S4SF*#, 3W4*#, 3Z4*#
3Q5	3Q5G, GT, GT/G, 3B5, G5, 3C5GT
3Q5GT	3Q5, G, GT/G, 3B5, GT, 3C5GT
3Q5GT/G	3Q5, G, GT, 3B5, GT, 3C5GT
3S4	3S4SF*, 3Q4#, 3W4*, 3Z4*
3V4	3C4*, 3E5*
3W4	3S4SF, 3Z4, 3S4*, 3Q4*#
5AR4	-
5U4G	5U4GA, GB, 5AR4, 5AS4, A, 5AU4, 5DB4, 5R4G, GTY, GY, GYA, GYB, 5T4, 5V3, 5V3A
5U4GA	5U4GB, 5AS4, A, 5AU4, 5DB4, 5R4G, GTY, GY, GYA, GYB, 5V3, A
5U4GB	5AS4, A, 5AU4, 5DB4, 5V3, A
5Y3GT	5Y3G, GA, GT/G, 5AR4, 5AX4GT, 5CG4, 5R4G, GTY, GY GYA, GYB, 5T4, 5V4, GA, GY, 5Z4, G, GT, GT/G, MG
5Y3GT/G	5Y3G, GA, GT, 5AR4, 5AX4GT, 5CG4, 5R4G, GTY, GY, GYA, GYB, 5T4, 5V4G, GA, GY, 5Z4, G, GT, GT/G, MG
6AB4	-
6AC7	6AC7A, Y, 6AB7, Y, 6AJ7, 6SG7*#, GT*#, Y*#, 6SH7*#, GT*#
6AF6	6AF6G/GT, 6AD6G
6AG5	6BC5, 6CE5, 6AK5*, 6AU6#, A#, 6AW6#, 6CB6#, A#, 6CF6#, 6DC6#, 6DE6#, 6CY5*#, 6EA5*#, 6EV5*#
6AJ8	-
6AK6	-
6AL5	6EB5
6AL7GT	6AL7
6AN4	6AF4#, 6AF4A#, 6T4#
6AQ5	6AQ5A, 6BM5, 6HG5
6AQ5A	6HG5, 6AG5*!, GBM5*!
6AQ6	6AT6*, 6AV6*, 6BK6*, 6BT6*6AQ7GT
6AQ8	6DT8*
6AS5	6CA5*
6AT6	6AV6, 6BK6, 6BT6, 6AQ6*
6AU6	6AU6A, 6BA6, 7543, 6AW6#, 6AG5#, 6BC5#, 6CB6#, A#, 6CE5#, 6CF6#, 6DE6#, 6DK6#
6AU6A	6AU6*!, 7543*!, 6BA6*!, 6CB6A#, 6CE5#
6AV6	6AT6, 6BK6, 6BT6, 6AQ6*
6BA6	6AU6, A, 6BD6, 6CG6, 7543, 6BZ6#
6BA7	-
6BE6	6BY6, 6CS6
6BH6	6AW6*, 6CB6*, A*, 6CF6*, 6DC6*
6BJ6	6BJ6A
6BJ7	-
6BL7GT	6BL7GTA, 6BX7GT, 6DN7
6BQ5	7189, 7189A
6BQ7A	6BQ7, 6BC8, 6BS8, 6BX8, 6BZ7, 6BZ8, 6HK8, X155, 6BK7*, A*, B*
6BR5	6DA5
6C4	-
6C5	6C5G, GT, GT/G, MG, 6J5, G, GT, GT/G, GTX, GX, MG, 6L5G
6C5GT	6C5, G, GT/G, 6C5MG, 6J5, G, GT, GT/G, GTX, GX, MG, 6L5G
6CA4	-
6CB6	6CB6A, 6AW6, 6CF6, 6DC6, 6DE6, 6DK6, 6HQ6, 6AG5#, 6AU6#, 6AU6A#, 6BC5#, 6CE5#, 6BH6#, 6HS6*#
6CB6A	6CB6*!, 6CF6*!, 6DC6*!, 6DE6*!, 6DK6*!, 6HQ6*!, 6AU6A#, 6CE5#, 6HS6*#
6CG7	6FQ7
6DA6	-
6DJ8	6ES8, 6FW8*, 6KN8*
6E5	6G5#
6E6	-
6EA8	6CQ8#, 6GH8#, 6KD8#, 6MQ8*, 6U8A*
6EQ7	6KL8#
6H6	6H6G, GT, GT/G, MG
6H6GT	6H6, G, GT/G,MG
6J5	6J5G, GT, GT/G, GTX, GX, MG, 6C5, G, GT/G, MG, 6L5G
6J5GT	6J5, G, GT/G, GTX, GX, MG, 6C5, G, GT, GT/G, MG, 6L5G
6J6	6J6A
6J7	6J7G, GT, GTX, MG, 6W7G*
6K6GT	6K6, G, GT/G, MG
6K6GT/G	6K6, G, GT, MG
6K8	6K8G, GT, GTX
6L6G	6L6, A, GA, GAY, GB, GC, GT, GX, Y, 5881, 7581, 7581A
6L6GB	6L6, A, G, GA, GAY, GC, GT, GX, Y, 5881, 7581, 7581A
6SA7	6SA7G, GT, GT/G, GTX, GTY, Y, 6SB7, Y, GTY
6SA7GT	6SA7, G, GT/G, GTX, GTY, Y, 6SB7, Y, GTY
6SB7Y	6SB7, GTY
6SC7	6SC7GT, GTY
6SF5	6SF5GT
6SF7	6SF7GT

* OK for parallel-filament circuits # May not work in all circuits ! OK for series circuits not requiring controlled warm-up

TUBE	MAY BE REPLACED BY
6SG7	6SG7Y, 6SH7, GT, 6AB7*#, Y*#, 6AC7*#, A*#, Y*#, 6AJ7*#
6SH7	6SH7GT, L, 6SG7, GT, Y, 6AB7*#, Y*#, 6AC7*#, A*#, Y*#, 6AJ7*#
6SJ7	6SJ7GT, GTX, GTY, Y
6SJ7GT	6SJ7, GTX, GTY, Y
6SK7	6SG7, 6SK7GT, GT/G, GTX, GTY, Y, 6SS7*, GT*
6SK7GT	6SG7, 6SH7, 6SK7, G, GT/G, GTX, GTY, Y, 6SS7*, GT*
6SL7GT	6SL7A, GTY, L
6SN7GT	6SN7A, GTA, GTB, GTY, L
6SN7GTB	6SN7GTA*!
6SQ7	6SQ7G, GT, GT/G, 6SZ7*
6SQ7GT	6SQ7, G, GT/G, 6SZ7*
6SQ7GT/G	6SQ7, G, GT, 6SZ7*
6SR7	6SR7G, GT, 6ST7*
6SS7	6SS7GT, 6SK7*, G*, GT*, GT/G*, GTX*, GTY*, Y*
6T8	6T8A, 6AK8
6U5	6G5, 6G5/6H5, 6H5, 6T5, 6U5/6G5
6U5/6G5	6U5, 6G5, 6G5/6H5, 6H5, 6T5
6U6GT	6Y6G*, GA*, GT*
6U8	6EA8, 6LN8, 6CQ8#, 6KD8*, 6MQ8*, 6U8A
6V6GT	6V6, G, GTA, GT/G, GTX, GTY, GX, Y, 7408
6V6GT/G	6V6, G, GT, GTA, GTX, GTY, GX, Y, 7408
6V8	-
6X4	6BX4, 6AV4*
6X5	6X5G, GT, GT/G, L, MG, 6AX5GT*, 6W5G*, GT*
6X5GT	6X5, G, GT/G, L, MG, 6AX5GT*, 6W5G*, GT*
6X8	6X8A, 6AU8*#, A*#, 6AW8*#, A*#, 6EH8#
6Y3G	6Y3
7A4	XXL
7A6	-
7A7	7A7LM, 7H7, 7B7*
7A8	7B8*, 7B8LM*
7AG7	7AH7, 7G7*, 7T7*
7AH7	7AG7,7B7, 7H7*
7B6	7B6LM, 7C6*
7B7	7AH7, 7A7*, 7A7LM*, 7H7*
7C5	7C5LT, 7B5*, 7B5LT*
7C6	7B6*, 7B6LM*
7F7	-
7F8	-
7H7	7A7, LM, 7AH7*, 7B7*
7Q7	-
7Y4	7Z4*
7Z4	-
117Z3	-
117Z6GT	117Z6G, GT/G
1273	-
12AD7	12AX7*, A*, 12BZ7*, 12DF7*, 12DM7*, 12DT7*, 7025*, A*
12AL5	-
12AT6	12AT6A, 12AV6, A, 12BK6, 12BT6
12AT7	12AZ7*, A*
12AU6	12AU6A, 12BA6, A, 12AW6#
12AU7	12AU7A, 12AX7, A#
12AU7A	12AU7, 12AX7#, A#
12AV6	12AV6A, 12AT6, A, 12BK6, 12BT6
12AX7	12AX7A, 12DF7, 12DT7, 7025, A, 12AD7*, 12AU7#, A#, 12BZ7*, 12DM7*
12AX7A	12AX7, 12DF7, 12DT7, 7025, A, 12AD7*, 12AU7#, A#, 12BZ7*, 12DM7*
12B7	12B7ML, 14A7, 14H7, 14A7/12B7, 14A7ML, 14A7ML/12B7ML
12BA6	12AU6, A, 12BA6A, 12BZ6#
12BA7	-
12BD6	12BA6, 12BA6A
12BE6	12BE6A, 12CS6
12BF6	12BU6
12DT8	-
12H6	-
12J5	12J5GT
12J5GT	12J5
12K7GT	12K7G, GT/G
12SA7	12SA7G, GT, GT/G, GTY, Y, 12SY7, GT
12SA7GT	12SA7, G, GT/G, GTY, Y, 12SY7, GT
12SF7	12SF7GT, Y
12SG7	12SG7GT, Y
12SJ7	12SJ7GT
12SJ7GT	12SJ7
12SK7	12SK7G, GT, GT/G, GTY, Y
12SK7GT	12SK7, G, GT/G, GTY, Y
12SQ7	12SQ7G, GT, GT/G
12SQ7GT	12SQ7, G, GT/G
144	-
14A7	12B7, ML, 14H7, 14A7/12B7, 14A7ML, 14A7ML/12B7ML
14B6	-
14F8	-
14Q7	-
18FW6	18FW6A, 18GD6, 18GD6A
18FX6	18FX6A
18FY6	18FY6A, 18GE6, 18GE6A
19T8	19T8A, 19C8
25F5	25F5A
25L6G	25L6, GT, GT/G, 25W6GT
25L6GT	25L6, G, GT/G, 25W6GT
25Z5	25Z5MG, 25Y5
25Z6G	25Z6, GT, GT/G, MG
25Z6GT	25Z6, G, GT/G, MG
32ET5	32ET5A, 34GD5, A
35A5	35A5LT
35B5	-
35C5	35C5A, 30A5
35L6GT	35L6G, GT/G
35W4	35W4A
35Y4	-
35Z4GT	35Z4
35Z5	35Z5G, GT, GT/G
35Z5GT	35Z5, G, GT/G
35Z5GT/G	35Z5, G, GT
36AM3	36AM3A, B
50A5	-
50B5	-
50C5	50C5A
50L6	50L6G, GT
50L6G	50L6GT
50L6GT	50L6G
5881	6L6GC, 7581, A
7025	7025A, 12AX7, A, 12DF7, 12DT7, 12AD7*, 12BZ7*, 12DM7*
EABC80	6AK8
EC92	6AB4
ECC85	6AQ8
ECH81	6AJ8
EF89	6DA6
EL84	6BQ5
EL95	6DL5
EM80	6BR5
EM81	6DA5

* OK for parallel-filament circuits

\# May not work in all circuits

! OK for series circuits not requiring controlled warm-up

Manufacturers and Their Brands

MANUFACTURER/SUPPLIER	CITY	BRANDS
A.E. Dufenhorst Co.	Rockford, IL	Saba
Admiral	Chicago, IL	Admiral
Aermotive Equipment Corp.	Kansas City, MO	Aeromotive
Affiliated Retailers	New York, NY	Artone
Air King Products Co.	Brooklyn, NY	Air King
Airadio Inc.	Stanford, CT	Airadio
Alamo Electronics	San Antonio, TX	Radioette
Algene Radio Corp.	Brooklyn, NY-	Algene
Allen B. Dumont Laboratories	Passaic, NJ	Dumont
Allied Radio Corp.	Chicago, IL	Knight
American Communications Co.	-	Liberty
American Elite	New York, NY	Telefunken
American Geloso Electric	New York, NY	Geloso
Andrea Radio Corp.	Long Island City, NY	Andrea
Ansley Radio	Trenton, NJ	Ansley
Apex Electric Mfg. Co.	Chicago, IL	Apex
Arthur Ansley Mfg.	Doylestown, PA	Arthur Ansley
Arvin Industries	Columbus, IN	Arvin
Associated Merchandising Corp.	-	AMC
Atomic Heater/Radio	New York, NY	Atlas
Audar, Inc.	Argos, IN	Audar
Audio Industries	Michigan City, IN	Ultratone
Auto Musical Instruments	Grand Rapids MI	AMI
Automatic Radio Mfg. Co.	Boston, MA	Automatic
Aviola Radio Corp.	-	Aviola
B.F. Goodrich Tire & Rubber Co.	Akron, OH	Mantola
Bell Sound Div.	Columbus, OH	Pacemaker
Belmont Radio Corp.	Chicago, IL	Belmont
Bendix Radio Div., Bendix Corp.	Baltimore, MD	Bendix
Benrus Watch Co.	New York, NY	Benrus
Blonder-Tongue Laboratories	-	Blonder-Tongue
Burstein-Applebee Co.	-	Calrad
Butler Brothers	Chicago, IL	Air Knight, Sky Rover
Capehart-Farnsworth Corp.	Ft. Wayne, IN	Capehart
Capital Appliance Distributor Div.	-	Robin
CBS Electronics	-	Columbia
CBS Columbia Div.	Long Island City, NY	CBS
CBS-Hytron	-	Columbia
Channel Master Corp.	Ellenville, NY	Channel Master
Cities Service Oil Co.	New York, NY	Cisco
Collins Radio Co.	Cedar Rapids, IA	Collins
Columbia Phonographs	New York, NY	Columbia
Columbia Records	New York, NY	Columbia
Concord Radio Corp.	Chicago, IL	Concord
Continental	-	Sharp
Continental Electronics Ltd.	New York, NY	Continental Electronics
Continental Merchandise Co.	New York, NY	Continental
Coronado	Minneapolis, MN; Los Angeles, CA	Coronado
Crescent Industries	Chicago, IL	Crescent
Crosley Corp.	Cincinnati, OH	Crosley
Crystal Products Co.	-	Coronet
Curtis Mathes Manufacturing Co.	-	Curtis Mathes
Dalbar Manufacturing Co.	Dallas, TX	Dalbar
David Bogen Co.	Paramus, NJ	David Bogen
Dearbom	Chicago, IL	Dearbom
Delco Radio, Div. of GM Corp.	Kokomo, IN	Delco
Delmonico International	Long Island City, NY	Delmonico, Emud, Sony
DeWald Radio Mfg. Co.	Long Island City, NY	DeWald
Dynamic Electronics	Richmond Hill, NY	Dynamic
Dynavox	Long Island City, NY	Dynavox
Electronic Corp. of America	Brooklyn, NY	ECA
Eckstein Radio & Television Co.	-	Karadio
Electro Appliances Mfg. Co.	New York, NY	Electro
Electro-Tone Corp.	Hoboken, NJ	Electro-Tone
Electro-Voice	Buchanan, MI	Electro-Voice
Electromatic	New York, NY	Electromatic
Electronic Laboratories	Indianapolis, IN	Electronic Labs
Electronic Specialty	Los Angeles, CA	Ranger
Electronic Utilities	Chicago, IL	Hitachi, Braun
Electronics Guild	Long Island, NY	Bulova
Emerson Radio & Phonograph Corp.	Jersey City, NJ; New York, NY	Emerson
Empire Designing Corp.	New York, NY	Empress
Espey Mfg. Co.	New York, NY	Espey, Philharmonic
Esquire Radio Corp.	New York, NY	Esquire
Excel Corp. of America	New York, NY	Excel, Toshiba
Fada Radio & Electric Co.	Long Island City, NY; Belleville, NJ	Fada
Fanon Electronic Sales Corp.	-	Fanfare
Farnsworth Television & Radio Corp.	Ft. Wayne, IN	Capehart, Farnsworth
Federal Telephone & Radio	Newark, NJ	Federal
Ferrar Radio & Television	New York, NY	Ferrar
Firestone Tire & Rubber Co.	Akron, OH	Air Chief, Firestone
Fisher Radio Corp.	Long Island, NY	Fisher
Flush-Wall Radio Mfg.	Newark, NY	Flush-Wall
Galvin Mfg. Corp.	Chicago, IL	Motorola
Gamble-Skogmo	Minneapolis, MN	Coronado
Garod Radio Corp.	Brooklyn, NY	Garod

MANUFACTURER/SUPPLIER	CITY	BRANDS
General Electric (GE)	Utica, NY; Syracuse, NY; Bridgeport, CN	General Electric
General Implement	Cleveland, OH	General Implement
General Television & Radio Corp.	Chicago, IL	General Television
Gilfillan Brothers	Los Angeles, CA	Gilfillan
Globe Electronics	New York, NY	Globe
Gonset	Burbank, CA	Gonset
Granco Products	Long Island City, NY	Granco
Grossman Music Co.	-	Stratovox
Guild Radio & Television Co.	Inglewood, CA	Guild
Harvey-Wells Electronics	Southbridge, MA	Harvey-Wells
Hallicrafters Co.	Chicago, IL	Echophone, Hallicrafters
Hammarlund Mfg.	New York, NY	Hammarlund
Harman-Kardon	Westbury, NY; Long Island, NY	Harman-Kardon
Heath Co.	Benton Harbor, MI	Heath
H.H. Scott	Maynard, MA; Cambridge MA	H.H. Scott
Hinners-Galanek Radio Corp.	Long Island City, NY	Cavalier
Hoffman Radio Corp.	Los Angeles, CA	Hoffman
Howard Radio	Chicago, IL	Howard
I.D.E.A.	Indianapolis, IN	Regency, Monitoradio, Policalarm
Industrial Electronic	New York, NY	Simplon
Intercontinental Industries, Inc.	-	Minute Man
International Detrola	Detroit, MI	Aria, Detrola
J.W. Davis Co.	-	Watterson
Jackson Industries	Chicago, IL	Jackson
Jefferson-Travis	New York, NY	Jefferson-Travis
Jewel Radio Corp.	Newark, NJ; Long Island City, NY	Belltone
John Meck Industries	Plymouth, IN	Meck, Meck Trail Blazer, Mirrortone
J.W. Miller	Los Angeles, CA	Miller
Kanematsu	New York, NY	NEC
Kappler Co.	Los Angeles, CA	Kappler
La Magna Mfg. Co.	E. Rutherford, NJ	Lamco
Lafayette Radio	Jamaica, NY	Kowa, Lafayette
Lawrence Co., Radio Div.	Cincinnati, OH	Bagpiper
Lear Inc.	Grand Rapids, MI	Learadio
Leopold Sales Corp.	-	Nanola (Nanao)
Lewyt Corp.	New York, NY	Lewyt
Lindex	New York, NY	Swank
Maco Electric Corp.	-	Maco
Madison Fielding	New York, NY	Madison Fielding
Magnavox Co.	Ft. Wayne, IN; Oakland, CA	Magnavox, Spartan
Maguire Industries Inc.	Greenwich, CT	Maguire
Majestic International Sales Corp.	Chicago, IL	Grundig Majestic
Majestic Radio & Television	Elgin, IL; St Charles, IL	Majestic

MANUFACTURER/SUPPLIER	CITY	BRANDS
Mark Simpson Co.	Long Island, NY	Masco
Mason Radio Products	Kingston, NY	Mason
Meissner Mfg. Co.	Mt. Carmel, IL	Meissner
Midwest Radio & Television	Cincinnatl, OH	Midwest
Minerva Corp. of America	New York, NY	Minerva
Mitchell Mfg. Co.	Chicago, IL	Mitchell
Molded Insulation	Philadelphia, PA	VIZ
Monitor Equipment Co.	New York, NY	Monitor
Monitoradio Div. I.D.E.A.	Indianapolis, IN	Monitoradio, Policalarm
Monitro	New York, NY	Monitor
Montgomery Ward & Co.	Chicago, IL	Airline
Motorola Inc.	Chicago, IL	Motorola
Muntz TV	Evanston, IL	Muntz
N. Pickens Import Co.	-	Blaupunkt
National Co-op	Chicago, IL	Co-op
National Union Radio Corp.	Newark, NJ	National Union
National Co.	Malden, MA	National
Newcomb Audio Products	Hollywood, CA	Newcomb
Noblitt Sparks Industries	Columbus, IN	Arvin
North American Philips Co. Inc.	-	Norelco
Northeastern Engineering Inc.	-	Electone
Olson Radio Corp.	-	Olson
Olympic Radio & Television	Long Island City, NY	Olympic
Olympic, Div. of Siegler Corp.	-	Olympic-Continental
Olympic-Opta	Long Island City, NY	Cremona, Magnet
Packard-Bell Electronics Corp.	Los Angeles, CA	Packard-Bell
Pedersen Electronics	Lafayette, CA	Pederson
Pentron Corp.	Chicago, IL	Astra-Sonic, Pentron
Philco Corp.	Philadelphia, PA	Philco
Philharmonic Radio Corp.	New York, NY	Philharmonic
Phillips Petroleum Co.	Bartelsville, OK	Woolaroc
Phillips Radio	Kokomo, IN	Phillips
Pilot Radio	Long Island City, NY	Pilot
Polyrad	Cincinnati, OH	Standard
Porto Products	Chicago, IL	Porto Baradio
Precision Electronics Inc.	Franklin Park, IL	Grommes
Premier Crystal Laboratories Inc.	New York, NY	Premier
Pure Oil Co.	Chicago, IL	Puritan
Radiaphone	Los Angeles, CA	Mayfair
Radio & Television Inc.	New York, NY	Brunswick
Radio Apparatus Co.	Indianapolis, IN	Monitoradio, Policalarm
Radio Corp. of America	Camden, NJ	Radiola, RCA Victor
Radio Craftsmen	Chicago, IL,	Radio Craftsmen, Kitchenaire
Radio Development & Research Corp.	New York, NY	Magic Tone

MANUFACTURER/SUPPLIER	CITY	BRANDS
Radio Manufacturing Engineers, Inc.	Peoria, IL	RME
Radio Wire & Television Inc.	New York, NY	Lafayette
Radionic	New York, NY	Chancellor
Rauland-Borg	Chicago, IL	Rauland
Rauland Corp.	-	Lyric
Rayenergy Radio & Television Corp.	New York, NY	Rayenergy
Raytheon Television & Radio Corp.	Chicago, IL	Raytheon
RCA Victor	Camden, NJ	RCA Victor
RCA Engineering Products	Camden, NJ	RCA Victor
RCA Home Instrument Division	Camden, NJ	Radiola
Realtone Electronics	-	Realtone
Regal Electronics Corp.	New York, NY	Regal
Regency Division I.D.E.A. Inc.	Indianapolis, IN	Regency
Remler Co.	San Francisco, CA	Remler
Renard Radio Mfg. Co.	-	Renard
Revere Camera Co.	Chicago, IL	Revere
Roland Radio Corp.	Mt. Vernon, NY	Roland
Royal	New York, NY	Royal
Sarkes Tarzian, Inc.	-	Sarkes Tarzian
SC Ryan	Minneapolis, MN	Darb
Scott Radio Laboratories Inc.	Chicago, IL	Scott
Sears & Roebuck & Co.	Chicago, IL	Silvertone
Sentinel Radio Corp.	Evanston, IN; Ft. Wayne, IN	Sentinel
Setchell-Carlson Inc.	New Brighton, MN; St. Paul, MN	Setchell-Carlson
Sheridan Electronics	Chicago, IL	Vogue
Sherwood Electronic Laboratories	Chicago, IL	Sherwood
Shriro, Inc.	New York, NY	Crown, Linmark
Signal Electronics, Inc.	New York, NY	Signal
Sonic Industries Inc.	Lynbrook, NY	Sonic
Sonora Radio & Television Corp.	Chicago, IL	Sonora
Sound	Chicago, IL	Sound
Sparks-Withington Co.	Jackson, MI	Sparton
Spartan	Ft. Wayne, IN	Spartan
Speigel Inc.	Chicago, IL	Aircastle, Continental
Stark	Ft. Wayne, IN	Cromwell, Plymouth, Stark

MANUFACTURER/SUPPLIER	CITY	BRANDS
Steelman Phonograph & Radio Co.	Mt. Vernon, NY	Steelman
Stewart-Warner Electric Corp.	Chicago, IL	Stewart-Warner
Stewart-Warner Radio Corp.	Chicago, IL	Stewart-Warner
Stromberg-Carlson Co.	Rochester, NY	Stromberg-Carlson
Superex Electronics Corp.	-	Superex
Sylvania Home Electronics	Batavia, NY	Sylvania
Sylvania Electric Products	Buffalo, NY	Sylvania
Symphonic Radio & Electronics Corp.	-	Symphonic
Tech-Master Corp.	-	Tech-Master
Telesonic Corp. of America	-	Telsonic (Medco)
Tele King Corp.	New York, NY	Tele King
Tele-Tone Radio Corp.	New York, NY	Tele-Tone
Televox, Inc.	Mt. Vernon, NY	Televox
Templetone Radio Mfg.	New London, CT	Temple
Transistor World Corp.	-	Toshiba
Trav-ler Karenola Radio & Television Corp.	Chicago, IL	Trav-ler
Union Electric	Long Island City, NY	Unitone
United Motors, GM Building	Detroit, MI	Delco
US Televison Mfg. Co.	New York, NY	Clearsonic
V-M Co.	Benton Harbor, MI	V-M
Van Camp Hardware & Iron Co.	Indianapolis, IN	Van Camp
Videola-Erie	Brooklyn, NY	Fonovox, Tonfunk
W.T. Grant Co.	New York, NY	Grantline
Warwick Mfg. Corp.	Chicago, IL	Clarion
Waterproof Electric Co.	Burbank, CA	Gon-set
Watterson Radio Mfg. Corp.	Dallas, TX	Watterson
Webcor Inc???	-	Webcor
Webster???	Chicago, IL	Webcor
Wells-Garner & Co.	-	Wells-Garner
Westinghouse Electric Co.	Sunbury, PA	Westinghouse
Westinghouse Home Radio Div.	Sunbury, PA	Westinghouse
Western Auto Supply Co.	Kansas City, KS	Truetone
Whitney & Co.	-	Arcadia
Wilcox-Gay Corp.	Brooklyn, NY	Majestic
Wilcox-Gay Corp.	Charlotte, MI	Recordio
Wilmak Corp.	-	Wilmak
Zenith Radio Corp.	Chicago, IL	Zenith

POPULAR HIT SONGS
1950-1960

1950 Autumn Leaves *Johnny Mercer*

1951 Unforgettable *Nat King Cole*

1952 Your Cheatin' Heart *Hank Williams Sr.*

1953 Rock Around the Clock *Bill Haley and the Comets*

1954 Shake, Rattle, and Roll *Joe Turner*

1955 Maybelline *Chuck Berry*

1956 Blueberry Hill *Fats Domino*

1957 Jailhouse Rock *Elvis Presley*

1958 Whole Lotta Shakin Goin On *Jerry Lee Lewis*

1959 Put Your Head on My Shoulder *Paul Anka*

1960 The Twist *Chubby Checker*

FIXING UP YOUR OLD RADIO

Here are some basic troubleshooting hints to help you identify and solve typical problems you might encounter when working on your radio.

First Step: DON'T PLUG IT IN!

For your own safety and to avoid circuit damage, make a visual inspection of the radio before powering up:

- Check the power cord for dried-out, cracked insulation and bare spots.
- Check the circuit wire insulation for bare spots. If the radio has been stored in an attic or shed, mice and other animals may have nested in it and chewed the wires.
- Check the dial cord, if your set has one, for frayed or broken wires.
- Check the speaker for cracks and tears. Because of the age of the radio you're working on, the speaker has probably deteriorated over time. You'll likely need to replace or repair it.
- Check to make sure the correct tube type is in each socket. If you need a schematic to verify placement, order the appropriate PHOTOFACT.
- Check all fuses for obvious breaks.

Then use an AC-DC variable isolation transformer to power the set up *gradually.* This gradual power-up avoids a sudden current surge which could damage defective circuits. It can also allow old, dried-out electrolytic capacitors to regenerate themselves, making replacement unnecessary.

Troubleshooting the Symptoms

Once you've identified and corrected any of the above-mentioned problems, you may encounter less obvious problems to bringing the radio up to full working order. On the next few pages are nine common complaints, the areas to check, the nature of the problem, and several possible causes listed in order of probability.

Set doesn't play at all

CHECK	PROBLEM	POSSIBLE CAUSE
Power supply	Not all tubes glow or warm up	Defective tube
		Tubes in wrong sockets
		Defective line cord
		Defective line switch
	Abnormal or no hum level	Shorts or opens in filter capacitors
	Abnormal voltage on B plus line	Defective rectifier tube
		Open filter resistor
Audio frequency section	Abnormal or no test tone response	Defective tube
		Defective first AF tube
		Shorted plate bypass capacitor
		Open coupling capacitor
		Open cathode resistor
		Open first AF plate resistor
		Open output transformer
		Shorted AF grid
		Shorted first AF plate
		Shorted first AF grid
		Defective speaker
Intermediate frequency and detector section	No signal note heard	Defective IF tube
		Defective detector tube
		Short or open in volume control
		Defective IF transformer
		Shorted IF grid circuit
		Antenna/oscillator misaligned

Set doesn't play at all (cont'd)

CHECK	PROBLEM	POSSIBLE CAUSE
Converter	No signal note heard	Defective converter tube
		Open converter cathode circuit
		Short in oscillator tuning capacitor
		Open input IF transformer
		Open oscillator coil
		Short in oscillator-grid circuit
		Shorted converter-grid circuit

Set doesn't pick up enough stations

CHECK	PROBLEM	POSSIBLE CAUSE
Power supply	Low sensitivity due to set location or insufficient gain	Poor set location
Audio frequency section		No antenna
Intermediate frequency and detector section		Weak tube
Converter		Defective loop antenna
		Misalignment
		Shorted or open AVC bypass capacitor
		Conductive dust in the gang tuning capacitor

Set fades

CHECK	PROBLEM	POSSIBLE CAUSE
Connections	Radio operates intermittently	Thermal tube
Tubes		House voltage change
Wiring		Loose connection anywhere in radio

Set doesn't play loudly enough

CHECK	PROBLEM	POSSIBLE CAUSE
Power supply Audio frequency section Intermediate frequency and detector section Converter	Low volume due to weak response or defective components	Weak rectifier tube Weak tube(s) in audio stages Defective filter capacitors Defective coupling capacitor Resistors changing in value Defective output transformer Defective speaker

Poor tone quality

CHECK	PROBLEM	POSSIBLE CAUSE
Speaker Filter system Tubes Capacitors	Poor tone due to hum or distortion in audio circuits	Defective tubes Incorrect cathode voltage Defective filter capacitors Leaking coupling capacitor AVC bypass capacitor leakage Open volume control Defective cathode-bias resistor Lower-than-normal plate voltage Positive voltage on control grid Rubbing voice coil Rattling cone

Hum at all points of tuning range

CHECK	PROBLEM	POSSIBLE CAUSE
Filter capacitors Tubes Grid circuits Capacitor blocks	Hum interferes with tone quality	Tube cathode-heater leakage Old or defective filter capacitors Leakage between capacitor block sections Open grid circuit

Hum only at specific points of tuning range

CHECK	PROBLEM	POSSIBLE CAUSE
Capacitors	Hum interferes with tone quality	Overload on strong signal due to defective AVC capacitor Defective capacitor

Squealing and motorboating

CHECK	PROBLEM	POSSIBLE CAUSE
Output filter capacitor Grid circuits Bypass capacitor Shield cans Wire dress	Interference with audio output across the entire tuning range	Noisy defective tubes Old output filter capacitor Open bypass capacitor Open grid circuit Ungrounded shield can Incorrectly dressed wiring
	Interference on a specific station only	Image-frequency interference

Set makes crackling noises

CHECK	PROBLEM	POSSIBLE CAUSE
Set location Connections Tubes Wiring	Noise interferes with audio quality	Set may receive outside electrical noise Defective tube(s) Corrosion in transformer windings Loose or poorly soldered connections Wires may be touching

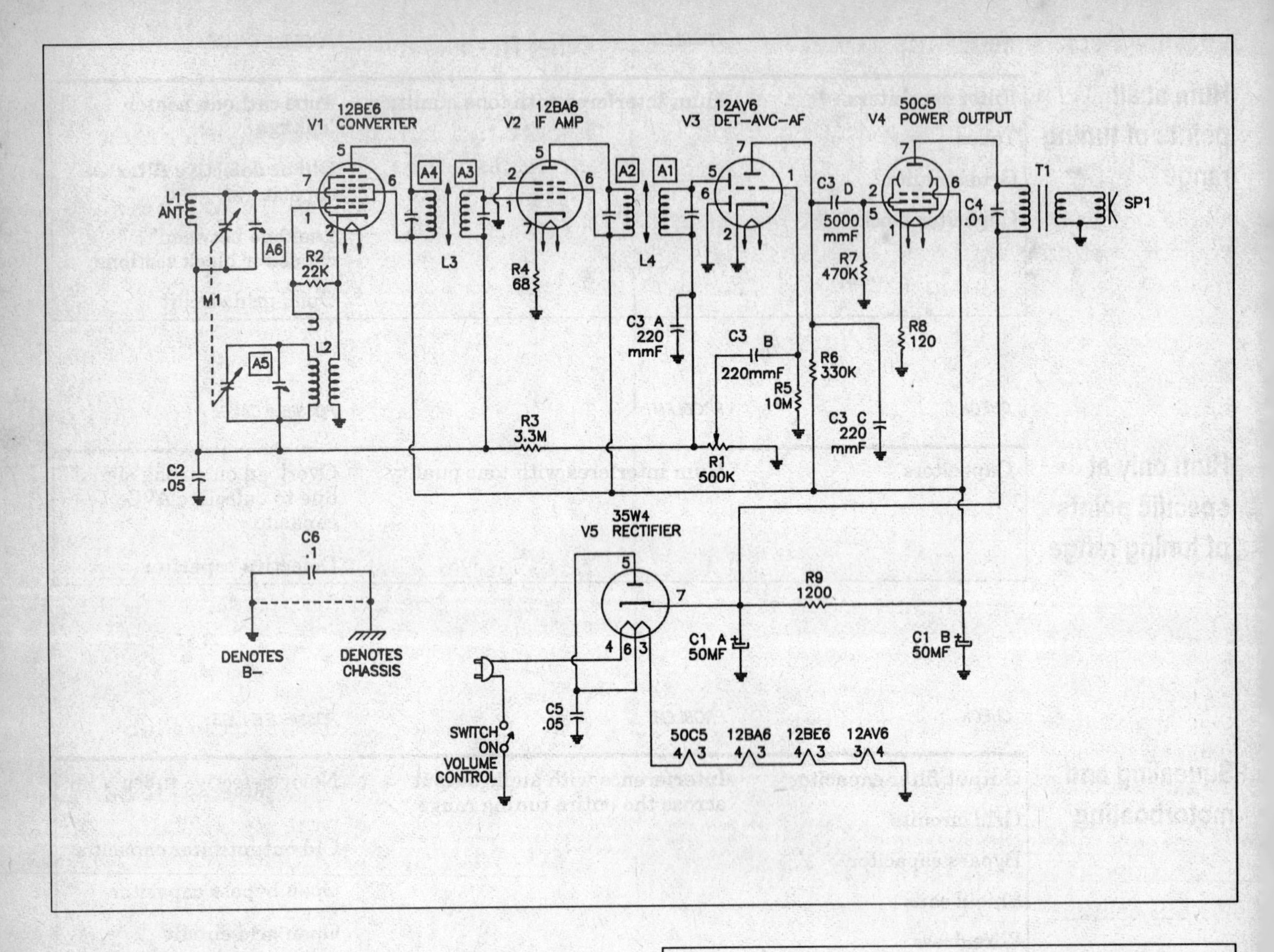

Typical Schematic

This 5-tube schematic represents the basic superheterodyne receiver. Tubes added to this basic configuration enhance the radio's performance by providing power regulation, audio gain, other band frequencies, better sensitivity, and other uses.

If you want a complete schematic for your specific radio, order the appropriate PHOTOFACT.

Tubes

V1	Converter
V2	IF Amp
V3	DET. -AVC-AF
V4	Power Output
V5	Rectifier

Capacitors

C1A	Filter-Red
C1B	Filter-Green
C2	AVC Filter
C3A	Diode RF Filter
C3B	Audio Coupling
C3C	AF Amp Plate Bypass
C3D	Audio Coupling
C4	Output Plate Bypass
C5	Line Filter
C6	Line Isolation

Controls

R1A	Volume Control
R1B	Power Switch

Resistors

R2	Oscillator Grid
R3	AVC Network
R4	IF Cathode
R5	AF Amplifier Grid
R6	AF Amplifier Plate
R7	Output Grid
R8	Output Cathode
R9	Filter

Locating Replacement Parts

Worn or defective parts can be replaced with new pieces available through electronic parts distributors.

Call Sams at 800-428-7267 for the name of your nearest distributor. Or order unusual pieces from specialty supply companies such as:

Antique Electronic Supply
6221 S. Maple Avenue.
Tempe, AZ 85283

Vintage TV & Radio
3498 W. 105th Street
Cleveland, OH 44111

You can also find a lot of still-functioning pieces at your local radio club swap meet. Using the indexes in this book, you'll discover models similar to yours from which you can salvage parts, and you'll find appropriate tube substitutions.

If you order a PHOTOFACT for your radio, it comes with a complete parts list to help you in your search.

Other Resources

A wealth of technical and collecting information is available from other radio enthusiasts through special-interest clubs and publications. Here are a few to get you started.

Magazines

Radio Age
636 Cambridge Road
Augusta, GA 30909

Antique Radio Classified
498 Cross Street
P.O. Box 2
Carlisle, MA 01741

National Clubs

Antique Wireless Assn.
Main Street
Holcomb, NY 14469
716-657-7489

Antique Radio Club
of America
81 Steeplechase Road
Devon, PA 19333
215-688-2976

If you enjoyed this volume and want to collect all six in the Radios of the Baby Boom Era series, look for them at your Sams authorized distributor or your favorite bookstore.

TO ORDER SAMS PHOTOFACT® SERVICE DATA

If you'd like to work on a radio but need more information than we provide here, simply order a Sams PHOTOFACT for that model.

PHOTOFACT service data includes schematics, parts lists, alignments/adjustments, testing procedures, and troubleshooting guidelines. A PHOTOFACT is available for every radio listed in this book.

To identify the PHOTOFACT for your particular radio, find the radio among the pictures, or locate the model number in the index entitled "Pictured Radios and Similar Models." A PHOTOFACT set number is provided there for each model. This is the number to order.

Your local Sams authorized distributor carries a large stock of PHOTOFACT service data. Call Sams customer service at

800-428-7267

for the name of your local distributor or to place a phone order. When you call, ask for Operator RB-1.

Howard W. Sams & Company is a recognized leader in technical publishing. Nearing our first half-century in the service documentation business, we have covered more than 150,000 models of products varying from radios and TVs to ovens and helicopters. When Sams provides technical information, you can count on it being accurate and complete.